W0225708

Springer Advanced Texts in Life Sciences

David E. Reichle, Editor

Springer Advanced Texts in Life Sciences
Series Editor: David E. Reichle

Dale W. Johnson   Robert I. Van Hook
Editors

Analysis of Biogeochemical Cycling Processes in Walker Branch Watershed

With 191 Figures

Springer-Verlag
New York Berlin Heidelberg
London Paris Tokyo

Dale W. Johnson
Oak Ridge National Laboratory
Division of Environmental Sciences
Oak Ridge, Tennessee 37831-6038, USA

Robert I. Van Hook
Oak Ridge National Laboratory
Division of Environmental Sciences
Oak Ridge, Tennessee 37831-6038, USA

Series Editor
David E. Reichle
Oak Ridge National Laboratory
Division of Environmental Sciences
Oak Ridge, Tennessee 37831-6038, USA

Library of Congress Cataloging-in-Publication Data
Analysis of biogeochemical cycling process in Walker Branch Watershed
/ Dale W. Johnson and Robert I. Van Hook, editors.
    p. cm.—(Springer advanced texts in life sciences)
  Bibliography: p.
  Includes index.
  ISBN 978-1-4612-8134-4   ISBN 978-1-4612-3512-5 (eBook)
  DOI 10.1007/978-1-4612-3512-5

    1. Biogeochemical cycles—Tennessee—Oak Ridge. I. Johnson, D.
W. (Dale W.), 1946– . II. Van Hook, Robert I. III. Title: Walker
Branch Watershed. IV. Series.
  QH344.A52 1989
  551.9—dc19
                                                      88-39675

Printed on acid-free paper

Typeset by David E. Seham Associates, Inc., Metuchen, New Jersey.

9 8 7 6 5 4 3 2 1

Preface

The Oak Ridge National Laboratory's Environmental Sciences Division initiated the Walker Branch Watershed Project on the Oak Ridge Reservation in east Tennessee in 1967, with the support of the U.S. Department of Energy's Office of Health and Environmental Research (DOE/OHER), to quantify land-water interactions in a forested landscape. It was designed to focus on three principal objectives: (1) to develop baseline data on unpolluted ecosystems, (2) to contribute to our knowledge of cycling and loss of chemical elements in natural ecosystems, and (3) to provide the understanding necessary for the construction of mathematical simulation models for predicting the effects of man's activities on forested landscapes. In 1969, the International Biological Program's Eastern Deciduous Forest Biome Project was initiated, and Walker Branch Watershed was chosen as one of several sites for intensive research on nutrient cycling and biological productivity. This work was supported by the National Science Foundation (NSF). Over the next 4 years, intensive process-level research on primary productivity, decomposition, and belowground biological processes was coupled with ongoing DOE-supported work on the characterization of basic geology and hydrological cycles on the watershed. In 1974, the NSF's RANN Program (Research Applied to National Needs) began work on trace element cycling on Walker Branch Watershed because of the extensive data base being developed under both DOE and NSF support. Budgets of cadmium, lead, and zinc were developed and compared with budgets of carbon and nitrogen, which significantly increased our understanding of the cycling of these elements in watershed ecosystems.

In 1978, the baseline DOE funding for basic biogeochemical cycling research was supplemented by two major new projects. The first was the experimental verification of the stream spiraling concept, using phosphorus and nitrogen, and was supported by the NSF's Ecosystem Studies Program; the second addressed atmospheric deposition and the effects of acidic precipitation on forests, and was sponsored by the Electric Power Research Institute. The baseline funding by DOE/OHER's Ecological Research Division has provided a unique and continuous program that has been supplemented over the years by funding from different agencies for research on specific topics relating to forested landscapes. The Walker Branch Watershed Project is one of only a few long-term intensive ecosystem studies in the United States. Numerous collaborators from universities, national and international institutes, and federal agencies (particularly the National Oceanic and Atmospheric Administration's Atmospheric Turbulence and Diffusion Division in Oak Ridge) have been involved in the project. This book synthesizes 18 years of research effort in one document, addressing, in addition to a comprehensive site description, in-depth analyses of biogeochemical cycling, carbon cycling and productivity, water balances, forest micrometeorology, stream spiraling, and mathematical simulation modeling.

Significant scientific accomplishments have been made over the past 18 years in several areas: dry deposition; the role of macropores in hydrologic transport; the development of biogeochemical cycles of carbon, nitrogen, sulfur, and several trace elements; and the particularly important long-term data bases on atmospheric deposition, hydrology, streamflow, biomass, and soils. Atmospheric deposition research has resulted in identification of the significance of dry deposition in the overall atmospheric input of elements such as sulfur (e.g., dry deposition represents 50% of the overall input). Detailed studies of hydrology have identified the important role of macropores in subsurface transport, which in turn has led to the development of additional research activities and improved simulation models for belowground water transport. The development of detailed biogeochemical cycles of nitrogen, sulfur, carbon, and several trace elements has pointed out the significant features of each of these cycles and contributed to our understanding of the basic ecological processes governing cycling and response to manmade stress. Long-term trends, which are obvious only in data sets that have been maintained over long periods of time, have pointed out changes in species composition and in the biogeochemical cycles of various elements. These long-term data sets are providing baselines against which to assess future change.

The results of Walker Branch research are applied in three different ways. First, our understanding of the fundamental behavior of forest ecosystems developed over the course of our research has allowed us to answer specific questions concerning the effects of stress on the watershed as well as on other forested landscapes. As particular problems arise (e.g., air pollution stress, heavy metal input, etc.) either on adjacent watersheds or on forested landscapes elsewhere in the eastern deciduous forest, our knowledge of basic processes derived from Walker Branch research allows us to investigate these new problems cost-effectively; that is, the numbers and kinds of samples required and the analyses needed on those samples can be minimized. Second, mathematical modeling, which has been a central theme of the Walker Branch Watershed Project throughout its history, has focused primarily on hydrologic transport, element cycling, and biological processes as they relate to the development of advanced models such as TEHM (Terrestrial Ecosystem Hydrology Model), PROSPER (evapotranspiration by vegetation), and UTM (Unified Transport Model). Third, the combination of long-term data sets for examining changes in ecosystem composition and behavior as a function of changing environmental conditions and the basic knowledge gained on specific processes (such as anion mobility in soils) has provided a unique tool for extrapolating Walker Branch Watershed results to the eastern deciduous forests. For example, the sensitivity of soils in the eastern United States to acidic precipitation has been developed and mapped, based on basic research results on ion mobility in Walker Branch Watershed soils coupled with specific measurements on soil samples obtained from other sites in the region.

The Walker Branch Watershed Project was initiated in the mid-1960s with the goal of understanding the basic biological, chemical, and physical processes and the interactions of them that govern the cycling of materials in forested landscapes. The project was begun before public concerns about acidic precipitation, increases in atmospheric carbon dioxide, emissions of trace elements from industrial sources, and shallow land burial of hazardous wastes, etc. had arisen. The principles governing watershed ecosystem dynamics and the data obtained from these studies have provided the basis for quickly responding to, and assisting in, the resolution of these problems.

Walker Branch Watershed research results have demonstrated the importance and value of long-term basic research in providing the knowledge needed for understanding and resolving environmental problems. Continuing fundamental investigation on (1) watershed input-output and inventories of elements in vegetation and soils, (2) processes important in biogeochemical cycling (belowground and stream processes), (3) elements of particular importance in aquatic and terrestrial ecosystems (phosphorus, calcium), and (4) extrapolation of the Walker Branch results to the eastern deciduous forests will not only be of significance in their own right, but will also provide data, approaches, and insights applicable to the solution of future energy-related problems associated with air pollution, contaminant transport, and waste management.

D.W. Johnson and R.I. Van Hook

Acknowledgments

Financial support for the research conducted on Walker Branch Watershed was provided by the U.S. Department of Energy's Office of Health and Environmental Research, Division of Ecological Research, and Division of Physical and Technological Research (under contract DE-AC05-84OR21400 between the U.S. Department of Energy and Martin Marietta Energy Systems, Inc., for the operation of Oak Ridge National Laboratory); the Electric Power Research Institute's Ecological Studies Program (under contracts RP1813-1 and RP1907-1 between Electric Power Research Institute and Martin Marietta Energy Systems, Inc.); and the National Science Foundation's Ecosystem Studies Program (DEB-7803012, BSR-8103181), International Biological Program (AG-199), and the Research Applied to National Needs Program (AG-389), with all contracts between the National Science Foundation and the U.S. Department of Energy/Oak Ridge National Laboratory. Additional contributions from the U.S. Army's Research Office and the Corps of Engineers Waterways Experiment Station and Engineering Topographic Laboratory, the U.S. Department of Defense, the U.S. Environmental Protection Agency, and the National Oceanic and Atmospheric Administration are greatly appreciated.

The authors are especially indebted to the originators of the Walker Branch Watershed Project, S.I. Auerbach of Oak Ridge National Laboratory, J.W. Curlin of the Office of Technology Assessment, and the late D.J. Nelson, who administratively and technically directed the initial stages and development of this facility. The authors are also indebted to T. Tamura and J.S. Olson, who reviewed the entire manuscript and offered many helpful comments and suggestions.

The research reported in Chapter 3 by B.A. Hutchison and D.D. Baldocchi was greatly assisted by D.R. Matt, R.T. McMillen, and J.D. Womack of the Atmospheric Turbulence and Diffusion Laboratory's scientific staff. Scientific assistance was also provided by L.J. Gross, University of Tennessee, and S.J.

Tajchman, West Virginia University. Technical assistance was provided by J.L. Sharp, J. Gholston, and J. Wynn, along with J.T. Dean, M.E. Hall, J.M. Taylor, and G. Venable. Various phases of the research effort were furthered by the services of P. Camara, the late J.R. Fowler, K. (Jamruz) Heffner, K.A. Parker, J.F. Stiefel, and J.L. Trimble. Computer assistance was provided by W.R. Martin and M.A. Mitchell. This assistance is gratefully acknowledged.

S.E. Lindberg wishes to acknowledge the guidance provided by R.C. Harriss of NASA's Langley Research Laboratory in the early stages of the research presented in Chapter 4. Also, thanks are due L.K. Mann, J.M. Coe, and M. Levin for field assistance, and N.M. Ferguson for performing chemical analyses.

R.J. Luxmoore and D.D. Huff thank the following persons for technical contributions in obtaining and summarizing some of the field data presented in Chapter 5: T. Grizzard, A.E. Hunley, J.R. Jones, W.J. Selvidge, and D.E. Todd, Jr.

N.T. Edwards wishes to thank B.M. Ross-Todd for technical support in the collection of respiration data presented in Chapter 6. He and S.B. McLaughlin thank R.K. McConathy for his help in collecting the photosynthate allocation data. D.W. Johnson wishes to thank D.E. Todd, Jr., and L.M. Stubbs of Oak Ridge National Laboratory and A.J. Pieper and B.A. Yelczyn of the University of Wisconsin at Stevens Point for collection of the Walker Branch Watershed inventory data.

D.W. Johnson and G.S. Henderson, the authors of Chapter 7, gratefully acknowledge the technical assistance provided by D.E. Todd, Jr., L.M. Stubbs, K.C. Dearstone, W.J. Selvidge, A.E. Hunley, and T. Grizzard in collecting much of the data for Chapter 7 and in maintaining the facilities on Walker Branch Watershed over the years, and by N.M. Ferguson in performing the chemical analyses. Thanks are also extended to two summer students, J.J. Pieper and B.A. Yelczyn of the University of Wisconsin at Stevens Point, for their help in the forest inventory in 1979.

The synthesis of research results presented in Chapter 8 by J.W. Elwood and R.R. Turner represents the contribution of numerous individuals who have worked on various studies on the streams draining Walker Branch. For their technical assistance in collecting, processing, and analyzing samples in both the field and laboratory and in preparing data for publication, the authors thank S. Rucker, R.M. Cushman, R.W. Stark, J.C. Barmier, P.T. Singley, L.A. Ferren, and J.D. Story. For their contributions of ideas and data on various stream studies, thanks and gratitude are extended to several colleagues: J.D. Newbold, Stroud Water Research Center, P.J. Mulholland and R.V. O'Neill, Oak Ridge National Laboratory, for their contributions to the nutrient spiraling studies; C.E. Comiskey, Science Applications International Corporation, for his contributions to data on organic matter inputs and transport; G.S. Henderson, University of Missouri, for his contributions on episodic patterns in stream water chemistry; J.E. Segars, graduate student from the University of Tennessee, for her contributions to the work on phosphorus forms and concentrations in soil solutions, groundwater, and stream water; and S.E. Lindberg, Oak Ridge National Laboratory, for his contributions to the work on trace element geochemistry.

The preparation of this book would not have been possible without the able assistance provided by the publications staff of the Environmental Sciences Division of Oak Ridge National Laboratory: J.E. Holbrook, R.R. Adams, and R.E. Booker of Graphic Arts; D.D. Rhew, G.R. Carter, V.M. Davidson, D.H. Deaton, P.G. Epperson, K.N. Gibson, C.A. Kappelmann, M.A. Kirby, J. Saffell, and J.L. Seiber of the Word Processing Center; and A.L. Ragan and N.E. Tarr of the Editorial Office. Their expertise and dedication to excellence are gratefully acknowledged.

Contents

Contributors

Dennis D. Baldocchi
Oak Ridge Associated Universities, Oak Ridge, Tennessee 37831, USA

Nelson T. Edwards
Environmental Sciences Division, Oak Ridge National Laboratory, Oak Ridge, Tennessee 37831, USA

Jerry W. Elwood
Environmental Sciences Division, Oak Ridge National Laboratory, Oak Ridge, Tennessee 37831, USA

W. Franklin Harris
National Science Foundation, Division of Biotic Systems and Resources, Washington, D.C. 20545, USA

Robert C. Harriss
National Aeronautics and Space Administration, Langley Research Laboratory, Hampton, Virginia 23665, USA

Gray S. Henderson
Forestry Department, University of Missouri, Columbia, Missouri 65201, USA

William A. Hoffman, Jr.
Department of Chemistry, Denison University, Granville, Ohio 43023, USA

Dale D. Huff
Environmental Sciences Division, Oak Ridge National Laboratory, Oak Ridge, Tennessee 37831, USA

Boyd A. Hutchison
Air Resources Laboratory, Atmospheric Turbulence and Diffusion Division, National Oceanic and Atmospheric Administration, Oak Ridge, Tennessee 37831, USA

Dale W. Johnson
  Environmental Sciences Division, Oak Ridge National Laboratory, Oak Ridge,
  Tennessee 37831, USA

Steven E. Lindberg
  Environmental Science Division, Oak Ridge National Laboratory, Oak Ridge,
  Tennessee 37831, USA

Gary M. Lovett
  Institute of Ecosystem Studies, Carey Arboretum, Millbrook, New York 12545,
  USA

Robert J. Luxmoore
  Environmental Sciences Division, Oak Ridge National Laboratory, Oak Ridge,
  Tennessee 37831, USA

Samuel B. McLaughlin
  Environmental Sciences Division, Oak Ridge National Laboratory, Oak Ridge,
  Tennessee 37831, USA

Ralph R. Turner
  Environmental Sciences Division, Oak Ridge National Laboratory, Oak Ridge,
  Tennessee 37831, USA

Robert I. Van Hook
  Environmental Sciences Division, Oak Ridge National Laboratory, Oak Ridge,
  Tennessee 37831, USA

Chapter 1
Introduction

R.I. Van Hook

The Walker Branch Watershed Project was initiated in 1967 under sponsorship of the U.S. Atomic Energy Commission with three primary objectives: (1) providing base-line values for unpolluted natural waters, (2) contributing to our knowledge of cycling and loss of chemical elements in natural ecosystems, and (3) enabling the construction of models for predicting the effects of man's activities on the landscape. Walker Branch Watershed is located in the Ridge and Valley province of east Tennessee on the Oak Ridge Reservation. The initial focus of the project centered primarily on the geologic and hydrologic processes that govern the quantity and quality of water moving through the watershed. The research focus expanded through time as new projects were initiated: intensive ecological process studies under the National Science Foundation (NSF) in 1970, trace contaminant studies under NSF in 1973, forest micrometeorology research jointly under the Department of Energy and the National Oceanic and Atmospheric Administration in 1976, atmospheric deposition studies under the National Atmospheric Deposition Program in 1980, and detailed acidic deposition effects research on canopy processes and soil chemistry in 1981. These studies of varying lengths have all contributed to a more complete understanding of the biological, chemical, geological, and physical processes governing watershed behavior.

The watershed-level approach to the study of ecosystem dynamics (Fig. 1.1) has provided a means of focusing the results from process-level studies, including productivity, decomposition, hydrology, and element cycling, on the understanding of the Walker Branch Watershed ecosystem. Detailed measurements of wet and dry deposition inputs (Fig.1.2), coupled with throughfall and stemflow, have yielded estimates of chemical inputs to the soil. These data have been utilized, along with empirical information

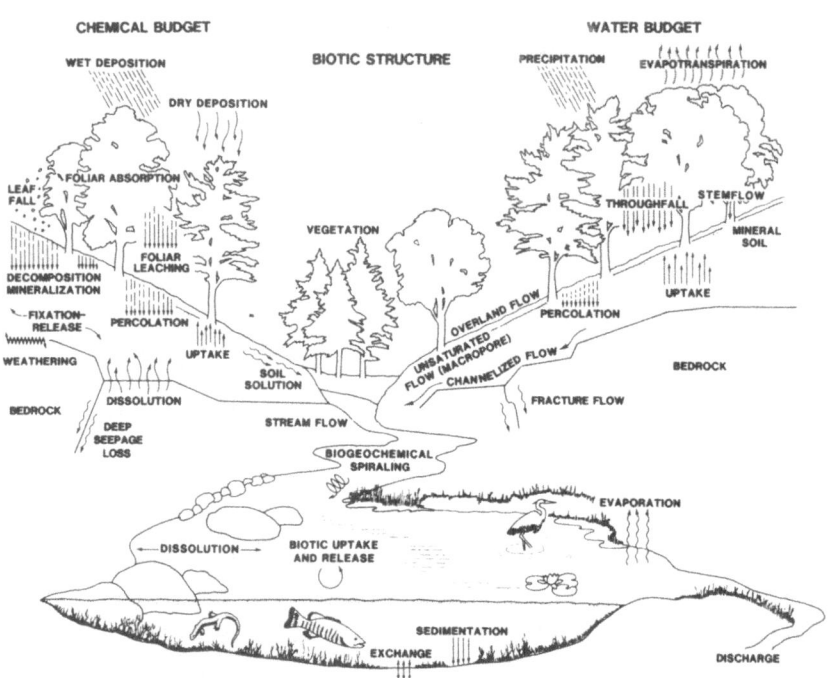

Figure 1.1. The watershed-level approach to the study of ecosystem dynamics includes the coupling of chemical and water inputs with biological and geochemical cycling of these materials to form an integrated biogeochemical cycle for the watershed system.

on the soil's hydrologic characteristics (Fig. 1.3) and chemical composition, in the development of hydrologic transport models, which account for subsurface flow, variable source areas, transport across the plant-soil water interface, evapotranspiration, and outflow (Fig. 1.4). Detailed studies of the chemical composition of wet and dry deposition, aboveground and belowground biomass, and various soil horizons have been combined with the hydrologic transport models to estimate the amounts and fluxes of materials in various components of the ecosystem as well as the total input and output. The watershed approach is essential for developing these types of analyses. To achieve a watershed-level approach, one must first conduct stand-level research (Fig. 1.5), which provides a suitable scale for detailed mechanistic studies. Detailed measurements of deposition, translocation, cycling, and leaching in individual deciduous and coniferous stands provide the information necessary for integrating up to the watershed level of organization. A thorough understanding of the physiological and ecological processes governing plant productivity, plant uptake of nutrients and water, decomposition, and biogeochemical cycling at the

Figure 1.2. Wet and dry deposition measurements are used in developing input-output budgets of water and chemical constituents in the landscape. This site, located on Walker Branch Watershed, is a component of the National Atmospheric Deposition Program.

Figure 1.3. Measurements of the hydrologic characteristics of the soil, including infiltration and conductivity, in field soil blocks provide data for hydrologic transport models.

Figure 1.4. Calibrated V-notch weirs are used in determining the hydrologic output and chemical composition of stream water from Walker Branch Watershed.

stand level provides the framework for developing process models based on biological and physical principles. These models may then be used for the synthesis of interrelated data sets as well as extrapolation of stand-level results up to the watershed level of organization. Stand-level investigations also provide data for correlating or verifying remotely sensed information on primary productivity and biogeochemical cycling.

The Walker Branch Watershed Project has expanded our basic understanding of the biological and physical mechanisms governing the ecological behavior of forested ecosystems. Long-term trends are now apparent in the data sets on hydrology, atmospheric deposition, streamflow, biomass dynamics, and soil composition that have been developed since 1967. In the following chapters, we discuss the significant results of 18 years of work on Walker Branch Watershed in the areas of forest meteorology, atmospheric chemistry, water, carbon dynamics, biogeochemical cycling, stream spiraling, and modeling. The biogeochemical cycles developed in this work provide unique descriptions of the state of health of this ecosystem as well as a base line for evaluating future changes. We

ORNL–DWG 85-10963

CONCEPTUAL DIAGRAM OF A STAND-LEVEL STUDY

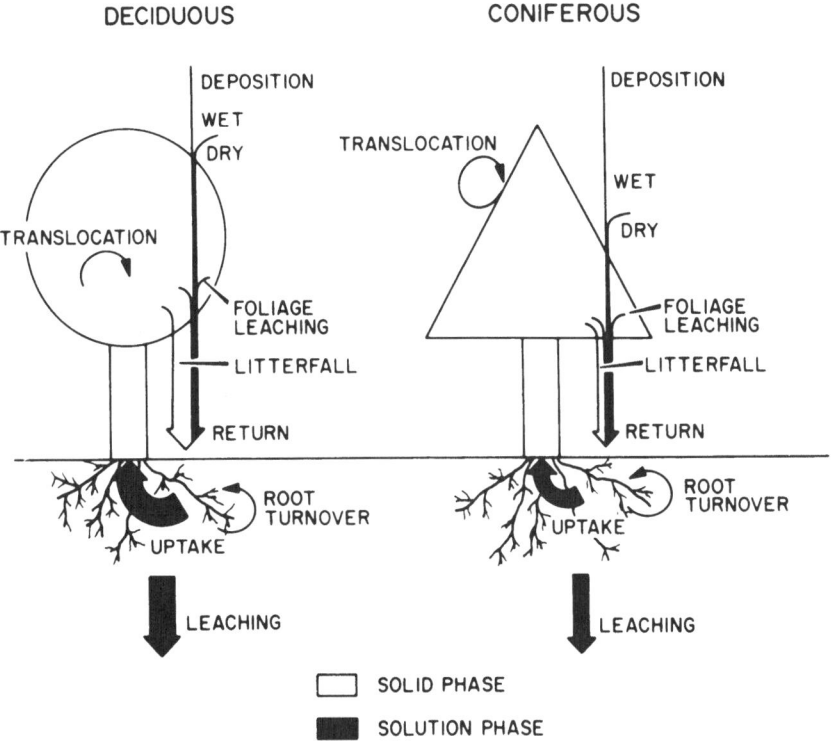

Figure 1.5. Stand-level analyses of basic biological, chemical, and physical processes for different forest types are essential components of watershed research.

also discuss the applications of Walker Branch Watershed data sets, knowledge, and models to relevant situations in which there is a need to understand the potential impact of man-made stress on a particular forested ecosystem. The hydrology, materials flow, forest growth, and stream spiraling models that have been developed as a result of these studies are unique tools for assisting in evaluating the effects of both natural and anthropogenic stresses on forest ecosystems and are now available for such applications.

Chapter 2
Site Description

D.W. Johnson

2.1 Location (Curlin and Nelson 1968)

Walker Branch Watershed is located on the U.S. Department of Energy's Oak Ridge Reservation in Anderson County, Tennessee, at latitude 35°58′ north, longitude 84°17′ west, near Oak Ridge National Laboratory (Fig. 2.1), and drains into an embayment of the Tennessee Valley Authority's (TVA) Melton Hill Reservoir, which is formed by the impoundment of the Clinch River.

The watershed occupies a total of 97.5 ha, consisting of two subwatersheds—the west catchment being 38.4 ha and the east catchment 59.1 ha (Fig. 2.2). The catchment basin is bounded on the north by Chestnut Ridge, which reaches an elevation of 350 m and slopes rapidly southward to an elevation of 265 m in the valley at the confluence of the two forks (Fig. 2.2). The watershed is almost completely encircled by a tertiary road system, which provides easy access to the perimeter and to the center ridge dividing the subwatersheds.

2.2 Climate (Curlin and Nelson 1968)

The climate of Oak Ridge is typical of the humid southern Appalachian region. The mean annual rainfall is ~139cm, and the mean median temperature is 14.5°C. Precipitation is predominately in the form of rainfall, although under unusual conditions snowfall can represent a significant portion of the total water income. Storm tracks appear to travel northwest to southeast. The precipitation pattern during the year is characterized by a wet winter and a comparatively dry spring, followed by a relatively

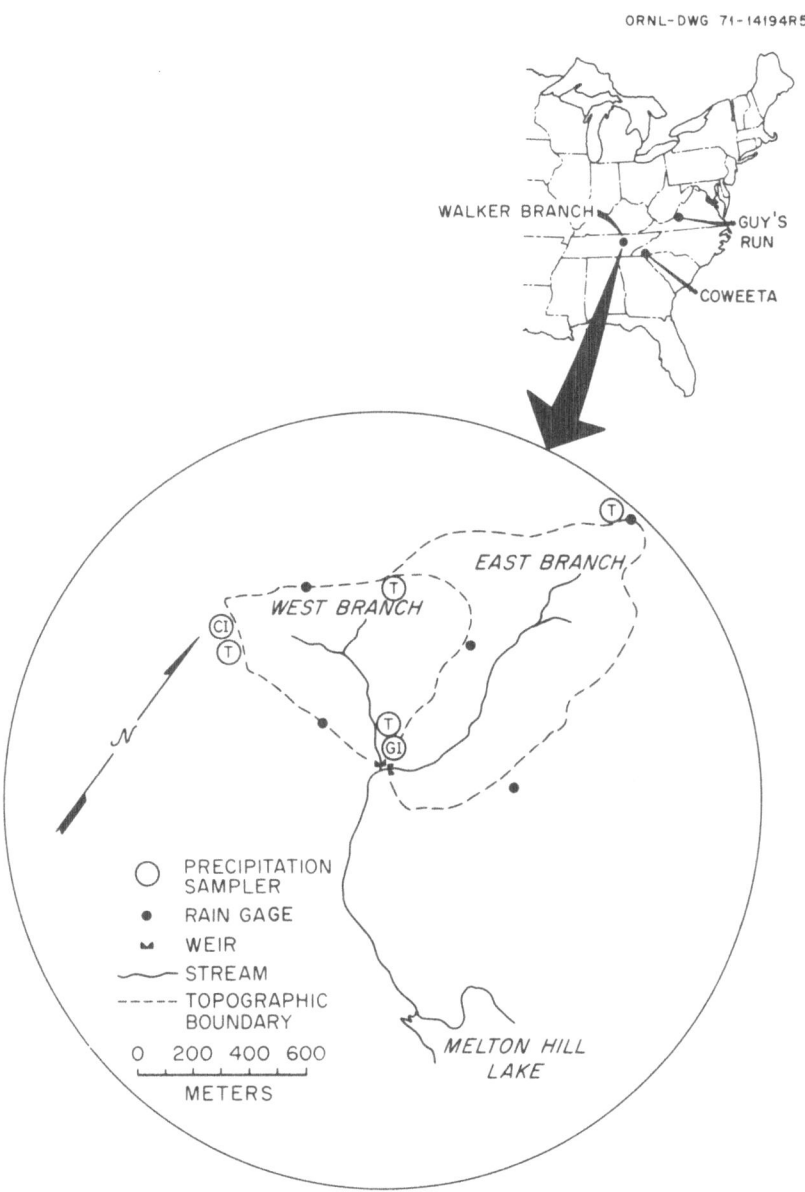

ORNL-DWG 71-14194R5

Figure 2.1. Location of Walker Branch Watershed Project.

wet summer and dry autumn. July rainfall (15 cm) normally approaches that of the wet winter months, whereas June (8 cm) is almost as dry as the autumn. July is generally the hottest month (average temperature 25.1°C), and January is the coldest (4.4°C). On the average, there are 200 frost-free days.

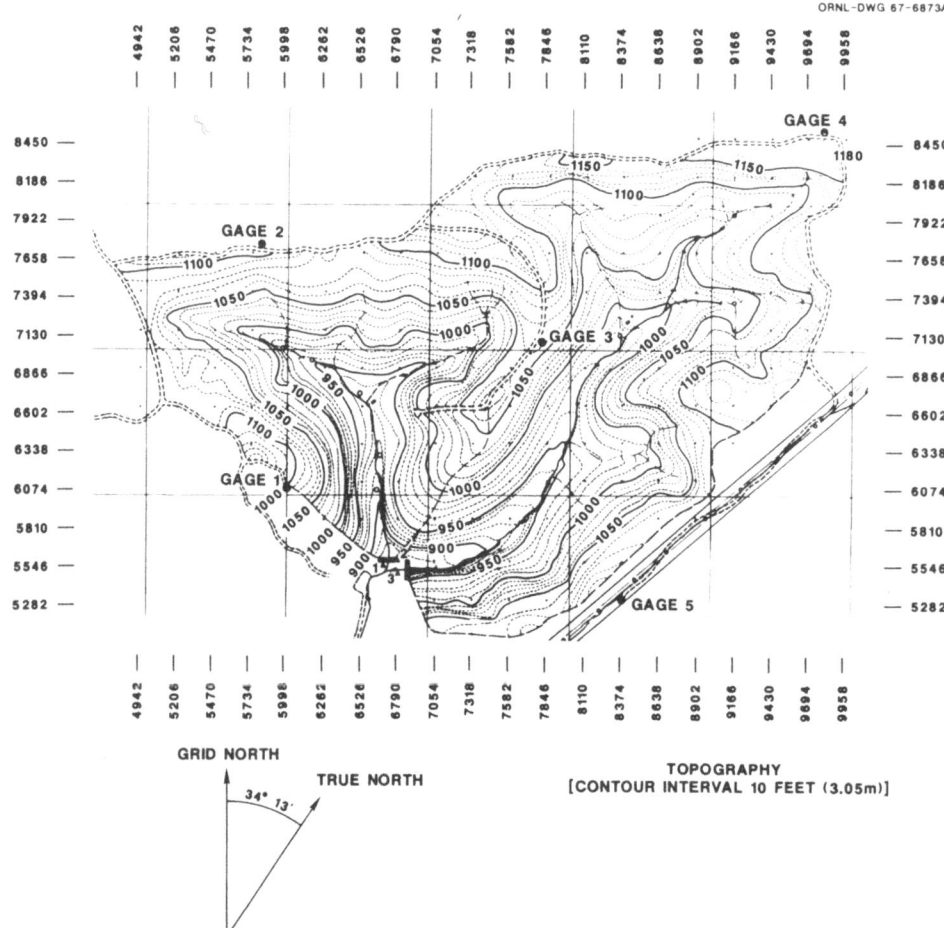

Figure 2.2. Topographic map of Walker Branch Watershed.

The striking feature of the OakRidge climate, as far as plant growth is concerned, is the development of comparatively early moisture deficits in the spring. However, the rainfall in July and August normally prevents the development of severe summer deficits, which often occur in other areas of the southern United States.

2.3 Air Quality (Lindberg et al. 1979)

The watershed provides an ideal field laboratory in which to examine deposition and cycling of airborne trace contaminants derived from atmospheric emissions related to energy technology. The catchment is located within ~20 km of two major coal-fired power plants and one small plant (Fig. 2.3). As indicated by the annual windrose, these point sources are frequently upwind of the watershed. The Kingston Power Plant (operated by TVA) is 22 km to the west-southwest. This facility consumes 12,000 Mg of coal daily, generating 1,200 MW of electricity. The plant was upgraded in 1976 by the addition of two 330-m stacks and best-available-technology electrostatic precipitators. About 13 km to the east-northeast of the watershed is TVA's Bull Run Power Plant. This facility was upgraded in 1977 by the installation of new, high-efficiency (99.9% mass removal) electrostatic precipitators in line with the single 270-m stack. It consumes 7,000 Mg of coal daily, generating 890 MW of electricity. The

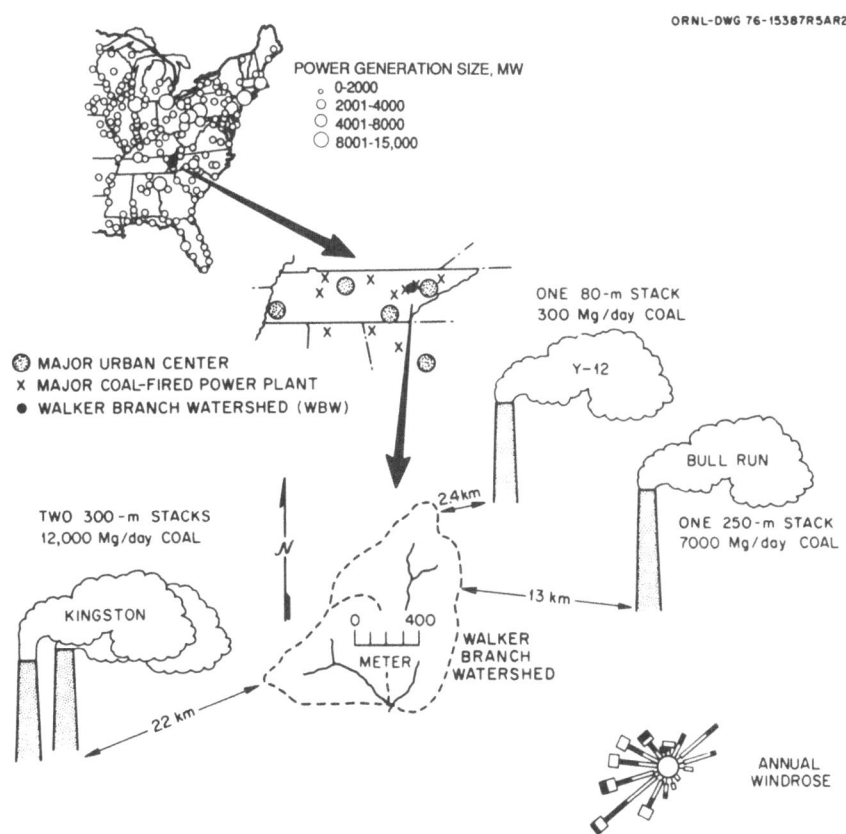

ORNL-DWG 76-15387R5AR2

POWER GENERATION SIZE, MW
- 0-2000
- 2001-4000
- 4001-8000
- 8001-15,000

ONE 80-m STACK
300 Mg/day COAL

⊗ MAJOR URBAN CENTER
X MAJOR COAL-FIRED POWER PLANT
● WALKER BRANCH WATERSHED (WBW)

Y-12

BULL RUN

TWO 300-m STACKS
12,000 Mg/day COAL

ONE 250-m STACK
7000 Mg/day COAL

2.4km

13 km

KINGSTON

0   400
METER

WALKER
BRANCH
WATERSHED

22 km

ANNUAL
WINDROSE

Figure 2.3. Geographical location of Walker Branch Watershed in relation to continental, regional, and local sources of atmospheric emissions due to fossil fuel utilization. Shown are the spatial distribution of centers of coal- and oil-fired power generation in the eastern United States, the location of specific regional coal-fired power plants and major urban centers, and the proximity of three local power plants to the watershed.

third plant, situated ~2.4 km to the northwest of the watershed, is a much smaller unit operated by the U.S. Department of Energy on the Oak Ridge Reservation. This unit is equipped with an 87-m stack and somewhat older, less efficient (98%) precipitators. The facility burns 300 Mg of coal daily for steam production only.

In addition, there are several other sources of atmospheric emissions within ~35km of the watershed. Numerous manufacturing industries are located in the Tennessee River Valley in the Rockwood-Harriman industrial corridor, ~35km to the west-southwest. The industrial corridor includes facilities for the production of steel, wood, paperboard, and textile products. The TVA Kingston Power Plant is also located in this corridor. Sources in this direction from the watershed may be particularly important because of the influence of the topography, which is dominated by parallel ridges. These ridges run west-southwest to east-northeast and strongly influence the channeling of winds through the parallel valleys. The annual windrose in Fig. 2.3 indicates that this industrial corridor is frequently upwind of the watershed. Total annual emissions during 1977 from coal combustion in this area are as follows (in megagrams): suspended particulates = 35.6×10^3, sulfur oxides = 157×10^3, nitrogen oxides = 30×10^3. The contributions from the Kingston Power Plant have been estimated to be 97, 98, and 99%, respectively. Of the total annual emissions in 1977 in the county due to combustion of all types of fossil fuel (coal, fuel oil, natural gas, and gasoline), the relative contribution of the Kingston Power Plant was 97% of the suspended particulates and sulfur oxides and 94% of the nitrogen oxides. The potential influence of this one point source on local air quality is obvious.

The possibility of regional effects on local air quality must not be neglected, however. In addition to these local sources, the watershed is located ~20 km west of greater Knoxville, within 260 km of three major regional urban centers (Chattanooga, Nashville, and Atlanta), and within 350 km of 22 coal-fired power-generating stations with a combined generating capacity of 2×10^4 MW (Fig. 2.3).

2.4 Geology (Henderson et al. 1971)

The Walker Branch Watershed basin is underlain by the Knox Group, a 610-m-thick sequence of siliceous dolomite, divided into four formations on the basis of minor variations in lithology and chert characteristics. Approximate contacts between formations are shown in Fig. 2.4, and the stratigraphic sequence, thicknesses, and general characteristics are summarized in Table 2.1. Bedding planes in the rocks generally strike north 60° east, and dip is 30 to 40° to the southeast. Although it is likely that minor structural anomalies exist in the basin as small folds or faults, no evidence of such anomalies was observed in adjacent basins. Two prominent sets of joints were observed, one striking about north 55° west and the other nearly east-west.

The bedrock geology is largely obscured by a mantle of weathering products from the underlying dolomite and is thickest along drainage divides. Chert is interspersed throughout the weathered mantle, but is concentrated at the ground surface and the bedrock-weathered material interface. The depth to bedrock is highly variable, but often reaches 30 m on ridgetops.

2.5 Soils (Peters et al. 1970)

The soils formed over the dolomitic substrate are primarily Ultisols. These soils develop in humid climates of the temperate to tropical zones on old or highly weathered parent material under forest or savannah vegetation. Small areas of Inceptisols are found in alluvial areas adjacent to the streams. The soils are generally well drained and have a high infiltration capacity. The predominant clay mineral found in these soils is kaolinite, with lesser amounts of vermiculite, hydrous micas, and quartz forming the complement.

Soils of the Fullerton and Bodine Series, both typic Paleudults, occupy over 90% of the watershed (Fig. 2.5 and Table 2.2). Fullerton soils occupy the ridgetops and upper-slope positions, and Bodine soils are found on intermediate and lower slopes. Claiborne soils occupy minor areas in the major stream bottom of the east catchment on alluvial or colluvial deposits washed from the uplands. Stonyland and Rockland occur on lower slopes near the weirs where the dolomite substrate tends to outcrop or lie near the soil surface.

Both the Bodine and Fullerton soils are acidic (pH 4.2–4.6) and low in exchangeable bases, nitrogen, and phosphorus. The relatively thin A1 horizons have moderate base saturation (25%), whereas the lower, A and B, horizons have approximately 10 and 7%, respectively. The Bodine and Fullerton soils are very permeable and very well drained.

The minor soil series on the watershed are widely different from the Bodine and Fullerton soils. The colluvial origin of the Linside soil is apparent from the relatively high content of sand and chert in the lower horizons. The fragipan in the profile description of the Tarklin soil is indicated by the higher bulk density in the lower horizon of that soil. Claiborne, though morphologically similar to the Bodine and Fullerton soils, is chemically more similar to the Linside and Tarklin soils. It is more acid than the latter two soils, but has a moderate base saturation. These three soils (Linside, Tarklin, and Claiborne) are cherty, moderately fertile, and permeable. Though minor in extent, their proximity to the stream make them extremely important in terms of both their ability to supply moisture to mesophytic and riparian vegetation and their possible influence on the quality of water flowing into the stream.

In summary, the soils of Walker Branch Watershed can be described as very cherty, infertile, and very permeable. They are ill suited for agriculture, and as such, are characteristic of many forested soils of the Ridge and Valley Physiographic Province.

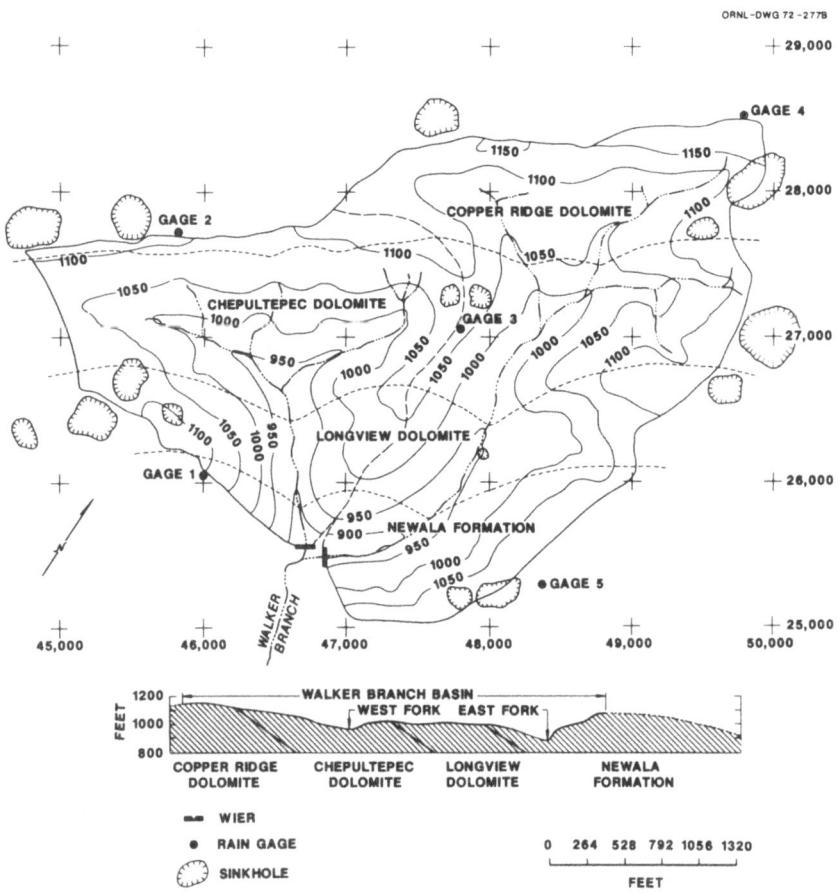

Figure 2.4. Topographic map of Walker Branch Watershed, showing bedrock types and approximate boundaries. Sinkhole depressions that possibly are tributary to Walker Branch are outlined. Mapping was done using aerial photographs at a scale of 1 in. to 264 ft and a 10-ft contour interval map (1 ft = 0.3048 m).

ORNL-DWG 87-12230

SYSTEM	SERIES	GROUP	FORMATION	THICKNESS IN FEET (APPROX)	DESCRIPTION
ORDOVICIAN	LOWER ORDOVICIAN	KNOX	NEWALA FORMATION	1100	UPPER PART IS DOLOMITE, LIGHT GRAY, VERY FINELY CRYSTALLINE, SILICEOUS, MOSTLY IN BEDS 2 TO 4 FEET THICK. OUTCROP AREA OF UPPER PART IS OUTSIDE WALKER BRANCH BASIN ALONG SOUTHEAST SLOPES OF SOUTH-EAST DRAINAGE DIVIDE. LOWER PART IS DOLOMITE, LIGHT GRAY, COMMONLY CONTAINING ABUNDANT MAROON AND GREENISH MOTTLINGS; FINELY CRYSTALLINE: ABUNDANT JASPER CHERT IN NODULES AND LAYERS AND ON GROUND SURFACE. A FEW THIN BEDS OF OLIVE-GRAY LIMESTONE. WEATHERED MATERIAL OVER BEDROCK APPEARS TO BE THINNER THAN OVER OTHER FORMATIONS IN BASIN AND OUTCROPS ARE COMMON.
			LONGVIEW DOLOMITE	400	DOLOMITE, LIGHT GRAY TO GRAY-TAN FINELY TO COARSELY CRYSTALLINE, VERY SILICEOUS. ABUNDANT CHERT IN WEATHERED MATERIAL IS GENERALLY LIGHT-COLORED, DENSE, ANGULAR, CHERT OCCURS IN BLOCKS LOCALLY OF LARGE SIZE. WEATHERED MATERIAL APPEARS TO BE FAIRLY THICK AND OUTCROPS ARE SPARSE.
			CHEPULTEPEC DOLOMITE	550	DOLOMITE, LIGHT GRAY FINELY TO MEDIUM CRYSTALLINE. LESS SILICEOUS THAN OTHER FORMATIONS IN WALKER BRANCH BASIN, BUT LOCAL CHERT CONCENTRATIONS OCCUR ON GROUND SURFACE. CHERT MOSTLY LIGHT GRAY, POROUS OR CAVERNOUS, ROUNDED SURFACES, AND CONTAINS SMALL WHITE OOLITES. PRESENCE OF SANDSTONE BEDS IN LOWER PART INDICATED BY 3- TO 6-INCH THICK BLOCKS OF LIGHT-BROWN SANDSTONE FLOAT. WEATHERED MATERIALS ARE FAIRLY THICK AND OUTCROPS ARE VERY FEW.
CAMBRIAN	UPPER CAMBRIAN		COPPER RIDGE DOLOMITE	1000	DOLOMITE, MEDIUM TO DARK GRAY, COARSELY CRYSTALLINE, ASPHALTIC, VERY SILICEOUS. DOES NOT CROP OUT IN WALKER BRANCH BASIN. RESIDUAL CHERT IS TYPICALLY LIGHT TO MEDIUM GRAY, DENSE, AND CONTAINS PROMINENT BANDED OOLITES. WEATHERED MATERIALS APPEAR TO BE THICK.

Figure 2.4. *Continued.*

Table 2.1. Characteristics of geologic formations underlying Walker Branch Watershed basin

System	Series	Group	Formation	Thickness (m)
Ordovician	Lower Ordovician	Knox	Newala	100
Ordovician	Lower Ordovician	Knox	Longview Dolomite	34
Ordovician	Lower Ordovician	Knox	Chepultepec Dolomite	50
Cambrian	Upper Cambrian	Knox	Copper Ridge Dolomite	93

Source: G.S. Henderson et al. 1971. Walker Branch Watershed: A study of terrestrial and aquatic system interactions. pp. 30–48. IN Ecological Sciences Division Annual Report, 1971. ORNL-4359. Oak Ridge National Laboratory, Oak Ridge, Tennessee.

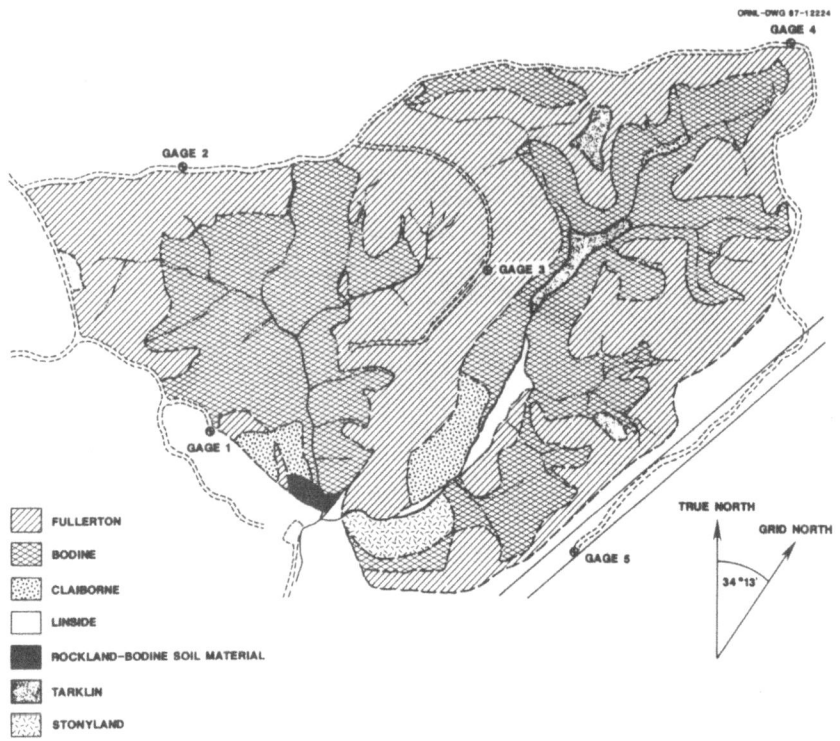

ORNL-DWG 87-12224

GAGE 4

GAGE 2

GAGE 3

GAGE 1

TRUE NORTH

GRID NORTH

34°13'

GAGE 5

▨ FULLERTON

▨ BODINE

▨ CLAIBORNE

☐ LINSIDE

■ ROCKLAND-BODINE SOIL MATERIAL

▨ TARKLIN

▨ STONYLAND

Figure 2.5. Soil map of Walker Branch Watershed.

2.6 Vegetation (Harris 1977)

The vegetation is primarily oak-hickory (*Quercus* spp.–*Carya* spp.), with scattered pine (*Pinus echinata* Mill. and *P. virginiana* Mill.) on the ridge-tops and mesophytic hardwoods (predominantly *Liriodendron tulipifera* L. and *Fagus grandifolia* Ehrh.) in protected coves and stream bottoms (Fig. 2.6). The mean basal area is ~23 m²/ha. Detailed descriptions of Walker Branch Watershed and the four originally defined major forest types are given by Grigal and Goldstein (1971). The four major forest types originally defined were (1) oak-hickory (*Quercus-Carya* spp.), (2) chestnut oak (*Q. prinus*), (3) yellow-poplar (*Liriodendron tulipifera*), and (4) pine (*Pinus echinata* and *P. virginiana*).

The oak-hickory forest type consists primarily of hickory (*Carya glabra* and *C. tomentosa*), chestnut oak (*Quercus prinus*), white oak (*Q. alba*), and red maple (*Acer rubrum*), with lesser amounts of sourwood (*Oxydendrum arboreum*), blackgum (*Nyssa sylvatica*), yellow-poplar (*Liriodendron tulipifera*), and (in descending order of importance) occasional

Table 2.2. Soil series and mapping units observed on Walker Branch Watershed

Soil series	Texture	Slope range (%)	Degree of erosion	Comprehensive classification (7th approximation)
Tarklin	Sil to cherty sil	2–12	Slight	Typic Fragiudults; fine-loamy, siliceous, thermic
Fullerton	Sil, cherty sil to cherty sicl	5–30	Slight to severe	Typic Paleudults: clayey, kaolinitic, thermic
Bodine	Cherty sil	5–30	Slight	Typic Paleudults; loamy-skeletal, siliceous, thermic
Claiborne	Cherty sil to cherty sicl	12–30	Slight to servere	Typic Paleudults: fine-loamy, siliceous, mesic
Linside	Sil	0–5	Slight	Aquic Fluventic Eutrochrepts; fine, silty, mixed, mesic
Stonyland (Fullerton soil material)	Dolomite outcrop	20	—	Not classified
Rockland (predominately chert)	Chert fragments	20	—	Not classified

Source: Adapted from L.N. Peters et al. 1970. Walker Branch Watershed Project: Chemical, physical, and morphological properties of the soils of Walker Branch Watershed. ORNL/TM-2968. Oak Ridge National Laboratory. Oak Ridge, Tennessee.

shortleaf pine (*Pinus echinata*), red oak (*Q. rubra*), black oak (*Q. Velutina*), with dogwood (*Cornus florida*) in the understory. These forests occur both on ridgetops and on slopes. The chestnut oak type consists primarily of chestnut oak with (in descending order of importance) lesser amounts of white oak, hickory, red maple, shortleaf pine, blackgum, and sourwood. This forest type occupies relatively dry sites near ridgetops. The yellow-poplar type consists primarily of yellow-poplar with lesser amounts of hickory and (in descending order of importance) occasional white oak, red maple, red oak, and sourwood. These forests occur primarily along streams and in valleys (Grigal and Goldstein 1971) but are invading dry, ridgetop pine-type forests as well. The pine type consists primarily of shortleaf pine with lesser amounts of yellow-poplar and (in descending order of importance) very occasional hickory, red maple, and dogwood. These forests occur primarily on steep slopes. A 2.5-ha loblolly pine (*Pinus taeda*) plantation, which was planted in 1949 on a ridgetop in the northwest corner of the watershed, was later designated a fifth forest type. The stand consists almost entirely of loblolly pine with an occasional invading yellow-

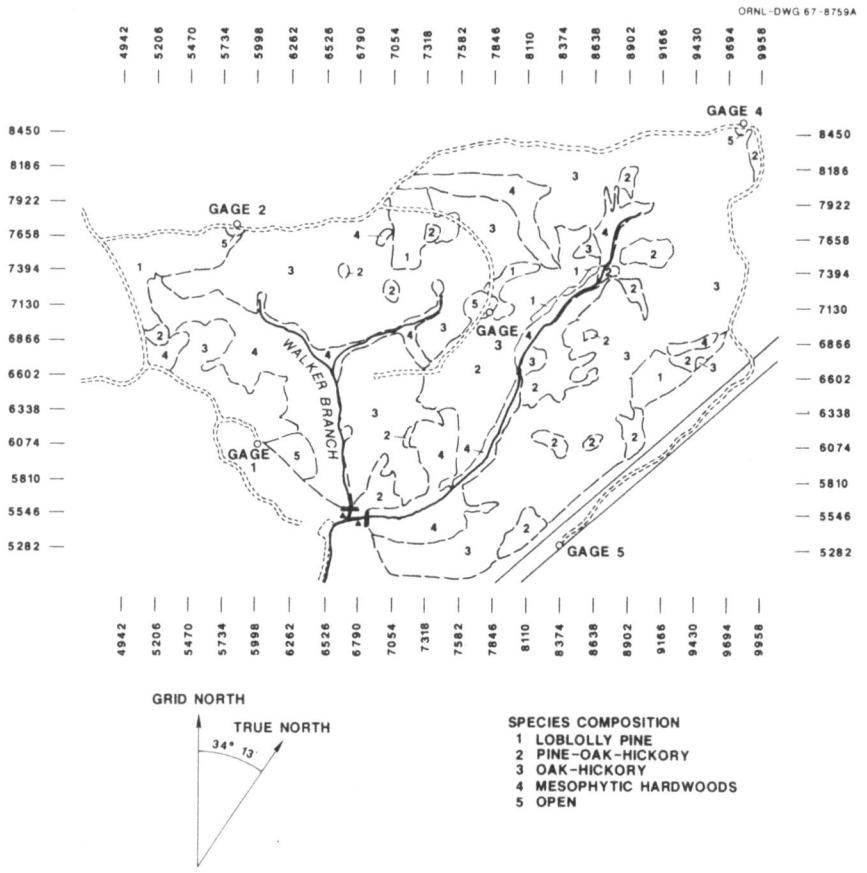

Figure 2.6. Forest cover map.

poplar and dogwood in the understory. The biomass pools and dynamics of these forests are described in Chapter 6.

2.7 Land-Use History (Henderson et al. 1971)

Prior to acquisition by the U.S. government in 1942, the Atomic Energy Commission's OakRidge Reservation was primarily inhabited by agriculturally oriented families. Land settlement progressed from the major river bottoms and terraces up the minor stream courses to the ridgetops. Thus, population densities and intensity of agricultural activity decreased from river bottoms to ridgetops, reflecting the suitability of the land for cultivation. Walker Branch Watershed lies on the southern slope of Chestnut Ridge and was less intensively managed than adjacent land along the Clinch

River. In spite of the steep slopes (average ~30%) and shallow soils, cultivation and other agricultural and forestry practices were employed in the area and influenced subsequent vegetation and soil characteristics.

Land management in the area immediately prior to 1942 varied greatly, depending on the topography and fertility of the land and the needs of individual owners. Seven major land-use areas are delineated in Fig. 2.7, and descriptions of the areas are given in Table 2.3. Agriculture was practiced on 21% of the watershed with varying degrees of intensity. Most farming was of the subsistence type, the notable exception being area 5A, where some vegetables and flowers were grown for sale at local markets. The gardens provided food for family consumption, and the fields yielded hay and corn for livestock feed, with some of the corn undoubtedly diverted to the production of whiskey. A technique commonly used in subsistence agriculture was practiced on the watershed, particularly in areas 4A and 4B. Land on steep slopes was cleared and subsequently cultivated until decreased fertility and erosion limited crop yields. These areas were allowed to revegetate naturally and were then used as pasture, with burning used to control brush and sage grass. Fields were burned in the winter and pastured the following spring and summer. Occasionally, a field was allowed to lie fallow 2 or 3 years to increase the organic matter content of the soil and then cultivated again for 1 or 2 years. Both of these practices

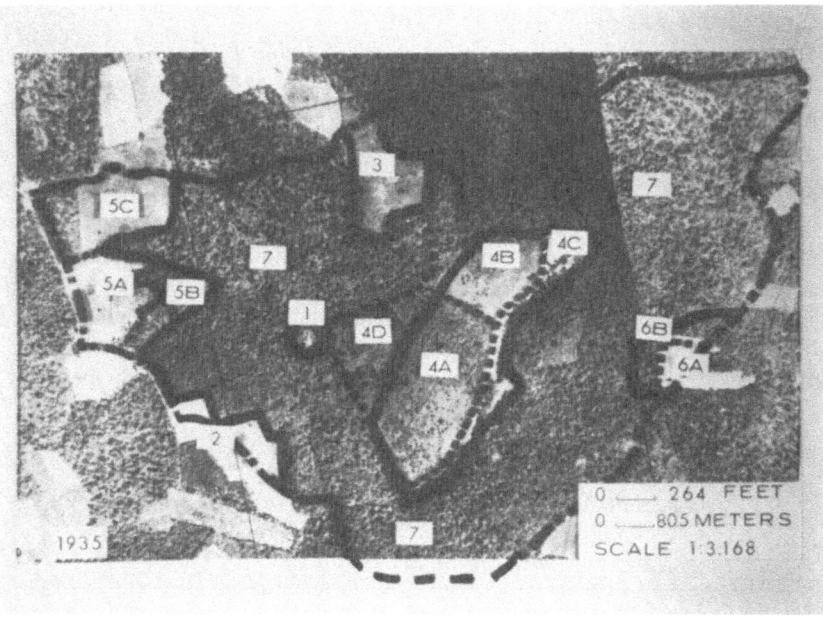

Figure 2.7. Land-use areas on Walker Branch Watershed superimposed on 1935 photograph of the area. Descriptions of land-use areas are given in Table 2.3.

Table 2.3. Descriptions of land-use areas on Walker Branch Watershed as delineated in Fig. 2.7

Area	Description
1	Homesite with small cabin in the bottom of the west catchment; rock walls and flower beds; lawn and garden maintained immediately north and west of cabin; no large-scale farming practiced owing to narrowness of valley.
2	Two large fields on gently sloping ridgetop extending beyond boundary of watershed; primarily corn grown without addition of fertilizer; no evidence of severe erosion.
3	Pasture on gently sloping ridgetop extending onto upper slopes, which are somewhat steeper; some erosion on steeper slopes; area severely burned in 1967, causing complete seedling and sapling mortality as well as some mortality in the overstory.
4	This area represents the agriculture complex of the owner and is broken down into the following sections:
4A, 4B	Areas on steep lower and middle slopes alternately used for pasture and cultivated crops; burn-and-graze type of agriculture; advanced sheet and gully erosion active in 1942; mostly burned in 1967.
4C	Level bottom in east catchment; homesite with outbuildings in northwest portion; extensive gardens and several fruit and walnut trees; some better pasture; a road ran through this area.
4D	Pastured woodlot on steep-to-gentle slopes; probably better timber selectively removed; this area was subjected to burning from fire escaping from areas 4A and 4B.
5	Extensively cultivated area on gentle ridgetops grading to steeper upper-position slopes; divided into following areas:
5A	Predominately gently sloping ridgetop; large commercial vegetable garden and hay and corn fields; erosion evidence slight; a barn occupied southwest portion of this area.
5B	Pasture on farily steep slopes; heavily used prior to 1930, then partially abandoned owing to severe erosion.
5C	Western portion of this area was a woodlot until 1930 when it was partially cleared for pasture; eastern portion was a productive corn and hay field, and fertilization and crop rotation were apparently practiced; entire area was planted to loblolly pine in 1951 or 1952.
6A, 6B	Predominately a pastured area, which changed repeatedly between 1924 and 1942; gently sloping ridgetop with small cabin; A is the oldest cleared area; B is the most recently cleared.
7	Includes the bulk of the watershed; steep topography with relatively undisturbed deciduous forest; oak trees selectively cut for railroad ties; certain areas may have been used more extensively for agriculture prior to the 1900s.

caused severe erosion and soil degradation, which eventually led to abandonment or conversion to woodlot pasture.

The landowners practiced little forest management prior to 1942. They cut small quantities of timber for their own use for building construction, fence rails, and firewood. They also sold some timber to neighbors, who selectively harvested larger oaks for use as railroad ties.

It is difficult to assess the total impact of fire on Walker Branch Watershed. It is assumed that most of the area has burned periodically throughout its history. Fires probably escaped from pastures where fire was used to control brush. These fires were not severe, consuming only ground cover and some of the soil's organic matter layer; however, they undoubtedly contributed to some soil degradation due to increased erosion and susceptibility of nutrients being leached from the soil.

In 1967 two fires burned parts of the watershed shown in Fig. 2.8. One of these burned only 1.8 ha on the west catchment of the watershed and had a limited effect on the vegetational structure. The other was more severe, burning a total of 39 ha, mostly on the east catchment. This fire was extremely hot on the ridgetops, as evidenced by tree trunks' being charred as high as 4 m above the ground, but it decreased in intensity as it burned downslope. Since 1967, vegetation patterns in the more severely

Figure 2.8. Aerial extent of two fires that burned on Walker Branch Watershed during 1967 superimposed on 1967 photograph of the area.

burned areas have changed appreciably: most of the understory and some overstory trees have died, with a resultant upsurge in herbaceous and woody undergrowth. Adjacent burned and unburned acres are compared in Fig. 2.8. The two fires combined burned 38% of the total watershed area, with about 15% burned severely.

References

Curlin, J.W., and D.J. Nelson. 1968. Walker Branch Watershed Project: Objectives, facilities, and ecological characteristics. ORNL/TM-2271. Oak Ridge National Laboratory, Oak Ridge, Tennessee.

Grigal, D.F., and R.A. Goldstein. 1971. An integrated ordination-classification analysis of an intensively sampled oak-hickory forest. J. Ecol. 59:481–492.

Harris, W.F. 1977. Walker Branch Watershed: Site description and research scope. pp. 4–17. IN D. L. Correll (ed.), Watershed Research in Eastern North America. A Workshop to Compare Results. Chesapeake Bay Center for Environmental Studies, Smithsonian Institution, Edgewater, Maryland.

Henderson, G.S., R.M. Anderson, L. Boring, J.W. Elwood, T. Grizzard, W.F. Harris, A. Hunley, W. McMasters, W.J. Selvidge, J.L. Thompson, and D.E. Todd. 1971. Walker Branch Watershed: A study of terrestrial and aquatic system interactions. pp. 30–48. IN Ecological Sciences Division Annual Report, 1971. ORNL-4359. Oak Ridge National Laboratory, Oak Ridge, Tennessee.

Lindberg, S.E., R.C. Harriss, R.R. Turner, D.S. Shriner, and D.D. Huff. 1979. Mechanisms and rates of atmospheric deposition of trace elements and sulfate to a deciduous forest watershed. ORNL/TM-6674. Oak Ridge National Laboratory, Oak Ridge, Tennessee.

Peters, L.N., D.F. Grigal, J.W. Curlin, and W.J. Selvidge. 1970. Walker Branch Watershed Project: Chemical, physical, and morphological properties of the soils of Walker Branch Watershed. ORNL/TM-2968. Oak Ridge National Laboratory, Oak Ridge, Tennessee.

Chapter 3
Forest Meteorology

B.A. Hutchison and D.D. Baldocchi

3.1 Introduction

Interactions between a forest canopy and the ambient atmosphere above create the unique microclimate within the forest and govern the exchanges of materials and energy between the forest and the atmosphere in which it is immersed. Consequently, many of the physical, chemical, and biological phenomena operating in forested watersheds are functions of these canopy-atmosphere interactions.

With the support of the U.S. Department of Energy (DOE) and its predecessor agencies, the Atmospheric Turbulence and Diffusion Laboratory (ATDL) of the National Oceanic and Atmospheric Administration (NOAA) (now the Atmospheric Turbulence and Diffusion Division, Air Resources Laboratory, NOAA) established a forest meteorology research site immediately adjacent to the Walker Branch Watershed. This research facility, which became operational in 1978, permits above- and within-canopy studies of the interactions between a deciduous forest canopy and the atmosphere along with characterization of the boundary conditions that govern those interactions. Since the inception of this facility, support from a variety of sources, in addition to continuing DOE sponsorship, has permitted a wide range of research.

The interactions between vegetation, such as a forest, and the atmosphere take the form of exchanges of energy (solar and thermal radiation, latent and sensible heat, and momentum) and mass (e.g., water vapor, CO_2, nutrients, pollutants). These exchanges occur by means of several distinctly different but closely coupled mechanisms: (1) radiative energy transfers; (2) turbulent transfers of mass, momentum, and sensible heat; and (3) gravitational mass transfer from the atmosphere to the forest (i.e.,

precipitation or sedimentation of supermicron-size particles). Because of
fundamental differences in these transfer mechanisms, different measurement techniques are required for studies of the transfer mechanisms
and for the quantification of exchange rates. Parallel research efforts at
the Walker Branch Watershed have addressed a variety of exchanges occurring by means of these phenomena. This chapter summarizes the ATDL
research efforts directed toward canopy radiation transfer and turbulent
exchange mechanisms and rates. Precipitation and sedimentation phenomena are addressed in Chapters 4, 5, and 7.

One of the more significant unifying concepts of contemporary micrometeorology derives from the classical studies of evaporation through porous films conducted by Brown and Escombe (1900). They calculated the
rate of diffusion through a stomatal tube, using Fick's diffusion law, an
analog of Ohm's law, which can be expressed as follows:

$$\text{Flux} = \frac{\text{concentration difference}}{\text{resistance to transfer}}$$

This was later extended by Raschke (1956) to predict the temperature and
transpiration of a leaf from its energy budget. Waggoner and Reifsnyder
(1968) applied the Ohm's law analogy to an entire plant canopy, leading
to the conceptualization of canopy-atmosphere exchange (under steady-
state conditions) as a network of resistances to exchange, as shown in
Fig. 3.1. The utility of this conceptualization is that consideration of the
nature of the resistances identified in Fig. 3.1 enables identification of the
boundary conditions and forcing functions operating in canopy-atmosphere
exchange.

For example, the aerodynamic resistance (r_a) is a function of the wind
speed (u) and atmospheric stability (ϕ) in the free atmosphere above the
canopy. Both u and ϕ are affected by large-scale surface characteristics
such as albedo and topography. The aerodynamic resistance is also a
complex function of the local canopy structural features, including crown
shape and dimension and canopy element spatial distribution and density,
which define the aerodynamic roughness (z_0) of the canopy. This leads to
the conclusion that the turbulent exchange between a forest and the atmosphere is governed by ambient properties of the atmosphere such as u,
ϕ, net radiation (R_n), and canopy architecture as it affects surface roughness (z_0).

At the surfaces of canopy elements as well as at soil, litter, rock, or
water surfaces on the forest floor, another set of resistances to transfer
must be considered. Diffusive boundary layer resistances exist by virtue
of the fact that very near surfaces immersed in a moving fluid a quasi-
laminar boundary layer develops in which turbulence is damped out by
the viscosity of the air. Thus, the only mechanism operating to transfer
entities through such layers is molecular diffusion. While the molecular

ATDL-M86/478

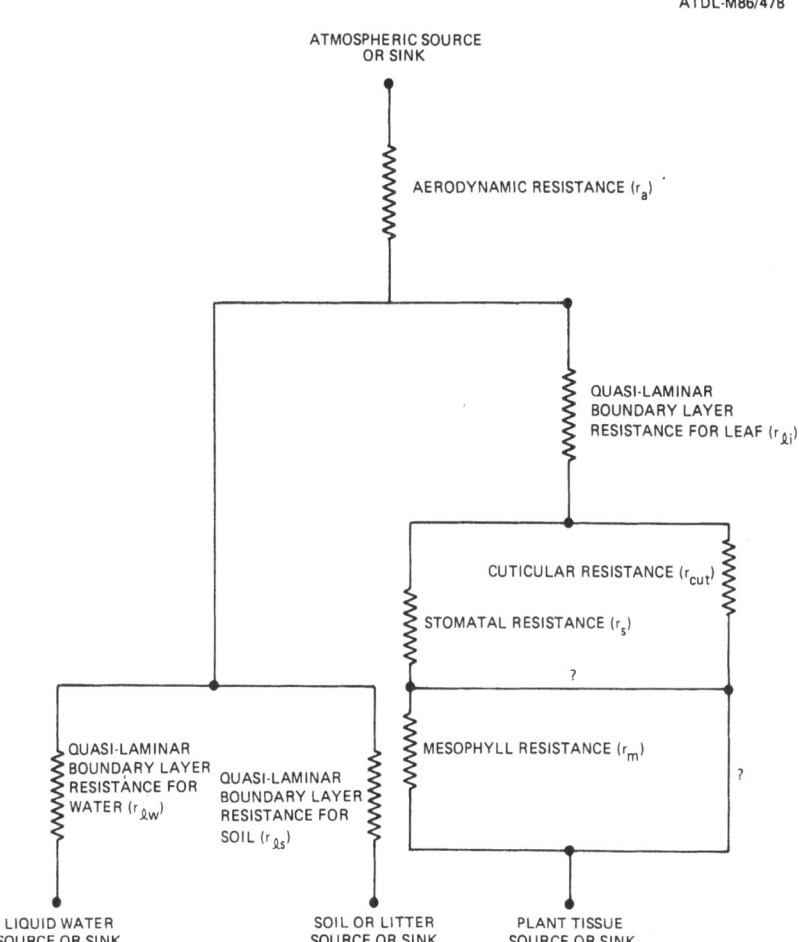

Figure 3.1. Schematic representation of canopy-atmosphere exchanges as an analog to Ohm's law for electrical current flow.

diffusivity of an entity is only a function of its molecular weight and temperature, the diffusive resistances (r_ℓ's) for the variety of surfaces encountered within a forest are functions of other factors in addition to molecular diffusivity. Since the thickness of the quasi-laminar layer profoundly affects the length of time required for diffusion of a given amount of an entity through the layer, the various r_ℓ's are functions of both the molecular diffusivity of the entity under consideration and the thickness of the quasi-laminar boundary layer. Because the thickness of the quasi-laminar layer is affected by u and ϕ in the vicinity of the surface and by microscopic surface roughness features, r_ℓ is a function of ambient micrometeorological conditions in the vicinity of the surfaces under consideration as well as of the characteristics of the surfaces in question.

For biologically active materials whose sources or sinks lie within plant tissues, other resistances come into play. For pathways into the plant involving the stomata and the mesophyll tissues, resistances to molecular diffusion through the stomata (r_s) and in the cellular interstices of the mesophyll (r_m) must be considered. Stomatal geometry is implicated in defining r_s, but since stomatal behavior is environmentally dependent and physiologically mediated, r_s is also a function of the status of water in the soil-plant-atmosphere continuum, atmospheric concentrations of CO_2, O_3, and other materials affecting stomatal physiology, as well as air temperatures, vapor pressure deficits, and amounts of photosynthetically active radiation (PAR) incident upon the leaf under scrutiny. The mesophyll resistance (r_m) is a function of the molecular diffusivity of the entity being diffused, just as are the various r_l's and r_s. In addition, r_m is a function of the source or sink strengths within the leaf of the entity in question. With r_m, the distance over which molecular diffusion must occur, which is a function of the cellular structure of the leaves in question, is also implicated. Since the cellular surfaces of tissue cells in leaves are moist, the degree of affinity between the diffusing entity and water is also important in defining r_m. If the leaf cuticle is not impervious to an entity, another pathway between the leaf surface and plant tissues exists. Resistance to transfer through the cuticle ($r$cut) is determined by factors such as the thickness of the cuticular layer and the chemical and physical characteristics of the entity being transferred.

The point of this introduction to the Ohm's law analogy of canopy-atmosphere exchange is that a variety of conditions and processes operate to define and control such exchange. The turbulent transport of energy and mass to and from a forest canopy is affected by interactions among ambient, above-canopy conditions, canopy architecture, the physiological functioning of canopy elements as related to water use and stomatal behavior, and the transfer and partitioning of solar energy, which is the forcing function for all these interrelated processes. Since it is the interactions among the boundary conditions and the forcing function that define the exchange processes and rates, forest canopy-atmosphere exchanges cannot be studied in isolation. To produce generalizable knowledge of exchange mechanisms and quantities, holistic and integrated studies of forest meteorology are required.

This chapter summarizes our attempts to conduct such forest meteorology research at the Walker Branch Watershed. Following a description of the research site and facilities, the characterization of relevant boundary conditions is discussed. Results of intensive experimental and theoretical studies of deciduous forest canopy radiation transfer are presented, and the status and preliminary results of canopy-atmosphere turbulent exchange are described.

3.2 Site Description

The ATDL deciduous forest meteorological research site is situated to the west of the Walker Branch Watershed on the DOE Oak Ridge Reservation. The site is on a spur ridge, which slopes about 3% to the west-southwest and which originates at Chestnut Ridge (see Fig. 3.2). Average elevation is 365 m above MSL, roughly 35 m above the valley floors on either side of Chestnut Ridge.

The soil is cherty silt-loam (Fullerton series, Typic Paleudult), derived from dolomitic limestone, and is characteristically infertile and acidic. It exhibits a high infiltration capacity and a moderate water-retention capacity, and it is well drained. Since the Ridge and Valley Physiographic Province in which this site is embedded is geologically very old, weathering processes have created a deep regolith, exceeding 30 m in depth at some points on the watershed and averaging 13 m in depth. The forest trees, therefore, experience water stress only during extended and severe droughts, since moisture remains available in the regolith after soil moisture nearer the surface is depleted. Such water may be taken up by the deeply penetrating support roots of the trees; some water also diffuses upward to the rooting zone, where it becomes available for uptake by finer tree roots.

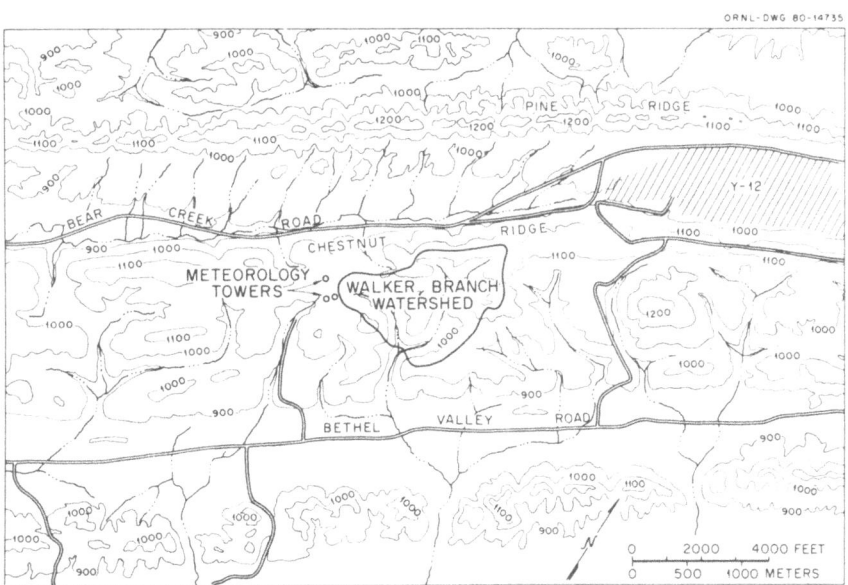

Figure 3.2. Contour map showing location of the Atmospheric Turbulence and Diffusion Laboratory's forest meteorology research facility adjacent to the Walker Branch Watershed on the U.S. Department of Energy's Oak Ridge Reservation. (Contour interval is 100 ft.)

Figure 3.3. General view of the east Tennessee deciduous forest at the Atmospheric Turbulence and Diffusion Laboratory's Walker Branch Watershed site in winter. The decreasing density of woody biomass with depth to a relatively open trunk space is indicative of the leaf distribution in the canopy during the growing season. Also visible is the 33-m tower that serves as a tail spar supporting pairs of cables at seven levels within and one level above the canopy. Instrumented trams (five are clearly visible) traverse these cables to obtain horizontally and vertically replicated measurements of incoming and outgoing fluxes of solar, photosynthetically active, and all-wave radiation. (ATDL photo.)

The meteorology research site is covered by a 40-or-more-year-old oak-hickory stand, which originated with the abandonment of mountain pastures at the advent of the Manhattan Project (ca. 1940). Major codominant species in the forest canopy include *Quercus alba* L., *Q. prinus* L., *Acer rubrum* L., *Q. velutina* Lam., *Q. falcata* Michx., *Liriodendron tulipifera* L., and *Carya glabra* (Wangenh.) Sarg. Understory vegetation consists of seedlings and saplings of these species along with suppressed *Cornus florida* L., *Nyssa sylvatica* Marsh., and *Oxydendrum arboreum* (L.) DC. Further details of the forest stand are presented in Sect. 3.3.2.

Research facilities at this site include a 44-m-tall stairway tower and two 33-m triangular towers that serve as head and tail spars for an instrument-support system. This instrument-support system consists of eight parallel pairs of cables mounted at seven levels within and one level above the canopy. These allow spatial replication of measurements in planes parallel to the forest floor, using a moving-sensor approach. Vertical replication of measurements is achieved through simultaneous operation of the eight levels. In addition, portable buildings located nearby provide field laboratory-workshop space and house a variety of computer-centered data acquisition facilities. A winter view of the research site is shown in Fig. 3.3.

3.3 Characterization of Boundary Conditions

3.3.1 Climatological Monitoring

As noted above, among the boundary conditions governing canopy-atmosphere interactions are the ambient meteorological conditions. Hence, understanding a particular component of these interactions requires knowledge of the relevant above-canopy meteorological conditions obtaining during the event under study. Moreover, since many micrometeorological and ecological phenomena have "memories" (i.e., the phenomena reflect antecedent as well as concurrent conditions), a site climatology is required. In view of this, the research site was instrumented, and basic climatological monitoring began in 1978.

Variables monitored above the canopy include incoming and outgoing fluxes of global solar (R_g), photosynthetically active (PAR), and all-wave (R_a) radiation, diffuse R_g and PAR, wind speed and direction, air temperature, and relative humidity. Within the canopy, the vertical distribution of air temperature and soil temperature is monitored along with the surface soil heat flux. Precipitation inputs to the site are measured with a weighing-bucket rain gage located on the ground in a nearby clearing (Walker Branch Watershed rain gage 2), and an estimate of throughfall amounts is obtained as an event average of records from three other weighing-bucket rain gages on the forest floor at the research site.

With the exception of the various rain gages, all climatological monitoring instrumentation is hardwired to a dedicated data acquisition system for data recording and preliminary processing. After further processing and editing, these data are archived on magnetic tape and microfiche as hourly averages.

Mean daily totals (5 years) of global solar, photosynthetically active, and net radiation incident upon the Walker Branch Watershed over the

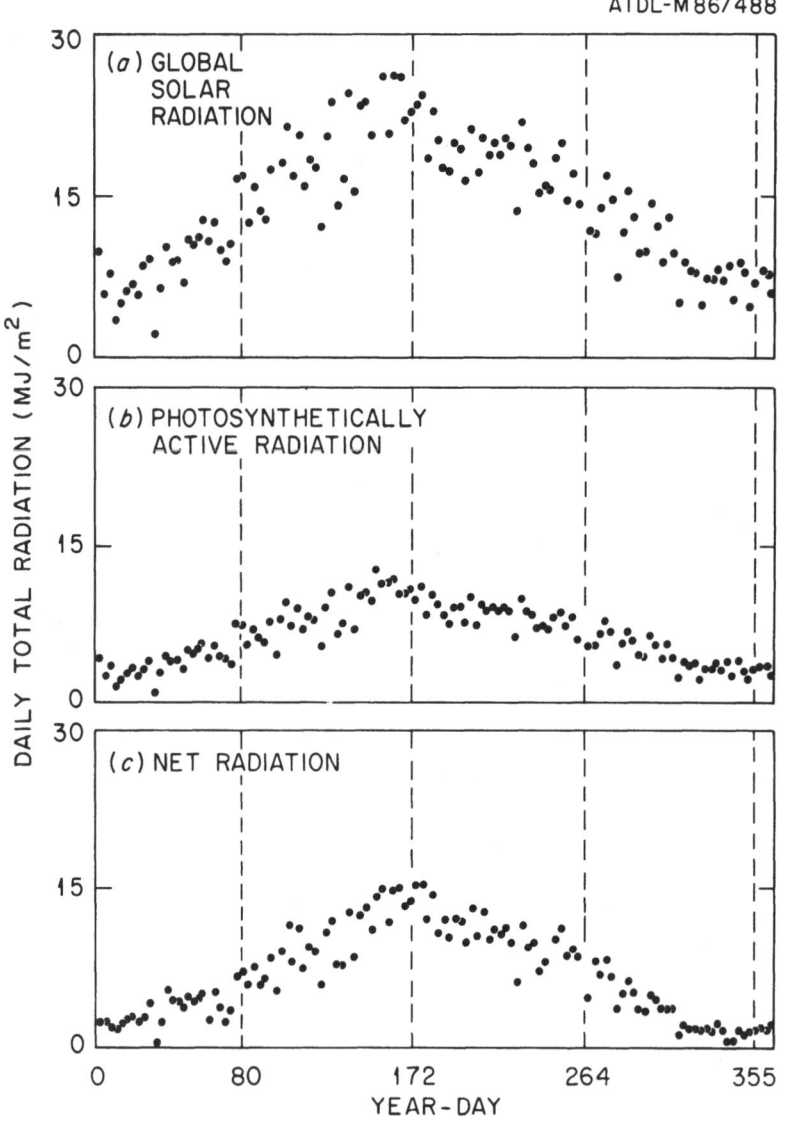

Figure 3.4. Annual course of mean daily radiation totals incident upon the Walker Branch Watershed. Vertical dashed lines indicate the vernal and autumnal equinoxes (year-days 80 and 264, respectively) and the summer and winter solstices (year-days 172 and 355, respectively).

course of the year are shown in Fig. 3.4. Because of seasonal variations in cloudiness and in atmospheric turbidity, the time traces of mean daily global solar and photosynthetically active radiation are asymmetrical about the solstices. Maximal daily totals for both these wave bands occur in the period just prior to the summer solstice. Following the summer solstice, increasing levels of atmospheric turbidity associated with frequent stagnating high-pressure systems and high atmospheric relative humidities reduce insolation more than earlier in the year. In Fig. 3.4c, the effects of the phenological dynamics of canopy structural development are evident in the time course of the daily net radiation total. Because of higher albedos in the leafless canopy phenological phase, more radiation is reflected by the forest about the time of the vernal equinox than about the autumnal equinox, when the forest is still fully leaved. As a result, daily total net radiation is lower around the vernal than around the autumnal equinox.

The data of Fig. 3.4 are further combined as mean daily totals and frequency distributions of daily total solar radiation by phenoseason in Fig. 3.5. The phenoseasons of Fig. 3.5 are defined in Table 3.1. Winter

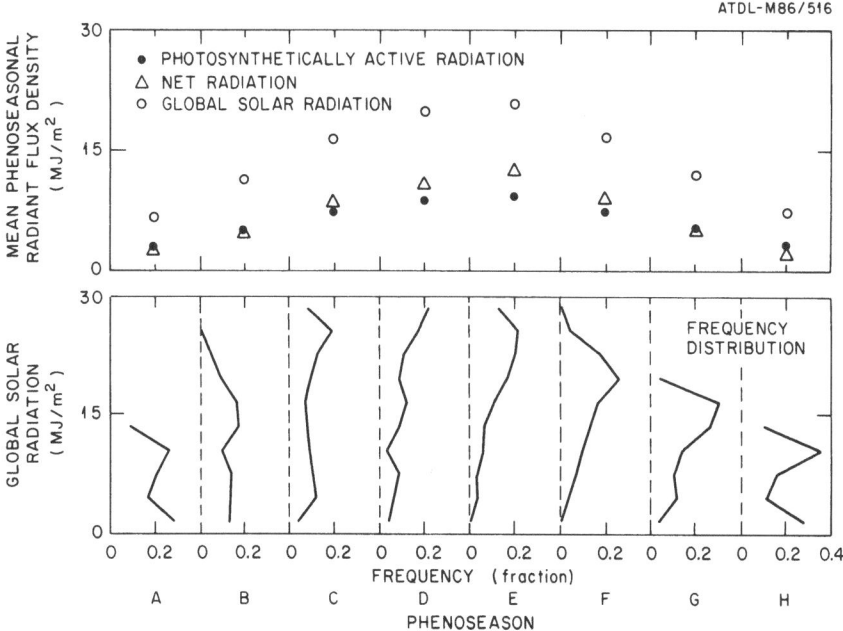

Figure 3.5. Mean daily total photosynthetically active (●), net (△), and global solar (○) radiation by phenoseason. The frequency distributions of mean daily global solar radiation for each phenoseason are shown in the lower plot to illustrate the temporal variation in incident solar radiation resulting from seasonal changes in earth-sun geometry and in atmospheric transparency to solar radiation. A = winter leafless; B = spring leafless; C = spring leaf expansion; D = summer leaf expansion; E = summer fully leaved; F = autumn fully leaved; G = autumn leaf abscission; and H = winter leaf loss phenoseasons.

Table 3.1. Duration of Phenoseasons

Code	Phenoseason	Period of year		
		Year–day	to	Year–day
A	Winter leafless	346		035
B	Spring leafless	036		090
C	Spring leaf expansion	091		125
D	Summer leaf expansion	126		150
E	Summer full leaf	151		220
F	Autumn full leaf	221		275
G	Autumn leaf abscission	276		310
H	Winter leaf loss	311		345

distributions of daily total solar radiation tend to be bimodal, reflecting the alternation of mostly clear and mostly overcast days resulting from winter frontal systems passing over east Tennessee. In summer phenoseasons, convective activity dominates in this region, resulting in high modal daily totals and distributions that are highly skewed toward lower daily total values.

3.3.2 Canopy Architecture

As noted earlier, the forest at the meteorology research site is of the oak-hickory type. The mean diameter at breast height (dbh) of the codominant overstory trees was 0.32 m and basal area was 26 m^2/ha when last surveyed in 1976. The mean height of the trees forming the overstory was 21.5 m, ranging from 17 to 25 m, when measured in 1978.

The architectural features of this canopy, as defined by phytoactinometric theory, have been the subject of intense study. Results of this study are reported in Hutchison et al. (1986). These features include leaf inclination angle distributions and vertical distributions of silhouette area densities of leaves and woody canopy elements. As is discussed in Sect. 3.4, studies of radiative transfer in this forest have shown that the degree of foliage clumping is a critical architectural characteristic in defining forest radiation regimes. However, a technique for direct measurement of this characteristic is yet to be devised.

Figure 3.6 shows the cumulative proportion of leaf area density falling into 10° inclination angle classes for the three major canopy strata in this forest. In the subcanopy, the inclination-angle distributions were planophile (i.e., the leaves were displayed nearly horizontally), according to the de Wit classification scheme (de Wit 1965). Higher above the forest floor, the inclination angle distributions became plagiophile (i.e., most leaves were displayed at angles midway between horizontal and vertical). Mean

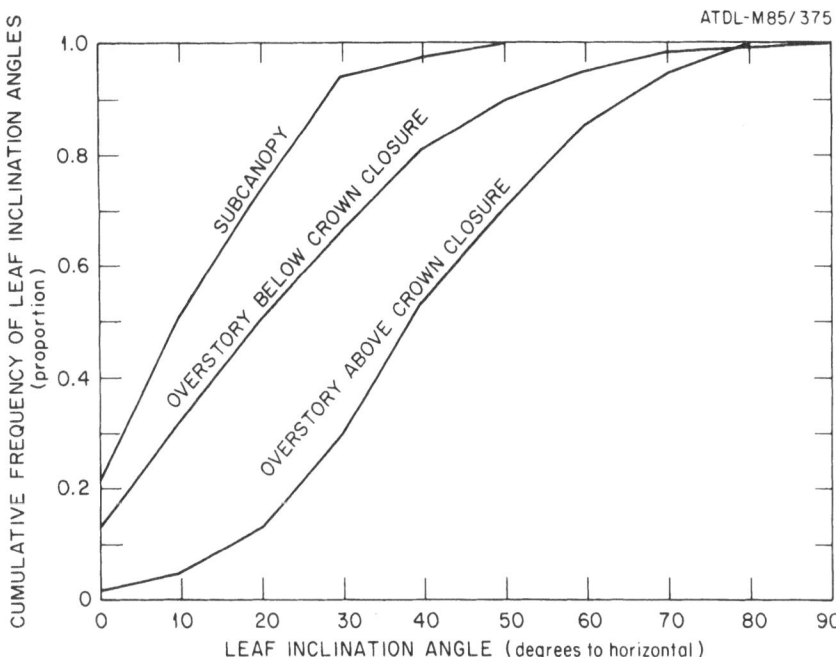

Figure 3.6. Cumulative area-weighted frequency distributions of leaf inclination angles in the three major canopy strata of an east Tennessee deciduous forest. Source: B.A. Hutchison, D.R. Matt, R.T. McMillen, L.J. Gross, S.J. Tajchman, and J.M. Norman. 1986. The architecture of a deciduous forest canopy in eastern Tennessee, U.S.A. J. Ecol. 74:635–676.

leaf inclinations were 10, 20, and 38° in the subcanopy, overstory below crown closure, and overstory above crown closure, respectively.

The vertical distribution of leaf area density in the fully leaved forest canopy is shown in Fig. 3.7. As expected, the greatest leaf area densities were found in the overstory above crown closure. Plotting the data of Fig. 3.7 in integral form yields Fig. 3.8, the cumulative vertical distribution of leaf area. Figure 3.8 also includes the cumulative vertical distribution of the silhouette area of woody canopy elements. All data on Fig. 3.8 are expressed as nondimensional area indexes (i.e., element area per unit ground area, m^2/m^2). As indicated on Fig. 3.8, the total leaf area above the level where crown closure occurs (about 15 to 17 m above the forest floor) was ~3.8. Slightly over one additional unit of leaf area was present in the subcanopy, yielding a total leaf area index of ~4.9 in this forest. The total woody canopy element (silhouette) area index in the stand was ~0.6, giving a total plant area index of ~5.5 for this forest in full leaf.

The vertical distribution of normalized dry leaf weight per unit of surface area is shown in Fig. 3.9 for three common tree species in this forest. These data were normalized by dividing the average unit area dry leaf

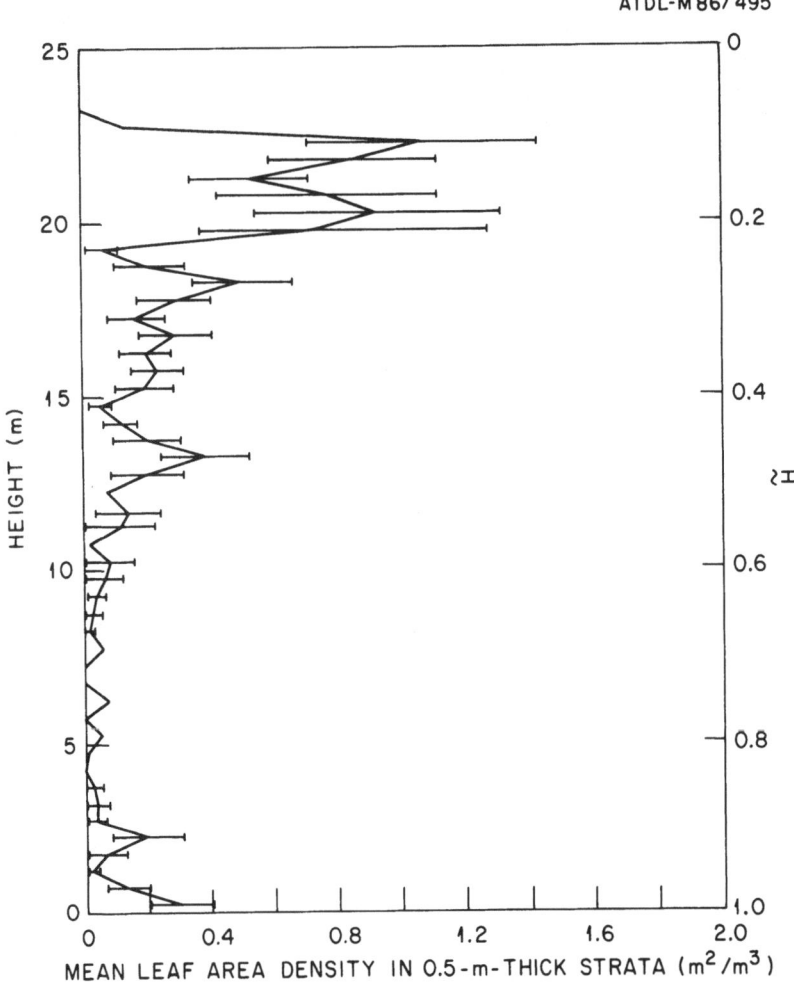

Figure 3.7. Vertical distribution of average leaf area density in an east Tennessee deciduous forest. Horizontal bars represent ±1 SE. The right-hand ordinate represents height in the forest in the nondimensional form: $\bar{H} = 1 - H/\hat{H}$, where H = height in the forest and H = maximum forest tree height. Source: B.A. Hutchison, D.R. Matt, R.T. McMillen, L.J. Gross, S.J. Tajchman, and J.M. Norman. 1986. The architecture of a deciduous forest canopy in eastern Tennessee, U.S.A. J. Ecol. 74:635–676.

weight of individual canopy strata by the maximum value encountered in samples from all strata. The values used for this normalization were 11.83, 12.82, and 10.11 mg/cm² for *Quercus alba*, *Q. prinus*, and *Acer rubrum*, respectively. Insufficient information was obtained to determine whether the uniform increases in the normalized weights per unit area with height in the forest of Fig. 3.9 were the result of increased leaf thickness or tissue density.

ATDL-M 85/372

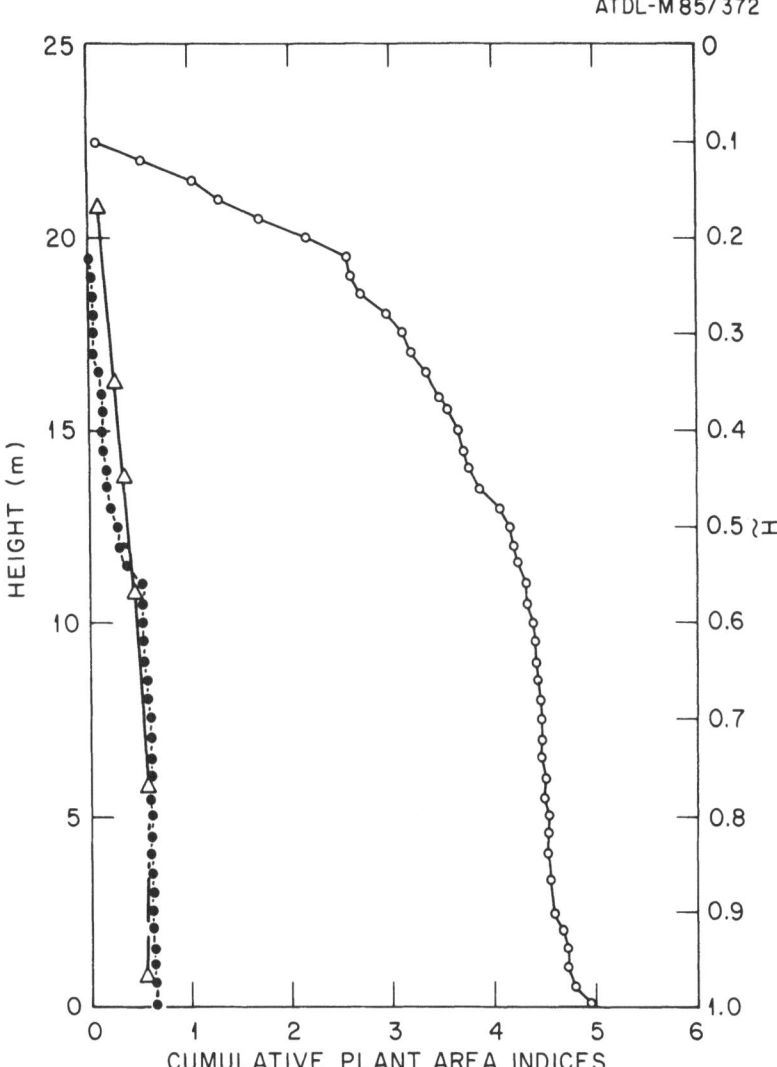

Figure 3.8. Cumulative vertical distribution of leaf area index (○) and woody element silhouette area index (●, direct measurement; △, indirect measurement) in an east Tennessee deciduous forest. $\bar{H}$ is the nondimensional height defined in Fig. 3.7. Source: B.A. Hutchison, D.R. Matt, R.T. McMillen, L.J. Gross, S.J. Tajchman, and J.M. Norman. 1986. The architecture of a deciduous forest canopy in eastern Tennessee, U.S.A. J. Ecol. 74:635–676.

If forest canopy architecture were static, the data presented in Figs. 3.6 through 3.9 would constitute a reasonably complete characterization of a canopy in terms of radiation transfer. However, the structure of a deciduous canopy varies drastically from season to season and more subtly from year to year. Thus, these data must be accepted only as approximations of canopy structure at the phenological extremes of the full-leaf

ATDL-M 85/378 R

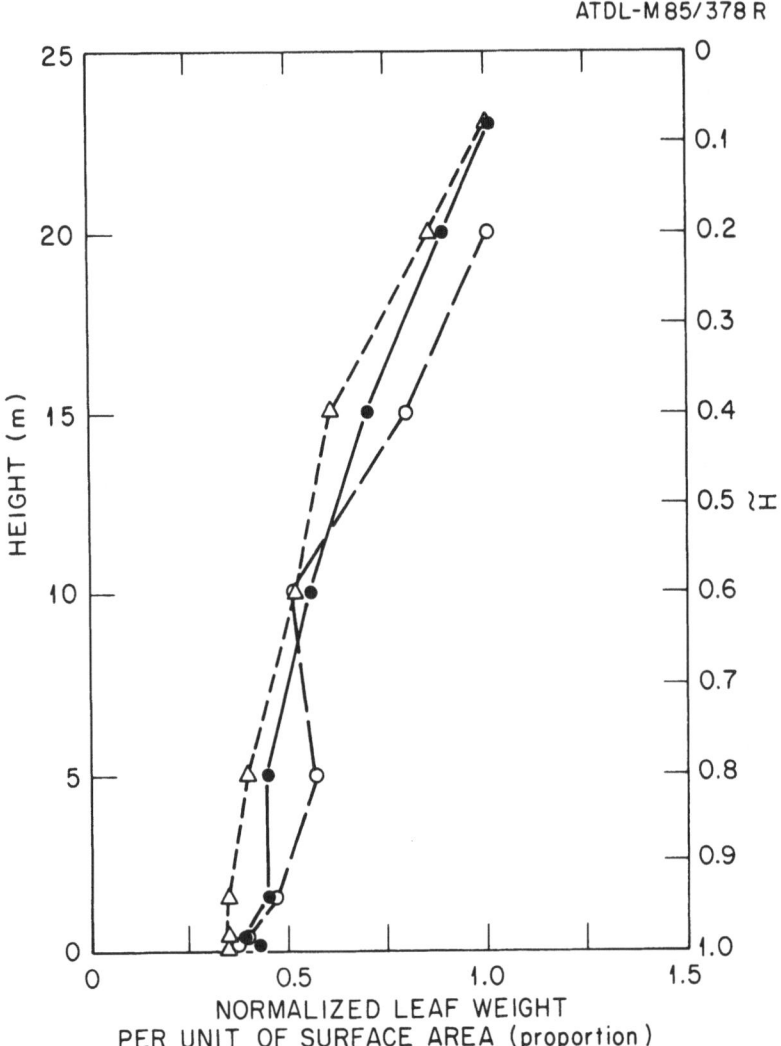

Figure 3.9. Mean normalized leaf weight per unit surface area of three species of trees as a function of height in an east Tennessee deciduous forest canopy (●, *Quercus alba*; △, *Q. prinus*; and ○, *Acer rubrum*). $\bar{H}$ is the nondimensional height defined in Fig. 3.7. Source: B.A. Hutchison, D.R. Matt, R.T. McMillen, L.J. Gross, S.J. Tajchman, and J.M. Norman. 1986. The architecture of a deciduous forest canopy in eastern Tennessee, U.S.A. J. Ecol. 74:635–676.

and leafless phases. More efficient, indirect canopy structure assessment techniques are being developed and tested to allow characterization of the phenological and successional dynamics of canopy architecture. We noted earlier that our test of phytoactinometric theory in this forest indicated that foliage clumping was an additional structural characteristic of importance in terms of canopy radiation transfer. Techniques for assessments of canopy element clumping must be developed. In addition,

the architectural characteristics of forest canopies of importance in turbulent mass and energy exchange have been little studied. Thus, two major
tasks at the Walker Branch site for the future must be to identify and
define such characteristics and to develop techniques for their quantification.

Canopy radiative transfers are also governed by the radiative characteristics of canopy elements. Leaf reflectivities and transmissivities as
functions of tree species, canopy level, upper vs. lower leaf surfaces,
phenology, and spectral wave band have been measured for common tree
species at the Walker Branch site. Data for well-watered, fully expanded
leaves in the photosynthetically active radiation wave band are summarized
in Table 3.2. While lower surface reflectivities (ρ_ℓ's) are from two to three

Table 3.2. Reflectivities and transmissivities of canopy leaves in the photosynthetically
active radiation (PAR) (0.4- to 0.7-μm) wave band of the electromagnetic spectrum

Species			Reflectivity (ρ)		Transmissivity (τ)	
Common name	Scientific name	Canopy position	Upper surface	Lower surface	Upper surface	Lower surface
Overstory species						
Chestnut oak	*Quercus prinus*	Crown	0.069	0.21	0.060	0.045
Chestnut oak	*Quercus prinus*	Subcanopy	0.074	0.17	0.066	0.068
Tulip poplar	*Liriodendron tulipifera*	Crown	0.084	0.26	0.070	0.059
Tulip poplar	*Liriodendron tulipifera*	Subcanopy	0.092	0.21	0.12	0.12
White oak	*Quercus alba*	Crown	0.087	0.19	0.077	0.068
White oak	*Quercus alba*	Subcanopy	0.076	0.18	0.078	0.066
Red maple	*Acer rubrum*	Crown	0.059	0.30	0.071	0.041
Red maple	*Acer rubrum*	Subcanopy	0.083	0.22	0.093	0.10
Shagbark hickory	*Carya ovata*	Subcanopy	0.081	0.12	0.076	0.084
Understory species						
Sassafras	*Sassafras albidum*	Subcanopy	0.062	0.22	0.10	0.11
Sourwood	*Oxydendrum arboreum*	Subcanopy	0.11	0.15	0.12	0.12
Dogwood	*Cornus florida*	Subcanopy	0.10	0.22	0.12	0.10
Devil's walking stick	*Aralia spinosa*	Subcanopy	0.073	—	0.073	—

times greater than upper surface reflectivities (ρ_u's), the transmissivities (τ's) of the two surfaces are similar.

3.3.3 Canopy Physiology

Canopy physiology operates as a further boundary condition affecting canopy-atmosphere exchanges of mass and energy. This influence is effected by the mediating role of stomatal and mesophyll resistance distributions in the uptake and evolution of biologically active materials. Thus, an understanding of canopy-atmosphere exchange presupposes knowledge of the spatial and temporal dynamics of those physiological processes affecting r_s and r_m and of the nature of the process effects on r_s and r_m distributions in space and in time.

Because of the vertical extent and spatial heterogeneity of the uneven-aged, mixed species forest canopy under scrutiny at the Walker Branch site, routine direct measurement of stomatal and mesophyll resistance distributions is infeasible. Consequently, our approach in the case of r_s has been to develop a physically and physiologically realistic model of bulk canopy stomatal resistance (r_{sc}). By coupling a canopy radiation transfer model (Norman 1979, 1982) and a leaf stomatal resistance model (Jarvis 1976), individual leaf r_s is calculated on the basis of canopy irradiance distributions and observed variations in r_s with PAR flux densities (Baldocchi et al. 1987). An example of the observed variation in r_s of white oak (*Quercus alba*) as a function of incident PAR is shown in Fig. 3.10.

In using the model, the individual leaf r_s's are averaged to obtain r_{sc}. Thus, the nonlinear response of stomatal aperture to irradiance is accounted for. The canopy stomatal resistance model was tested for water vapor exchange against estimates of r_{sc} derived from steady-state diffusion porometer measurements of r_s on an individual leaf basis in a soybean canopy (Baldocchi et al. 1985a) and from computations using the Penman-Monteith equation for latent heat exchange (Monteith 1973). Individual leaf r_s data were normalized by the leaf area index (LAI) of the soybean canopy to maintain dimensional comparability. Stomatal resistance data and environmental parameters required for the Penman-Monteith calculations were observed over a wide range of climatic and crop-water-stress conditions. Consequently, test results are thought to be fairly general.

Figure 3.11 presents the comparison of bulk canopy stomatal resistance as derived by these three approaches. Values of r_{sc} derived using the Jarvis-Norman coupled model approach are highly correlated with values derived both from leaf r_s measurement ($r = 0.95$) and the Penman-Monteith formulation ($r = 0.89$). Furthermore, as indicated on Fig. 3.11, the magnitudes of the r_{sc} values agree quite well with the magnitudes of r_s/LAI measurement data. Poorer agreement with the Penman-Monteith model is indicated. Differences between estimates obtained by these two approaches

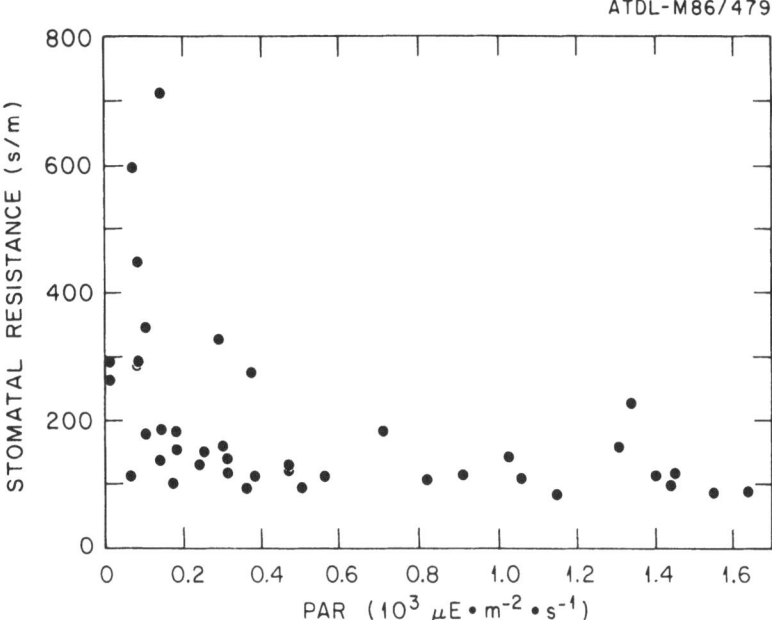

Figure 3.10. Measured stomatal resistances of white oak leaves as a function of incident photosynthetically active radiation (PAR).

are of the order of 30 to 50%. Finnigan and Raupach (1987) have concluded that the r_{sc} predicted by the Penman-Monteith approach is not a parallel, area-weighted sum of individual leaf r_s; rather, it incorporates an aerodynamic resistance component into the r_{sc} derivation. Furthermore, the Penman-Monteith r_{sc} term also contains a dependency upon individual leaf net radiation balances that is not present in the r_{sc} values derived from the Jarvis-Norman model. As a result, an overestimate of r_{sc} by nearly a factor of 2 by the Penman-Monteith model is to be expected. With these comments in mind, the comparisons of Fig. 3.11 strongly support the conclusion that the coupled Jarvis-Norman model of Baldocchi et al. (1987) adequately characterizes bulk canopy stomatal behavior in terms of canopy-atmosphere exchanges of latent heat in the case of a soybean canopy. Similar tests, using r_s/LAI and environmental data from the Walker Branch forest canopy, await acquisition of suitable data sets.

3.3.4 Edaphic Conditions

A variety of other boundary conditions associated with canopy-atmosphere exchange that are also operating can be conveniently grouped as edaphic, or site, conditions. Included are those features of the site and of its upwind fetch that affect the flow of air in the surface boundary layer over the

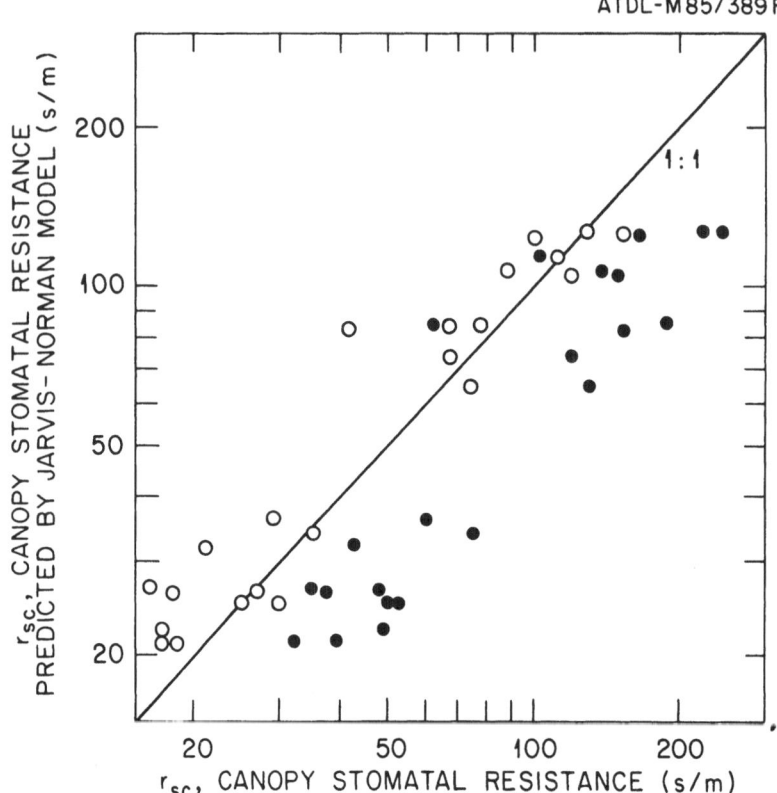

Figure 3.11. Comparison of bulk canopy stomatal resistances (r_{sc}) for water vapor transfer as computed with the Jarvis-Norman model and the Penman-Monteith equation (●) and as measured in a soybean canopy (○). The measured bulk canopy stomatal resistance was obtained by dividing the observed mean individual leaf stomatal resistance by the leaf area index (i.e., as r_s/LAI). Source: D.D. Baldocchi, B.B. Hicks, and P. Camara. 1987. A canopy stomatal resistance model for gaseous deposition to vegetated surfaces. Atmos. Environ. 21:91–101.

site. Because of the role of plant water potential in establishing r_s values, soil moisture availability is another edaphic condition requiring characterization for studies of forest meteorological phenomena involving exchanges of water or in which r_s plays a mediating role. In the longer term, the availability of soil moisture and nutrients affects canopy-atmosphere exchange by virtue of their influence on stand structure, species composition, and ecological succession.

As with canopy architecture, the role of fetch topography in governing canopy-atmosphere exchange is only beginning to be considered. Consequently, the topographic characteristics of importance remain to be defined at the Walker Branch site and elsewhere.

Because of the chain of reactions along the soil-plant-atmosphere continuum that link canopy r_s with soil moisture availability, soil moisture distributions with depth are measured at approximately biweekly intervals at this site. Eventually, the canopy r_s model will have to be expanded to account for soil moisture in addition to irradiance. These routine soil moisture observations will provide data for both developing and preliminary testing of such an expanded model.

No attempt has been made to characterize the relationships among soil moisture, site fertility, forest growth, succession, and canopy architecture at this site. However, such characterizations will eventually be needed so that research results from the Walker Branch Watershed can be extrapolated to dissimilar forests growing on other sites.

3.3.5 Summary and Synthesis

The mean daily total global solar radiation incident upon the Walker Branch forest canopy throughout the year is well within the range of values that are characteristic of continental temperate zone climates in the northern hemisphere, as cited by Jarvis and Leverenz (1983) and Galoux et al. (1981). The values at the Walker Branch site (lat. 35°57'30" N) are somewhat higher than the daily irradiances observed by Anderson (1964) at the more northerly Maddingly Wood site in England (lat. 52°13' N) and by Galoux et al. (1981) at Virelles, Belgium (lat. 50°4' N).

With an LAI of 4.9, the Walker Branch forest falls toward the upper limit of 6 or so for deciduous forests reported by Jarvis and Leverenz (1983). LAI data for other deciduous stands are provided in studies by Miller (1967), Burgess and O'Neill (1975), and Rauner (1976). The data on leaf area, woody element area, and total plant area from these stands are shown in Table 3.3. Comparison of these data indicates that the Walker Branch forest LAI may be intermediate relative to other deciduous stands, whereas its woody element area index (WAI) is relatively high. Despite the variability in LAI and WAI between stands, the total stand plant area indexes for the few deciduous forests for which data are available appear more uniform. The implications of this uniformity in terms of ecosystem productivity should be further investigated.

Studies of the canopy architecture of Colorado oak and aspen stands have also been made by Miller (1967). Similarly, Ford and Newbould (1971) conducted extensive studies of the architecture of a coppiced *Castanea sativa* forest in Ham Street Woods, England. As with the Walker Branch canopy, leaf inclination angles increased with height in these three forest types. Cumulative frequency distributions of leaf inclination in the form of Fig. 3.6 are provided at five canopy depths in the *Castanea* stand by Ford and Newbould (1971). As with the Walker Branch oak-hickory stand, distributions in the upper canopy are plagiophile but become increasingly planophile with increasing depth in the canopy.

Table 3.3. Leaf, woody canopy element, and total plant area indexes of some deciduous forest stands

Forest types	LAI	WAI	PAI	Source
Aspen (*Populus tremula*)	7.13	0.27	7.40	Rauner 1976
Mixed oak	7.0	2.0	9.0	Galoux et al. 1981
Tulip poplar (*Liriodendron tulipifera*)	6.0	—	—	Burgess and O'Neill 1975
Birch (*Betula verrucosa*)	5.30	0.74	6.04	Rauner 1976
Maple (*Acer platanoides*)	5.02	0.16	5.18	Rauner 1976
Oak-hickory (*Walker Branch*)	4.9	0.6	5.5	Hutchison et al. 1986
Linden (*Tilia cordata*)	4.78	0.32	5.10	Rauner 1976
Oak (*Quercus robur*)	4.60	0.48	5.08	Rauner 1976
Oak (*Quercus gambellii*)	4.2	—	—	Miller 1967
Aspen (*Populus tremuloides*)	2.0	—	—	Miller 1967

Abbreviations: LAI = leaf area index; WAI = woody canopy element area index; PAI = plant area index.

Ford and Newbould (1971) also present data on the variation in leaf dry weight per unit of surface area with height in the canopy. As in the Walker Branch stand (Fig. 3.9), leaf dry weight per unit area increased with height. However, Ford and Newbould did not normalize their weight per area data, so direct comparisons are not possible. Variations in leaf weight per area similar to those of Fig. 3.9 have been reported in other forests (e.g., Kira et al. 1969) and have been shown to be correlated with variations in photosynthesis and respiration rates (Kira et al. 1969). The observed vertical variation in leaf size and weight agrees with predicted variation based upon thermal regulation, water-use efficiency, and photosynthetic potential (Givnish 1979). The fact that the rate of increase in mean unit-area dry leaf weight with height was relatively constant among species warrants further study of this variable. If such constancy were found among species in a variety of forests, the height variation in unit-area dry leaf weight might be a useful characterization of forest canopies at particular sites, or it may represent an ecological constant that is maintained over a range of deciduous canopy types.

We have attempted to parameterize the leaf stomatal resistance vs. PAR response curve for derivation of bulk canopy stomatal resistances by means of the coupled Norman-Jarvis model described earlier. Minimal stomatal resistance values observed in the Walker Branch forest canopy are of the order of 100 to 150 s/m. Such values are comparable to those reported by Hinckley et al. (1978) and Dougherty et al. (1979) for various species of oak growing in Missouri and by Elias (1979) for oaks in Czechoslovakia. Values of b, a parameter that is equal to the PAR irradiance at stomatal resistances twice the minimum value and defines the degree of curvature of the parabolic relationship between r_s and PAR, are of the order of 22 W/m². While b values for other deciduous forests are not available in the literature, this value for the Walker Branch forest does compare favorably with those reported for other C-3 crop species (see Baldocchi et al., 1987).

3.4 Canopy Radiation Transfers and Regimes

3.4.1 Experimental Studies

The penetration of radiation into a forest canopy is controlled by the amount and spatial distribution of canopy elements and their radiative characteristics, along with the angles between the solar beam and canopy element surfaces and the sky-brightness distribution. Consequently, the radiation climate of a deciduous forest varies temporally as a result of diurnal and seasonal changes in earth-sun geometry and of phenological variations in both canopy architecture and the radiative characteristics of canopy elements. Since the spatial distribution of canopy elements and their geometrical and radiative properties is nonuniform, canopy radiation distributions vary spatially as well.

The global downward flux of radiation incident upon the earth's surface comprises a diffuse and a solar beam component. The beam component originates at the solar disk. Since the solar disk, as viewed from earth, subtends a quite small angle ($\sim$0°30'), for many purposes the solar beam may be considered to emanate from a point source. Diffuse radiation, on the other hand, emanates from the entire hemisphere of sky. Thus, a rigorous treatment of the penetration of diffuse radiation requires consideration of the directional distribution of diffuse radiant flux densities (i.e., the sky-brightness distribution). Consideration of such distributions involves considerable complexity, however. With clear to partly cloudy skies, the sky brightness is maximal adjacent to the solar disk and falls off rapidly with angular distance from the solar disk (e.g., see Hutchison et al. 1980). Consequently, under such sky conditions, the apparent position of the sun strongly controls not only the penetration of beam radiation into a forest but also the penetration of the diffuse component as well. For overcast skies, the sky-brightness distribution tends to be a function of the optical density of the cloud cover.

Global and diffuse incoming fluxes and global outgoing fluxes of solar, photosynthetically active, and all-wave radiation have been measured in and above the Walker Branch forest canopy for periods of several days in all seasons and phenological canopy phases. The moving-sensor technique was utilized to characterize the spatial variations in these fluxes in planes parallel to the forest floor. Vertical variability was accounted for by replicating the moving-sensor measurements at seven levels within the forest. Fluxes incident upon the canopy were measured with a single, stationary instrumented tram situated about 10 m above the canopy. Further details of the measurement system are provided by Baldocchi et al. (1984a). These observations, along with the canopy architecture and canopy element radiative characteristics assessments described earlier, provide the information necessary to quantify the interactions that control canopy radiative transfers of broad-band components of the solar and terrestrial electromagnetic spectra. The framework used to quantify these interactions is provided by a variety of phytoactinometric theories extant in the literature. However, before discussing our tests of phytoactinometric theory, it is useful to qualitatively consider the effects of interactions between earth-sun geometry, canopy architecture, canopy element radiative properties, and sky-brightness distribution upon canopy radiative transfer (see Baldocchi et al. 1984a, 1984b).

The role of seasonal and diurnal changes in the geometry of the solar beam and forest canopy elements in forest radiation regimes under clear to partly cloudy skies becomes evident from a consideration of Figs. 3.12 and 3.13. Figure 3.12 is a hemispherical-field-of-view photograph of the winter leafless forest taken from near the forest floor, and Fig. 3.13 is the same type of photograph taken under summer, full-leaf conditions. Superimposed on these two figures are the apparent paths of the solar disk across the hemisphere of sky at the winter solstice (southernmost line), the vernal and autumnal equinoxes (middle line), and the summer solstice (northernmost line).

In the winter forest (Fig. 3.12), the sky near the horizon is almost completely occluded. Furthermore, the annulus of occluded sky is asymmetrical because of the 3% southwest slope of this site. Consequently, penetration of the solar beam just after sunrise and just before sunset is minimal, and, because of the slope, the period of minimal beam penetration persists longer in the morning than in the afternoon. Above the zone of nearly total occlusion is another annulus of moderate optical density extending to an angular elevation of ~30°. (For reference, the solar noon elevation of the sun at the winter solstice is 30.6°.) Throughout this annulus, the solar beam encounters several to many tree crowns along its path to the forest floor. Thus, attenuation of the beam is still substantial during winter at midday. Above 30° elevation, the optical density of the canopy decreases with further increases in angular elevation. As a result, midday solar beam penetration increases from the winter solstice until just after the vernal equinox. After that time, bud break and leaf expansion rapidly close the canopy, and Fig. 3.12 no longer applies.

Typically, this canopy attains full leaf expansion about year-day 150 (i.e., ~3 weeks prior to the summer solstice). At the latitude of the Walker Branch Watershed, growing seasons are long (frost-free periods average

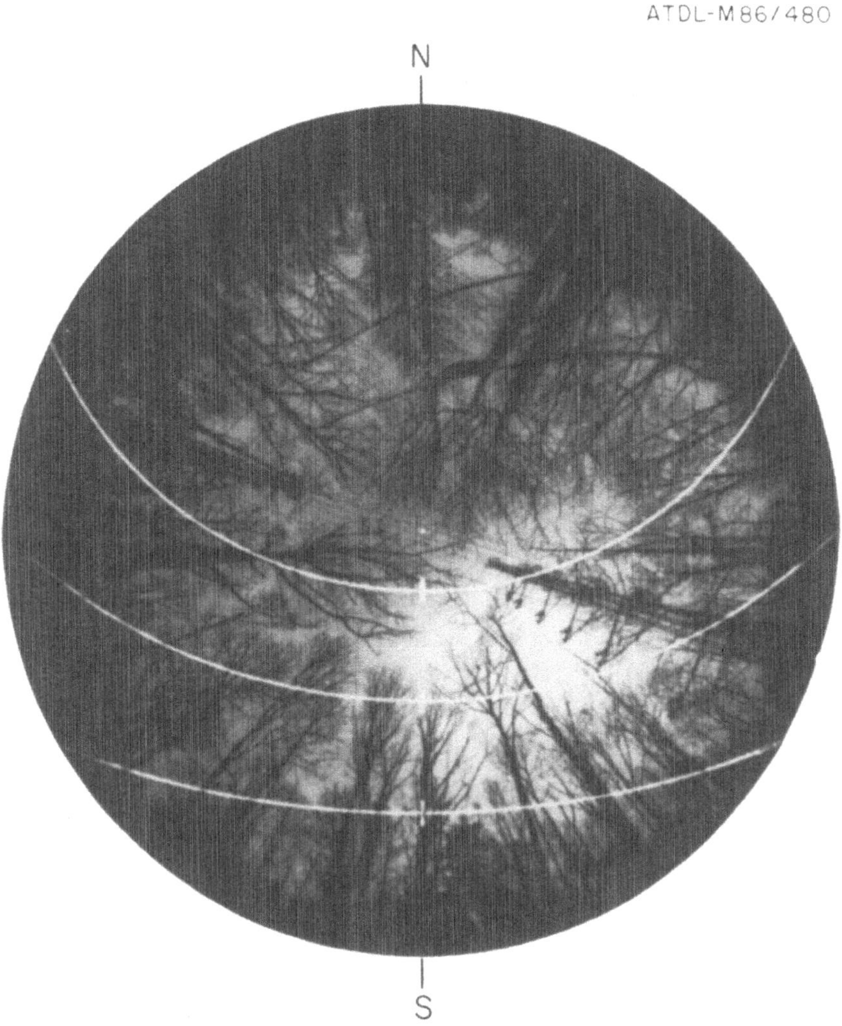

ATDL-M86/480

Figure 3.12. Hemispherical field-of-view photograph of the leafless forest canopy in winter. Despite the leafless canopy, the optical density of woody canopy elements along angular elevations corresponding to winter solar paths is considerable, implying significant attenuation of the solar beam in the winter, leafless forest canopy. (ATDL photo.)

well over 180 days), and the oak species that predominate in this forest tend to hold their senescent leaves well into the winter. As a result, the canopy characteristics shown in Fig. 3.13 persist with only minor variation from late spring until early winter (roughly from year-day 120 until year-day 325).

ATDL-M86/481

N

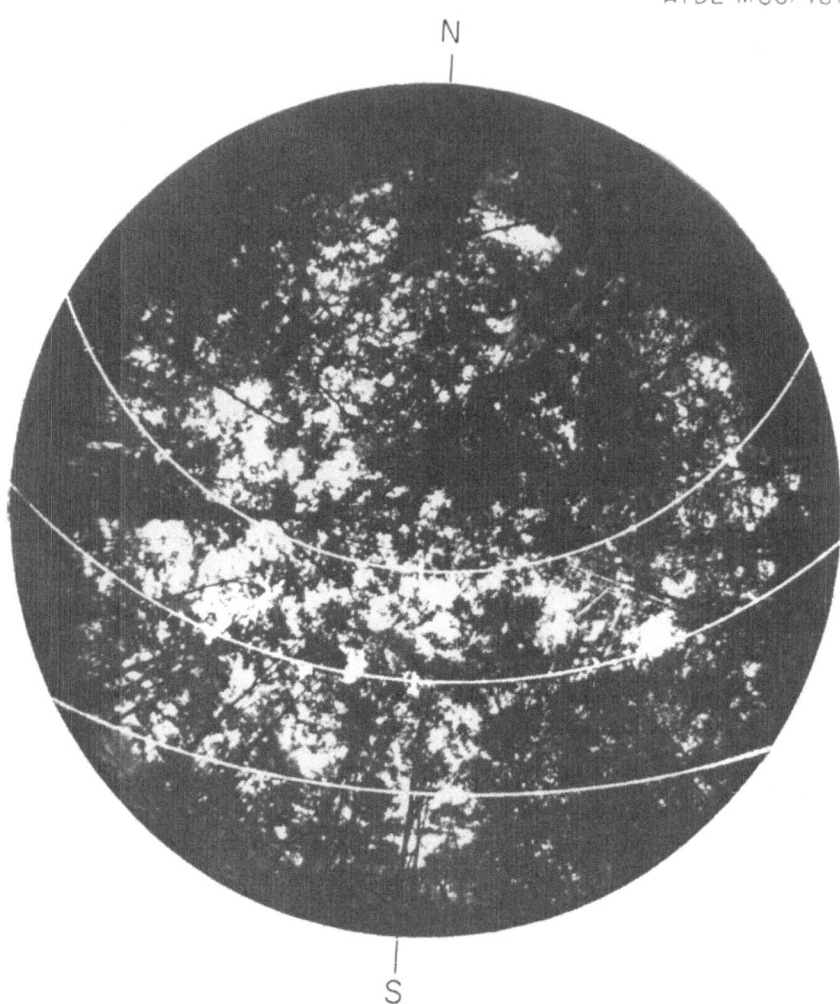

S

Figure 3.13. Hemispherical field-of-view photograph of the fully leaved forest canopy in summer. The absence of consistent changes in the optical density of the canopy with angular elevation implies that beam penetration into this canopy is not a strong function of solar elevation. (ATDL photo.)

In the fully leaved forest, the optical canopy-density dependence upon angular elevation is much less pronounced than in the winter, leafless canopy (Fig. 3.13). Thus, despite the high midday solar elevations of summer, the penetration of beam radiation is low.

The results of these interactions are quantitatively shown in Fig. 3.14. The penetration of the photosynthetically active component of solar radiation (PAR) to the floor of the winter forest is shown in Fig. 3.14a. On

ATDL-M 83/214R

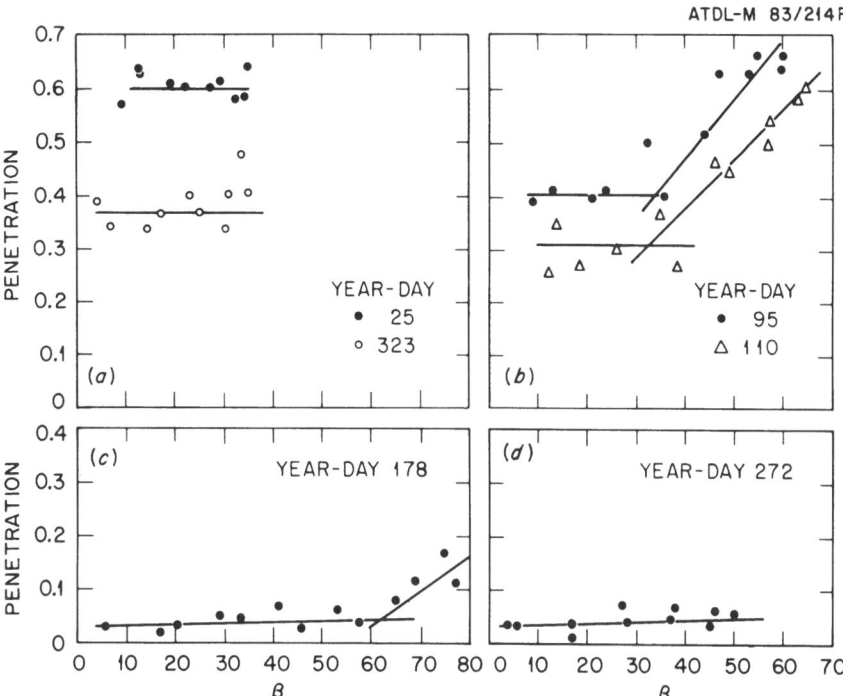

Figure 3.14. Penetration of photosynthetically active radiation (PAR) (defined as the ratio of PAR below the canopy to that above the canopy) as a function of solar elevation (β) for (a) year-days 25 (leafless canopy) and 323 (autumnal leaf fall); (b) year-days 95 and 110 (spring leafing period); (c) year-day 178 (fully leaved summer period); and (d) year-day 272 (fully leaved early autumn period). Source: Adapted from D.D. Baldocchi, B.A. Hutchison, D.R. Matt, and R.T. McMillen. 1984. Seasonal variations in the radiation regime within an oak-hickory forest. Agric. For. Meteorol. 33:177–191.

this figure, year-days 25 and 323 represent the winter solstice $+36$ and -30 days, respectively. Since the position of the sun is symmetrical about the solstices, and since the positions of the solar paths vary only slowly from day to day near the solstices, the solar paths on the 2 days of Fig. 3.14a are similar and fall about one-third of the distance between those of the winter solstice and the equinoxes. However, the optical densities of the canopy on these days are distinctly different. On year-day 323, senescent leaves persist in the canopy, yielding a canopy density approaching that of Fig. 3.13. On year-day 25, however, the canopy is leafless, as in Fig. 3.12. Thus, the absolute values of PAR penetration are reduced on year-day 323. Since the optical densities along solar paths on these days are relatively invariant in both the leafless (Fig. 3.12) and the fully leaved (Fig. 3.13) forest, no variation in PAR penetration with changing solar elevation throughout these days is apparent in Fig. 3.14a.

As the vernal equinox approaches, the apparent rate of movement of solar paths to the north increases. Hence, the sun appears to rise higher in the sky each day, and the solar beam paths become ever shorter and less optically dense. Consequently, a break appears in the relationship between beam PAR penetration and solar elevation at roughly 30° (Fig. 3.14b), the elevation where a marked decrease in the optical density of the leafless canopy is visible in Fig. 3.12. Bud break had occurred by year-day 95, and sufficient leaf expansion had occurred to reduce the PAR penetration to levels well below those of year-day 25 (Fig. 3.14a). Continuing leaf expansion between year-days 95 and 110 further decreases PAR penetration on year-day 110, but the spatial distribution of canopy elements apparently remains similar to that shown on Fig. 3.12, since the break in the PAR penetration curve occurs at about the same solar elevation, and the rate of increase in PAR penetration with increasing solar elevation following the break in curvature is nearly the same as on year-day 95.

With full leaf expansion (about year-day 150), the increased optical density of the canopy further displaces the fractional PAR penetration curves downward. A week or so before the summer solstice (year-day 178), the fractional penetration of PAR is <10% over most of the solar elevations present on that day (Fig. 3.14c). Only when the sun reaches maximal midday elevations does the PAR penetration show a definite, direct relationship with solar elevation.

By autumn (year-day 272 of Fig. 3.14d), the solar paths have moved far southward once again, and midday solar elevations no longer exceed the 60° or so elevation where the frequencies of canopy gaps increase in the fully leaved canopy. As a result, the penetration of PAR on year-day 272 is nearly identical to that of year-day 178 but without the break in the curve.

Other information about the nature of the interactions among apparent solar position, canopy architecture, and canopy radiative properties involved in canopy radiation transfer can be inferred from consideration of mean radiation profiles in the canopy. Figures 3.15 and 3.16 show the seasonal variations in the penetration of mean daily global solar and PAR radiation in the forest at the Walker Branch site. In these figures, the global components within the canopy (R_g, PAR) are normalized by those incident upon the canopy (R_{g0}, PAR$_0$) for ease of comparison. Data used to generate these figures are from clear to mostly clear days.

Radiation is least attenuated in the winter leafless forest (year-day 25) in both the total solar spectrum and the PAR wave band. Despite limited leaf expansion and significantly higher solar elevations throughout much of year-day 95, the penetration of both R_g (Fig. 3.15) and PAR (Fig. 3.16) radiation is reduced from that of year-day 25. Continued leaf expansion further reduces radiation penetration by year-day 110. However, comparison of the data for solar radiation in Fig. 3.15 with that for PAR in

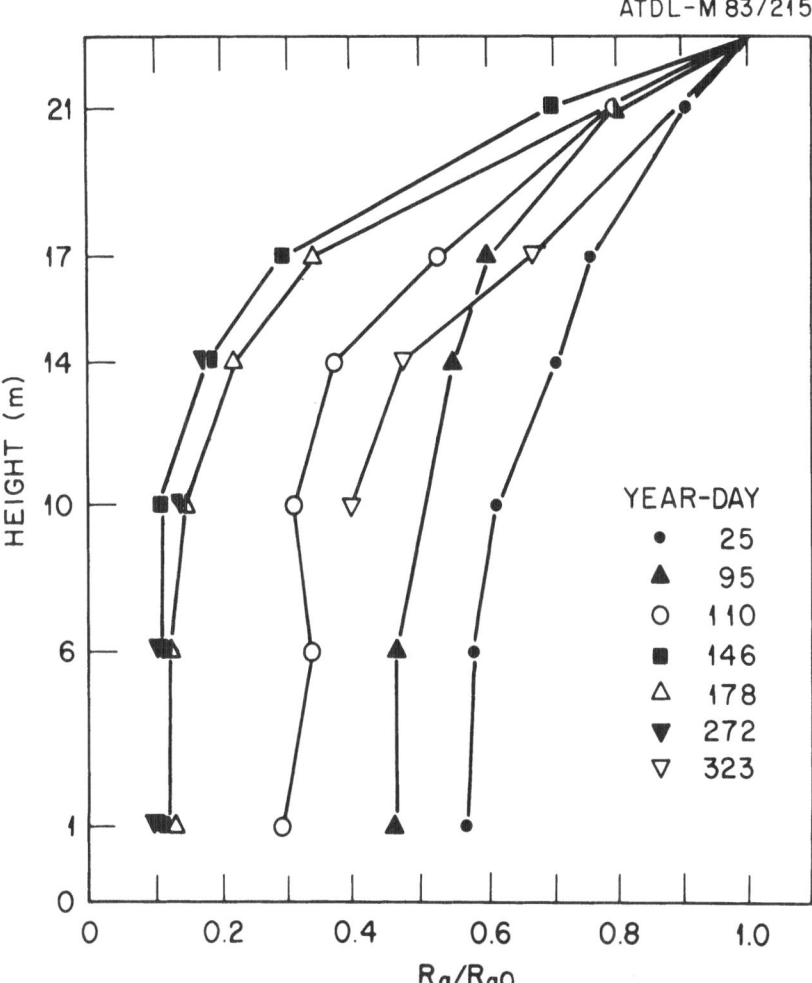

Figure 3.15. Normalized incoming solar radiation [incoming solar radiation within the canopy (R_g)/incoming solar radiation incident upon the top of the canopy (R_{g0})] as a function of height in the forest canopy and of time of year. Source: Adapted from D.D. Baldocchi, B.A. Hutchison, D.R. Matt, and R.T. McMillen. 1984. Seasonal variations in the radiation regime within an oak-hickory forest. Agric. For. Meteorol. 33:177–191.

Fig. 3.16 on this day indicates that in the 15-day period from day 95 to 110, solar radiation penetration was reduced more than that of PAR. Earlier in the year, the penetration of solar radiation and that of its PAR component are nearly identical.

This divergence is related to the spectral radiative characteristics of leaves and the spectral composition of total solar radiation. Solar radiation is composed of nearly equal proportions of PAR (0.3 to 0.7 μm) and near

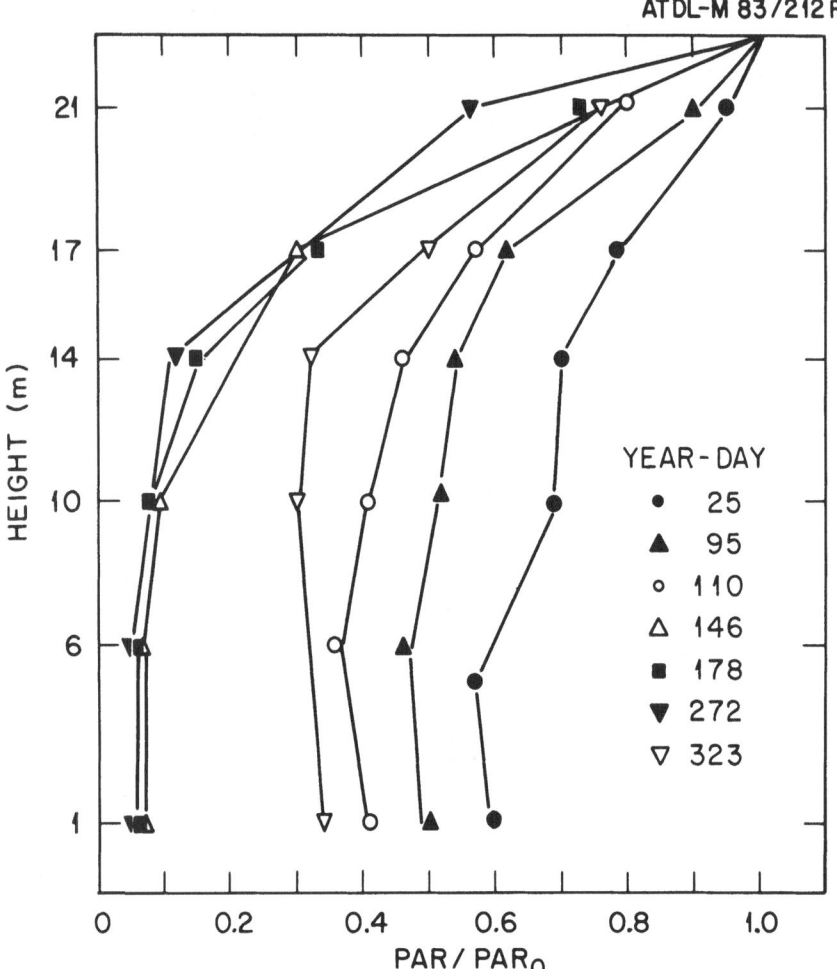

ATDL-M 83/212R

Figure 3.16. Normalized incoming photosynthetically active radiation (PAR/PAR$_0$) as a function of height in the forest canopy and of time of year. Source: Adapted from D.D. Baldocchi, B.A. Hutchison, D.R. Matt, and R.T. McMillen. 1984. Seasonal variations in the radiation regime within an oak-hickory forest. Agric. For. Meteorol. 33:177–191.

infrared radiation (NIR) (0.7 to 3.0 μm). Thus, in the winter leafless forest, the absence of spectrally selective leaf pigmentation allows nearly equal penetration of PAR and NIR, and little difference in the profiles of mean daily total R_g and PAR radiation is observed. This is further illustrated by comparing the daily mean ratios of PAR/R_g for year-days 25 and 323 on Fig. 3.17. Little difference in the height profiles of the ratios on these days is indicated on this figure. With the onset of leaf expansion and the synthesis of leaf pigments in spring, this situation changes drastically.

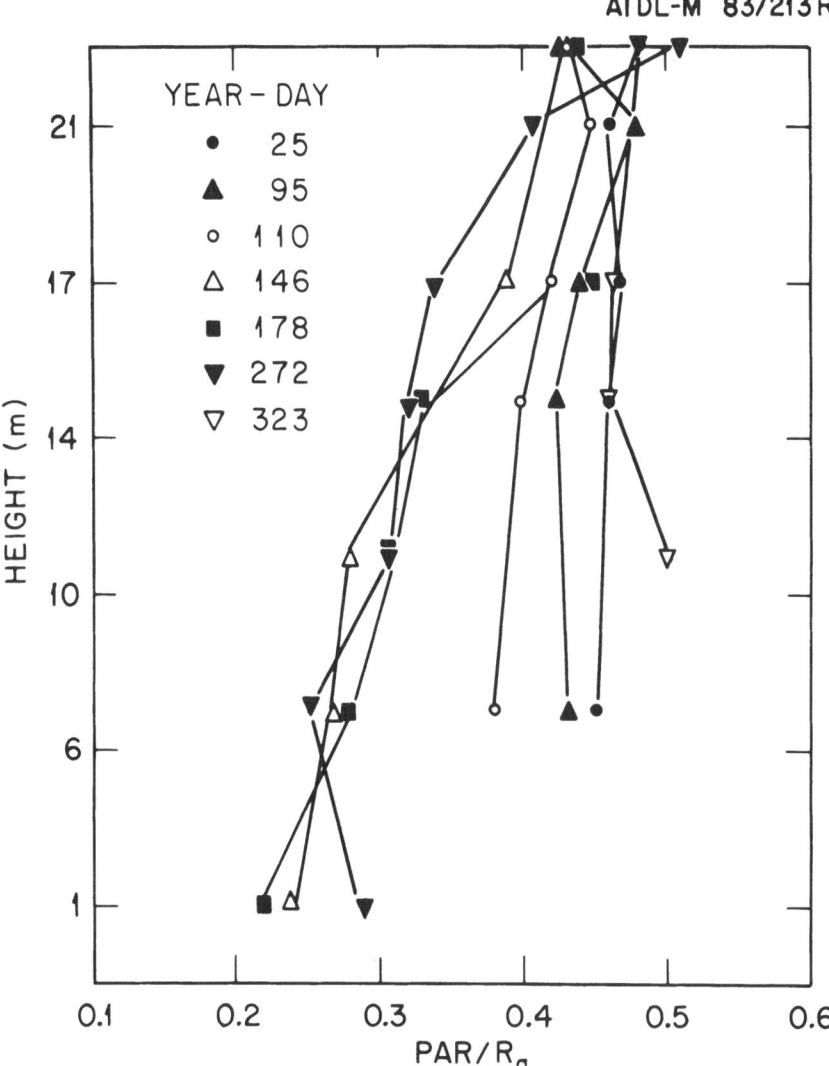

ATDL-M 83/213R

Figure 3.17. The ratio of photosynthetically active radiation (PAR) to global solar radiation (R_g) as a function of height in the forest canopy and time of year. Source: Adapted from D.D. Baldocchi, B.A. Hutchison, D.R. Matt, and R.T. McMillen. 1984. Seasonal variations in the radiation regime within an oak-hickory forest. Agric. For. Meteorol. 33:177–191.

With leaf expansion, phytomass is added to the canopy. This increases the attenuation of radiation of all wavelengths. However, as phytomass in the canopy increases, spectrally selective pigments are synthesized within canopy leaves. Thus, differential absorption and transmission of PAR and NIR by plant pigments leads to altered radiation climate qualities in plant stands. Chlorophyll, for example, strongly absorbs PAR but is nearly transparent to NIR. Hence, radiation within leafy canopies tends to be rich in NIR and deficient in PAR relative to the more equal mix of components above. As a comparison of the data of Figs. 3.15, 3.16, and 3.17 shows, throughout the full-leaf and leaf senescence phases of canopy development (year-days 146, 178, and 272), PAR is attenuated more strongly than total solar radiation. However, during leaf expansion (year-day 110), R_g is more strongly attenuated. The reasons for this are not clear from these data. Sanger (1971) reports that expanding oak leaves contain less chlorophyll than fully expanded ones; thus, we speculate that leaf pigments and phytomass during the leaf expansion period somehow absorb more NIR relative to PAR than in other phenological phases of canopy development, leading to a brief springtime period when R_g is attenuated more strongly than PAR radiation.

From the time of full leaf expansion until abscission begins, there is little temporal variation in the attenuation of either R_g or its PAR component. With the onset of senescence and abscission at about year-day 270, some reduction in attenuation occurs. Because of the later abscission of leaves of oaks and the persistence of senescent oak leaves in the canopy, the attenuation of both R_g and PAR is much greater on year-day 323 than on year-day 25, despite similar solar paths.

While the distributions of radiant flux densities with height in a forest canopy (as shown in Figs. 3.15, 3.16, and 3.17) are useful descriptors of energy transfer, they are less than satisfactory for explaining the functional relationships involved. Such explanation must have a theoretical, or at least a hypothetical, foundation. The earliest attempt to mathematically explain the penetration of radiation into vegetation canopies was made by Monsi and Saeki (1953). They postulated that the attenuation of radiation in a vegetation canopy was analogous to that in a perfectly turbid, uniform medium. Adapting the Beer-Bouguer expression for radiation attenuation in turbid media to the canopy situation, Monsi and Saeki proposed that

$$I/I_0 = \exp^{-KF},$$

where I is the radiant flux density at some level in the canopy, I_0 is the radiant flux density incident upon the top of the canopy, K is an extinction coefficient, and F is the leaf or plant area index above the canopy level in question. This negative exponential expression can be rewritten as

$$\ln(I/I_0) = -KF,$$

yielding a straight-line semilogarithmic plot of I/I_0 vs. F. Thus, a simple test of the validity of this theory for canopy radiative transfer is to plot $\ln I/I_0$ as observed in and above the Walker Branch forest, over the plant area index present above each level. If the observational data produce straight lines, the theory is acceptable, and the slopes correspond to the

K's for the various radiative components observed. Figure 3.18 shows such a plot for incoming R_g, Rn, PAR, and R_a. As shown in Fig. 3.18, the relationship between the relative irradiance within the canopy and the amount of canopy surface area above is linear to a rather high level of significance for each of these global radiative components.

An extensive analysis of extinction coefficients from the fully leaved period was performed by Baldocchi et al. (1984a). The mean extinction coefficients derived from this study are given in Table 3.4. Given the

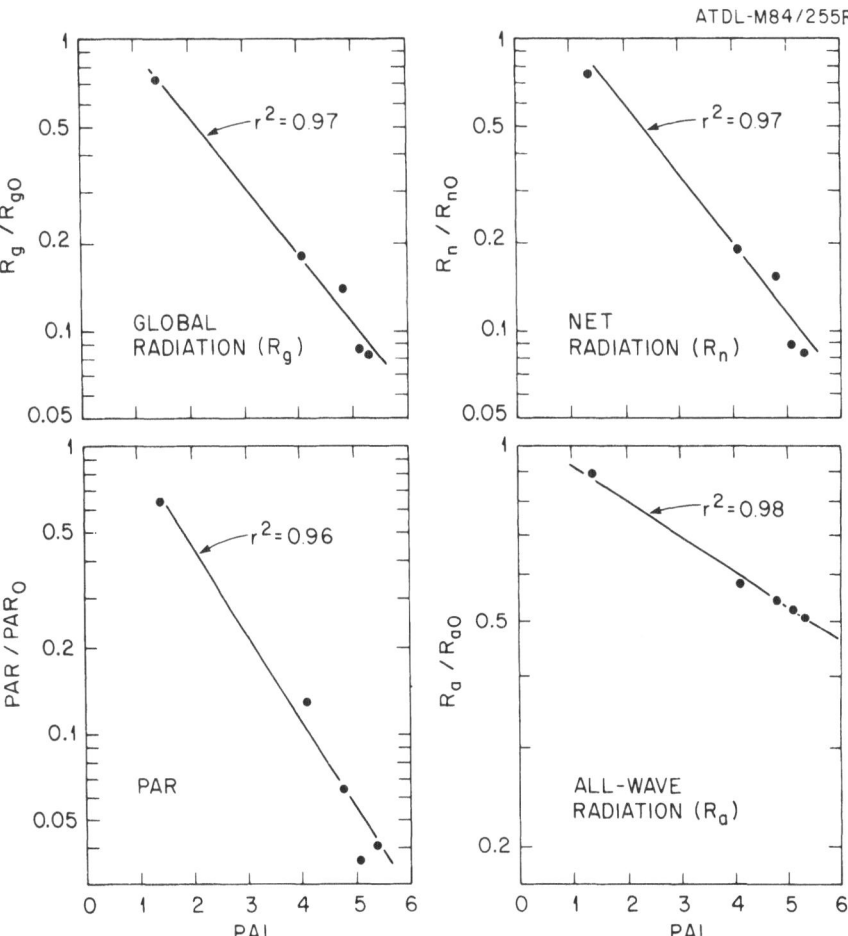

Figure 3.18. Normalized incoming radiation components as a function of plant area index (PAI) in the fully leaved forest canopy at the Walker Branch site. Observations made from 1100 to 1200 h on September 30, 1981. Source: Adapted from D.D. Baldocchi, D.R. Matt, B.A. Hutchison, and R.T. McMillen. 1984. Solar radiation in an oak-hickory forest: An evaluation of extinction coefficients for several radiation components during fully leafed and leafless periods. Agric. For. Meteorol. 32:307–322.

Table 3.4. Extinction coefficients
observed in the fully leaved forest

	Wave band	K	SD
R_g	(0.3 to 3 µm)	0.506	0.133
PAR	(0.3 to 0.7 µm)	0.655	0.139
R_a	(0.3 to 60 µm)	0.115	0.057

spectrally selective absorption of PAR, NIR, and thermal radiation by
plant pigments, the K's vary depending upon the spectral wave band under
consideration.

Because chlorophyll strongly absorbs PAR, K_{PAR} is high. On the other
hand, chlorophyll does not absorb NIR strongly. Also, leaves are thin,
maintain good thermal contact with the surrounding air, and have high
infrared emittance. Hence, upper and lower leaf surface temperatures are
nearly identical, resulting in nearly equal thermal radiation from those
surfaces.

This results in more or less isotropic thermal radiation fields inside the
canopy and translates to very low values of $K_{all-wave}$.

Since global radiation originates from the hemisphere of space above
the measurement point, and since the fully leaved canopy limits the penetration of the solar beam at nearly all solar elevations, the vertical variation in leaf inclination has little effect on the extinction of R_g, as indicated
by the linearity of the R_g vs. PAI plot of Fig. 3.18a. However, consideration
of the solar beam component alone suggests that the decreasing inclination
of leaves with depth in this canopy does affect the extinction of the solar
beam (R_{beam}). Figure 3.19 shows the mean normalized solar beam radiation
during the hour centered on 0930 for year-day 272 as a function of PAI.
Since the PAI of the overstory canopy above crown closure is ~4.0 (see
Fig. 3.8), the break in the curve of Fig. 3.19 at that PAI indicates an
abrupt change in K_{beam} at about the level of crown closure, the level where
leaf inclination angle distributions also change from plagiophile to planophile (Fig. 3.6).

Furthermore, since the angle of incidence of R_{beam} varies throughout
the day, the K_{beam}'s vary temporally as well as vertically in this canopy.
The diurnal variation in K_{beam} in the summer fully leaved forest is shown
in Fig. 3.20, along with the theoretical K_{beam}'s for canopies of constant
leaf inclinations (θ) of 0, 30, and 60°. Given the more steeply inclined
leaves in the overstory canopy, the observed K_{beam}'s are lower there than
in the understory, where leaves tend toward the horizontal. However,
the mean inclination in the overstory was observed to be 38°; yet the
observed K_{beam} values are consistently lower than the theoretical ones for
a canopy having a 60° leaf inclination. As is evident on Fig. 3.13, the fully
leaved canopy does not exhibit uniformly distributed canopy elements.

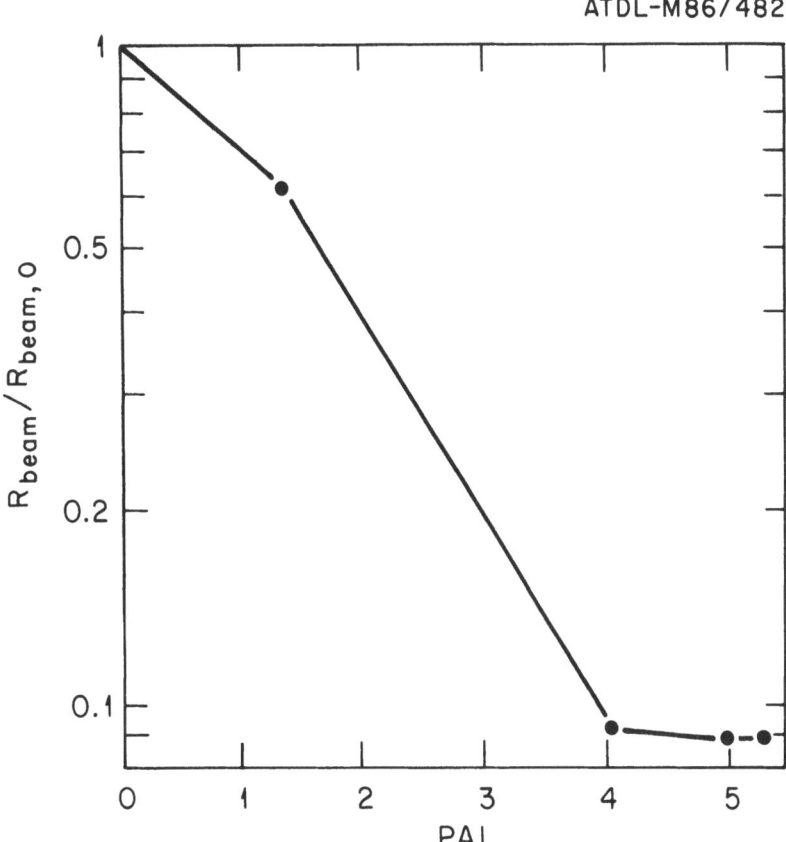

ATDL-M86/482

Figure 3.19. Normalized incoming solar beam radiation ($R_{beam}/R_{beam,0}$) as a function of plant area index (PAI). Observations made from 1100 to 1200 h on September 30, 1981. Source: Adapted from D.D. Baldocchi, D.R. Matt, B.A. Hutchison, and R.T. McMillen. 1984. Solar radiation in an oak-hickory forest: An evaluation of extinction coefficients for several radiation components during fully leaved and leafless periods. Agric. For. Meteorol. 32:307–322.

Rather, canopy elements appear to be clustered. This clumping of foliage has the effect of increasing the probability of radiation penetration for any combination of foliage area, inclination, and radiative properties. Thus, observed K_{beam}'s are lower than those predicted for a canopy having uniformly distributed elements inclined at 60°. In the understory, closer agreement between observed and predicted K_{beam}'s for a planophile canopy is evident during morning and midday hours. Agreement is poor later in the day, however.

The lack of symmetry about solar noon in the observed K_{beam}'s in both the overstory and understory are probably further effects of the spatial heterogeneity in canopy structure. For example, because of the topography

ATDL-M84/259R

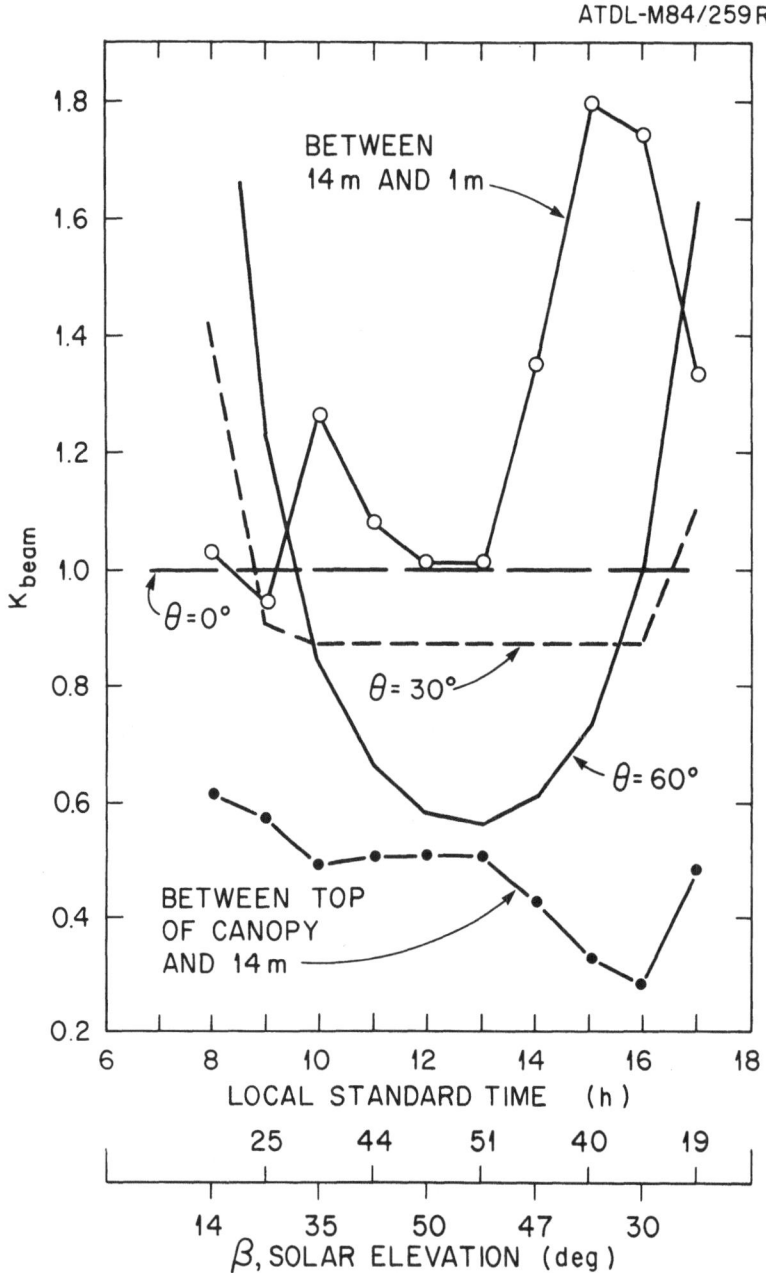

Figure 3.20. Diurnal course of the hourly mean extinction coefficient of beam solar radiation (K_{beam}) in the fully leaved forest canopy at the Walker Branch site. Data from 6 days is included in each hourly mean value. (θ = leaf inclination angle.) Source: Adapted from D.D. Baldocchi, D.R. Matt, B.A. Hutchison, and R.T. McMillen. 1984. Solar radiation in an oak-hickory forest: An evaluation of extinction coefficients for several radiation components during fully leaved and leafless periods. Agric. For. Meteorol. 32:307–322.

surrounding this site, the horizon to the east is higher than that to the west. Thus, the solar beam encounters more phytomass along early morning paths into the overstory than in late afternoon. In the understory, the opposite situation obtains. Because of codominant tree mortality, a rather sizable area of forest to the east and southeast of this site contains few overstory trees. It is this overstory-crown depauperate portion of the forest canopy that the morning solar beam traverses to reach the understory and forest floor at the research site. Thus, the solar beam reaching the forest floor encounters less phytomass in the morning than in the afternoon. Since we have not attempted to document the variation in PAI in horizontal space, our derivations of K_{beam}'s by means of the Beer-Bouguer approximation assume constant PAI. Thus, K_{beam}'s are overestimated when our assumed PAI is less than actual and underestimated when our assumed PAI is greater than actual. The problems of quantitatively assessing canopy element clumping and spatial variability remain to be resolved.

Because of the spatially heterogeneous structure of the Walker Branch forest canopy, frequency distributions of insolation components tend to deviate from normal (i.e., Gaussian) ones. Consequently, while space or time mean irradiances are useful in assessing energy transfers within the forest canopy, these means may be less than useful descriptors of the radiation environment for processes that vary nonlinearly with irradiance. For nonlinear light-dependent processes such as photosynthesis, consideration of actual frequency distributions may be required. Such distributions observed in the three major canopy strata on mostly clear days in the Walker Branch forest are shown in Figs. 3.21, 3.22, and 3.23. Figure 3.21 shows the frequency distribution of normalized PAR irradiance for one midday hour on a winter day in the leafless forest. Similarly Figs. 3.22 and 3.23 represent midday periods in the spring leafless (Fig. 3.22) and summer fully leaved (Fig. 3.23) forest. As comparison of these three figures indicates, interactions of earth-sun geometry and phenological canopy structure development profoundly alter irradiance frequency distributions.

In the winter leafless forest, the relative PAR irradiance distributions approach normal throughout the depth of the canopy but are slightly skewed and somewhat kurtotic (see Baldocchi et al. 1986a). Modal relative irradiances decline with depth, as expected. However, only in the overstory canopy above crown closure do relative irradiances approach those incident upon the canopy. Lower in the forest, virtually no sun flecks contain incident PAR flux densities despite the leafless canopy. This graphically demonstrates the importance of penumbral scattering in the attenuation of radiation in a leafless forest canopy. With the low solar elevations of winter and the long optical paths that therefore obtain, the fine twigs and branches, in sum, act as a neutral-density gray filter. The result is a reduction of radiant flux densities in what appear to be full-intensity sun flecks at all depths in the canopy.

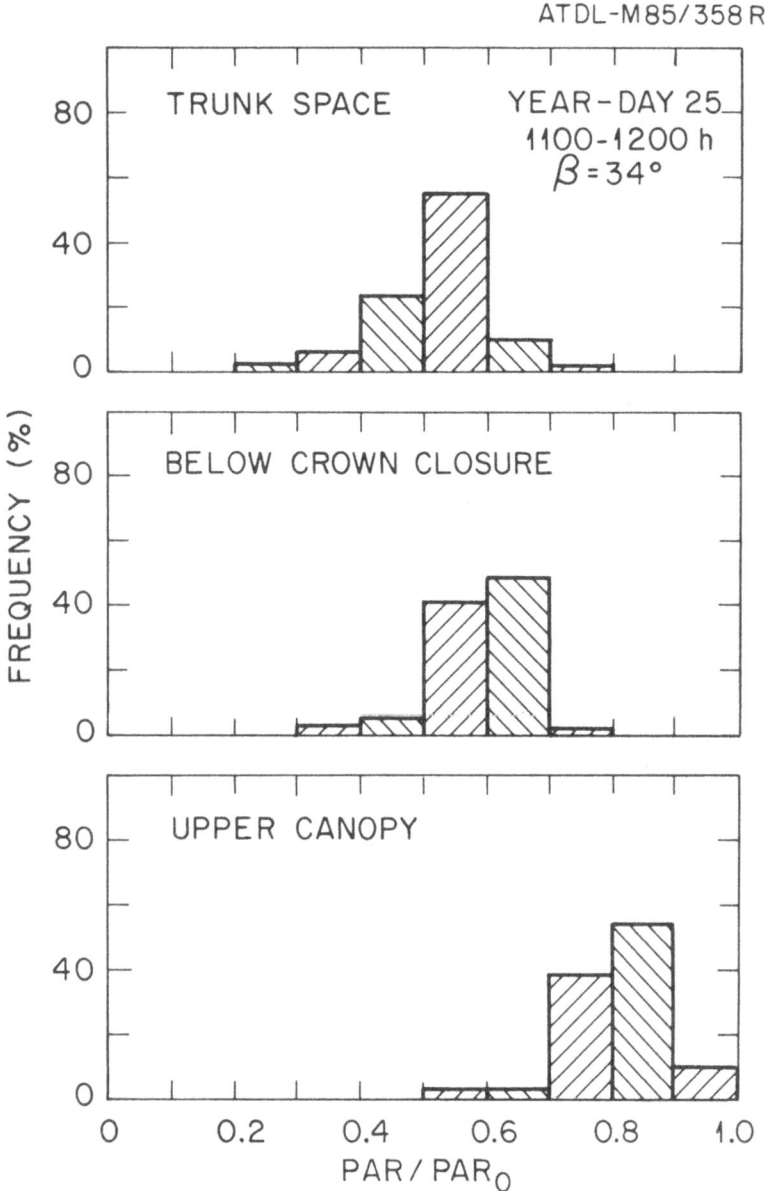

Figure 3.21. Frequency distributions of relative photosynthetically active radiation (PAR/PAR₀) in the three major canopy strata of the Walker Branch forest on a mostly clear day in the winter leafless phenoseason. Source: Adapted from D.D. Baldocchi, B.A. Hutchison, D.R. Matt, and R.T. McMillen. 1986. Seasonal variation in the statistics of photosynthetically active radiation penetration in an oak-hickory forest. Agric. For. Meteorol. 36:343–361.

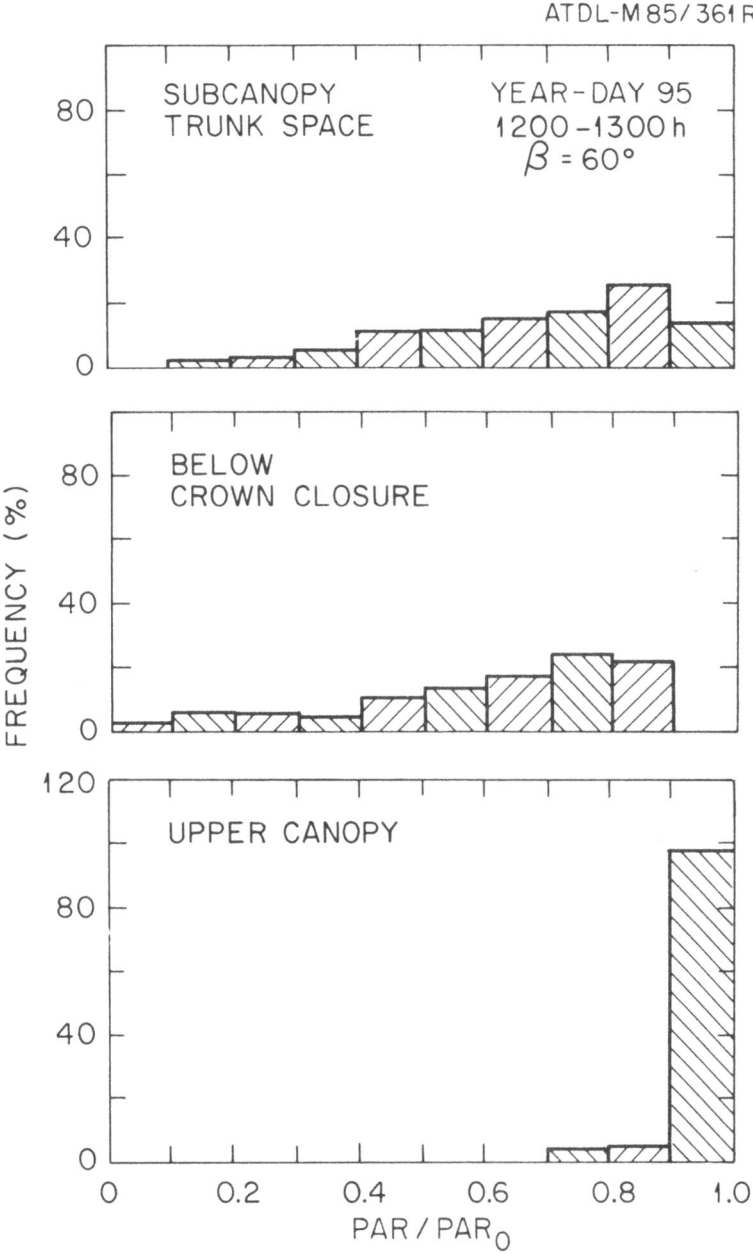

Figure 3.22. Frequency distributions of relative photosynthetically active radiation (PAR/PAR$_0$) in the three major canopy strata of the Walker Branch forest on a mostly clear day in the spring leafless phenoseason. Source: Adapted from D.D. Baldocchi, B.A. Hutchison, D.R. Matt, and R.T. McMillen. 1986. Seasonal variation in the statistics of photosynthetically active radiation penetration in an oak-hickory forest. Agric. For. Meteorol. 36:343–361.

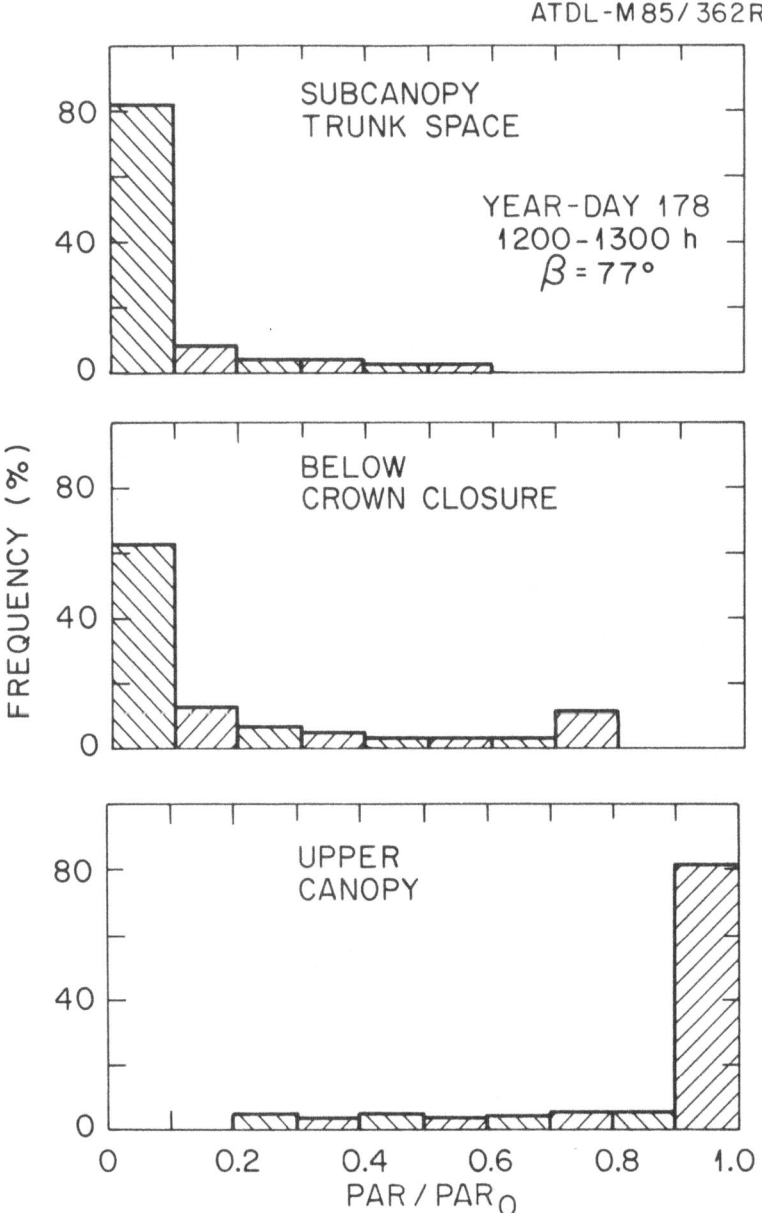

ATDL-M 85/ 362R

Figure 3.23. Frequency distributions of relative photosynthetically active radiation (PAR/PAR$_0$) in the three major canopy strata of the Walker Branch forest on a mostly clear day in the summer fully leaved phenoseason. Source: Adapted from D.D. Baldocchi, B.A. Hutchison, D.R. Matt, and R.T. McMillen. 1986. Seasonal variation in the statistics of photosynthetically active radiation penetration in an oak-hickory forest. Agric. For. Meteorol. 36:343–361.

With the increasing solar elevations of spring, this situation changes dramatically. In the spring leafless forest represented by Fig. 3.22, frequency distributions of relative PAR irradiances depart strongly from normal. With the higher solar elevations, the modal relative irradiances at all levels in the forest increase from those of winter days (Fig. 3.21). In the overstory above crown closure, nearly all irradiance occurs at flux densities approaching those incident upon the canopy. Below crown closure, while modal irradiances are higher than in winter, the scattering effects of penumbra cast by woody biomass and the yet tiny but numerous newly expanding leaves smear the frequency distributions over many relative-irradiance classes. As a result, modal frequencies are strongly reduced, and the frequency distributions become highly skewed. Because modal relative irradiances are high and because relative irradiances cannot exceed 1.0, the distributions are negatively skewed (i.e., toward lower values) throughout the canopy.

With continued expansion of leaves to full leaf area in the summer forest and the maximal solar elevations of the summer solstice, further evolution of the relative irradiance frequency distributions occur (Fig. 3.23). As in the leafless canopy, modal relative irradiance in the upper canopy remains near incident values, but the expanded leaves result in a reduced modal frequency. Because of radiation interception by the fully expanded leaves, the upper-canopy frequency distribution of relative PAR irradiances becomes more skewed to lower relative values. Deeper in the canopy, the dense shade produced by the fully expanded leaf canopy reverses the frequency distributions of spring. Modal relative irradiances at all levels below crown closure reflect the very low background levels of diffuse radiation present there. Penetrating beam radiation (sun flecks) skews the distributions to higher relative-irradiance values, but frequencies of occurrence are low. As in the winter forest, penumbra cast by the canopy elements reduce the beam radiant flux densities in sun flecks occurring below crown closure to values lower than those incident upon the canopy.

While the effects of the interactions of above-canopy radiation fields and canopy characteristics on within-canopy radiation regimes are of considerable importance, these interactions also affect the manner in which the forest and the atmosphere are coupled. Therefore, the above-canopy effects of the interactions between radiation fields and canopy characteristics are also of interest. Figure 3.24 shows the daily course of bulk canopy albedo for relatively clear days in the various phenoseasons experienced by the Walker Branch forest. The U shape of the diurnal albedo curves is typical of vegetation in general and illustrates the effect of the angle of incidence of radiation upon the amount reflected from canopy surfaces. With low solar elevations early and late in the day, optical paths through the canopy are long and dense; thus, impinging radiation has little likelihood of penetrating canopy gaps, and reflected radiation amounts are maximal. Consequently, bulk canopy albedos are higher early and late

ATDL-M84/288R

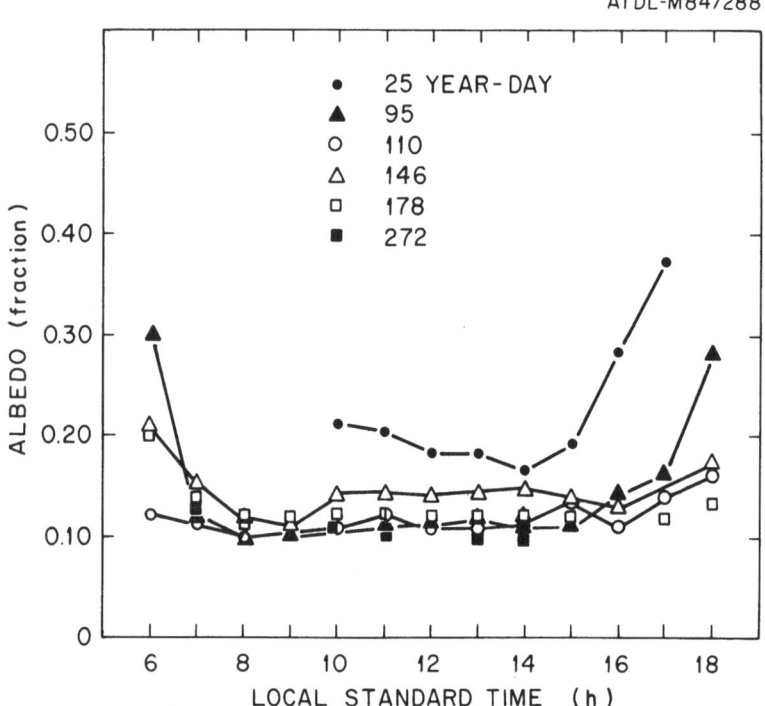

Figure 3.24. Diurnal trends in shortwave radiant albedos of the Walker Branch forest canopy as a function of year-day.

on all days. With the higher solar elevations of midday, more radiation penetrates canopy gaps. Within the canopy, multiple reflection from canopy elements results in increased radiation absorption. Consequently, bulk canopy albedo declines with higher solar elevations.

The data of Fig. 3.24 show that the reflectivity of the leafless forest is highest with the low solar paths of winter (year-day 25). By year-day 95, the higher solar elevations of spring have reduced bulk leafless canopy albedos to values near those present in the fully leaved canopy. With the development of the leafy canopy, bulk canopy albedos vary only slightly through the day and over seasons.

Although the phenological changes in the spectral characteristics of the canopy have little impact upon bulk canopy albedo, these changes greatly affect forest canopy reflectivity in the PAR wave band. Figure 3.25 shows the daily and phenoseasonal course of PAR reflectivities of the Walker Branch forest. As with albedo, PAR reflectivities are highest in the winter leafless phenoseason. The PAR reflectivity of the leafless canopy declines with rising solar elevations just as does the albedo. However, with leaf expansion, PAR reflectivities continue to decline. Following full leaf expansion, little effect of seasonal variations in solar path elevations is ev-

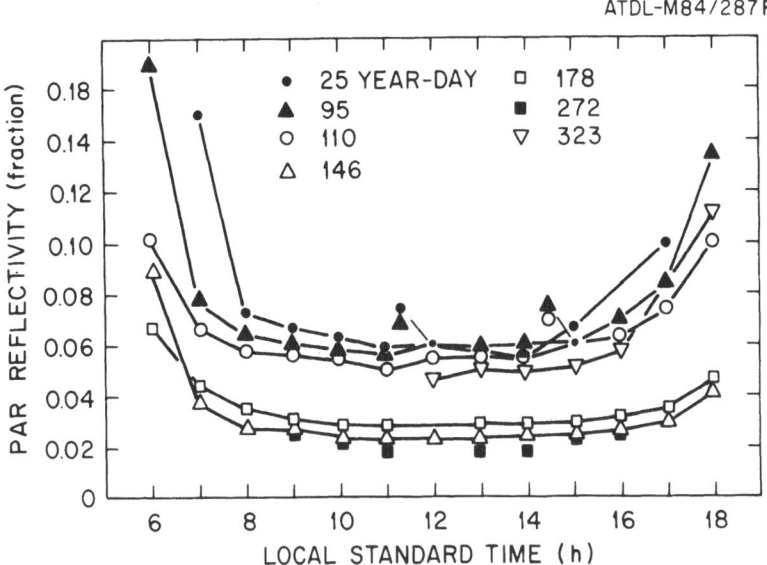

ATDL-M84/287R

Figure 3.25. Diurnal trends in the reflectivities of photosynthetically active radiation (PAR) in the Walker Branch forest canopy as a function of year-day.

ident in the data of Fig. 3.25, although the daily effects remain. It seems likely that solar paths are sufficiently high in the sky from the vernal to the autumnal equinox that those seasonal changes in midday elevations that do occur have little effect upon either PAR reflectivity or albedo. This hypothesis is supported by the relatively flat bottoms of the U-shaped reflectivity and albedo time traces of both Figs. 3.24 and 3.25. However, since the sun rises and sets at the horizon throughout the year, extremely low solar elevations are present twice each day, solar elevations sufficiently low that effects upon surface reflection are present. Thus, PAR reflectivities are higher about sunrise and sunset.

3.4.2 Tests of Phytoactinometric Theory

A number of models of radiation transfer in vegetation canopies have been developed, mostly for agronomic crop canopies. These models relate irradiance at a level in a canopy to the area index of canopy elements above the level in question, the radiative characteristics of those elements, their inclination, and the incident beam and diffuse radiative flux densities at the top of the canopy. We have compared simulations by two well-known canopy radiative transfer models with the observed radiant flux densities as a test of their applicability in deciduous forests (Baldocchi et al. 1985b). The models selected were those of Norman (1979) and de Wit (de Wit 1965; Idso and de Wit 1970).

The Norman model assumes that the canopy is horizontally homogeneous with a spherical leaf inclination angle distribution. The penetration of both diffuse and beam radiation is approximated by means of Poisson probability distributions, using the Ross-Nilson formulation for the beam component (Ross and Nilson 1971). A unique feature of the Ross-Nilson approach is that it allows treatment of canopies with nonrandom vertical foliage distributions. Because of the differences in scattering of PAR and NIR by chlorophyll, these two wave bands are treated separately in the Norman model. The de Wit model, on the other hand, is an extension of the Monsi and Saeki (1953) model, allowing specification of the leaf inclination angle distribution.

Figure 3.26 shows a comparison of observation and simulation for R_g

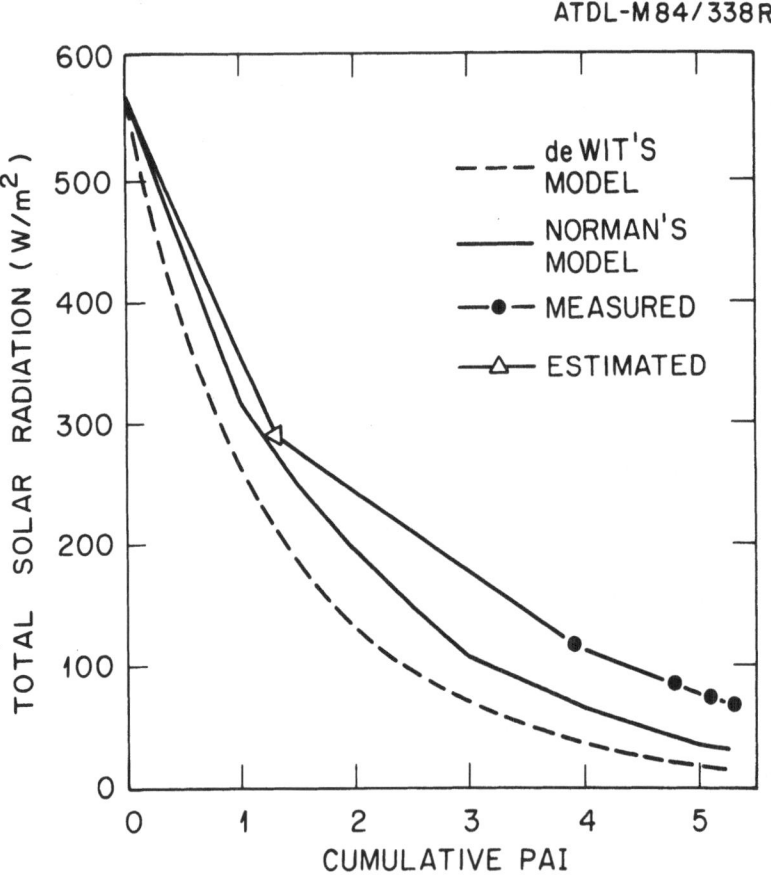

Figure 3.26. Simulated and observed flux densities of global solar radiation in the fully leaved forest canopy as a function of plant area index (PAI). Source: Adapted from D.D. Baldocchi, B.A. Hutchison, D.R. Matt, and R.T. McMillen. 1985. Canopy radiative transfer models for spherical and known leaf inclination angle distributions: A test in an oak-hickory forest. J. Appl. Ecol. 22:539–555.

in the fully leaved forest canopy (year-day 272). Both the Norman and the de Wit models underestimate the penetration of R_g into this canopy, with the de Wit model showing the greatest underestimation. Reasons for this underestimation can be inferred from tests of the beam and diffuse radiation penetration submodels. While both models underestimate the penetration of beam radiation (Fig. 3.27), the Norman approach simulates observed diffuse radiation penetration with fair agreement (Fig. 3.28). The de Wit model underestimates diffuse irradiance throughout the depth of the canopy, whereas the Norman model overestimates this quantity in the upper canopy (Fig. 3.28).

Both models assume randomly distributed foliage in horizontal canopy strata. Since the actual foliage distribution is strongly clumped, greater penetration of beam radiation occurs at any given PAI than if the same

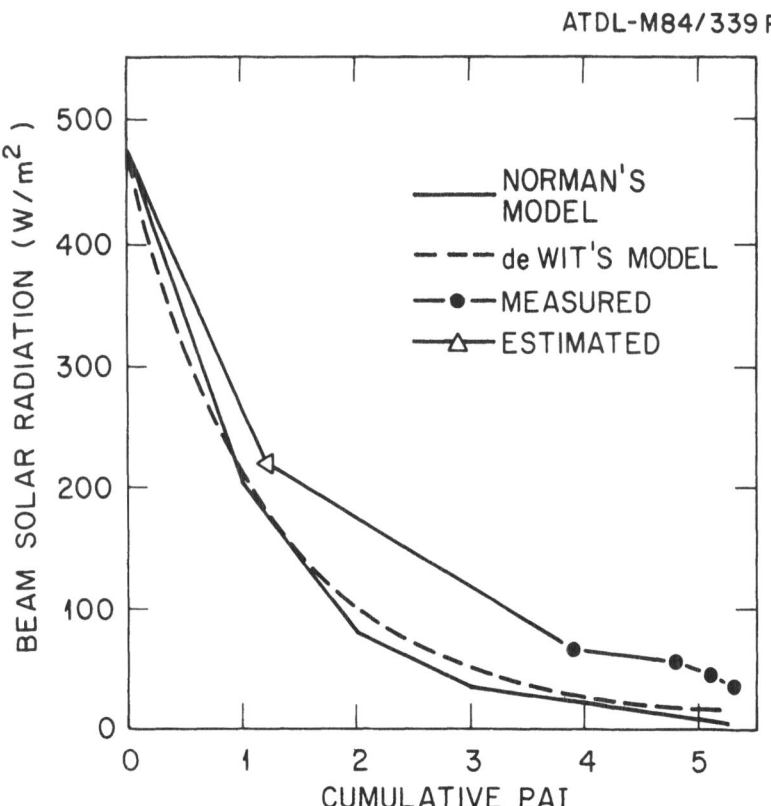

Figure 3.27. Simulated and observed flux densities of beam solar radiation as a function of plant area index (PAI) in the fully leaved forest canopy. Source: Adapted from D.D. Baldocchi, B.A. Hutchison, D.R. Matt, and R.T. McMillen. 1985. Canopy radiative transfer models for spherical and known leaf inclination angle distributions: A test in an oak-hickory forest. J. Appl. Ecol. 22:539–555.

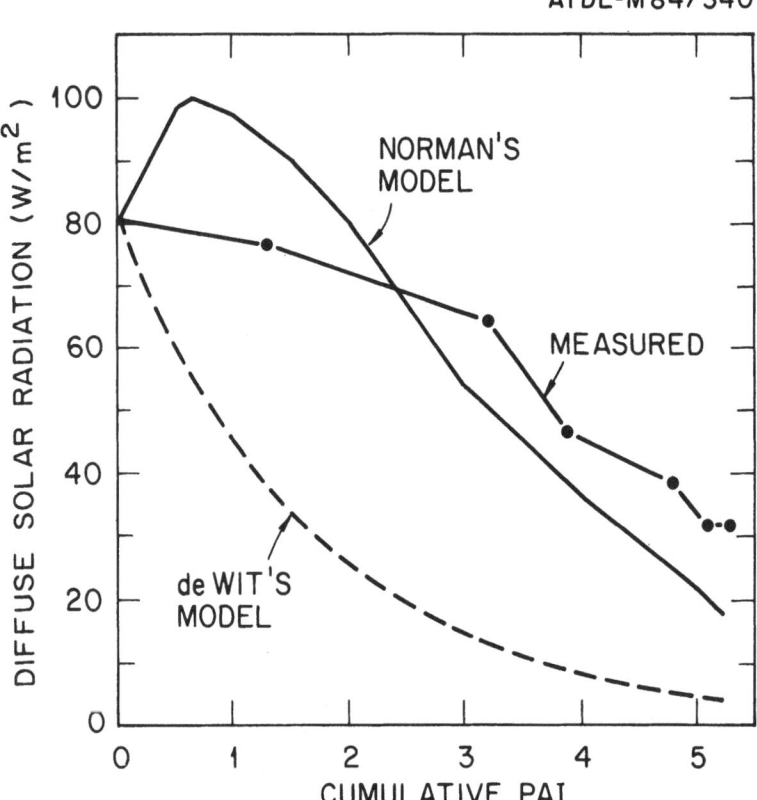

Figure 3.28. Simulated and observed flux densities of diffuse solar radiation as a function of plant area index (PAI) in the fully leaved forest canopy. Source: Adapted from D.D. Baldocchi, B.A. Hutchison, D.R. Matt, and R.T. McMillen. 1985. Canopy radiative transfer models for spherical and known leaf inclination angle distributions: A test in an oak-hickory forest. J. Appl. Ecol. 22:539–555.

amount of foliage were truly randomly distributed. Thus, both models underestimate the penetration of beam radiation. Similarly, the greater gap frequency in canopies with clumped foliage admits more diffuse radiation to subcanopy levels. Consequently, both models underestimate subcanopy diffuse irradiance as well. However, the underestimation of diffuse irradiance by the de Wit model cannot be completely ascribed to foliage clumping. The de Wit model considers only first-order scattering, whereas the Norman model iteratively incorporates multiple scattering of radiation by canopy elements. Thus, some of the underestimation of diffuse irradiance by the de Wit approach results from its inability to account for the effects of multiple reflections. The Norman model, on the other hand, overestimates upper-canopy diffuse irradiance even though it does account for multiple reflection effects. The problem is that in canopies with randomly dispersed foliage, greater amounts of beam radiation are converted to diffuse radiation by multiple reflections than in the Walker Branch canopy with clumped foliage. Thus, the Norman model overestimates the conversion of beam to diffuse radiation.

Although the actual leaf inclination angle distribution in the Walker Branch Watershed forest canopy does not approximate a spherical distribution, analyses of the Norman model indicates that this is of little consequence. Indeed, given the overall plagiophile nature of this canopy, the assumption of a spherical distribution should result in an overestimation of both diffuse and beam irradiance within the canopy. Thus, it appears that the assumption of spherical foliage inclination angle distribution can be made without serious error in canopies having inclination angle distributions other than erectophile or planophile.

It has been suggested that a negative binomial probability distribution would better simulate beam interception in canopies having clumped foliage than the Poisson approach of Ross and Nilson (Acock et al. 1970; Nilson 1971). Thus, simulations of beam irradiance using the Norman/Ross-Nilson (Poisson), the de Wit, and the negative binomial approaches are compared with observations in Fig. 3.29. The negative binomial simulation exhibits better agreement with observation than the other two approaches. This suggests that the problem of radiation penetration into clumped canopies is largely resolved; however, the negative binomial approach requires a parameter that characterizes the degree of foliage clumping present, and at present this parameter cannot be evaluated analytically. The value used to generate the curve of Fig. 3.29 was derived empirically from observed beam radiation penetration in the Walker Branch Watershed forest. Thus, the agreement shown in Fig. 3.29 may be spurious.

A major impetus for the study and simulation of canopy radiation transfer is the need for a better understanding of photosynthesis and stomatal conductance in plant canopies. Both these light-dependent phenomena operate as nonlinear functions of irradiation. Thus, the use of radiative transfer models developed for canopies with spherically distributed canopy elements to simulate irradiation in a canopy having clumped foliage such as the Walker Branch forest may result in serious errors in the prediction or understanding of photosynthesis and stomatal conductance. To determine the magnitude of such errors, canopy photosynthesis and stomatal conductance were computed [with the models of Marshall and Biscoe (1980) and Turner and Begg (1973), respectively] using three sets of vertical irradiation-profile data. Profiles of PAR irradiance observed in the Walker Branch forest in late summer, 1981, were used as the control. Test profiles of PAR irradiance were computed using the Poisson (Norman/Ross-Nilson) and the negative binomial approaches (see Baldocchi and Hutchison 1986). Comparison of the hourly total photosynthesis and hourly average stomatal conductance computed using these three sets of solar

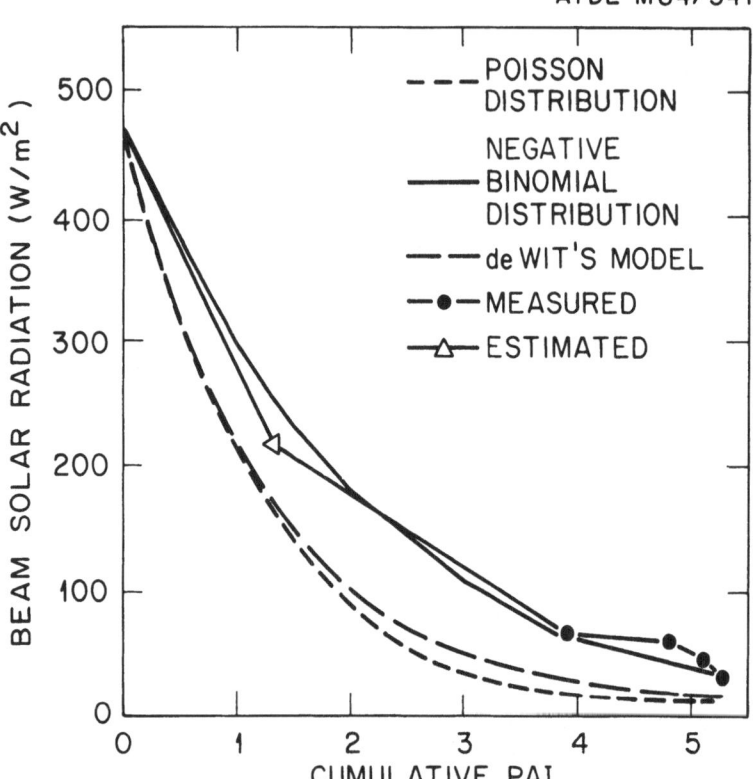

Figure 3.29. Flux densities of beam solar radiation as a function of plant area index (PAI), simulated using the Poisson and negative binomial probability distributions and de Wit's model, along with the flux densities observed in the fully leaved Walker Branch forest. Source: Adapted from D.D. Baldocchi, B.A. Hutchison, D.R. Matt, and R.T. McMillen. 1985. Canopy radiative transfer models for spherical and known leaf inclination angle distributions: A test in an oak-hickory forest. J. Appl. Ecol. 22:539–555.

radiation data indicates that the Norman model (Poisson distribution, spherical leaf inclination angle distribution assumption) underestimates both daily total photosynthesis (by 17%) and mean stomatal conductance (by 10%) as compared with values computed using the observed PAR irradiance profiles. The negative binomial approach yielded estimates of photosynthesis that exceeded those computed from observed PAR distributions by 8% and estimates of stomatal conductance that overestimated those computed from observed radiation data by 9%. Thus, the negative binomial model of canopy radiation transfer is shown to yield slightly better estimates of canopy photosynthesis in a forest having clumped foliage than a model developed for crop canopies with more uniformly dispersed canopy elements. Little improvement in the estimation of stomatal conductance is indicated, however. Preliminary analyses indicate that the overestimates of the negative binomial model predictions result from the model's inability to adequately treat penumbral reduction of penetrating beam radiation flux densities in the canopy.

3.4.3 Thermal Radiation Regimes

The exchanges of thermal radiant energy between canopy elements and the atmosphere are of interest, on the one hand, from the standpoint of energy partitioning in a forest canopy and, on the other, in terms of the thermal radiant signatures of forest canopies. In collaboration with colleagues from the Colorado State University, EG&G Energy Measurements, Inc., and the U.S. Army Engineers Waterways Experiment Station, we have made preliminary measurements of the directional thermal exitance from the leafless canopy and, more recently, from the fully leaved canopy. In addition, a series of effective radiant temperature (ERT) measurements were made to provide data for tests of a thermal vegetative canopy model (TMOD) derived by Kimes et al. (1981) and Smith et al. (1981).

Figure 3.30 shows an example of the ERT distribution (8–14 μm) of the leafless canopy at midday (1200–1210 h) on a clear winter day. No attempt was made to contour temperatures in the azimuth sector, ~40 to 80°, since towers and buildings were in the field of view (FOV) of the infrared thermometers at those azimuths. Considerable ERT variation with azimuth and nadir angle is evident owing to the varied surfaces within the 2° FOV of the multiple sensors and to the directional variations in amounts of sunny or shaded surfaces in the sensor FOVs. Because the canopy was leafless, the nadir-looking sensor (2° FOV) mostly "saw" the forest floor; thus, the ERT indicated by this sensor is that of a circular area on the forest floor of the order of 1.2 m diam. As the nadir angle of the thermal infrared (TIR) sensor increases, the amount of woody biomass in the FOV increases. Temperatures over northerly azimuths are highest, because, with the sun to the south, the northerly-looking sensors see the sunlit sides of boles, branches, and twigs. With southerly-looking angles, the shaded portions of the woody biomass are viewed and ERTs are correspondingly lower. For southerly-looking angles, temperatures show strong variation with nadir angles above 40°.

The thermal vegetation canopy model (TMOD) of Kimes et al. (1981) and Smith et al. (1981) computes canopy temperature by means of a set of steady-state energy balance equations which simulate the interactions of shortwave and long-wave radiation within a canopy. The TMOD was developed for coniferous forest canopies; hence, for application to a broadleaved canopy, the parameterization of convective heat exchange of the leaves had to be modified. The convective heat exchange coefficient used for the deciduous canopy leaves was derived from a theory developed by

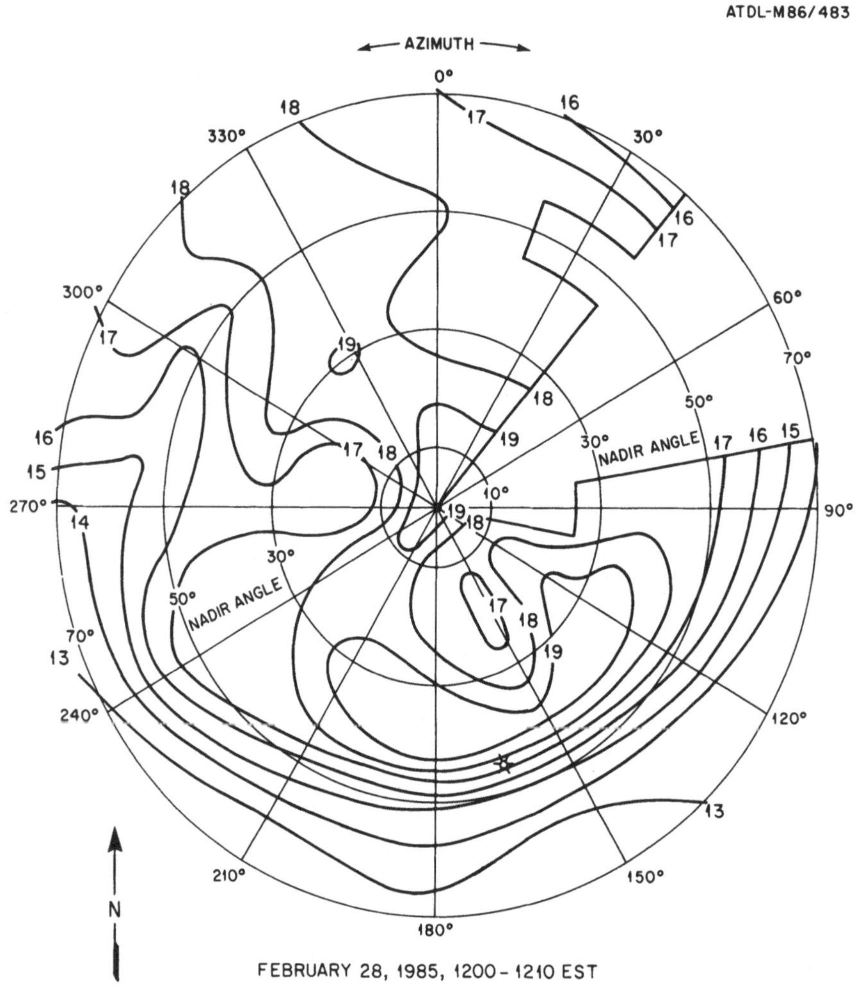

ATDL-M86/483

FEBRUARY 28, 1985, 1200-1210 EST

Figure 3.30. Directional effective radiant temperature distribution for 1200 to 1210 h, February 18, 1985. Contours are labeled in degrees Celsius, and the position of the sun is indicated at its median azimuth and zenith angle during this time period. Source: L.K. Balick and B.A. Hutchison. Directional thermal infrared exitance distributions from a leafless deciduous forest. IEEE Trans. Geosci. Remote Sensing GE-24:693–698.

Grace and Wilson (1976). Shortwave (solar) radiation transfer within the canopy is simulated by the solar radiation in vegetated canopies model (SRVC) of Smith and Oliver (1974). This model uses the de Wit (1965) model to calculate the probability that a light ray will encounter a canopy gap in its direction of travel and performs the actual light ray tracing using a Monte Carlo approach.

Figure 3.31 compares measured and TMOD simulated canopy element

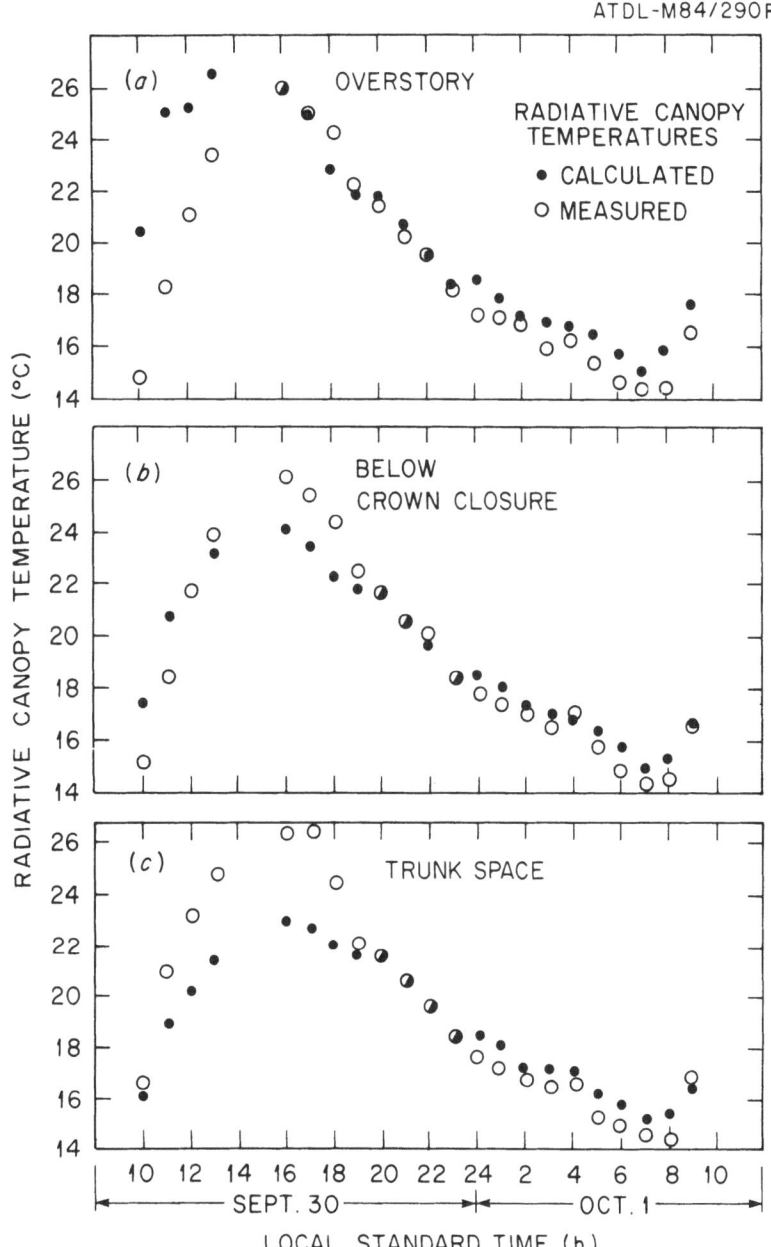

Figure 3.31. Observed and simulated effective radiant temperatures of major canopy strata on a clear, early autumn day with a fully leaved canopy.

temperatures for the three major canopy strata in the fully leaved phenological canopy phase. In the overstory, TMOD overestimates observed ERTs by ~2°C during daylight hours. At night, in the absence of insolation, simulated values are in good agreement with observation. In that portion of the canopy below crown closure and in the trunk space, the calculated temperatures overestimate observed ERTs in the morning, underestimate them in the afternoon, and agree with them at night.

The daytime differences between simulation and observation are ascribed to limitations in the parameterization and conceptualization of energy balance components by TMOD and its components. For one thing, the fact that foliage is clumped in the deciduous forest canopy increases the penetration of solar radiation into the canopy. Since the model assumes randomly dispersed foliage, the increased penetration into the clumped canopy results in higher than predicted element temperatures in the middle and lower canopy with the higher solar elevation angles of midday. Furthermore, in the energy balance equations used in TMOD to partition incident solar energy into latent and sensible heat transfers, these transfers are simulated using the electrical analog approach. In TMOD the main resistances involved are r_a and r_s. Both these resistances must be specified; TMOD does not allow their simulation from canopy radiant energy and wind speed distributions. Since r_a and r_s tend to vary widely with the changing radiation and wind environments in a forest canopy throughout a diel cycle, a single value parameterization of each of these quantities must lead to errors in the computation of energy partitioning, resulting in disagreement between simulated and observed ERTs.

3.4.4 Summary and Synthesis

We have shown that the radiation regime within a deciduous forest is a function of solar position, spatial and temporal variations in canopy structure, and optical properties of the canopy elements. Maximal radiation penetration occurs during the winter leafless period when ~60% of incoming PAR reaches the forest floor. This value agrees well with measurements by Federer (1971) and Rauner (1976) made in leafless deciduous forests. This value also agrees well with theoretical values computed with the model of Federer (1971). Lesser values of R_g penetration have been reported by Hutchison and Matt (1977) for a *Liriodendron tulipifera* canopy. The greater radiation attenuation in the leafless *L. tulipifera* canopy results from the greater biomass density of that canopy and from topographic shading. (The *L. tulipifera* stand studied was situated in a karst topographic depression.) Galoux et al. (1981) reported 30 to 40% penetration of solar radiation through the leafless oak forest at Virelles, Belgium. This reduced penetration results from the high latitude of the Belgian site and from the greater woody biomass area index (WAI) of that forest (see Table 3.3).

The penetration of R_g and its PAR component decreases with expanding leaf biomass in the spring. This decrease is somewhat offset by increased solar elevations near the summer solstice. Less than 5% of the incoming R_g penetrates the fully leaved forest. Similar results are reported by Horn (1971), Rauner (1976), Floyd et al. (1978), and Galoux et al. (1981). Radiation penetration again increases in autumn with the loss of foliage, despite reduced solar elevations.

The frequency distributions of PAR penetration also exhibit seasonality. In the winter leafless forest, the frequency distribution is unimodal and skewed. This finding agrees with the observations of Hutchison and Matt (1977) in a tulip poplar forest. In a leafing forest, the frequency distribution of relative PAR irradiances becomes broader and strongly skewed. This results from higher solar elevations, which enhance PAR penetration in the overstory, and from expanding leaves, which have the opposite effect at lower levels through absorption and scattering of the incoming radiation. Ovington and Madgwick (1955) report that the frequency distribution of radiation in a *Quercus rubra* forest is not skewed. Their results may be suspect, since the spatial replication of their measurements may not have been adequate to characterize the spatial variability in radiant flux densities within the forest. The frequency distribution of PAR in the fully leaved canopy is unimodal and negatively skewed in the upper canopy, bimodal below crown closure, and unimodal and positively skewed in the sub-canopy trunk space. Similar observations have been reported by Ovington and Madgwick (1955), Norman and Jarvis (1974), Hutchison and Matt (1977), and Sinclair and Knoerr (1982) for coniferous and deciduous forests. Low solar elevation angles and leaf senescence and abscission cause the frequency distribution in the autumnal forest to approach uniformity.

The canopy reflectivity for PAR and shortwave radiation exhibits seasonality as well. PAR reflectivity and solar albedo are maximal when the canopy is leafless and minimal when it is fully leaved. The range of midday values between these two extreme phenological phases is 0.06 and 0.02 for PAR and 0.20 and 0.10 for shortwave radiation. This seasonal pattern is distinctly different from those reported for other, higher-latitude deciduous forests. Galoux et al. (1981) reported mean shortwave radiation albedos of 0.178 for a foliated oak forest in Belgium and 0.122 for the leafless phase. Similarly, DeWalle and McGuire (1973) found growing season albedoes from 0.159 to 0.176 and dormant season albedos (with no snow cover) from 0.125 to 0.145. The reversal in seasonal maximum and minimum between the Walker Branch forest and the more northerly forests should be further investigated.

The radiation regime data were used to test extant phytoactinometric theory. The penetration of total radiation (direct plus diffuse) decreases with increasing leaf area, in accordance with the theory of Monsi and Saeki (1953). No effect of solar elevation on these relationships is discernible. However, the extinction coefficient in the Monsi-Saeki (1953) relationship is markedly affected by the presence or absence of leaves. Unlike total radiation, the penetration of the beam component does not decrease exponentially in the fully leaved forest. This is a consequence of the more steeply inclined leaves in the overstory and clumped foliage, a condition that enhances beam penetration, and the more horizontal leaves in the subcanopy, which inhibit beam penetration. Thus, phytoactinometric models based on the Poisson distribution (e.g., de Wit 1965; Norman 1979; Ross 1981) fail in this canopy, whereas models based on the negative binomial distribution (e.g., Acock et al. 1970; Nilson 1971; Baldocchi et al. 1985b) are more successful. Use of the Poisson-based phytoactinometric model to compute the rates of nonlinear, light-dependent biological processes, such as canopy photosynthesis and stomatal conductance, can lead to a substantial underestimation of these quantities.

Preliminary measurements of thermal infrared radiation exitance show a strong nadir and azimuthal dependence in the leafless forest. Radiant temperatures decrease from northerly to southerly azimuth angles. Radiant temperatures also decrease as the nadir angle decreases from horizontal.

3.5 Canopy-Atmosphere Turbulent Exchange

3.5.1 Background

Canopy-atmosphere exchanges of entities other than electromagnetic radiation involve a variety of mechanisms. Large (supermicron-size) particles settle out of the atmosphere by gravitational attraction as precipitation or dry-deposited material; exchanges of gases and small particles involve other processes. In the free atmosphere, vertical transport of mass and momentum is almost entirely the result of turbulence. Molecular diffusion and phoretic mechanisms operate there but contribute little to vertical transport. Turbulent fluid motions near surfaces are damped out by molecular viscosity, resulting in a quasi-laminar boundary layer immediately adjacent to those surfaces. Thus, entities being exchanged between surface elements and the atmosphere must be transported through this boundary layer by processes other than turbulence—primarily by molecular diffusion and impaction.

Although processes other than turbulent transport affect the exchanges between surfaces and the atmosphere, such processes tend to operate at microscopic levels and at rates dependent upon the nature of the surface and its location relative to other surfaces. Given the complex geometry of a forest canopy and the myriad surfaces presented to the atmosphere by such a canopy, the determination of stand-level exchange rates and amounts by measurements of molecular diffusion and impaction would be a formidable task. In many cases it can be assumed that the rate of vertical turbulent exchange of an entity in the vicinity of a surface feature, such as a forest canopy, is limited by the rate of evolution or uptake of that entity by the surface feature. That is, the collective rate of exchange of any conservative entity across the quasi-laminar boundary layers surrounding the surface elements making up the "surface" must equal the rate of turbulent transport of that entity to or from the "surface."

Micrometeorological techniques for the measurement of air-surface exchange make use of this fact to measure exchange rates from measures of vertical turbulent transport. In practice, however, the restriction of no flux divergence implies difficulty in relating measured turbulent transport to surface exchange rates in other than the simplest situations at meteorologically ideal sites, that is, over level, smooth surfaces under steady-state atmospheric conditions. Thus, accurate measurement of turbulent transport over forests is especially difficult because of the extreme roughness of forest canopies and the tendency for contemporary forests to occur largely on terrain too irregular for agricultural use or commercial development. Obviously, the constraints associated with flux measurement techniques that require meteorologically ideal sites must be relaxed to study the problem of forest canopy-atmosphere exchange. Thus, the first task undertaken with regard to canopy-atmosphere exchange at the Walker Branch Watershed site was an evaluation of flux measurement techniques.

The approach that held greatest promise, *a priori*, was the eddy correlation technique. This technique operates on the concept that the vertical flux of an entity is equal to the mean covariance between turbulent fluctuations in the atmospheric concentration of that entity and in the vertical component of the wind velocity over the time period of interest. Unlike most other flux measurement techniques, the covariance approach measures the flux under any circumstance. Nevertheless, the areal representativeness of this flux measurement depends upon the absence of horizontal flux divergence and the presence of steady-state atmospheric conditions. Despite this limitation, the eddy correlation technique has attributes that offer the possibility that areally representative fluxes can be determined, at least part of the time, at such meteorologically nonideal sites as the Walker Branch site.

With irregular terrain and rough surfaces, the unequal irradiance of northerly and southerly exposed slope facets implies that surface energy and mass balances are not everywhere uniform. Thus, horizontal gradients in mass or energy concentrations may be established along which advective exchange may occur. Consequently, flux measurements over nonideal sites are suspect because a point measure of a flux may not be representative of the flux to the entire surface feature of interest.

Another problem of irregular surfaces has to do with the production of the turbulence that drives the vertical transport process. Turbulence is generated mechanically, as with fluid flow over a rough surface. Turbulence is also generated by buoyancy; warm or moist air is less dense than cool or dry air, and, therefore, warm air tends to rise. Since the frame of reference of mechanical turbulence is oriented perpendicular to the stream lines of the local flow and buoyancy operates along the geopotential vertical, the corresponding "verticals" may not be coincident. Consequently, an instrument array designed to measure a flux in the geopotential vertical may "see" only a part of the flux in the surface normal direction and vice versa when used at nonideal sites.

Furthermore, terrain-induced deformations of the flow field over a portion of the earth's surface will create zones of convergence, divergence, or, in certain situations, separation of the flow from the surface. In terms of flux measurements, such deformation creates further problems with regard to the representativeness of a point flux measure for the surface area of interest.

The properties of the eddy correlation technique having bearing on this situation are the following: (1) Because of the turbulent mixing in boundary layer flow over the earth's surface, the wind structure at any point in the boundary layer represents the integrated effect of some upwind area of surface upon the boundary layer flow at that point (see Pasquill 1972). Thus, the flux indicated by eddy correlation at that point is an areally integrated measure of the exchange between that area of surface and the atmosphere. Furthermore, the area of integration, which is a function of atmospheric stability, wind speed, and surface roughness, is definable (Pasquill 1972). (2) Eddy correlation flux measurements are independent of energy balance considerations. (3) Since eddy correlation measurements of vertical fluxes involve direct measurement of the vertical wind velocity, mathematical rotation of the coordinate system can be performed. Now, in what way do these properties have a bearing on the problem of flux measurement at nonideal sites?

First, since the eddy correlation measurement of a vertical flux is areally integrated, the problem of spatial representativeness is partially resolved. The area over which the flux measurement is integrated is a function of wind speed and direction, surface roughness, and atmospheric stability. In the most restrictive cases with high wind speeds and unstable, well-mixed atmospheres, this area of integration is much larger than the typical scale of spatial canopy heterogeneity within most continuous forest types. Thus, spatial variations in vertical fluxes resulting from canopy irregularity within continuous forest types are accounted for in the eddy correlation flux measurements. Those spatial flux variations resulting from larger-scale variations in forest type or terrain irregularity are not accounted for. Hence, spatially replicated flux measures would be required to assess their significance.

Second, the fact that eddy correlation flux measurements are independent of energy balance considerations provides a test for the presence of errors due to advection or other sources of flux divergence. The eddy correlation technique allows direct measurement of the sensible (H) and latent (λE) heat fluxes. The sum of $H + \lambda E$ can also be inferred as the residual of the energy balance; that is,

$$H + \lambda E = R_n - S - G - P - A$$

where R_n is the net all-wave radiation, S is the enthalpy storage in the air and biomass within the forest volume, G is the soil heat flux, P is the energy fixed by photosynthesis, and A is an advection term. R_n is fairly easily measured, and S and G can be measured or inferred. P is generally so small that it is ignored, but it can be included if the situation warrants. Agreement between $H + \lambda E$ and $R_n - S - G$ indicates the absence of flux divergence (i.e., $A \rightarrow O$), provided that errors in R_n, S, and G are negligible. By extension, if $H + \lambda E$, as determined by eddy correlation, is free of significant error due to flux divergence, then other fluxes measured at the same time by means of eddy correlation can be accepted with confidence as well.

Finally, the capacity to mathematically rotate the coordinate system provides for the resolution of mechanically driven and buoyancy driven fluxes to a common vertical. In practice, mathematical rotation also eliminates nonzero vertical velocities that appear as artifacts of advection. While such rotation of coordinate axes is computationally trivial, given contemporary computer facilities, the potential for error propagation in the computation is sufficiently great that careful tests of the procedure had to be conducted to ensure that the results were of acceptable accuracy.

3.5.2 Applicability of Eddy Correlation at the Walker Branch Site

A set of wind component velocities was collected above the Walker Branch forest over multiday periods throughout all seasons and phenological phases of the canopy. Despite the variable terrain, the wind component velocity standard deviations [σ_u, σ_v, and σ_w for the horizontal mean wind (u), horizontal transverse wind (v), and vertical wind (w) directions] normalized by the friction velocity (u_*) are remarkably stable over all wind directions, as shown in Fig. 3.32 (McMillen 1983). (Data for northerly winds are not included on Fig. 3.32, since winds from that direction pass directly through the tower.) Thus, wind component velocity deviations (σ's) do not appear to be affected by wind direction or, by extension, topography. This implies that the turbulence characteristics over the moderately complex terrain of the Walker Branch Watershed are not significantly affected by the topography. Consequently, the areal representativeness of point flux measurements does not appear to be significantly degraded by topography at this site.

Figure 3.33 is a test of energy balance closure during typical midsummer

ATDL-M86/484

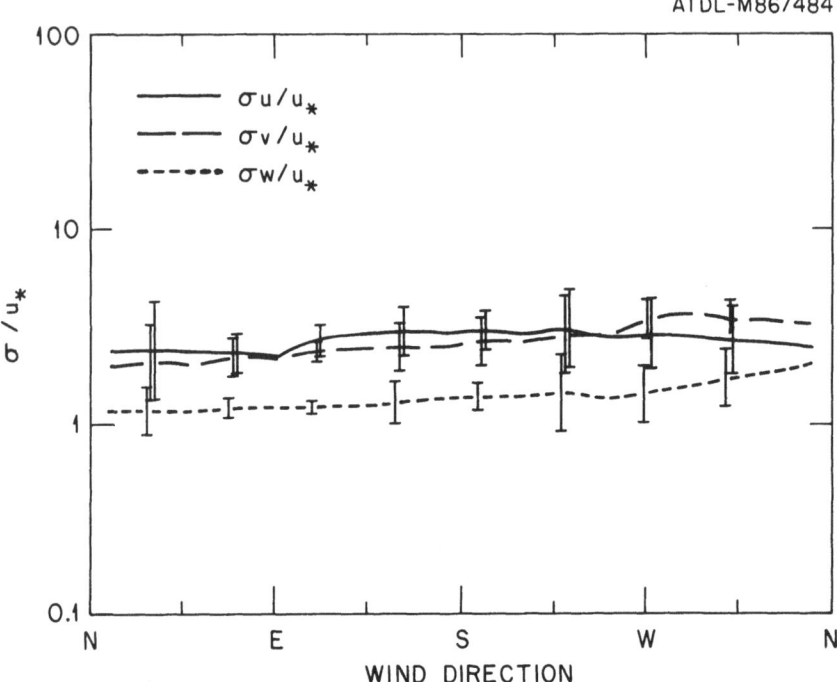

Figure 3.32. Standard deviations (σ) [normalized by the friction velocity (u_*)] of the horizontal mean wind (u), the horizontal transverse wind (v), and the vertical wind (w) components as a function of wind direction. Source: R.T. McMillen. 1983. Eddy correlation calculations done in real-time. pp. 111–114. IN Preprint Volume, Seventh Conference on Fire and Forest Meteorology, American Meteorological Society, Boston, Massachusetts.

conditions at the Walker Branch site. The data of Fig. 3.33 exhibit scatter of the order of ±30% about the 1:1 correspondence line. Error analyses of the measurements involved indicate that scatter up to about ±20% is attributable to the eddy correlation measurement of $H + \lambda E$ and the radiometric measurement of R_n.

Canopy heat storage (S) was estimated for purposes of this test, assuming that the time rate of the temperature change of the biomass in the forest was equal to that of the within-canopy air mass. Since the thermal mass of boles and branches in this forest is substantial, the thermal inertia of the forest biomass is considerably greater than that of the air within the forest volume. Hence, there may be large errors in the values of S computed in this manner, especially near sunrise and sunset, when air temperatures change far more rapidly than biomass temperatures.

Another source of the scatter in the data of Fig. 3.33 is the comparison of areally integrated measures of $H + \lambda E$ with essentially point estimates or measures of other energy balance components. Since the upwind areas

ATDL-M 86/485

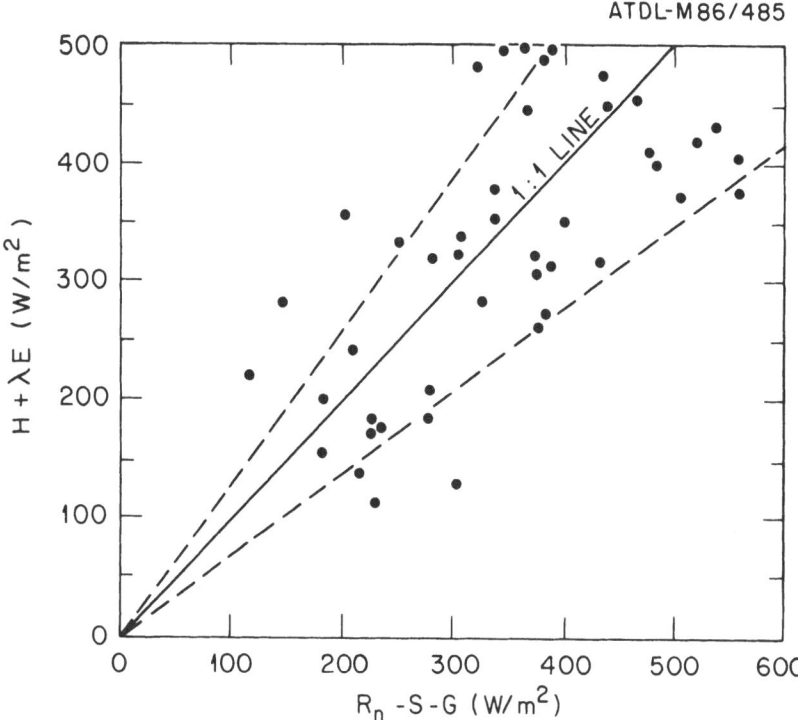

Figure 3.33. Comparison of sensible plus latent heat flux ($H + \lambda E$), as measured above the fully leaved forest canopy, with the estimated residual of the energy balance ($R_n - S - G$). Dashed lines represent the ±30% envelope around the 1:1 correspondence line. Source: S.B. Verma, D.D. Baldocchi, E.E. Anderson, D.R. Matt, and R.J. Clement. 1986. Eddy fluxes of CO_2, water vapor, and sensible heat over a deciduous forest. Boundary-Layer Meteorol. 36:71–91.

of integration for all wind directions at this site contain variable amounts of northerly and southerly slope exposures, the solar irradiance of any area of integration is nonuniform. Thus, the amount and partitioning of R_n is spatially variable. This variability is illustrated in Fig. 3.34, where the normalized energy balance residual is plotted as a function of wind direction. This normalized residual of the energy balance is defined as ($R_n - H - \lambda E - S - G)/R_n$, with H and λE measured by means of eddy correlation, R_n measured at a single point above the Walker Branch canopy, and S and G inferred from point measurements of pertinent meteorological variables within the canopy at the research site (see Baldocchi et al. 1985c). For upwind fetches in the azimuth sector 60 to 150°, the terrain is predominately south facing (see Fig. 3.2). With winds from that sector, the observed residual tends to be less than zero, because the areally integrated R_n for that sector is greater than that observed at the research site. For winds from the sector 210 to 240° of azimuth, where north-facing

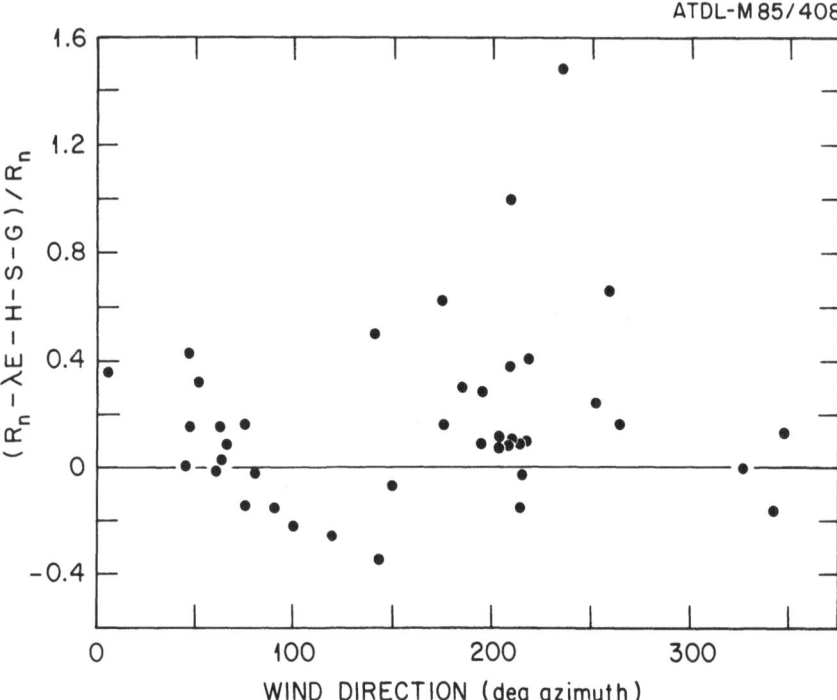

ATDL-M 85/408

Figure 3.34. Effects of wind direction on the normalized residual of the energy balance [$(R_n - \lambda E - H - S - G)/R_n$] on mostly clear summer days at the Walker Branch forest meteorology research site. The high positive values (>0.6) of the normalized energy balance residual were observed during periods of intermittent obscuration of the sun by small clouds. Thus, the estimates of canopy heat storage (S) for these periods are highly questionable, as are the derived energy balance residuals. Source: D.D. Baldocchi, D.R. Matt, R.T. McMillen, and B.A. Hutchison. 1985. Evapotranspiration from an oak-hickory forest. pp. 414–422. IN Proceedings of the National Conference on Advances in Evapotranspiration, December 16–17, 1985, Chicago, Illinois. American Society of Agricultural Engineers, St. Joseph, Michigan.

slopes predominate, the areally integrated R_n must be less than that observed at the Walker Branch site, with the result that the normalized energy balance residual exceeds zero. The upshot of this situation is that areally integrated measures or estimates of R_n, S, and G must be obtained for the same area of integration "seen" by the eddy correlation instrumentation to rigorously test for flux divergence errors in the flux measurements.

Mathematical rotation of the wind component axes has been performed on velocity data collected above the Walker Branch forest. This procedure not only references the data to a common "vertical," but also reduces the time mean cross and vertical wind velocities to zero, a condition required by the principle of mass conservation. The resultant rotated velocity data produce derived turbulence parameters that are stable and possess acceptable levels of scatter and whose magnitudes compare favorably with those observed at ideal sites.

In practice, it is necessary to produce real-time measures of the fluxes of interest so as to allow monitoring of the performances of eddy flux measurement system components. A digital, recursive-filter technique with a 200-s time constant is employed to estimate the real-time running means, which allows on-line derivations of the turbulent fluctuations as required for real-time flux calculations. The filtered data approach was tested by comparing the covariances produced with filtered data with the same co-variances computed with wind component velocity data preprocessed with a three-dimensional coordinate-axis rotation scheme derived by Tanner and Thurtell (1969). Results of this comparison are shown in Fig. 3.35, where agreement is generally good. Points falling outside the $\pm 10\%$ confidence limits represent data obtained under near-calm conditions. We conclude that a digital, recursive-filter technique can be used to compute reliable flux covariances at moderately complex sites such as the Walker Branch Watershed.

3.5.3 Canopy-Atmosphere Exchange

As confidence in the use of eddy correlation techniques at the Walker Branch site developed, flux measurement programs were initiated. Because of the current concern about acid deposition, capabilities for routine measurements of total S, SO_2, O_3, and particulate SO_4 were developed in addition to those for momentum, H, and λE. Collaborative measurements with university scientists allowed expansion of flux measurement capabilities to include NO_x and CO_2, though not on a routine basis.

Pollutant Exchange

An example of observed SO_2 deposition velocities (v_d) and atmospheric concentrations is presented in Fig. 3.36. The deposition velocity (v_d) is defined as

$$v_d = \frac{\text{flux}}{\text{concentration}}$$

At the time of these measurements, the leaves in the forest canopy were fully expanded.

Both the atmospheric concentrations and the deposition velocities for SO_2 during this period tend to be highest during the day and near zero during the night. (By convention, positive values of v_d indicate a flux to the canopy.) During the early morning hours of July 10, however, plume impaction occurred, causing SO_2 concentrations to rise sharply to

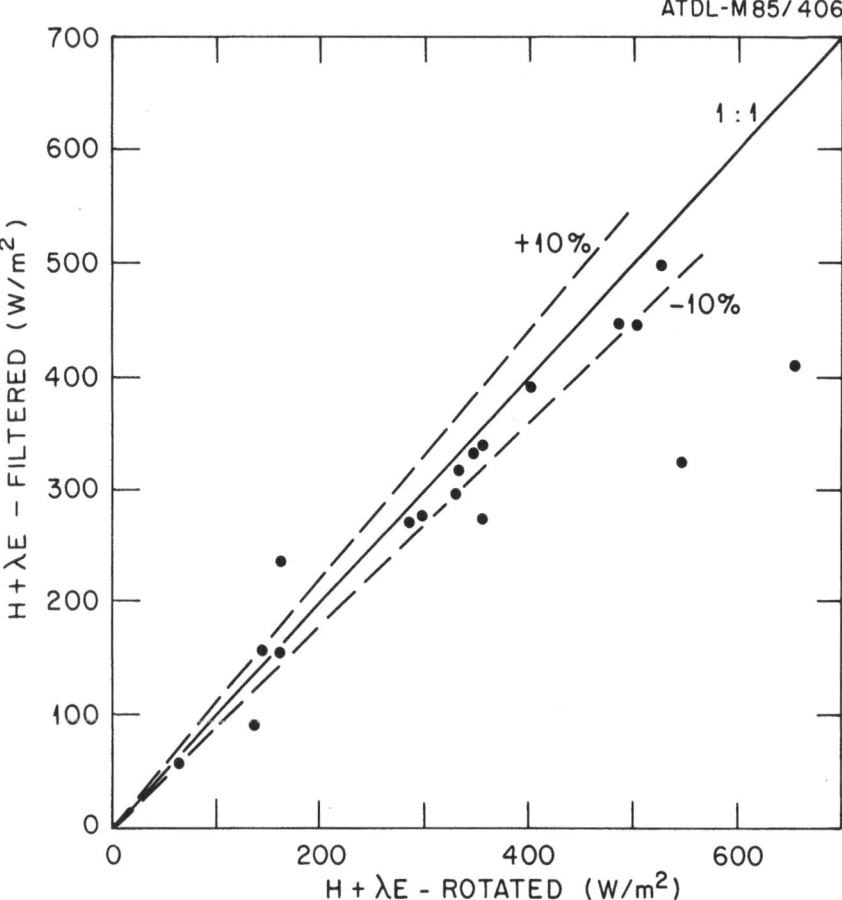

ATDL-M85/406

Figure 3.35. A comparison of sensible plus latent heat flux $(H + \lambda E)$ computed using digitally filtered data and data preprocessed with a three-dimensional co-ordinate rotation. All data presented were selected from periods of southerly winds. Source: D.D. Baldocchi, D.R. Matt, R.T. McMillen, and B.A. Hutchison. 1985. Evapotranspiration from an oak-hickory forest. pp. 414–422. IN Proceedings of the National Conference on Advances in Evapotranspiration, December 16–17, 1985, Chicago, Illinois. American Society of Agricultural Engineers, St. Joseph, Michigan.

extremely high values. Unfortunately, the flux measurement system failed just after midnight and did not resume operation until 0430 h. As the plume moved off site, slightly negative v_d's were observed as the SO_2-laden within-canopy air diffused upward and was replaced by cleaner air penetrating from above. Later that morning, a fumigation event occurred in which SO_2-rich air trapped within the planetary boundary layer by an elevated temperature inversion was mixed downward toward the surface. The increasing SO_2 concentrations caused flux divergence, because some

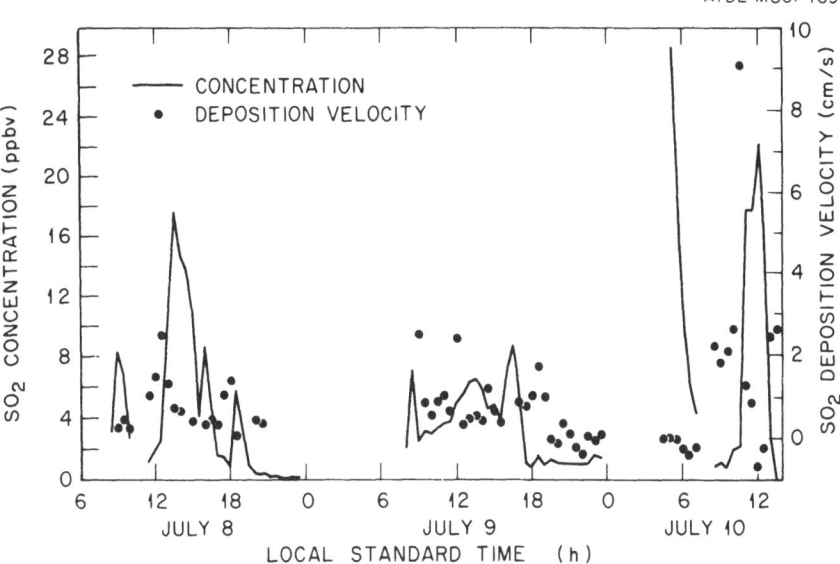

ATDL-M86/489

Figure 3.36. Atmospheric concentrations of SO_2 and concurrent deposition velocities to the deciduous forest canopy derived from eddy correlation measures
of the vertical SO_2 flux. Source: D.R. Matt, R.T. McMillen, J.D. Womack, and
B.B. Hicks. 1987. A comparison of estimated and measured SO_2 deposition velocities. Water Air Soil Pollut. 36:331–347.

of the SO_2 that was transported downward was not deposited; rather, the
SO_2 concentration in the air below the height of measurement increased.
Thus, the measured v_d's cannot be accepted as indicative of true surface
values. As the fumigation event weakened (~1100 h), SO_2-rich air was
flushed out of the canopy by cleaner air from above, resulting in the computed negative v_d values shown in Fig. 3.36.

Similar data on NO_x exchange across the canopy-atmosphere interface
along with the concurrent H and λE fluxes are presented in Table 3.5
(from Hicks et al. 1983). These data were observed over the fully leaved
forest canopy in July 1983. As with SO_2, daytime v_d's for NO_x are moderate
to high and operate toward the canopy. At night, the v_d's are mostly negative but near zero, indicating limited rates of exchange. Outliers were
statistically removed from the data of Table 3.5, and the resultant mean
v_d's were plotted as a function of time of day, yielding Fig. 3.37. Despite
high daytime variability, a strong diel cycle in NO_x v_d's is evident.

Latent Heat Exchange

Analyses of eddy correlation measurements of H and λE fluxes above the
early summer, nearly fully leaved canopy (late May 1983) conducted by

Table 3.5. Fluxes observed above a deciduous forest canopy in eastern Tennessee, July 1983

Date and time	R_n (W/m²)	H (W/m²)	λE (W/m²)	u. (cm/s)	$v_d(NO_x)$ (cm/s)	NO_x (ppbv)
July 14						
1000	572	458	202	26.7	0.48	10.5
1100	626	484	328	26.9	0.29	21.1
1200	637	435	438	13.5	0.28	10.6
1300	680	472	399	32.1	0.39	4.9
1400	627	448	306		0.05	4.6
1500	492	402	354		−1.08	4.2
1600	355	135	145	17.8	0.61	4.8
1700	210	11	30	15.9	0.02	5.1
1800	58	−11	2	11.9	−0.03	5.8
1900	−33	0	174	16.8	−0.18	6.0
July 15						
0900	283	224	122	24.7	2.70	7.7
1000	479	310	176	27.8	1.46	7.2
1100	625	290	233	19.3	4.90	8.4
1200	672	496	365	32.0	0.76	9.3
1300	663	353	338	27.4	−4.46	6.9
1400	547	309	227		0.34	2.4
1500	344	93	75	13.2	1.60	3.4
1600	284	64	26	14.1	0.47	2.9
1700	164	7	26	13.5		
1800	36	−2	39	19.3		
1900	−43	−3	54	23.4		
2000	−59	−2	36	17.3	0.01	18.4
2100	−57	−7	43	18.8	−0.11	12.5
2200	−58	−11	37	13.6	−0.06	10.3
2300	−51	−14	40	16.9	−0.05	7.0
2400	−52	−5	38	17.1	−0.12	7.4
July 16						
0100	−41	−12	42	18.7	−0.16	4.5
0200	−54	−21	42	22.9	−0.18	9.2
0300	−50	−4	34	8.7	−0.05	6.2
0400	−48	−2	35	8.0	−0.05	4.1
0500	−6	−4	38	9.3	−0.11	5.2
0600	51	4	37	18.5	−0.22	4.1
0700	132	20	64	24.4	−1.88	3.6
0800	387	182	115	30.4	−0.02	6.0
0900	351	141	89	22.9	0.11	5.2
1000	496	122	98	19.8	0.30	6.3

Times (EST) represent the ends of the hourly averaging periods.

Abbreviations: R_n = net radiation; H = sensible heat flux; λE = latent heat flux; u. = friction velocity; v_d = deposition velocity.

Source: Adapted from B.B. Hicks, D.R. Matt, R.T. McMillen, J.D. Womack, and R.E. Shetter. 1983. Eddy fluxes of nitrogen oxides to a deciduous forest in complex terrain. pp. 189–201. IN P.J. Sampson (ed.), Meteorology of Acid Deposition, Conference Proceedings. Air Pollution Control Association, Hartford, Connecticut.

ATDL-M83/152

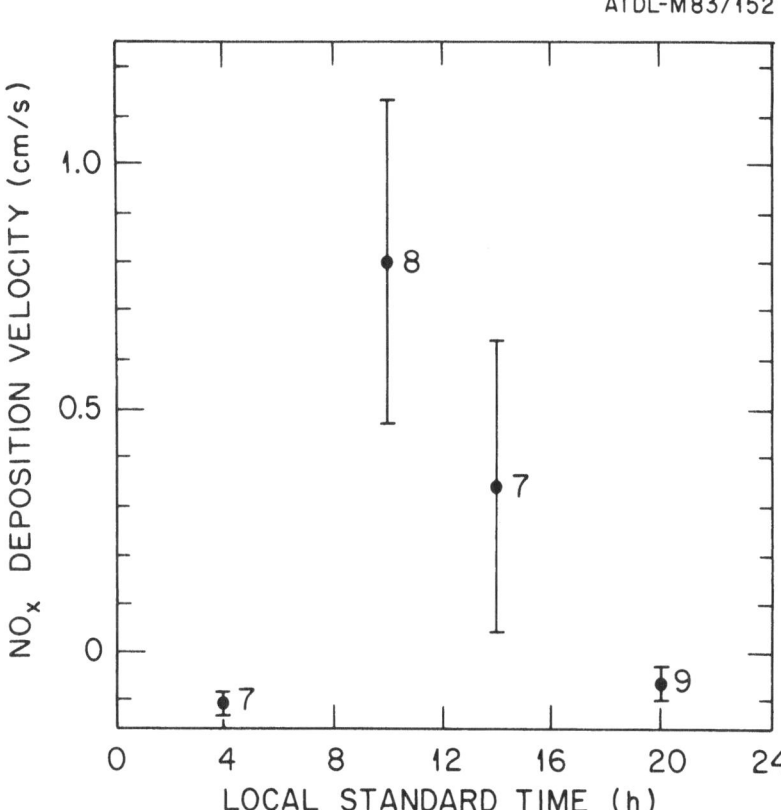

Figure 3.37. The diel variation in the velocities of NO_x deposition to the deciduous forest canopy. Numbers next to the data points indicate the number of hourly averages used to compute the four hourly means as plotted. Source: Adapted from B.B. Hicks, D.R. Matt, R.T. McMillen, J.D. Womack, and R.E. Shetter. 1983. Eddy fluxes of nitrogen oxides to a deciduous forest in complex terrain. pp. 189–201. IN P. J. Sampson (ed.), Meteorology of Acid Deposition, Conference Proceedings. Air Pollution Control Association, Hartford, Connecticut.

Baldocchi et al. (1985c) support the contention of McNaughton and Jarvis (1983) that λE over aerodynamically rough forest canopies is relatively decoupled from R_n. Plotting measured λE as a function of concurrent values of R_n (Fig. 3.38) reveals that the ratio $\lambda E/R_n$ above the nearly fully leaved, well-watered canopy ranges from about 0.2 to 0.4. Ratios of these entities over well-watered agricultural crops are much higher (e.g., Baldocchi et al. 1981b, 1985a).

McNaughton and Jarvis (1983) have defined the coupling coefficient (Ω) as

$$\Omega = [(1+\lambda)/(\lambda+s)](r_{sc}/r_{av})^{-1}$$

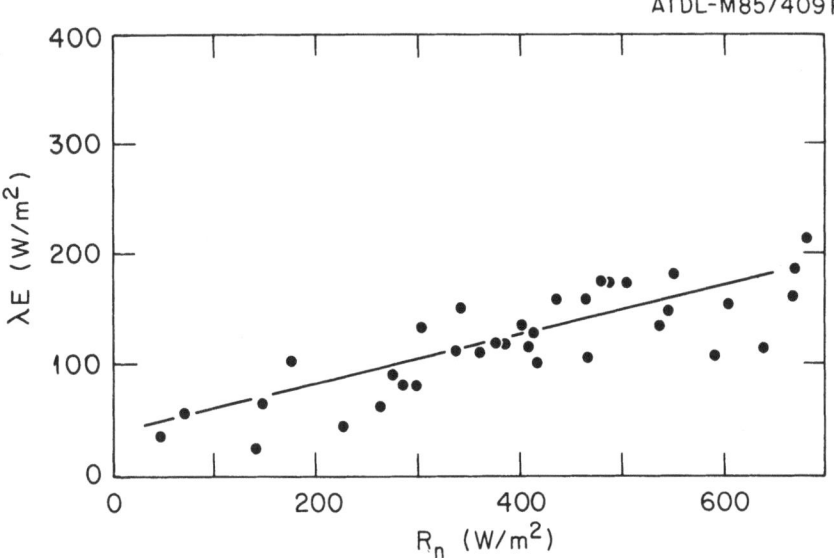

Figure 3.38. The relationship between latent heat flux (λE) and net radiation (R_n) over the nearly fully leaved forest canopy observed during periods of southerly winds. Source: Adapted from D.D. Baldocchi, D.R. Matt, R.T. McMillen, and B.A. Hutchison. 1985. Evapotranspiration from an oak-hickory forest. pp. 414–422. IN Proceedings of the National Conference on Advances in Evapotranspiration, December 16–17, 1985, Chicago, Illinois. American Society of Agriculture Engineers, St. Joseph, Michigan.

where λ is the psychrometric constant, s is the slope of the saturation vapor pressure vs. temperature curve, r_{sc} is the bulk canopy stomatal resistance, and r_{av} is the aerodynamic resistance for water vapor. This dimensionless coefficient ranges from 0 to 1, with lower values indicating weaker coupling between the canopy and the atmosphere. Values of Ω for conifer forests are typically of the order of 0.1 to 0.2, whereas for grass, values are ~0.8 (McNaughton and Jarvis 1983). Daytime values of Ω for the Walker Branch forest in nearly full leaf and under well-watered conditions ranged from 0.4 to 0.6—considerably larger than the values for conifer stands cited by McNaughton and Jarvis (1983). Nevertheless, Ω for the Walker Branch forest is less than has been reported for aerodynamically smoother grass and crop canopies.

The time traces of R_n, H, and λE on a reasonably clear day in late May above the Walker Branch forest are shown in Fig. 3.39. Prior to 1330 h, winds were from the northeast to east; afterward they were from the southeast to south. Wind speeds (u) in the morning at 7 m above the canopy were 2 to 3 m/s (Fig. 3.40a), declining to 1.5 to 2.0 m/s after midmorning, and then falling off toward zero later in the afternoon. The vapor pressure

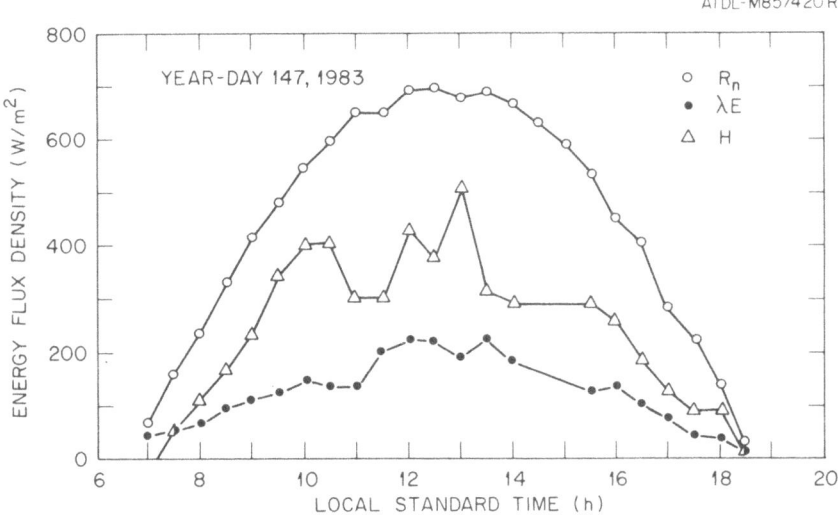

ATDL-M85/420 R

Figure 3.39. The diurnal variation of net radiation (R_n), sensible heat flux (H), and latent heat flux (λE) over the nearly fully leaved forest canopy on year-day 147, 1983. Source: Adapted from D.D. Baldocchi, D.R. Matt, R.T. McMillen, and B.A. Hutchison. 1985. Evapotranspiration from an oak-hickory forest. pp. 414–422. IN Proceedings of the National Conference on Advances in Evapotranspiration, December 16–17, 1985, Chicago, Illinois. American Society of Agricultural Engineers, St. Joseph, Michigan.

deficit (VPD, Fig. 3.40b) and air temperature (T, Fig. 3.40c) increased throughout the day.

The concurrent variations in r_{sc}, r_{av}, and Ω, as derived from the observed fluxes, are shown in Fig. 3.41. The r_{sc}'s increased throughout the day (Fig. 3.41a), as would be expected with the increasing VPD and T values of Figs. 3.40b and c. Values of r_{av}, on the other hand, were steady during the portion of the day with northeasterly to easterly winds. With the direction shift to southerly flows, r_{av} increased substantially as a result of the reduced wind speeds that accompanied the direction shift. Despite the sharp afternoon increase in r_{av} values, Ω (Fig. 3.41c) was relatively conservative. Apparently, the effects of higher r_{av} values on Ω were masked by less substantial but nevertheless marked increases in r_c following the shift in wind direction at ~1330 h.

Forest-Atmosphere CO₂ Exchange

In the summer of 1984, in collaboration with the Center for Agricultural Meteorology of the University of Nebraska, simultaneous eddy correlation measurements of CO_2 fluxes above (29 m) and within (1 m) the forest

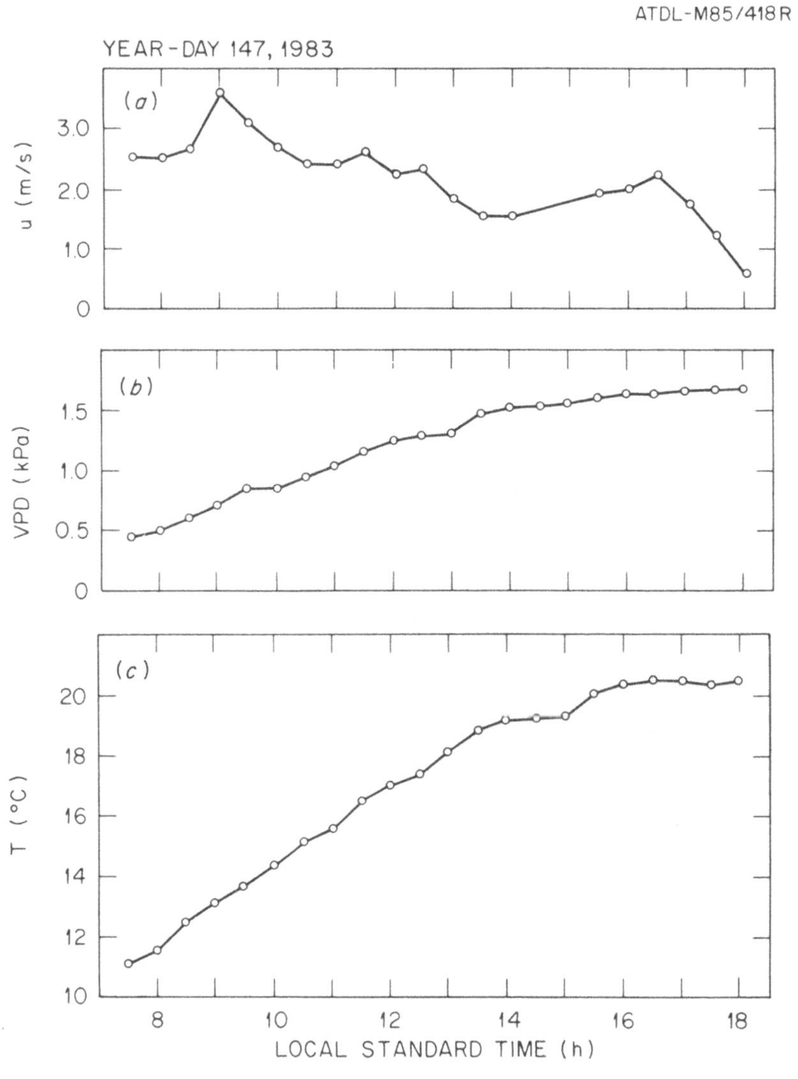

ATDL-M85/418R

Figure 3.40. The diurnal variation in wind speed (u), vapor pressure deficit (VPD), and air temperature (T) at the 29-m level (which is some 7 m above the nearly fully leaved forest canopy) on year-day 147, 1983. Source: Adapted from D.D. Baldocchi, D.R. Matt, R.T. McMillen, and B.A. Hutchison. 1985. Evapotranspiration from an oak-hickory forest. pp. 414–422. IN Proceedings of the National Conference on Advances in Evapotranspiration, December 16–17, 1985, Chicago, Illinois. American Society of Agricultural Engineers, St. Joseph, Michigan.

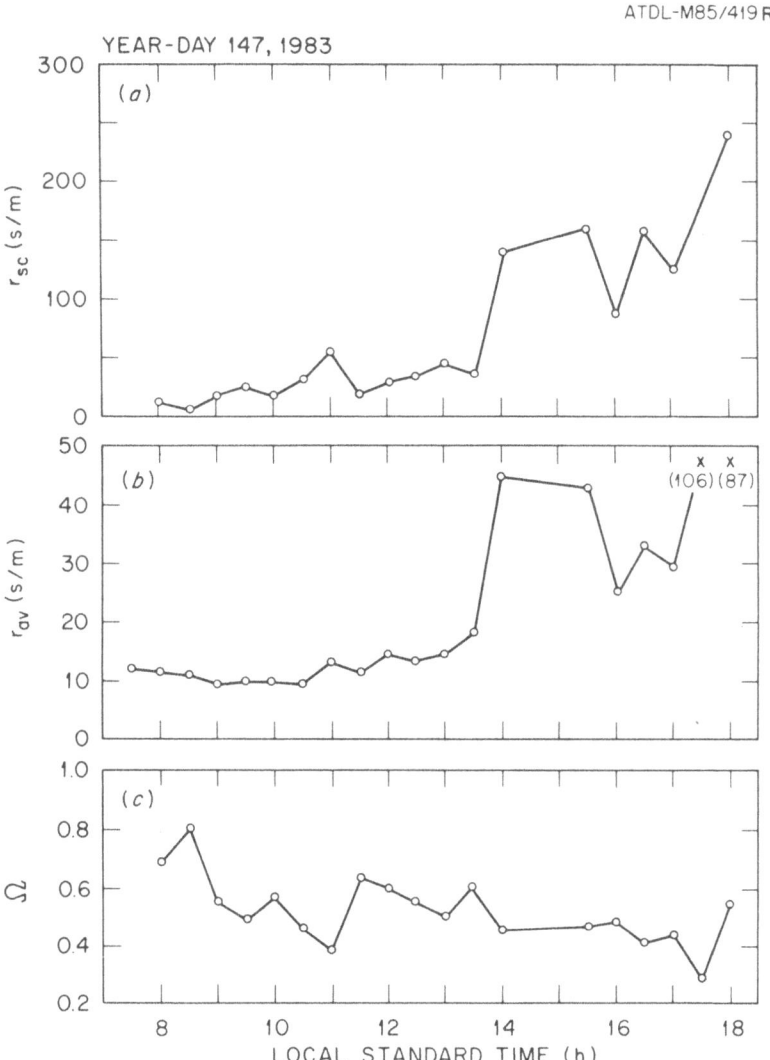

Figure 3.41. The diurnal variation in the bulk canopy stomatal resistances (r_{sc}), the aerodynamic resistances to water vapor transfer (r_{av}), and the water vapor transfer coupling coefficient (Ω) above the nearly fully leaved forest canopy on year-day 147, 1983. Source: Adapted from D.D. Baldocchi, D.R. Matt, R.T. McMillen, and B.A. Hutchison. 1985. Evapotranspiration from an oak-hickory forest. pp. 414–422. IN Proceedings of the National Conference on Advances in Evapotranspiration, December 16–17, 1985, Chicago, Illinois. American Society of Agricultural Engineers, St. Joseph, Michigan.

canopy were made at the Walker Branch site. Figure 3.42 shows the hourly mean CO_2 fluxes (F_c) along with the concurrent hourly mean solar insolation for 6 August days (from Verma et al. 1986).

In Fig. 3.43, the above-canopy F_c's are plotted as a function of R_g and VPD. Despite the considerable scatter of these data, it is clear that canopy F_c increases with insolation as expected, since photosynthesis, the sink for this flux, is a light-driven process. As with other nonplanophile canopies reported in the literature, no evidence of light saturation is evident in the data for the Walker Branch forest canopy. While the CO_2 flux rates observed by Denmead et al. (1983) over ponderosa pine were strongly dependent upon VPD, no such relationship is evident in Fig. 3.43. This probably reflects the well-watered status of the deciduous forest at the time of these measurements, although differences between conifer and deciduous forest photosynthesis and water-use strategies cannot be ruled out at this time.

The diurnal variation in forest floor CO_2 efflux, as determined by eddy correlation measurements at the Walker Branch site, is shown in Fig. 3.44. Maximum efflux rates are of the order of 0.3 to 0.4 mg m^{-2} s^{-1} and occur in afternoon hours. These values are somewhat higher than those reported from studies using enclosed chambers. The burst in early evening may reflect increased catabolism of carbohydrate in the roots and boles of the forest trees. However, further measurements will be required to verify the existence of such a phenomenon.

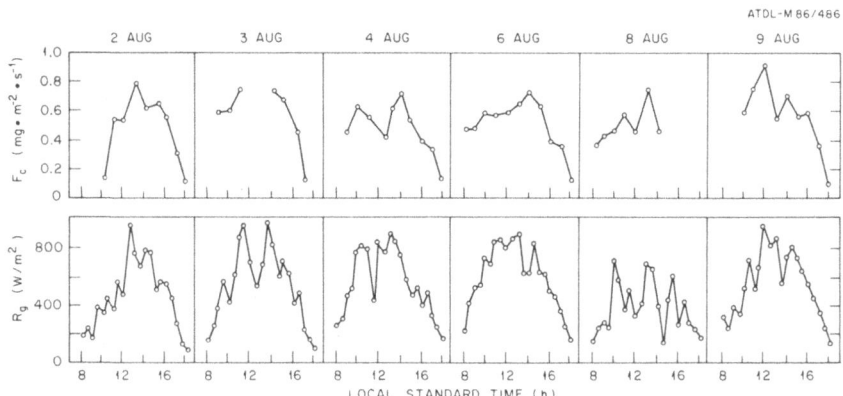

Figure 3.42. Diurnal patterns in CO_2 flux densities (F_c) and global solar radiation (R_g) above the fully leaved forest canopy in August 1984. Source: S.B. Verma, D.D. Baldocchi, D.E. Anderson, D.R. Matt, and R.J. Clement. 1986. Eddy fluxes of CO_2, water vapor, and sensible heat over a deciduous forest. Boundary-Layer Meteorol. 36:71–91.

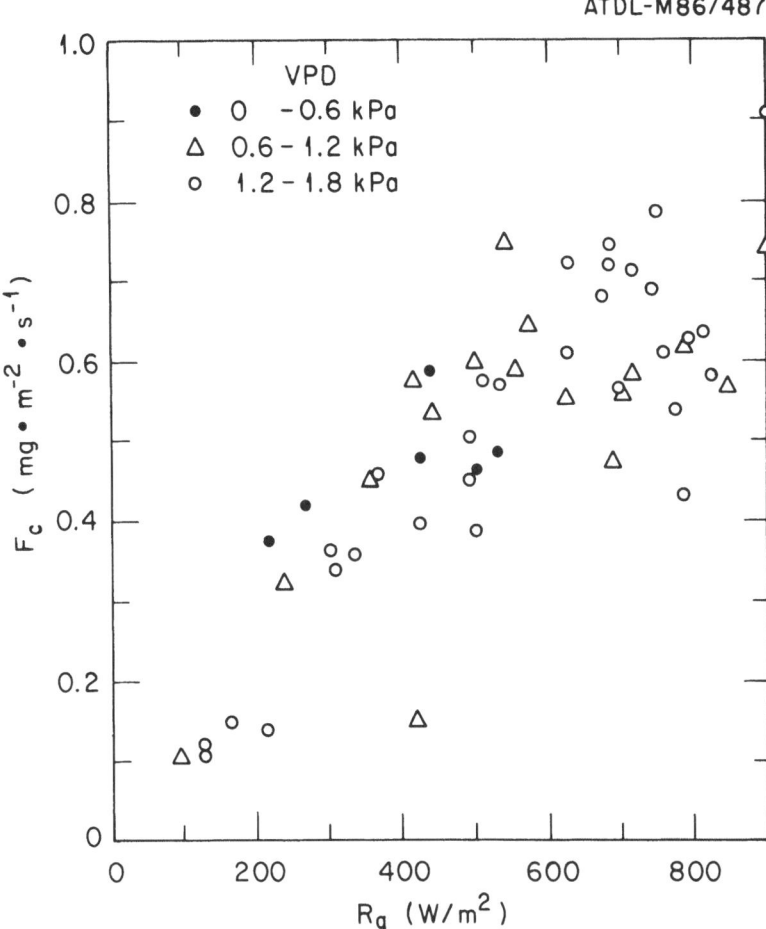

ATDL-M86/487

Figure 3.43. CO_2 flux densities (F_c) in relation to global solar radiation (R_g) and vapor pressure deficit (VPD) above the fully leaved forest canopy in summer. Source: S.B. Verma, D.D. Baldocchi, D.E. Anderson, D.R. Matt, and R.J. Clement. 1986. Eddy fluxes of CO_2, water vapor, and sensible heat over a deciduous forest. Boundary-Layer Meteorol. 36:71–91.

3.5.4 Summary and Synthesis

Studies of canopy-atmosphere turbulent exchange are recent and ongoing efforts of the Atmospheric Turbulence and Diffusion Laboratory (ATDL). These preliminary studies, however, have yielded new and interesting information regarding mass, momentum, and energy exchanges above the forest canopy.

Tests of the eddy correlation flux measurement technique suggest that the level of complexity of the underlying terrain at the Walker Branch

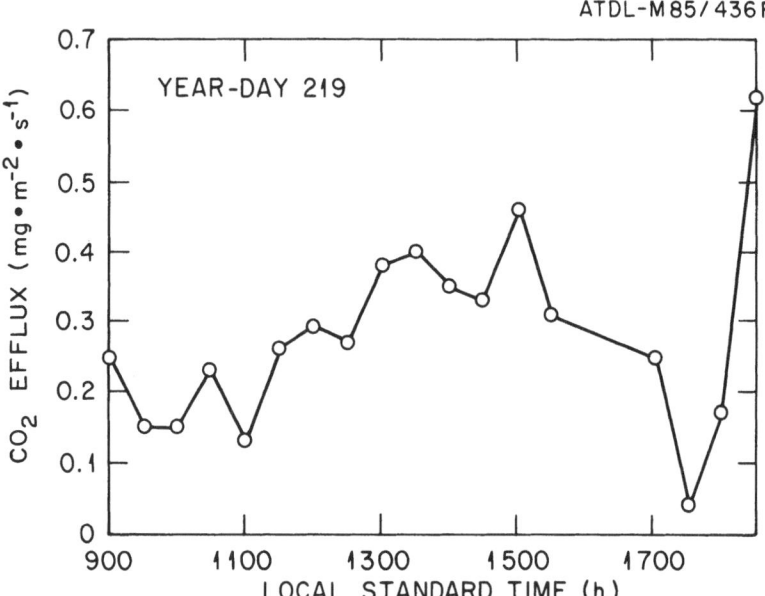

Figure 3.44. CO_2 flux densities at 1 m above the forest floor throughout a day in the fully leaved forest in summer. Source: Adapted from D.D. Baldocchi, S.B. Verma, D.R. Matt, and D.E. Anderson. 1986. Eddy-correlation measurements of CO_2 efflux from the floor of an oak-hickory forest. J. Appl. Ecol. 23:967–975.

Watershed field site does not strongly affect the turbulence characteristics of the wind at this site. Consequently, the eddy correlation technique can be applied to measure fluxes of water vapor, sensible heat, CO_2, and pollutants with moderate confidence.

Energy balances at this site often fail to close. However, this lack of closure is primarily an artifact of our inability to accurately estimate canopy heat storage (S) and of the influence of complex topography on the areally averaged, net radiation balance (R_n). Further work on these problems is needed.

The measurement of the deposition of pollutants implicated in the acid precipitation phenomenon is a major focus of current canopy-atmosphere exchange research at ATDL. Flux measurements are being made throughout the phenological year and used to parameterize the processes that control the dry deposition of pollutants to a deciduous forest canopy. Models developed for this work are currently being used to infer dry deposition on a continuing basis (e.g., Hicks et al. 1985; Baldocchi et al. 1987). The basic studies of canopy-atmosphere interactions discussed in this chapter have proved valuable in enhancing our ability to develop these inferential models and to apply them—a classic example of the application of basic research results. For example, the deposition of pollutant gases is strongly influenced by biology (e.g., stomatal conductance and leaf area) and surface wetness. Through past studies of canopy radiative transfer in this deciduous forest, we have been able to tailor extant phytoactinometric models for use in a deciduous forest. Subsequently, these models are being used in computing canopy stomatal conductance, a nonlinear, light-dependent entity, which operates as a critical input parameter in the dry deposition model.

The structural and physiological characteristics of a deciduous forest canopy are uniquely different from those of agricultural crop and coniferous forest canopies. These differences manifest themselves as differences in the partitioning of net radiation and in the magnitude of CO_2 and pollutant gas exchanges. For example, the minimum stomatal resistances of leaves of most tree species in a deciduous forest are much greater than those of leaves of agricultural crops (Jarvis 1980). On the other hand, the aerodynamic roughness of deciduous forests tends to be greater than that of agricultural crops but less than that of coniferous forests. These factors lead to a moderate decoupling of λE from R_n, as can be quantified by the omega theory of McNaughton and Jarvis (1983) and by the smaller values of canopy CO_2 exchange as compared with those measured over agricultural crops (see Baldocchi et al. 1981a, 1981b; Verma et al. 1986).

While no other stand-level CO_2 exchange data for deciduous forests are available for comparison, Dougherty et al. (1979) reported a maximum white oak (*Quercus alba*) leaf photosynthesis rate of 0.4 $mg \cdot m^{-2} \cdot s^{-1}$. Denmead et al. (1983) observed CO_2 fluxes over an Australian ponderosa pine plantation in the range of 0.4 to 1.2 $mg \cdot m^{-2} \cdot s^{-1}$. The midday values of F_c of Fig. 3.42 are well within those ranges, lending further credence to their validity.

To sum up the status of tests of eddy correlation flux measurement techniques at the Walker Branch site, strong, indirect evidence of the validity of eddy correlation flux measurements at this site has been obtained. Spatially replicated measurements of turbulence characteristics, vertical fluxes, and energy balance components over varying topography, forest type, and soil moisture availability gradients are needed to develop more unequivocal evidence of the representativeness of eddy correlation flux measurements at nonideal sites such as the Walker Branch Watershed. Such evidence should lead to better techniques for extending observations made at a single, nonideal location to other locations having different boundary conditions in the same general area.

References

Acock, B., J.H.M. Thornley, and J. Warren Wilson. 1970. Spatial variation of light in the canopy. pp. 91–102. IN Proceedings, IBP/PP Technical Meeting, Trebon, Czechoslovakia. Pudoc, Wageningen, Netherlands.

Anderson, M.C. 1964. Studies of the woodland light climate. II. Seasonal variation in light climate. J. Ecol. 52:643–663.

Baldocchi, D.D., and B.A. Hutchison. 1986. Effects of clumped foliage on the estimation of canopy photosynthesis and stomatal conductance in a deciduous forest. Tree Physiol. 2:155–168.

Baldocchi, D.D., S.B. Verma, and N.J. Rosenberg. 1981a. Environmental effects on CO_2 flux and CO_2-water flux ratio of alfalfa. Agric. Meteorol. 24:175–184.

Baldocchi, D.D., S.B. Verma, and N.J. Rosenberg. 1981b. Mass and energy exchange of a soybean canopy under various environmental regimes. Agron. J. 73:706–710.

Baldocchi, D.D., D.R. Matt, B.A. Hutchison, and R.T. McMillen. 1984a. Solar radiation in an oak-hickory forest: An evaluation of extinction coefficients for several radiation components during fully-leafed and leafless periods. Agric. For. Meteorol. 32:307–322.

Baldocchi, D.D., B.A. Hutchison, D.R. Matt, and R.T. McMillen. 1984b. Seasonal variations in the radiation regime within an oak-hickory forest. Agric. For. Meteorol. 33:177–191.

Baldocchi, D.D., S.B. Verma, and N.J. Rosenberg. 1985a. Water use efficiency in a soybean field: Influence of plant water status. Agric. For. Meteorol. 34:53–65.

Baldocchi, D.D., B.A. Hutchison, D.R. Matt, and R.T. McMillen. 1985b. Canopy radiative transfer models for spherical and known leaf inclination angle distributions: A test in an oak-hickory forest. J. Appl. Ecol. 22:539–555.

Baldocchi, D.D., D.R. Matt, R.T. McMillen, and B.A. Hutchison. 1985c. Evapotranspiration from an oak-hickory forest. pp. 414–422. IN Proceedings of the National Conference on Advances in Evapotranspiration, December 16–17, 1985, Chicago, Illinois. American Society of Agricultural Engineers, St. Joseph, Michigan.

Baldocchi, D.D., B.A. Hutchison, D.R. Matt, and R.T. McMillen. 1986a. Seasonal variation in the statistics of photosynthetically active radiation penetration in an oak-hickory forest. Agric. For. Meteorol. 36:343–361.

Baldocchi, D.D., S.B. Verma, D.R. Matt, and D.E. Anderson. 1986b. Eddy-correlation measurements of CO_2 efflux from the floor of an oak-hickory forest. J. Appl. Ecol. 23:967–975.

Baldocchi, D.D., B.B. Hicks, and P. Camara. 1987. A canopy stomatal resistance model for gaseous deposition to vegetated surfaces. Atmos. Environ. 21:91–101.

Balick, L.K., and B.A. Hutchison. 1986. Directional thermal infrared exitance distributions from a leafless deciduous forest. IEEE Trans. Geosci. Remote Sensing GE-24:693–698.

Brown, H., and F. Escombe. 1900. Static diffusion of gases and liquids in relation to the assimilation of carbon and translocation in plants. Philos. Trans. R. Soc. Lond., Ser. B 193:223–291.

Burgess, R.L., and R.V. O'Neill. 1975. Eastern Deciduous Forest Biome Progress Report, September 1, 1973—August 31, 1974. EDFB/IBP-7511. Environmental Sciences Division, Oak Ridge National Laboratory, Oak Ridge, Tennessee.

Denmead, O.T., E.F. Bradley, and E. Ohtaki. 1983. CO_2 exchange above and within a pine forest. Paper presented at the Forest Environmental Measurements Conference, October 23–28, 1983, Oak Ridge, Tennessee.

DeWalle, D.R., and S.G. McGuire. 1973. Albedo variations of an oak forest in Pennsylvania. Agric. Meteorol. 11:107–113.

de Wit, C.T. 1965. Photosynthesis of leaf canopies. Agricultural Research Report 663. Pudoc, Wageningen, Netherlands.

Dougherty, P.M., R.O. Teskey, J.E. Phelps, and T.M. Hinckley. 1979. Net photosynthesis and early growth trends of a dominant white oak (*Quercus alba*). Plant Physiol. 64:930–935.

Elias, P. 1979. Stomatal activity within the crowns of tall deciduous trees under forest conditions. Biol. Plant. 21:266–274.

Federer, C.A. 1971. Solar radiation absorption by leafless hardwood forests. Agric. Meteorol. 9:3–20.

Finnigan, J.J., and M.R. Raupach. 1987. Modern theory of transfer in plant canopies in relation to stomatal characteristics. pp. 385–429. IN E. Zeiger, G. Farquhar, and I. Cowan (eds.), Stomatal Function. Stanford University Press, Stanford, California.

Floyd, B.W., J.W. Burley, and R.D. Nobel. 1978. Foliar developmental effects on forest floor light quality. For. Sci. 24:445–451.

Ford, E.D., and P.J. Newbould. 1971. The leaf canopy of a coppiced deciduous woodland. I. Development and structure. J. Ecol. 59:843–862.

Galoux, A., P. Beneck, G. Gietl, H. Hager, C. Kayser, O. Kiese, K.R. Knoerr, C.E. Murphy, G. Schnock, and T.R. Sinclair. 1981. Radiation, heat, water, and carbon dioxide balances. pp. 87–205. IN D.E. Reichle (ed.), Dynamic Properties of Forest Ecosystems. Cambridge University Press, New York.

Givnish, T. 1979. On the adaptive significance of leaf form. pp. 375–407. IN O.T. Solbrig, S. Jain, G.B. Johnson, and P.R. Raven (eds.), Topics in Plant Population Biology. Columbia University Press, New York.

Grace, J., and J. Wilson. 1976. The boundary layer over a *Populus* leaf. J. Exp. Bot. 27:231–241.

Hicks, B.B., D.R. Matt, R.T. McMillen, J.D. Womack, and R.E. Shetter. 1983. Eddy fluxes of nitrogen oxides to a deciduous forest in complex terrain. pp. 189–201. IN P.J. Sampson (ed.), Meteorology of Acid Deposition, Conference Proceedings. Air Pollution Control Association, Hartford, Connecticut.

Hicks, B.B., D.D. Baldocchi, R.P. Hosker, Jr., B.A. Hutchison, D.R. Matt, R.T. McMillen, and L.C. Satterfield. 1985. On the use of monitored air concentration to infer dry deposition. NOAA Technical Memorandum ERL ARL-141. National Oceanic and Atmospheric Administration, Air Resources Laboratory, Silver Spring, Maryland.

Hinckley, T.M., R.G. Aslin, R.R. Aubuchon, C.L. Metcalf, and J.E. Roberts. 1978. Leaf conductance and photosynthesis in four species of the oak-hickory type. For. Sci. 24:73–84.

Horn, H.S. 1971. The Adaptive Geometry of Trees. Princeton University Press, Princeton, New Jersey.

Hutchison, B.A., and D.R. Matt. 1977. The distribution of solar radiation within a deciduous forest. Ecol. Monogr. 47:185–207.

Hutchison, B.A., D.R. Matt, and R.T. McMillen. 1980. Effects of sky brightness distribution upon penetration of diffuse radiation through canopy gaps in a deciduous forest. Agric. Meteorol. 22:137–147.

Hutchison, B.A., D.R. Matt, R.T. McMillen, L.J. Gross, S.J. Tajchman, and J.M. Norman. 1986. The architecture of a deciduous forest canopy in eastern Tennessee, U.S.A. J. Ecol. 74:635–676.

Idso, S.B., and C.T. de Wit. 1970. Light relations in plant canopies. Appl. Opt. 9:177–184.

Jarvis, P.G. 1976. The interpretation of the variations in leaf water potential and stomatal conductance found in canopies in the field. Philos. Trans. R. Soc. Lond., Ser. B 273:593–610.

Jarvis, P.G. 1980. Stomatal conductance, gaseous exchange and transpiration. pp. 175–204. IN J. Grace, E.D. Ford, and P.G. Jarvis (eds.), Plants and Their Atmospheric Environment. Blackwell Scientific Publications, Boston, Massachusetts.

Jarvis, P.G., and J.W. Leverenz. 1983. Productivity of temperate, deciduous and evergreen forests. pp. 234–280. IN O.L. Lange, P.S. Nobel, C.B. Osmond, and H. Zeigler (eds.), Encyclopedia of Plant Physiology New Series, Vol. 12D: Physiological Plant Ecology IV. Springer-Verlag, Berlin.

Kimes, D.S., J.A. Smith, and L.E. Link. 1981. Thermal IR exitance model of a plant canopy. Appl. Opt. 20:623–632.

Kira, T., K. Shinozaki, and K. Hozumi. 1969. Structure of forest canopies as related to primary productivity. Plant Cell Physiol. 10:129–142.

Marshall, B., and P.V. Biscoe. 1980. A model for C_3 leaves describing the dependence of net photosynthesis on irradiance. J. Exp. Bot. 31:29–39.

Matt, D.R., R.T. McMillen, J.D. Womack, and B.B. Hicks. 1987. A comparison of estimated and measured SO_2 deposition velocities. Water Air Soil Pollut. 36:331–347.

McMillen, R.T. 1983. Eddy correlation calculations done in real-time. pp. 111–114. IN Preprint Volume, Seventh Conference on Fire and Forest Meteorology, American Meteorological Society, Boston, Massachusetts.

McNaughton, K.G., and P.G. Jarvis. 1983. Predicting effects of vegetation changes on transpiration and evaporation. pp. 1–47. IN T. Kozlowski (ed.), Water Deficits and Plant Growth, Vol. 7. Academic Press, New York.

Miller, P.C. 1967. Leaf temperatures, leaf orientation, and energy exchange in quaking aspen (*Populus tremuloides*) and Gambell's oak (*Quercus gambelli*) in central Colorado. Oecol. Plant. 2:241–270.

Monsi, M., and T. Saeki. 1953. Über den Lichtfaktor in den Pflanzengesellschaften und seine Bedeutung für die Stoffproduktion. Jpn. J. Bot. 14:22–52.

Monteith, J.L. 1973. Principles of Environmental Physics. Edward Arnold, London.

Nilson, T. 1971. A theoretical analysis of the frequency of gap in plant stands. Agric. Meteorol. 8:25–38.

Norman, J.M. 1979. Modeling the complete crop canopy. pp. 249–277. IN B.J. Barfield and J.F. Gerber (eds.), Modification of the Aerial Environment of Plants. American Society of Agricultural Engineers, St. Joseph, Michigan.

Norman, J.M. 1982. Simulation of microclimates. pp. 65–99. IN J.L. Hatfield and I.J. Thompson (eds.), Biometeorology in Integrated Pest Management. Academic Press, New York.

Norman, J.M., and P.G. Jarvis. 1974. Photosynthesis in Sitka spruce (*Picea sitchensis* (Bong.) Carr. III. Measurements of canopy structure and interception of radiation. J. Appl. Ecol. 11:375–398.

Ovington, J.D., and H.A.I. Madgwick. 1955. A comparison of light in different woodlands. Forestry 28:141–146.

Pasquill, F. 1972. Some aspects of boundary layer description. Q.J.R. Meteorol. Soc. 98:469–494.

Raschke, K. 1956. Mikrometeorologisch gemessene Energie-umsatze eines Alocasiablattes. Arch. Meteorol. Geophys. Bioklimatol., Ser. B 7:240–268.

Rauner, J.L. 1976. Deciduous forest. pp. 241–264. IN J.L. Monteith (ed.), Vegetation and the Atmosphere, Vol. 2: Case Studies. Academic Press, New York.

Ross, J., and T.A. Nilson. 1971. A mathematical model of the radiation regime of vegetation. pp. 253–270. IN V.K. Pyldmaa (ed.), Actinometry and Atmospheric Optics. Israel Program for Scientific Translations, Jerusalem.

Ross, Y.L. 1981. The Radiation Regime and Architecture of Plant Stands. W. Junk, The Hague, Netherlands.

Sanger, J.E. 1971. Quantitative investigations of leaf pigment from their inception in buds through autumn coloration to decomposition in falling leaves. Ecology 52:1075–1089.

Sinclair, T.R., and K.R. Knoerr. 1982. Distribution of photosynthetically active radiation in the canopy of a loblolly pine plantation. J. Appl. Ecol. 19:183–191.

Smith, J.A., and R.E. Oliver. 1974. Effects of changing canopy directional reflectance on feature selection. Appl. Opt. 13:1599–1604.

Smith, J.A., K.J. Ranson, D. Nguyen, L. Balick, L.E. Link, L. Fritschen, and B.A. Hutchison. 1981. Thermal vegetation canopy model studies. Remote Sensing Environ. 11:311–326.

Tanner, C.B., and G.W. Thurtell. 1969. Anemoclinometer measurements of Reynold stress and heat transport in the atmospheric surface layer. Report RI-R10, U.S. Army Electronics Command, Ft. Huachuca, Arizona.

Turner, N.C., and J.E. Begg. 1973. Stomatal behavior and water status of maize, sorghum, and tobacco under field conditions, I. At high soil water potential. Plant Physiol. 51:31–36.

Verma, S.B., D.D. Baldocchi, D.E. Anderson, D.R. Matt, and R.J. Clement. 1986. Eddy fluxes of CO_2, water vapor, and sensible heat over a deciduous forest. Boundary-Layer Meteorol. 36:71–91.

Waggoner, P.E., and W.E. Reifsnyder. 1968. Simulation of the temperature, humidity, and evaporation profiles in a leaf canopy. J. Appl. Meteorol. 7:400–409.

Chapter 4
Atmospheric Chemistry, Deposition, and Canopy Interactions

S.E. Lindberg, R.C. Harriss, W.A. Hoffman, Jr.,
G.M. Lovett, and R.R. Turner

4.1 Introduction

One of man's impacts on geochemical cycles involves the atmospheric chemistry and deposition of several major and trace elements. The critical links between increased atmospheric emissions and effects on ecosystems are the rates at which, and the mechanisms by which, atmospheric constituents are transported to the earth's surface and made available to receptor organisms. Forest vegetation is a particularly important sink for atmospheric emissions; because of the reactivity and large surface area of forest canopies, they are effective receptors of airborne material delivered by both wet and dry deposition processes. In the eastern United States, forest canopies represent the initial point of interaction between the atmosphere and the biosphere for ~50% of the land surface area. On a global scale, we estimate that the combined surface area of all forest canopy leaves and needles is of the same order as that of the total global land surface area ($\sim 10^{14}$ m^2).

Dry deposition cannot be neglected in element cycling studies because of the potential for efficient interception by foliar canopies. This is particularly true when dry-deposited gases or particles contain water-soluble constituents. An important aspect of deposition and bioavailability is the water solubility of deposited material, which strongly influences its mobility in the environment. Elements transported to vegetation in an insoluble form may be of little ecological consequence. However, a constituent of relatively high solubility deposited on the vegetation surface can be readily mobilized by interaction with moisture films. Such wet-dry interactions are of particular importance due to (1) their episodic nature, (2) their ability to transport gas- or particle-associated elements to the forest canopy, partly in solution, and (3) their potential to mobilize previously dry-deposited material; all of these enhance the possibility of pollutant absorption by vegetation surfaces. This chapter is a review of recent research at Walker Branch Watershed on atmospheric deposition rates and processes, interactions of airborne material with the forest canopy, and the role of the atmosphere in forest element cycling.

4.2 Methodology

The locations of the specific sites used for research on atmospheric deposition are shown in Fig. 4.1. The detailed work on deposition processes described here was initiated in 1976 and is continuing (the most recent data summarized here were collected during 1983). Before 1977, deposition was sampled weekly using Wong wet and dry samplers at each of the five forest-clearing rain gage sites shown in the figure (Swank and Henderson 1976). During 1976–1977, deposition was sampled on an event basis at the circled sites (Fig. 4.1), as described below. From 1978 to 1983, bulk deposition was collected monthly in continuously open containers at the three westernmost rain gage sites (Johnson et al. 1982). Beginning in March 1980, wet-only deposition was sampled weekly at the rain gage site in the northwest corner of the watershed as part of the National Atmospheric Deposition Program (NADP 1982); beginning in January 1981, wet-only deposition was also sampled on an event basis at this same site as part of the regional MAP3S (Multistate Atmospheric Power Production Pollution Study) network (MAP3S 1980). From 1981 to 1983, both wet- and dry-deposited materials were collected above and below the canopy on an event basis as described below.

Methodology used in the detailed studies of deposition has been described in detail in the literature (Lindberg et al. 1979; Hoffman etal. 1980a; Lindberg and Harriss 1981, 1983; Lindberg 1982; Lindberg and Lovett 1983, 1985) and will be discussed only briefly here. The problems involved in the sampling and analysis of both wet and dry deposition are well documented (Galloway and Likens 1978; Lindberg and McLaughlin 1985). We employed a version of the HASL (Health and Safety Laboratory, New York) sampler for collection of precipitation as wetfall only (Fig. 4.2). We modified the sampler with selected construction materials and rearranged component parts to minimize trace-level contamination. The sampler was equipped with an automatic precipitation sensor such that the collection bottles remained tightly sealed to exclude dry deposition. Samples were collected directly in individual bottle/funnel arrangements, allowing different prewashing, sample preservation, and analytical methods to be used for a given element. Precipitation was collected primarily on an event basis at sites above the forest canopy (on a 46-m meteorological

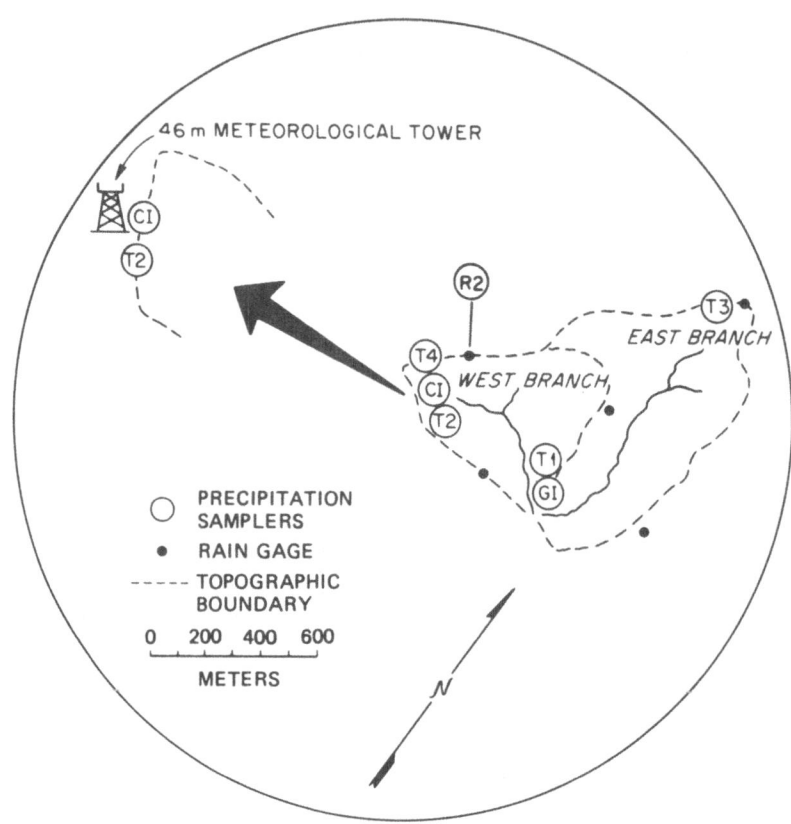

Walker Branch Watershed Deposition Network.

Figure 4.1. Detailed deposition networks operated at Walker Branch Watershed since 1970. The sites indicated by dots (rain gages) represent locations of hydrologic input measurements from 1970 to the present. These sites were also used to measure rain chemistry from 1970 to 1976. The circled sites were used for chemistry measurements of precipitation (CI, on a tower above the canopy; GI, in a clearing on the ground) and throughfall (T) from 1976 and 1977. Bulk and wet-dry chemistry measurements have been made at selected rain gage sites from 1978 to the present, as described in the text.

tower), at ground level in the open, and directly beneath the chestnut oak (*Quercus prinus*) canopy (throughfall). An event was defined as a rain period preceded and followed by continuous 6-h periods of no measurable precipitation (<0.25 mm).

Measurement of dry deposition presents a more complex problem because of its many component processes, the importance of each of which is dependent on the surface characteristics of the receptor (Hicks et al.

Figure 4.2. Wet deposition collectors used at Walker Branch Watershed for trace metal and major ion chemistry studies since 1976.

1980). We used two methods of sampling: collecting of material on inert, flat surfaces situated in the upper canopy (Fig. 4.3), and extraction of material from leaves collected at the same locations (in a mature chestnut oak stand with a 20- to 25-m canopy). Briefly, the procedures involved in-laboratory extraction of sequentially collected upper canopy leaves (*Quercus prinus*) to determine net surface material accumulation and extraction of polyethylene and polycarbonate plates following exposure in the canopy at the same foliage collection sites. Further details are described elsewhere (Lindberg and Lovett 1983, 1985). Samples were collected manually during dry weather intervals of 4- to 7-d duration during the 1977 growing season (April to October). From 1981 to 1983 we used an automatic rain shield device to expose plates for generally dry periods of 4 to 15 d (Fig. 4.3). Concurrent with these experiments, aerosols and gases were also collected above the forest canopy. Atmospheric vapor concentrations were measured above the canopy using a filter pack for HNO_3 (Huebert 1983), chemiluminescence for NO_x (Kelly and Meagher 1985), and, at a nearby clearing, flame photometry for SO_2 (TVA 1982). Suspended particle concentrations and size distributions were measured above the canopy using standard filtration and impactor methods (Lindberg and Harriss 1983). Dry deposition rates of large ($\geqslant 2$ μm) particles to individual inert surfaces were scaled to the full canopy using canopy/plate factors

Figure 4.3. Dry deposition collectors used in and above the forest canopy since 1977.

determined from statistical analysis of throughfall chemistry (Lovett and Lindberg 1984). These factors were 2.7 during the growing season (leaf area index $\sim$5 m^2 of foliage/m^2 of ground) and 1.0 during the dormant season. Vapor and small-particle ($<$2 μm) dry deposition fluxes were calculated from atmospheric concentration data collected above the canopy and deposition velocities measured on site or taken from the literature (Fowler 1980; Hicks etal. 1983; Huebert 1983). Deposition velocities were as follows: during the growing season, SO_2 = 0.5 cm/s, HNO_3 = 2 cm/s, and small particles = 0.2 cm/s; during the dormant season, SO_2 = 0.2 cm/s, HNO_3 = 0.4 cm/s, and small particles = 0.05 cm/s.

For the initial studies during 1976–1977, which included analyses of the trace metals Cd, Mn, Pb, and Zn in deposition, all apparatus was acid washed and rinsed prior to use in the field, and all field samples were stored in clear polyethylene bags and returned to the laboratory, where extractions were performed at a laminar-flow clean bench. Similar laboratory leaching procedures were applied to all samples to extract first the water-soluble (pH $\approx$5.6) and second the dilute acid-leachable fractions (pH $\approx$1.2, 0.1 N Ultrex® pure HNO_3), the sum of which was considered to represent the environmentally mobile material, termed the available fraction (Lindberg et al. 1979). Selected samples were also acid digested for determination of total metal content (concentrated HCl + HNO_3).

To minimize adsorption losses, precipitation samples were acidified to 0.1 N HNO_3 (Ultrex) following collection; thus, reported metal concentrations represent the water-leachable plus acid-leachable fractions. The low levels of suspended matter in rain (<5 mg/L) and the high water-soluble to acid-leachable concentration ratios of these constituents in aerosols (Lindberg and Harriss 1983) suggest that the concentrations were not influenced by particulate matter. Analyses involved standard methods (graphite furnace atomic absorption spectrometry for trace elements, using the method of standard additions; automated colorimetry for sulfate) and included studies of sampling reproducibility, trace-level contamination, and quality control (Lindberg et al. 1979). Precision and accuracy were checked by participation in a U.S. Department of Energy Environmental Measurements Laboratory interlaboratory comparison test using precipitation "control samples."

From 1981 to 1983 all samples were analyzed for SO_4^{2-}, NO_3^-, H^+, NH_4^+, Ca^{2+}, and K^+, using standard electrode, titrimetric, colorimetric, and chromatographic methods (Richter et al. 1983; Lindberg et al. 1984a). During 1976–1977 we collected 68 precipitation event samples, representing 76% of the total precipitation input (263 cm for 2 years), and sampled dry deposition of particulate metals for 800 h and suspended atmospheric metals for 1400h. From 1981 to 1983, we sampled 110 precipitation events, or 99% of the recorded precipitation (255 cm for 2 years); obtained continuous measurements of SO_2; measured HNO_3 for 2000 h; and sampled suspended and deposited particles for 4000 h.

4.3 Major Ions

4.3.1 Rates and Processes of Atmospheric Deposition to the Forest Canopy

Analysis of the atmospheric contribution to element cycles in forests is a complex task commonly approached in a simplistic manner. While some studies have reported detailed field measurements of either wet (Johannes et al. 1981) or dry (Weseley et al. 1983) deposition, few have intensively sampled precipitation, stemflow, throughfall, dry deposition, and the atmospheric concentrations of particles and vapors that are needed for accurate estimates of total input to the canopy and forest floor. Deposition is typically determined from samples of bulk precipitation collected periodically in continuously open containers (Likens et al. 1977). While this method may be useful in some remote locations where background aerosol and vapor concentrations are insignificant and wet deposition dominates input, it is generally acknowledged to produce results that are difficult to interpret (NAS 1983). This is particularly true for forests in industrialized areas, where high-surface-area canopies can interact with anthropogenic particles and vapors to produce significant dry deposition fluxes (Hosker and Lindberg 1982).

Atmospheric chemistry, deposition, and internal cycling of acid anions, base cations, and H^+ were intensively studied from 1981 to 1983 at Walker Branch Watershed, using the methods described above (Lindberg et al. in press). Dry deposition is a significant mechanism in the total annual flux of each of these ions to the forest (Table 4.1). The contribution of dry deposition ranges from 32% for NH_4^+ to ~50% for SO_4^{2-}, and H^+ and to ~60% for K^+ and NO_3^-, and is most important for Ca^{2+}, contributing 72% of the total input. Dry deposition is primarily controlled by vapor uptake for SO_4^{2-}, NO_3^-, and H^+ (70, 75, and 97% of the total dry input, respectively), by small-particle deposition for NH_4^+ (63%), and by large-particle deposition for K^+ and Ca^{2+} (~95% each). If total wet plus dry deposition is considered, these ions can be divided into three groups: those for which precipitation is the most important single process of deposition (SO_4^{2-} and NH_4^+), those for which vapor deposition is most important (NO_3^- and H^+), and those controlled by dry deposition of particles (K^+ and Ca^{2+}).

Table 4.1. Total annual atmospheric deposition of major ions to an oak forest at Walker Branch Watershed, Tennessee

Process	Atmospheric deposition[a] [mmol(+ or −)·m⁻²·year⁻¹]					
	SO_4^{2-}	NO_3^-	H^+	NH_4^+	Ca^{2+}	K^+
Precipitation	70 (5)	20 (2)	69 (5)	12 (1)	12 (2)	0.9 (0.1)
Dry deposition:						
Small particles	7 (2)	0.1 (0.02)	2.0 (0.9)	3.6 (1.3)	1.0 (0.2)	0.1 (0.05)
Large particles	19 (2)	8.3 (0.8)	0.5 (0.2)	0.8 (0.3)	30 (3)	1.2 (0.2)
Total particles	26 (3)	8.4 (0.8)	2.5 (0.9)	4.4 (1.3)	31 (3)	1.3 (0.3)
Vapors[b]	62 (7)	24 (4)	85 (8)	1.3[c]	0	0
Total dry	88 (8)	32 (4)	88 (8)	5.7 (1.3)	31 (3)	1.3 (0.3)
Total wet	70 (5)	20 (2)	69 (5)	12 (1)	12 (2)	0.9 (0.1)
Total deposition	160 (9)	54 (4)	160 (9)	18 (2)	43 (4)	2.2 (0.3)

[a]Values given are means and associated standard errors based on 2 years of data. Number of observations range from 15 (HNO_3) to 26 (particles) to 128 (precipitation) to 730 (SO_2). In comparing these deposition rates, it must be recalled that any such estimates are subject to considerable uncertainty. The standard errors given provide only a measure of uncertainty in the calculated sample means relative to the population means; hence, additional uncertainties in analytical results, hydrology, scaling factors, and deposition velocities must be included. We estimate the overall uncertainty in wet deposition fluxes to be on the order of 20 to 30%, and those for dry deposition to be ~50% for SO_4^{2-}, Ca^{2+}, K^+, and NH_4^+ and ~75% for NO_3^- and H^+.
[b]Includes SO_2, HNO_3, and NH_3. We assumed 100% conversion of deposited SO_2 to H_2SO_4 and of NH_3 to NH_4^+.
[c]NH_3 was not measured, but estimated from the literature (Tjepkema et al. 1981).

The relative contributions of these processes are not constant through-out the year (Fig. 4.4), as some have assumed (Mayer and Ulrich 1982). Dry deposition tends to dominate during the summer, when the canopy is fully developed, providing a significant surface area for interaction with suspended particles and vapors. Wet deposition is generally the most im-portant process during the winter when the canopy is barren and when

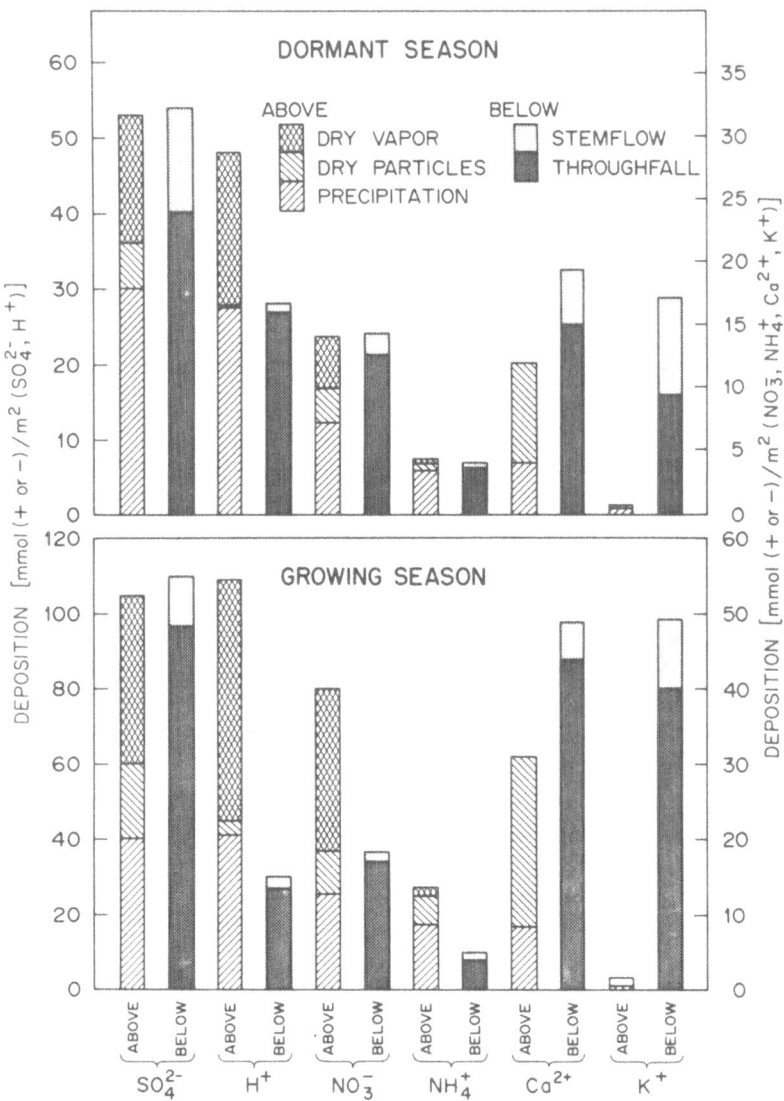

Figure 4.4. The contribution of several atmospheric deposition and internal transfer processes to the total ion flux above and below the forest canopy at Walker Branch Watershed for the growing (months 4–10) and dormant seasons during 1981–1983.

the air concentrations of most constituents are at a minimum. The only exceptions to these generalities are (1) predominance of wet deposition of NH_4^+ in the summer, which appears to have no significant vapor component (Tjepkema et al. 1981), and (2) predominance of large-particle deposition of Ca^{2+} in the winter, the size distribution of which is shifted upward during this period.

The contribution of various ions to the total deposition by each process reflects the chemistry of the contributing parent material. Precipitation is dominated by SO_4^{2-} and H^+, with sufficient SO_4^{2-} to account for all of the free acidity. However, correcting for sea-salt SO_4^{2-} and considering the probable contributions of $CaSO_4$ and $(NH_4)_2SO_4$ suggests that HNO_3 can contribute up to 30% of the free acidity. The small-particle dry deposition fraction is primarily a mixture of H_2SO_4 and $(NH_4)_2SO_4$, with at most 30% of the SO_4^{2-} occurring as acid sulfate. Large particles are apparently dominated by calcium salts of NO_3^- and SO_4^{2-}.

The atmospheric chemistry and behavior of these ions support the importance of dry deposition. The comparability between wet and dry deposition for SO_4^{2-} was suggested by atmospheric models (Shannon 1981), by large-scale budget approaches (Galloway and Whelpdale 1980), and for both SO_4^{2-} and NO_3^-, by theory (Fowler 1980). At Walker Branch, the mean annual air concentration of sulfur oxides is 10.4 μg S/m^3, 75% of which is SO_2. The predominance of the vapor and the size distribution of the particles (an average of 88% of the mass is associated with particles of ≤2 μm diam and 26% with particles ≤0.4 μm, Fig. 4.5) are conducive to removal of airborne sulfur by vapor dry deposition and precipitation scavenging. The importance of large-particle deposition for SO_4^{2-} (30% of dry deposition) has only recently been realized (Davidson et al. 1982), and emphasizes the importance of this particle fraction because of its efficient removal from the atmosphere by sedimentation and impaction. We calculate a mean annual dry deposition velocity for large-particle SO_4^{2-} to this forest of 0.4 cm/s, compared with submicron-particle deposition velocities for SO_4^{2-} of 0.1 to 0.7 cm/s measured in field studies and <0.1 cm/s indicated by wind tunnel studies (Hicks et al. 1983). Recent measurements of large-particle SO_4^{2-} flux, using our methods, in other areas of the United States (Davidson et al. 1985) and Europe (Lindberg et al. 1984b) support these results.

The importance of vapor and large-particle dry deposition is even more pronounced for NO_3^-. Nearly 75% of the dry deposition is attributable to HNO_3 vapor, a species not considered in deposition studies until recently (Huebert 1983), and small-particle NO_3^- deposition is negligible compared with the flux of large particles. The mean annual concentration of HNO_3 vapor in the atmosphere above this forest is 3.9 μg/m^3, and that of particle NO_3^- is 0.75 μg/m^3, approximately 70% of which apparently exists in the large-particle size range (Lindberg et al. 1984b) [note that all NO_3^- particle size data must be considered cautiously because of artifacts due to HNO_3

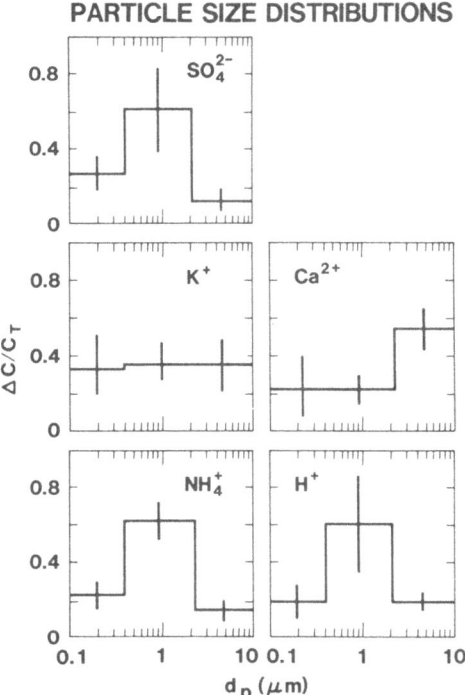

PARTICLE SIZE DISTRIBUTIONS

Figure 4.5. Particle size distributions of airborne SO_4^{2-}, K^+, and Ca^{2+}, NH_4^+, and H^+ collected above the forest canopy at Walker Branch Watershed during 1981–1983. Shown are mean normalized concentrations ($\Delta C/C_T$, where ΔC is the element concentration for a given size range and C_T is the total concentration of the element in all size ranges) and standard errors of the estimate of the means for five size classes. Particle diameters (d_p) were estimated from impactor data.

vapor and NH_4NO_3 volatilization (Appel and Tokiwa 1981)]. Nitric acid vapor, a major sink for atmospheric NO_x, is a highly soluble gas which is efficiently scavenged by precipitation. However, its solubility also results in very efficient deposition to plant canopies during dry periods (Huebert 1983). The deposition of other NO_x compounds is probably small relative to the total NO_3^- flux in Table 4.1 (Kelly and Meagher 1986).

For the major cations K^+ and Ca^{2+}, the dry deposition contribution falls at the upper end of this range, occurring primarily as large particles. This would be predicted from their aerosol size distributions and from their primary source in resuspended surface soil, road dust, and plant matter; 34% of the K^+ and 55% of the Ca^{2+} occur on particles greater than 2 μm diam (Fig. 4.5). The larger fraction of submicron K^+ in the atmosphere may explain the higher relative contribution of precipitation scavenging to its total deposition compared with that of Ca^{2+}. Because of their sources, dry-deposited Ca^{2+} and K^+ may not entirely represent new inputs to the forest, but rather recycled material of local origin. Evidence suggests that deposited Ca^{2+} does originate outside of the watershed, whereas K^+ may be derived primarily from internal sources. Profiles of Ca^{2+} aerosol concentrations and of Ca^{2+} deposition to inert surfaces each increase with increasing height above the forest floor, indicating an atmospheric source and a canopy sink. Profiles of K^+ concentration and deposition through the forest indicate highest levels within the canopy itself, decreasing both above and below the canopy, suggesting an in-canopy source of weathered leaf cuticle, trichomes, and pollen. This behavior was most pronounced in the spring and persisted throughout the growing season. Because most of the particle dry deposition occurred during the growing season (Fig. 4.4), it is possible that as much as 50% of the annual K^+ "deposition" represents an internal forest cycle. Potassium is known to cycle rapidly in the forest in the hydrologic medium; these results suggest that it behaves similarly in the atmospheric medium.

Wet deposition clearly dominates the input only of NH_4^+, as would be suggested by the atmospheric chemistry of the NH_3/NH_4^+ system. The vapor is highly soluble in acidic rain droplets or cloud water and has been reported to represent only 5% of the total atmospheric burden of NH_3 + NH_4^+ in the northeastern United States (Tjepkema et al. 1981), where summer NH_4^+ aerosol concentrations averaged 2.1 $\mu g/m^3$ compared with 2.0 $\mu g/m^3$ at Walker Branch. Because 85% of the airborne particulate NH_4^+ at our site exists on the ≤ 2-μm-diam particles (Fig. 4.5), removal by precipitation is favored over that by dry deposition. The particle deposition of NH_4^+ is dominated by the small-particle fraction (80% of particle input).

The deposition of H^+ is strongly influenced by the behavior of the acidifying gases and vapors. Nitric acid vapor and SO_2 represent 97% of the total dry deposition of potential free acidity, the remainder being attributable to small-particle deposition. This is reflected in air concentrations at this site; the mean total acidic vapor concentration as H^+ is 0.54 $\mu mol/m^3$, whereas that of particle-associated H^+ is 0.046 $\mu mol/m^3$, 80% of which occurs in the submicron size range (Fig. 4.5). Because of its reactivity, the total deposition of H^+ by wet or dry processes is difficult to quantify. For the vapors, we calculated potential H^+ assuming complete oxidation of deposited SO_2 and protonation of deposited NH_3. However, the chemistry of precipitation and of particle extracts necessarily represents the end result of any acid-base reactions occurring before analysis. Hence, actual H^+ input in particles could be higher, whereas the effective net deposition due to gases could be lower, primarily because of degasing of incompletely oxidized SO_2.

The apparent net deposition of free acidity to the Walker Branch ecosystem can be bracketed using the following assumptions: all Ca^{2+} was originally deposited as $CaCO_3$ and has already reacted completely with H^+, or, the other extreme, all Ca^{2+} still exists as $CaCO_3$ ready to react when solutions are mixed (e.g., incoming rain reacting with deposited particles on leaves); all NH_4^+ was originally deposited as NH_3 and has already reacted with H^+; and 20% of the deposited SO_2 is degased from the leaf before oxidation to H_2SO_4 [Taylor and Tingey (1983) measured the surface adsorption of SO_2 to foliage to be in the range of 10–30%]. The results of these assumptions yield an estimated range of total atmospheric deposition of free acidity of 90 to 200 mmol $H^+ \cdot m^{-2} \cdot year^{-1}$ (see Table 4.1), leaving little doubt that the net deposition to this forest is acidic.

4.3.2 Comparison with Other Data

Atmospheric deposition is commonly quantified in ecosystem studies by bulk collectors exposed continuously at ground level. Comparison of our results with bulk deposition indicates the potential error in such estimates of input to forests. Bulk deposition was collected at Walker Branch during this study by standard methods using replicate funnel/bottle collectors (Richter et al. 1983). A detailed analysis of these results (Lindberg et al. 1986) suggests that bulk deposition most significantly underestimates wet plus dry deposition for those ions with a major vapor or small-particle component in the atmosphere, as expected. Total deposition of H^+, NO_3^-, SO_4^{2-}, and NH_4^+ was underestimated by 50 to 70%. The underestimate of Ca^{2+} input was less (30%) because of its control by the dry deposition of large particles that are captured to some extent by bulk collectors. However, bulk K^+ deposition exceeded wet plus dry estimates (based on above-canopy sampling) by a factor of 2, reflecting the complication of the in-canopy source of K^+ discussed above. These comparisons suggest that biogeochemical cycling studies in which system inputs are based solely on measurements of bulk deposition must be interpreted cautiously.

Although our results are supported by published calculations, there are few if any directly comparable field studies. Dry deposition has been identified as being potentially important in the atmospheric flux of sulfur and nitrogen to forests in southern Sweden, using published air concentrations and deposition velocities (Grennfelt et al. 1980). Dry deposition was estimated to contribute 40% (NO_3^-) to 60% (SO_4^{2-}) of total input (30 and 140 mmol$(-) \cdot m^{-2} \cdot year^{-1}$, respectively), and was predicted to be dominated by HNO_3 and SO_2 (70 and 80% of dry deposition, respectively). In central Federal Republic of Germany (Mayer and Ulrich 1982), total SO_4^{2-} deposition was estimated to be 310 and 460 mmol$(-) \cdot m^{-2} \cdot year^{-1}$ to beech and spruce forests, dry deposition contributing 50 and 70% of the flux, respectively. However, wet and dry deposition were not measured separately but were estimated, assuming that soluble ions in bulk deposition represented the wetfall contribution and that the wet:dry deposition ratio estimated from winter throughfall beneath a leafless beech canopy was representative of summer conditions in both canopies. These assumptions are tenuous; both dry-deposited SO_2 and particle sulfur yield SO_4^{2-} ions to solution, and our data suggest that wet:dry deposition ratios are not constant between seasons (Fig. 4.4). In a coastal spruce forest in Scotland, total deposition was estimated from bulk collectors with and without overhanging mesh nets (Miller and Miller 1980). Total SO_4^{2-} deposition was estimated at 210 mmol$(-)\cdot$m$^{-2}\cdot$year^{-1}, 30 to 40% of which was attributed to the aerosol filtering action of the mesh. The filtered material was attributed primarily to relatively large sea-salt aerosols and did not include SO_2 or a significant submicron particle fraction. In addition, some of this filtered material may be derived from windblown mists that are not efficiently sampled by the bulk collectors.

In the United States, micrometeorological methods were used in a 1-month study to measure HNO_3 vapor deposition velocities to a grassy field (Huebert 1983). Measured values of 1 to 4 cm/s resulted in a dry deposition estimate of 2.4 mmol$\cdot$m$^{-2}\cdot$ month^{-1}, comparable to the wet deposition measured nearby. Particle deposition was not determined. Preliminary micrometeorological data from our forest suggest similar deposition velocities (B.B. Hicks, National Oceanic and Atmospheric Administration, and B.J. Huebert, University of Colorado, personal communications, 1984). In the northeastern United States, total SO_4^{2-} deposition to a deciduous forest was estimated using automatic wet-dry samplers for large particles and precipitation (Johannes et al. 1981), and SO_2 was estimated from literature values. Total deposition was similar to that measured in this study: 60% was wetfall and 30% was attributed to SO_2. The results of these studies, using widely varying approaches, suggest that the general contribution of dry removal processes to total atmospheric deposition of SO_4^{2-} and NO_3^- is in the range of 30 to 70%.

In a study in the southeastern United States, we compared input at Walker Branch with that at another regionally representative site and tested the hypothesis that deposition of sulfur to forests is influenced by local emissions (Lindberg and Kelly 1984). We sampled wet and dry sulfur deposition over a 1-year period at Walker Branch Watershed (WBW) and at Camp Branch Watershed (CBW) operated by the Tennessee Valley Authority. Recall that WBW is within 22 km of three coal-fired power plants of over 2500-MW(e) capacity, one of which [1600MW(e)] is generally upwind. Camp Branch is a mixed deciduous forest watershed on the Cumberland plateau ~100km to the southwest of WBW. In this comparison, CBW represents the "background" site, as it is much farther from local sources: one 1600-MW(e) plant is 75 km to the northeast, and another of 1960 MW(e) is 99 km to the southwest. Prevailing winds in this general region are from the southwest. From July 1981 to June 1982, wet and dry deposition and air concentration samples were collected above the canopy at CBW and WBW and used to estimate total input, as described earlier.

The results of our comparisons are summarized in Table 4.2, which shows mean sulfur concentrations in air and rain and annual deposition rates for several processes. Wet deposition during this period was com-

Table 4.2. Atmospheric concentrations and deposition of sulfur to two forested watersheds in eastern Tennessee from June 1981 to May 1982

	Site	
Parameter	Walker Branch	Camp Branch
Distance to nearest downwind power plant, km	22	99
SO_2 concentration, $\mu g/m^3$	15	12
Airborne SO_4^{2-}, $\mu g/m^3$	8.2	~3[a]
Particle SO_4^{2-} dry deposition rate to inert surfaces, $\mu g \cdot m^{-2} \cdot h^{-1}$	57	32
SO_4^{2-} in precipitation, mg/L	2.8	3.7
Precipitation amount, cm	116	104
Atmospheric deposition, kg $S \cdot ha^{-1} \cdot year^{-1}$		
Wet	11	13
Dry—large particles	2.6	0.9
Dry—small particles	1.5	~0.7[b]
Dry—gas	9.9	5.2
Dry—total	14	6.8
Total wet + dry	25	20

[a]Estimated using measured particle air concentrations at Walker Branch Watershed and measured dry deposition rates at both sites, and assuming that airborne sulfate concentration and large-particle dry deposition are proportional at each site.
[b]Estimated using measured SO_2 concentrations at both sites and small-particle air concentrations at Walker Branch Watershed, and assuming that SO_2 concentration and small-particle deposition are proportional at each site.

parable at the two sites, ranging from 11 to 13 kg $S \cdot ha^{-1} \cdot year^{-1}$, for WBW and CBW, respectively. However, according to our calculations, the dry deposition of sulfur at WBW exceeded that at CBW by a factor of 2 (14 vs. 6.8 kg $S \cdot ha^{-1} \cdot year^{-1}$, respectively). This is reflected in the lower mean SO_2 concentrations and mean particle dry deposition rates measured at CBW relative to WBW (SO_2 at CBW was 20% lower than that at WBW, and the dry deposition rate of airborne SO_4^{2-} particles to inert surfaces was ~40% lower). The total dry deposition of sulfur at the two sites differed by a larger factor because both the mean SO_2 concentration and the particle dry deposition rate during the summer months were substantially lower at CBW relative to those at WBW. Because of the increased canopy surface area during these months, differences in summer concentrations and deposition rates have a larger influence on the annual input than do similar differences in mean winter values.

Our measurements suggest that the proximity of these forested watersheds to commercial-scale power plants may influence the total annual deposition of sulfur. Total sulfur deposition to CBW, which is 99 km from the nearest upwind power plant, was ~20% lower than that at WBW, which is within 22 km of two power plants. Interestingly, the difference was primarily reflected in the dry deposition of sulfur at the two sites. This agrees with the general thinking that dry deposition is more sensitive to local emissions, whereas wet deposition is considered to reflect the regional mixture of emissions. However, because of the uncertainty associated with dry deposition measurements (Table 4.1), we cannot assess the significance of these differences.

Comparison of data on atmospheric deposition of major ions to Walker Branch with other regional and national data is limited by the lack of any large-scale dry deposition data. The only comparable extended studies of which we are aware have been discussed above. However, it is possible to compare atmospheric concentrations of selected constituents and both wet deposition and precipitation concentrations for several major ions, using existing SURE (Sulfur Urban Regional Experiment) and NADP (National Atmospheric Deposition Program) data.

Figures 4.6 and 4.7 show isopleths of annual wet deposition across the eastern United States for H^+, SO_4^{2-}, NO_3^-, NH_4^+, K^+, and Ca^{2+} during 1982, as measured by the National Atmospheric Deposition Program (NADP, 1985). The location of Walker Branch Watershed is shown by a large dot. Wet deposition of H^+ and SO_4^{2-} at Walker Branch is comparable to the highest values measured in the northeastern United States, whereas that of NO_3^- is ~30% lower than the highest deposition rates, but is generally comparable to values for the eastern states. Deposition of NH_4^+ and Ca^{2+} is ~50% lower than the highest values for deposition recorded (in the midwestern states) but is typical for the Southeast. Deposition of K^+ varies by an order of magnitude in the East; the value at Walker Branch is near the middle of the range.

Figure 4.8 illustrates regional trends of atmospheric particulate SO_4^{2-} in the East. These data are difficult to compare because of the different time periods considered. More recent regional data have not been published. The annual mean particle SO_4^{2-} concentration at Walker Branch from 1981 to 1983 was 8.2 $\mu g/m^3$, apparently higher than that recorded for this area during the 1977–1978 SURE study (OTA 1984). This more recent value for Walker Branch is comparable to the highest levels recorded in the Ohio River Valley during the earlier study (Fig. 4.8). All of these comparisons suggest that air quality and atmospheric deposition at Walker Branch are more typical of northeastern and midwestern states than of other southeastern states.

Comparison of current input estimates with historical data collected at Walker Branch Watershed is complicated by changes in methodology used over the years. Table 4.3 summarizes the historical record of deposition measurements published for Walker Branch Watershed for major ions, indicating the different methods used. These have included Wong wet/dry bucket samplers (1970–1976), bulk collectors (1979–1983), HASL-type wet-only collectors plus in-canopy particle collectors (1976–1977, 1981–1983), and the latter two plus above-canopy vapor samplers (1981–1983).

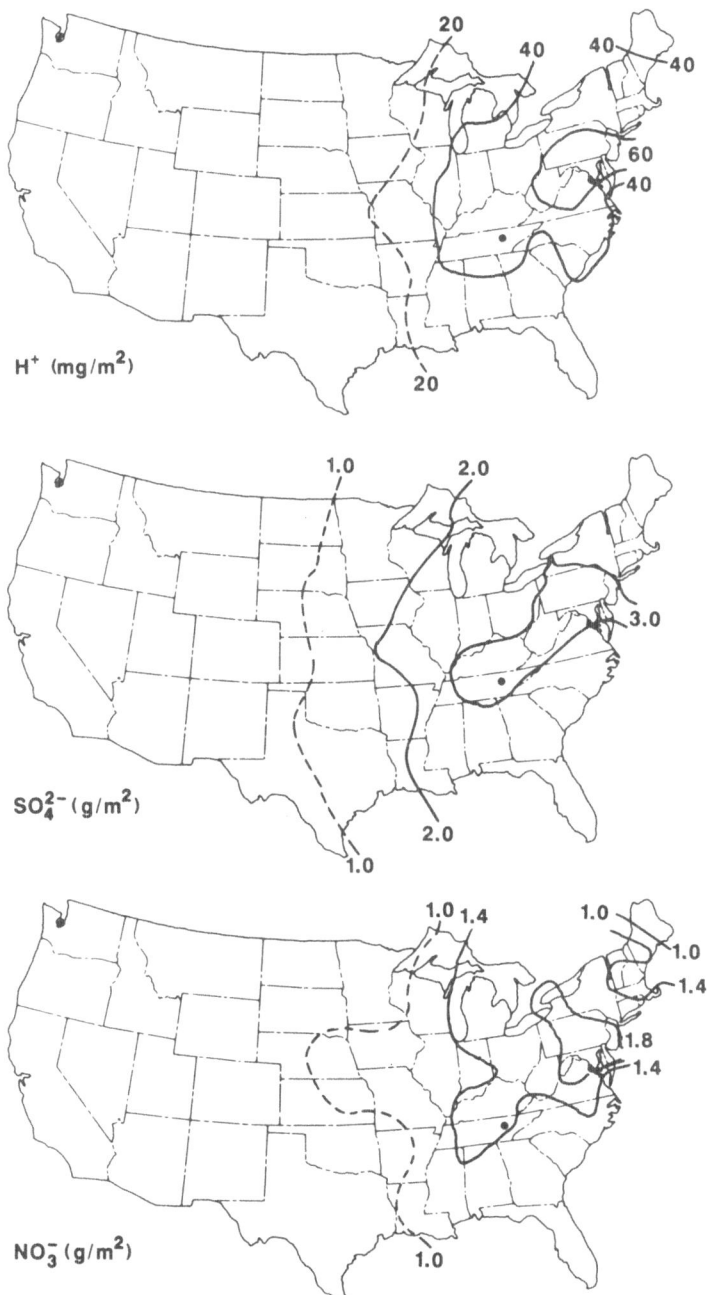

Figure 4.6. Isopleths of annual wet deposition of H^+, SO_4^{2-}, and NO_3^- during 1982, as measured by the National Atmospheric Deposition Program (NADP 1985). The location of Walker Branch Watershed (WBW) is shown by a large dot. The actual measured values for WBW are as follows: H^+ = 68 mg/m²; SO_4^{2-} = 3.7 g/m²; NO_3^- = 1.4 g/m².

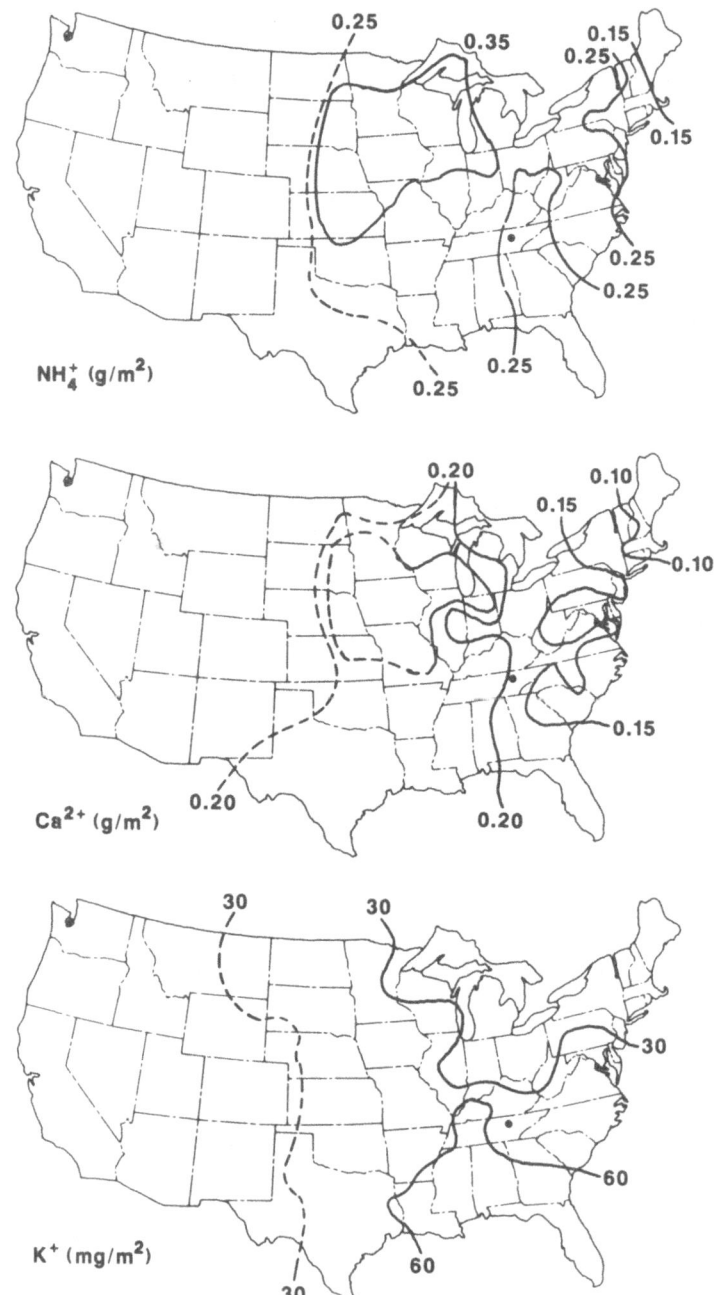

Figure 4.7. Isopleths of annual wet deposition NH_4^+, Ca^{2+}, and K^+ during 1982, as measured by the National Atmospheric Deposition Program (NADP 1985). The location of Walker Branch Watershed (WBW) is shown by a large dot. The actual measured values for WBW are as follows: $NH_4^+ = 0.23$ g/m²; $Ca^{2+} = 0.19$ g/m²; $K^+ = 32$ mg/m².

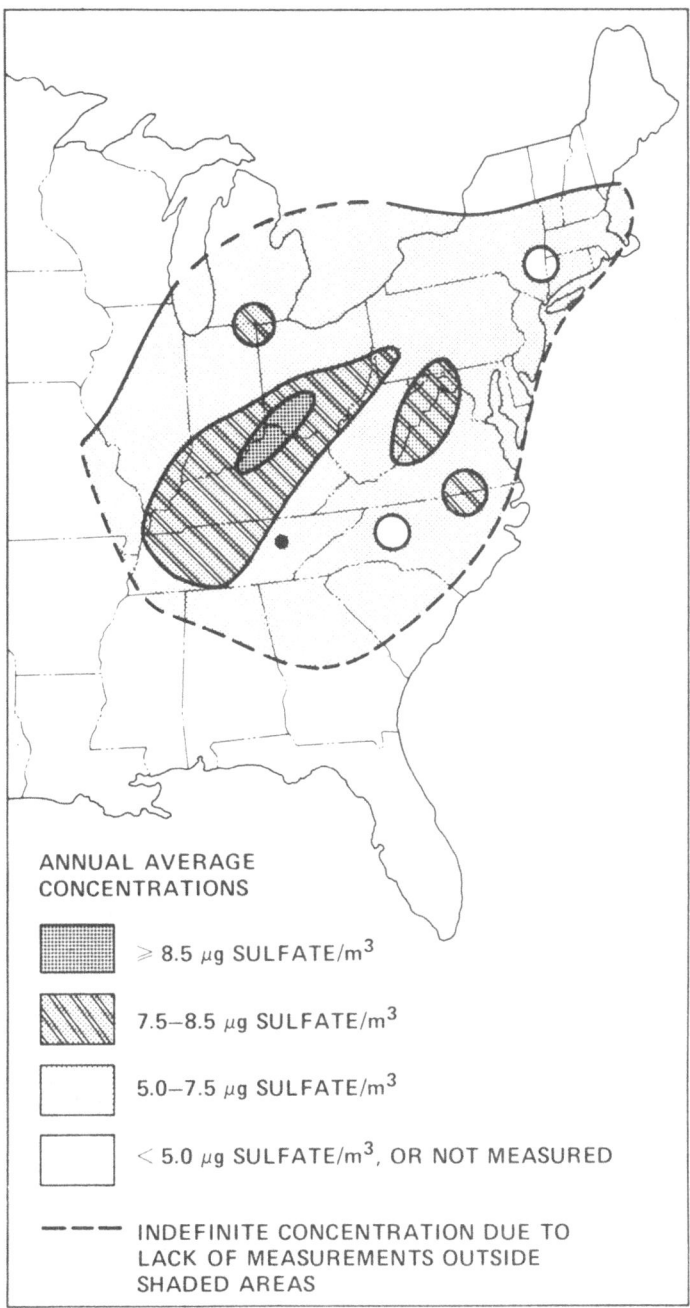

Figure 4.8. Isopleths of annual average airborne particle SO_4^{2-} concentrations during 1977–1978 (OTA 1984) and the annual mean value measured at Walker Branch Watershed during 1981–1983 (8.2 μg m^{-3}).

Table 4.3. Historical record of atmospheric deposition measured at Walker Branch Watershed

Atmospheric deposition (kg·ha^{-1}·year^{-1})

Date[a]	Methods[b]	Wet					Dry					Total or bulk				
		Ca	K	H	SO_4-S	NO_3-N	Ca	K	H	SO_4-S	NO_3-N	Ca	K	H	SO_4-S	NO_3-N
1970[c]	Wong[d]	—	—	—	—	—	—	—	—	—	—	17	6.2	—	—	—
1971[c]	Wong[d]	14	3.1	—	—	—	5.2	1.8	—	—	—	19	4.9	—	—	—
1972[c]	Wong[d]	14	2.4	—	—	—	6.6	1.8	—	—	—	21	4.2	—	—	—
1973[c,f]	Wong[d]	19	3.5	—	—	4.7	5.7	2.1	—	—	—	25	5.6	—	—	—
1974[f]	Wong[d]	9.5	0.9	—	16	3.1	4.3	2.7	—	4.0	—	14	3.6	—	20	—
1975[f]	Wong[d]	—	—	—	14	—	—	—	—	2.8	—	—	—	—	20	—
1976[f]	Wong[d]	—	—	—	14	—	—	—	—	2.9	—	—	—	—	17	—
1976[g]	W/D[h]	—	—	0.93	12	—	—	—	—	6.3	—	—	—	—	18	—
1977[g]	W/D[h]	—	—	0.77	13	—	—	—	—	6.3	—	—	—	—	19	—
1978	No data															
1979[i]	Bulk[i]	—	—	—	—	—	—	—	—	—	—	4.6	4.2	1.2	18	2.2
1980[k]	W/O[l]	1.3	0.17	0.44	7.2	2.2	—	—	—	—	—	—	—	—	—	—
1980[i]	Bulk[i]	—	—	—	—	—	—	—	—	—	—	3.4	1.4	0.38	6.4	0.2
1981[k]	W/O[l]	1.7	0.25	0.52	9.3	2.5	—	—	—	—	—	—	—	—	—	—
1981[i]	Bulk[i]	—	—	—	—	—	—	—	—	—	—	5.8	1.9	0.68	15	2.9
1982[m]	W/D/G[n]	2.0	0.24	0.71	11	2.8	6.2	0.51	0.90	14	4.6	8.2	0.75	1.6	25	7.4
1982[i]	Bulk[i]	—	—	—	—	—	—	—	—	—	—	6.1	1.7	0.83	13	0.92
1983[m]	W/D/G[n]	2.9	0.44	0.67	11	2.8	6.2	0.51	0.90	14	4.6	9.1	0.95	1.6	25	7.4
1983[i]	Bulk[i]	—	—	—	—	—	—	—	—	—	—	6.2	1.39	0.98	14	0.25

[a]Prior to 1977, Wong data were collected on a water-year basis (September–August); 1976 and 1977 W/D data were collected on a calendar-year basis; 1979–1981 data were collected on a March–February basis; 1982–1983 data were collected on a July–June basis.

[b]In each case, one method was used for a complete 12-month period.

[c]Elwood and Henderson (1975).

[d]Wong samplers allow separate collection of wetfall and dryfall in buckets using a rain-activated movable cover. Samples were collected in a forest clearing 1 m above the ground on a weekly schedule.

[e]Swank and Henderson (1976).

[f]Henderson et al. (1977).

[g]Lindberg et al. (1979).

[h]W/D = separate collection of incident precipitation as wetfall-only on an event basis on a tower above the forest canopy and of particle dry deposition using surrogate surface deposition plates situated in the actual canopy and collected on an event basis.

[i]Johnson et al. (1982); Richter et al. (1983).

[j]Bulk = precipitation and some particle dry deposition accumulated in continuously open funnel/bottle collectors situated in forest clearings 1 m above the ground and collected monthly.

[k]Data collected as part of the National Atmospheric Deposition Program.

[l]W/O = rain-activated wetfall-only collectors operated on a weekly basis as part of the National Atmospheric Deposition Program in a forest clearing 1.5 m above the ground.

[m]Lindberg et al. (1984b).

[n]W/D/G = wet-only precipitation, dry deposition of particles on flat plates, and estimated dry deposition of gases.

Several interesting points are apparent from these data. As discussed above, for comparable time periods, bulk estimates of total deposition are considerably lower than those determined from detailed analysis of all deposition processes. Historically, the measured deposition of Ca^{2+} and K^+ has decreased considerably since first reported in the early 1970s. These trends are attributable to major changes in methodology, however. Early samples were collected at ground level near gravel roads in Wong bucket collectors, which do not provide a positive seal over the wet samples. Entrainment of local soil and road dust plus plant debris in these samples, which were collected on a weekly basis, may explain the apparent high inputs of Ca^{2+} and K^+ in wet and total deposition from 1971 to 1974 compared with those from 1981 to 1983 (Table 4.3).

The only other ion with a sufficient record of analysis for comparison of temporal trends is SO_4^{2-}. Because of the role of rain, vapors, and particles in sulfur deposition and the changes in methodologies, the data are best compared graphically (Fig. 4.9). Shown in this plot are estimates of

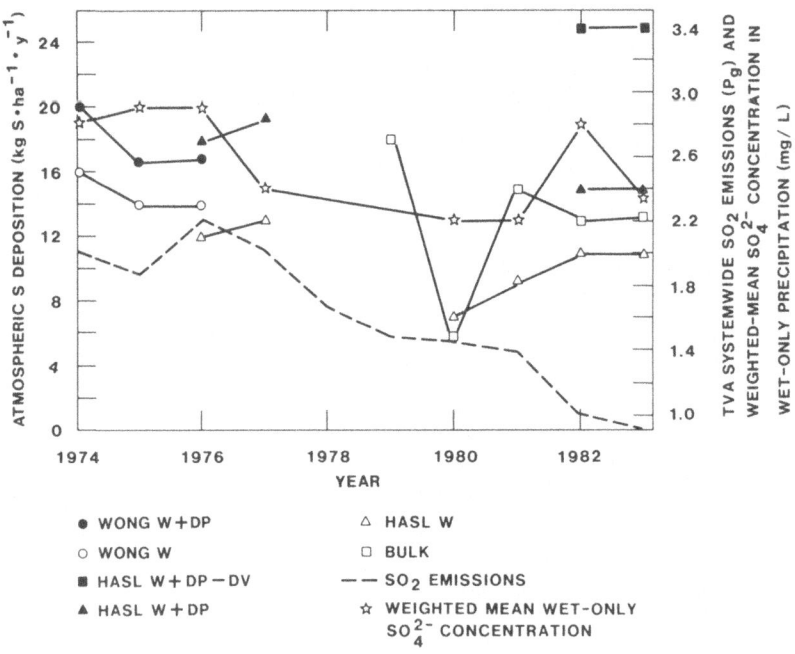

Legend:

- WONG W+DP   △ HASL W
- WONG W   □ BULK
- HASL W+DP−DV   — — SO₂ EMISSIONS
- HASL W+DP   ☆ WEIGHTED MEAN WET-ONLY SO_4^{2-} CONCENTRATION

Figure 4.9. Historical trends in atmospheric sulfur deposition as measured by several methods at Walker Branch Watershed. Also shown are trends in the weighted mean concentration of SO_4^{2-} in wet-only precipitation and in the Tennessee Valley Authority (TVA) systemwide SO_2 emissions. Deposition is indicated for wet-only precipitation by Wong and HASL collectors (Wong W, HASL W), for dry deposition of particles (DP) and vapors (DV), and for bulk precipitation.

annual deposition of sulfur by the methods described in Table 4.3 for bulk, wet, and dry deposition; annual weighted mean SO_4^{2-} concentrations in wet-only precipitation; and historical trends in Tennessee Valley Authority (TVA) systemwide emissions of SO_2. Deposition of sulfur by precipitation is a function of both air quality and rainfall amount, making trend analysis difficult. Normalizing the wet-only deposition estimates to annual rainfall amounts yields the weighted mean concentrations plotted near the top of the figure. Despite the scatter, the trend through 1981 generally reflects the decreasing trend in SO_2 emissions achieved by TVA, using a combination of coal washing, switching to low-sulfur coal, and installation of new electrostatic precipitators and sulfur scrubbers. A recent analysis suggests that over this time period there was also a significant and consistent decrease in both local and regional SO_2 levels in the Tennessee Valley (Parkhurst et al. 1984), suggesting that total wet plus dry deposition of sulfur to local forests has also decreased from a peak in the mid-1970s. The importance of analyses of such trends (NAS 1983) indicates the need for continuing long-term measurements of wet plus dry particle and vapor deposition to forested ecosystems.

4.3.3 Canopy Interactions with Deposited Material

The fate of material deposited on the forest canopy is reflected in the ion flux in throughfall plus stemflow collected below the canopy during the 1981–1983 study (Fig. 4.4). During the winter, when the canopy is leafless, the above- and below-canopy fluxes are generally comparable for most ions, indicating only a moderate influence of the exposed bark surface. This is not the case for Ca^{2+}, K^+, or H^+, the losses and gains of which are exchanged at nearly a 1:1 correspondence during the atmosphere-canopy interaction. Approximately 20 $mmol/m^2$ of H^+ is removed, and approximately 23 $mmol(+)/m^2$ of K^+ plus Ca^{2+} is released, whereas the major anions and other cations pass through the canopy with little change. The similarity in wet deposition fluxes of H^+ above (precipitation) and below (throughfall plus stemflow) the dormant canopy suggests that the net loss of H^+ occurs primarily from the dry-deposited component. The mean residence time of dry-deposited material in the canopy exceeds that of wet deposition significantly, allowing for a greater degree of canopy interaction with dry deposition.

During the growing season, the canopy has a significant influence on deposition reaching the forest floor for all ions except SO_4^{2-}. The canopy decreases the fluxes of H^+, NO_3^-, and NH_4^+ and increases those of Ca^{2+} and K^+. The chemistry of total atmospheric deposition to the canopy is acidic, consisting primarily of H_2SO_4 and HNO_3, with lesser amounts of NH_4^+ and Ca^{2+} salts of these strong acid anions. Below the canopy, the ion flux is dominated by Ca^{2+} and K^+ salts of SO_4^{2-}, with a minor contribution by the mineral acids. The removal of free H^+ from precipitation by deciduous forest canopies has been reported from studies of bulk (Cronan and Reiners 1983) or wet-only (Hoffman et al. 1980a) precipitation. These studies indicated a removal of 30 to 40% of the free H^+ in precipitation, whereas our most recent data suggest that the canopy can remove nearly 70% of the total deposition of free H^+ during the growing season. This reduction is thought to involve both ion exchange and weak base buffering reactions (Lovett et al. 1985; see also Sect. 4.3.4).

Deposited H^+ may remove canopy nutrients by cation exchange, as reflected in the growing season data for Ca^{2+} and K^+ (Fig. 4.4). The higher flux of these cations below the canopy compared with that above (~ 20 mmol$(+)$/m^2 for Ca^{2+} and ~ 50 mmol/m^2 for K^+) represents leaching of internal plant nutrients from the foliage. Using an ion charge balance of the interaction between deposition and the canopy and the measurement of mobile organic anions in throughfall, we reported that 40 to 60% of the nutrient cation leaching from this forest canopy is due to exchange for deposited free acidity (Lovett et al. 1985). The effects of nutrient leaching from the canopy depend on the ability of the tree to replenish these foliar pools.

The uptake of the nitrogen species is expected for a nitrogen-deficient forest ecosystem and primarily involves dry-deposited HNO_3 vapor and particulate NH_4^+ because of their long residence times on the foliage. However, foliar absorption of NO_3^- and NH_4^+ from precipitation must also occur to some extent. Sulfur exhibits no apparent retention by foliage in this sulfur-rich ecosystem (Johnson et al. 1982). Although 40% of the SO_4^{2-} deposition is from SO_2 uptake via leaf stomata, apparently all of this is available for foliar leaching during subsequent rain events (Fig. 4.4). Evidence suggests that throughfall SO_4^{2-} consists of two fractions exhibiting quite different leaching kinetics, one indicative of surface washoff and one of diffusion through leaf membranes (Lovett and Lindberg 1984). However, the actual source of any particular SO_4^{2-} ion in throughfall can only be determined by tracer studies.

4.3.4 Acid Exchange in the Canopy

As discussed in Sect. 4.3.3, the deciduous canopy at Walker Branch Watershed absorbs a considerable portion of the acidity deposited to it during the growing season. The forest canopy is known to have a major influence on precipitation acidity (Cronan and Reiners 1983), but this effect has not been quantified, and its mechanisms have not been studied in detail. Some work on rain chemistry at Walker Branch (Hoffman et al. 1980a) was designed to determine the influence of the oak canopy (*Quercus prinus* and *Q. alba*) on the nature and concentration of acidity in intercepted rain.

The contribution to the free acidity of precipitation, as measured by conventional pH electrodes, is from all protons in solution, regardless of source. Strong acids completely dissociate in solution, whereas weak acids, both organic and inorganic, only partially dissociate. The degree of weak-acid dissociation decreases with decreasing pH such that, at the pH of most precipitation samples at our site, they contribute negligibly to the free acidity and primarily to the total acidity (Galloway et al. 1976). For 29 storms sampled at Walker Branch Watershed during the 1977–1978 growing seasons, we found weak acids to account for ~30% of the total acidity in incident rain ($\overline{X} \pm SE = 32 \pm 3\%$). Total acidity averaged 137($\pm$8) μmol/L and strong acidity averaged 93($\pm$11) μmol/L. This is similar to the early data published on weak acids in rain by Galloway et al. (1976), who reported a contribution of 35% weak acids to the total acidity in one storm sampled during July in the Northeast.

In throughfall collected beneath the chestnut oak canopy, we found weak acids to contribute approximately twice as much to the total acidity [(55 $\pm$ 4)%] as was the case for incoming rain. The concentration of weak acids in throughfall has not been previously quantified. The interception of incoming rain by the canopy had little influence on total acidity [throughfall = 135($\pm$8) μmol/L compared with 137 in rain]. However, strong acidity was decreased significantly [61($\pm$9) μmol/L compared with 93 in rain]. Thus, the average net loss of strong acidity from the incoming rain to the canopy was approximately 30 μmol/L.

In general, the white oak canopy removed a greater amount of the incoming strong acidity than did the chestnut oak canopy (up to 100% in some events). In both canopies, the more leaf layers with which the rain interacted, the greater the degree to which the strong acids were scavenged. The primary physical factor influencing acid uptake was rain intensity. As intensity decreased (and rain residence time on the vegetation increased), strong-acid scavenging increased, suggesting the importance of rain-canopy reaction time.

Charge balance considerations suggest that uptake of H^+ in the canopy requires an equivalent release of balancing cations or uptake of associated anions. Our data do indicate a positive correlation between H^+ uptake and leaching of trace metals and major cations into intercepted rain (Lindberg et al. 1979). Although the composition of leaf cuticle is not well known, it is thought to be a polyester of C_{16}–C_{18} acids (Albersheim 1965). The behavior of the cuticular material as an exchange site following interception of rain could be expected to result in some degree of structural alteration (Garrells and Christ 1965) such that certain components of the negative framework would dissolve into the surrounding solution. Thus, the interaction between hydrogen ions in rain and the canopy may be responsible for both cation displacement and weak-acid leaching, as discussed in more detail by Hoffman et al. (1980a) and Lovett et al. (in press).

With this background, it is interesting to consider the event-by-event variations in rain and throughfall acidity during a 2-year period (Fig. 4.10). During the summer of 1976, the influence of the forest canopy was ap-

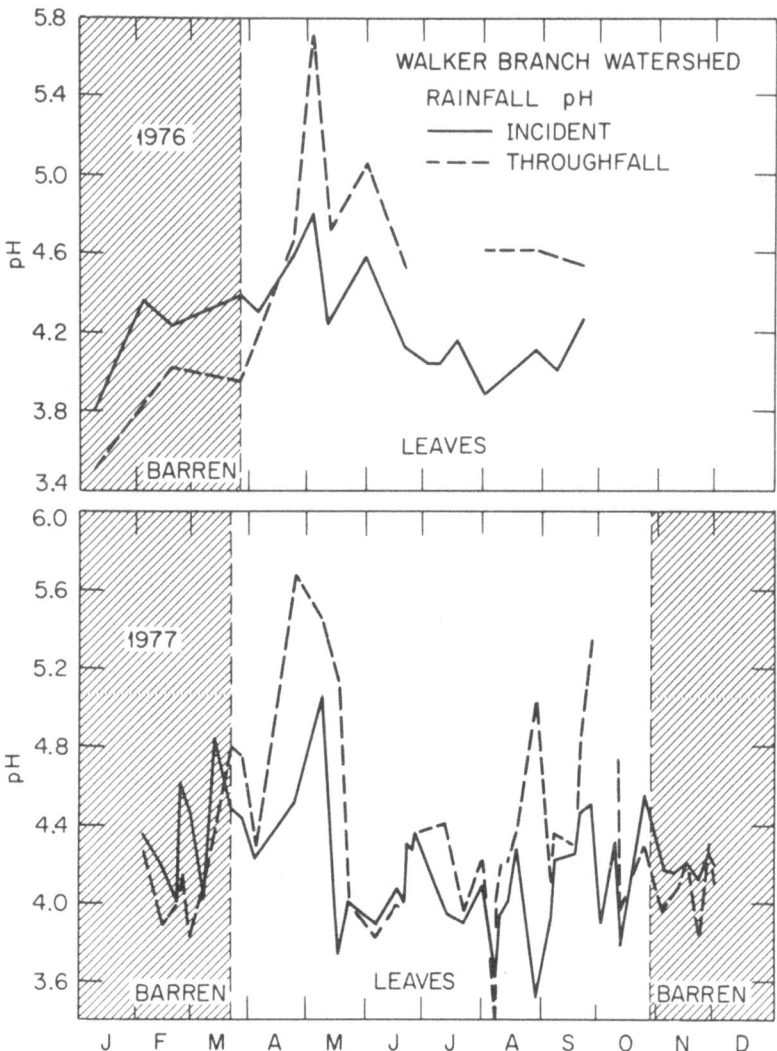

Figure 4.10. The pH of rain above and below the chestnut oak canopy during 1977–1978.

parent, with consistently higher pH values in throughfall than in rain above the canopy. Following leaf fall and during the entire dormant period, the throughfall pH values were similar to, but consistently lower than, the incident rain values. This has been attributed to leaching of weak organic acids and, in some cases, strong acids from bark and twigs (Hoffman et al. 1980a). Following budbreak and canopy refoliation, the pH again reversed such that the throughfall pH exceeded that of incoming rain, indicating hydrogen ion exchange at the newly formed leaf surface. However, this trend was not always maintained throughout the growing season. For four of five successive storms in May and June, the throughfall pH values were less than, or comparable to, those in rain above the canopy. This suggests that the fully developed canopy had temporarily lost the ability to remove hydrogen ions from incoming precipitation.

We found that this phenomenon of higher throughfall acidity relative to incident rain during full canopy development is caused by increases in both strong and weak acids in throughfall (Hoffman et al. 1980a). This indicates increased weak acid leaching from the leaves and possibly also solubilization of particles on the leaf surface having a strong acid reaction. Particles with a strong acid nature are abundant both in ambient air and on various surfaces at Walker Branch Watershed (Lindberg et al. 1979). Interestingly, the May-to-June pH reversal period was preceded by 15 d of essentially no precipitation, during which time several local air stagnation bulletins were issued. The period was characterized by calm, hazy conditions and frequent temperature inversion and fumigation events, during which time the concentrations of ozone, sulfur dioxide, total suspended particles, and nitrogen oxide in ambient air often exceeded air quality standards (Knoxville–Knox County Air Quality Board, Tennessee, personal communication, 1979).

The interaction of the strong oxidant ozone at, and internal to, the leaf surface is known to attack double bonds and affect membrane permeability (Mudd and Kozlowski 1975). The esterified C_{16}–C_{18} acids, which comprise the cuticle, include free carboxyl groups that would be likely sites of cation exchange. If these groups are points of reaction with ozone, as would be predicted by a strict chemical consideration, their ability to behave as hydrogen ion exchange sites would be significantly impaired. This may partially explain the subsequent pH reversal. In addition, the storm events occurring during the pH reversal period were generally small in volume (<2.5 cm) and high in intensity, leading to short residence times of the precipitation on the leaf surfaces. However, the first major storm following this period (>11 cm) resulted in the return of the throughfall-rain pH relationship to that expected for the developed canopy (throughfall pH > rain pH). We are not aware of any mechanism by which these exchange sites could have been restored, but the subsequent storms generally resulted in considerably higher throughfall pH values relative to incoming rain, as expected. This "normal" relationship was maintained throughout the remainder of the growing season. During the period following leaf fall, the throughfall pH was again consistently less than the pH of the incident precipitation, as during the previous year. Further information on the strong and weak acid content, sulfate and nitrate concentrations, and levels and identities of several organic compounds in rain and throughfall collected during these periods has been published (Hoffman et al. 1980a). Continuing examination of the chemistry resulting from the interaction between rain and leaves is both prudent and necessary to achieve a fuller understanding of internal and external mechanisms affecting leaf processes.

4.3.5 Summary and Implications

Our data support the hypothesis that dry deposition plays a major role in atmosphere-canopy interactions in forests of the eastern United States. Chronic exposure of the canopy to particles and gases increases the degree of interaction with foliage, which can result in both ion uptake and ion loss. The phenomenon of strong-acid scavenging from rain by the forest canopy is a well-known but poorly understood phenomenon resulting in the release of plant-related weak acids and nutrients (Eaton et al. 1973; Likens et al. 1977; Hoffman et al. 1980a). We estimated that ~50% of the cation leaching from the canopy at Walker Branch was due to exchange for deposited strong acidity (Lovett et al. in press). Accelerated nutrient leaching from foliage due to acid scavenging and cation exchange may increase the rate at which certain nutrients become limiting to forest growth. Future research in forests should concentrate on canopy effects in addition to soil effects, particularly on the role of hydrogen uptake and subsequent organic carbon and nutrient loss from foliage.

We estimate that deposition supplies from 10% (potassium) to 40% (nitrogen and calcium) to more than 100% (sulfur) of the needs of this forest for annual woody increment, either through foliar absorption or root uptake of deposited ions. For sulfur, even the total forest requirement (woody increment plus annual foliage production) is exceeded by deposition, and nitrogen input contributes 5 to 10%. As these proportions increase due to industrial emissions, forests may become increasingly dependent on the atmosphere for nutritional requirements at the same time that they are increasingly exposed to toxic trace contaminants (Lindberg et al. 1982b). The effects of excess sulfur, nitrogen, and metal deposition by atmosphere-dominated element cycles are possibly already being manifested in high-elevation forests in the eastern United States and Europe (Johnson and Siccama 1983).

4.4 Trace Metals

Man's activities may rival natural processes as sources of several constituents in the atmosphere (Lantzy and McKenzie 1979). Acid deposition is only one such manifestation; mobilization of trace metals also accompanies industrial activities. Forests may be particularly effective sinks for these emissions because of their high–surface area canopies and because of some aspects of their biogeochemical cycles. This section summarizes the results of an intensive 2-year study of trace metal deposition and cycling at Walker Branch Watershed from 1976 to 1977, using the techniques described earlier to sample and analyze Cd, Mn, Pb, and Zn in a variety of matrices.

4.4.1 Atmospheric Chemistry

Investigations of the atmospheric chemistry of metals at WBW included use of chemical mass balance models to identify the sources of metals in suspended particles (Andren and Lindberg 1977), an analysis of factors influencing metal concentrations in aerosols (Lindberg and Harriss 1981), and a study of metal solubility in airborne material (Lindberg and Harriss 1983). Solubility is a particularly important parameter in metal-cycling studies, because elements transported to vegetation in insoluble forms may be biologically inactive. A particle-associated metal of high solubility, however, may be readily mobilized on vegetation surfaces by interaction with moisture. We determined the solubility of metals in airborne particles, using selective extraction procedures to define two fractions: total extractable fraction (soluble in distilled water plus $0.1\ N\ HNO_3$) and soluble fraction (soluble in distilled water alone). Relative solubility was defined as (soluble)/(total extractable). Details of the methods have been published (Lindberg and Harriss 1983).

Of the metals included in this study, lead is the largest component of the total extractable fraction of the atmospheric particles sampled, occurring at concentrations up to 2 orders of magnitude above the other metals (Table 4.4). The combined contribution of these four metals to the total aerosol mass is 0.5%, ~70% of which is attributable to lead alone. Earlier research indicated that Al, Fe, Cl, Ca, and Na accounted for another 8%, and SO_4^{2-} accounted for ~25% of the average aerosol mass at this site (Andren and Lindberg 1977; Lindberg et al. 1979).

The air concentrations measured at Walker Branch Watershed are generally comparable to values reported for rural areas, with only lead approaching levels reported for urban environments (Table 4.5). Over the course of the sampling period, the concentrations of Cd, Mn, and Pb varied

Table 4.4. Statistical summary of total extractable[a] metal concentrations in air and in suspended particles based on total filter samples (n = 10); also shown are mass median diameters of each metal based on impactor data

Constituent	Air concentration (ng/m³)			Particle concentration[b] (µg/g)			Mass median diameter (µm)	
	Mean	SD	Range	Mean	SD	Range	Mean	Range
Cadmium	0.17	0.09	0.05–0.29	5.5	4.1	1.5–13	1.5	0.8–3.9
Manganese	10	4.1	2.0–18	340	150	31–580	3.4	2.0–6.1
Lead	110	38	63–170	3500	1800	2000–7300	0.5	0.45–0.70
Zinc	32	70	2.9–230	1100	2400	78–7800	0.9	0.50–1.5

[a]Extractable = water plus dilute-acid-leachable fractions.

[b]Expressed in relation to the mass of total suspended particles.

Table 4.5. Summary of representative aerosol composition studies at diverse locations

Location	Airshed type	Cd	Mn	Pb	Zn	Study
South Pole	Remote	—	0.01	0.63	0.03	Zoller et al. 1974
Greenland	Remote	—	—	0.15	<1.3	Davidson et al. 1981
Bolivia	Remote	—	0.8–7.1	3.6–7.1	1.2–5.3	Adams et al. 1977
Olympic National Park, Wash.	Remote	0.54	20	2.2	8.9	Davidson et al. 1983
White Mt., Calif.	Remote	—	—	8.0	—	Chow et al. 1972
Squaw Mt., Colo.	Remote	—	BD[a]	BD[a]	BD[a]	Lawson and Winchester 1978
North Atlantic	Marine	0.003–0.62	0.05–5.4	0.10–64	0.3–27	Duce et al. 1975
North Atlantic	Marine	—	2.2	9.9	4.3	Buat-Menard and Chesselet 1979
Chadron, Nebr.	Rural	0.57	5.7	45	16	Streumpler 1975
Belgium	Rural	6–16	—	212–594	218–799	Kretzschmar et al. 1977
Walker Branch Watershed	Rural	0.17	10	110	32	This study (mean values)
Pasadena, Calif.	Suburban	—	—	2140	76	Hammerle and Pierson 1975
Cleveland, Ohio	Suburban	1.6	66	451	264	King et al. 1976
Cleveland, Ohio	Urban	3.9	148	759	413	King et al. 1976
St. Louis, Mo.	Urban	—	—	~300	~150	Winchester et al. 1974
Belgium	Industrial	14–50	—	791–975	618–3054	Kretzschmar et al. 1977
Northwest Indiana	Industrial	—	63–390	400–3700	100–1540	Harrison et al. 1971

Concentrations (ng/m^3)

[a]BD = below detection (described as "less than a few ng/m^3" by the authors).

by factors of 3 to 9, whereas the concentration of zinc varied over nearly 2 orders of magnitude. The highest concentration of zinc (230 ng/m³) exceeded the next highest value by a factor of 7. The normalized particle size distributions for each element (Fig. 4.11) were less variable than the absolute concentrations, and indicate that the metals fall into two general size distribution categories: (1) monomodal distribution with maxima in the submicron-size particle mode (lead, zinc), and (2) relatively uniform size distribution with supermicron–particle size tendency (manganese) or submicron tendency (cadmium). The distributions are reflected in the calculated mass median diameters that are submicron for lead, near unity for zinc and cadmium, and supermicrometer for manganese (Table 4.4). The dominance of the coarse-particle mode for manganese suggests a surface soil dust source, and the strong fine-particle modes for lead and zinc suggest combustion sources. Both fine-particle combustion and coarse-particle dispersion sources apparently contribute to the cadmium size distribution.

The influence of combustion processes on the size distributions can be illustrated by normalizing element concentrations in each size class to the concentration of a typical lithophilic element (e.g., Al, Fe, Sc). These

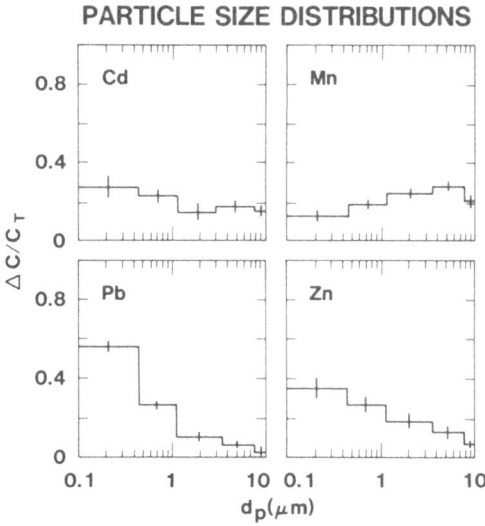

Figure 4.11. Normalized size distributions for Cd, Pb, Mn, and Zn, based on cascade impactor data collected during ten experimental periods. Shown are mean normalized concentrations ($\Delta C/C_T$, where ΔC is the element concentration for a given size range and C_T is the total concentration of the element in all size ranges) and standard errors of the estimate of the means for five size classes. Particle diameters (d_p) were estimated from impactor data.

ratios are further normalized by comparison with similar ratios for crustal material or surface soils and are referred to as enrichment factors: EF = $(X/Al)_{air}/(X/Al)_{soil.crust}$ (Zoller et al. 1974). Values near unity suggest a crustal or surface soil source. One sample was subjected to complete acid digestion and was analyzed for total concentrations of Cd, Mn, Pb, Zn, and Al. The EF values confirm our conclusions regarding possible sources to the atmospheric load of each element. With the exception of manganese, each element exhibited enrichment factors that generally increased with decreasing particle size. Manganese in all size classes was not significantly enriched over soil values (EF = 1–4), indicating suspended surface material as its primary source. Large-particle cadmium may be of crustal origin with EF = 3 to 4, whereas the submicron material was enriched over soil values by factors of 17 to 23. Both lead and zinc were considerably enriched in relation to soils in all size classes, particularly in the fine-particle range (EF = 400–13,000). The ratio of mean EFs for submicron particles to larger particles is ~10 for zinc and ~40 for lead, indicating an increased contribution of noncrustal, probably combustion-derived, material to the fine-particle concentrations.

In an attempt to identify factors influencing these metal concentrations, several meteorological characteristics were determined for each period during which we collected atmospheric particles; these include precipitation amount and duration, atmospheric stability (as measured by frequency of air stagnation events), frequency of wind direction from local sources, and atmospheric moisture status. The periods fall into three general air stagnation classes: high (100% of the sampling period under an air stagnation alert), medium (~30%), and low (0%). These classes were determined from the frequency of occurrence of regional air stagnation alerts issued by the National Weather Service for eastern Tennessee. As air stagnation frequency increased, there was a general increase in the air concentration of Cd, Pb, and Zn but a decrease in the concentration of manganese. This behavior supports the hypothesis of a combustion source for Cd, Pb, and Zn and a soil source for manganese. Air stagnation alerts are issued when vertical stratification and horizontal dispersion are such that pollutant particles and gases are likely to concentrate near the ground. Increased air stagnation also decreases surface wind resuspension of soil dust.

Wind direction exhibited no consistent influence on either air concentrations or particle size distributions despite the nearby power plants. This apparent lack of influence of wind direction on atmospheric concentrations suggests that particulate metals at this site result from regional transport phenomena rather than directly from local coal combustion emissions. Gordon (1975) reached a similar conclusion regarding the influence of a local power plant on airborne metal concentrations in rural Maryland. Application of a chemical mass balance model to aerosol composition at our sampling site suggested that the local power plants contributed <1%

of the cadmium and lead, ~2% of the manganese, and ~10% of the zinc to the total aerosol. The primary contribution to the lead burden in WBW aerosols was apparently automobile emissions (Lindberg et al. 1979; Andren and Lindberg 1977).

The relative solubilities of metals in the total aerosol are summarized in Table 4.6 for the 10 experiments. Relative solubility was defined earlier as being the water-soluble fraction expressed as a percentage of the water-soluble plus acid-leachable fractions. The relative solubilities indicate that substantial amounts (generally >50%) of the total extractable metals were soluble in distilled water. The pattern of mean relative solubilities (Zn > Mn ≃ Cd > Pb) is similar to the general order of the thermodynamic solubilities of the most probable salts of these elements (SO_4^{2-}, NO_3^-, and Cl^-) occurring in airborne particles in areas of fossil fuel combustion (Dams et al. 1975; Henry and Knapp 1980).

The water and dilute acid extractions used in this study did not result in complete recovery of total metal present in atmospheric particles. Results of concentrated acid wet digestion of dilute acid extraction residues for size-fractionated particles collected during one experiment indicated that the concentration ratios of the residual metal (that released by digestion) to the total extractable metal (that soluble in water plus dilute acid) could be significant (Lindberg and Harriss 1983). For manganese and cadmium, the ratios ranged from 0.003 to 0.44 for various size fractions but indicated a <15% increase in metal concentrations due to the additional wet digestion for the total aerosol. However, for lead and zinc, these ratios ranged from 0.22 to 250 for different size classes and indicated an ~100 to 200% concentration increase for the total aerosol. For all metals, there was a general increase in the ratio as particle size increased, suggesting that the refractory metals occurred primarily on larger particles. Com-

Table 4.6. Statistical summary of solubility characteristics of suspended particles collected above the forest canopy in Walker Branch Watershed during 1976–1977

Component	Atmospheric concentration[a]		Relative solubility[a,c] (%)	Solubility range (%)
	Acid-leachable[b] (ng/m³)	Water-soluble (ng/m³)		
Cadmium	0.17 ± 0.03	0.14 ± 0.03	82 ± 4	48–91
Manganese	9.4 ± 1.6	8.6 ± 1.2	83 ± 2	70–92
Lead	112 ± 12	84 ± 11	76 ± 3	55–88
Zinc	9.9 ± 4.0	8.9 ± 3.5	89 ± 2	74–95

[a]Mean and standard error of the estimate (n = 10).
[b]Acid-leachable concentration at pH of 1.2; includes water-soluble component.
[c]Defined as (water-soluble concentration)·100/(acid-leachable concentration).

parison of results of complete acid digestions with those of dilute acid plus water extractions indicated that relative solubility was a reasonable estimate of water solubility for all metals in the smallest size class and for cadmium and manganese in the total aerosol. However, relative solubility was a considerable overestimate of water solubility for lead and zinc in the total aerosol because of the relatively large additional quantities of lead and zinc released by the concentrated acid digestion. The ratios of relative solubility to water solubility for the total aerosol were as follows: $Cd = 1.1$, $Mn = 1.1$, $Pb = 2.0$, and $Zn = 3.2$. These factors can be used to modify the relative solubility values to reflect the more conventional definition of water solubility. Assuming the results of these complete acid digestions to be representative, the mean relative solubilities of the total aerosol (from Table 4.6) expressed in conventional water solubility terms would be as follows: Mn, 75%; Cd, 74%; Pb, 37%; and Zn, 28%.

These values are generally similar to those reported in the literature for coastal marine aerosols (Hodge et al. 1978) and for urban aerosols (Eisenbud and Kneip 1975), but are higher than values we previously determined by the same techniques for particles collected in a combustion plume 250 m downwind of a 2600-MW coal-fired power plant (Lindberg and Harriss 1980). Relative solubility increased during plume aging, however, to values comparable to those reported here for ambient aerosols. The trend is illustrated for lead in Fig. 4.12, and values for the other metals in plume aerosols collected 250 m and 7 km downwind of the power plant, respectively, were as follows: Cd, 55 and 90%; Mn, 55 and 90%; and Zn, 50 and 80%. Relative solubility measurements for size-fractionated particles indicated that the increase in the solubility of the total aerosol was due to significant increases in the relative solubilities of the finest particle size sampled.

The relationship between particle size and mean relative solubility in the ambient aerosols is also illustrated in Fig. 4.12 for lead. All elements exhibited a trend of increasing mean solubility and decreasing variability in solubility as particle size decreased. This trend is apparently not related to particle size distribution alone because of the differences in size distributions among elements. The trend of increased relative solubility with decreasing particle size can be explained by a combination of factors: (1) a basic difference in the chemistry of source material contributing to each size fraction; (2) the greater surface area-to-volume ratio of smaller particles combined with the surface predominance and overall elevated concentrations of metals on small particles (Keyser et al. 1978); and (3) the increased free H^+ and associated strong-acid anion concentrations in distilled water leachates of smaller particles (Winkler 1980). Increased water solubility of atmospheric particles with decreasing particle size was previously reported by Meszaros (1968), who found the major fraction of airborne water-soluble particles in the size range of 0.2 to 0.6 μm.

Of the meteorological parameters measured during each period, only

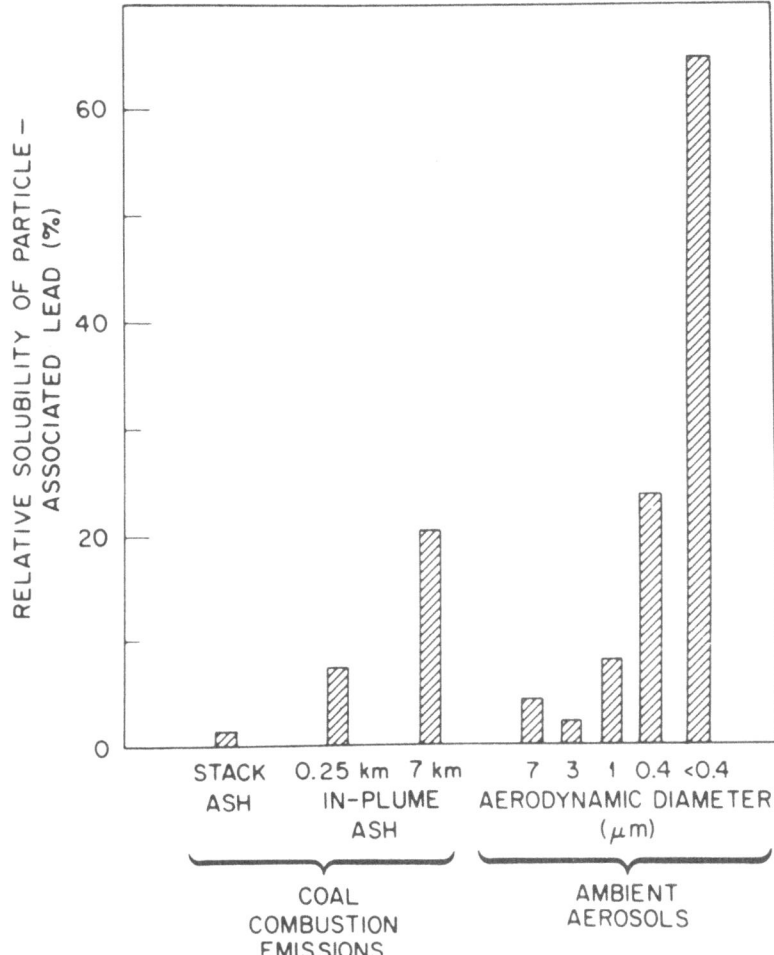

Figure 4.12. Relative solubility (water-soluble to water-soluble plus acid-leachable concentration ratio $\times$ 100) of lead in aerosols emitted from a modern coal-fired power plant, in the atmosphere over Walker Branch Watershed, and in particles deposited in the upper forest canopy.

atmospheric moisture had a significant influence on particle solubility. Although the correlation coefficient was significant only for lead ($r = 0.75$, $p < 0.05$), all elements exhibited a trend of increasing solubility with increased duration and frequency of atmospheric water-vapor saturation (Lindberg and Harriss 1983). The relationships between relative solubility of metals in ambient aerosols and atmospheric water-vapor saturation and between metal solubility in combustion aerosols and plume age suggest the importance of particle hydration in influencing metal solubility. Laboratory studies by Cofer et al. (1981) demonstrated that sorption and oxidation of SO_2 by soot particles are enhanced by high concentrations of atmospheric water vapor, O_3, and NO_2. We hypothesize that our field data showing increased aerosol metal solubility for high atmospheric water vapor conditions are consistent with these laboratory results. Ambient aerosols exposed to an atmosphere of high water vapor content will sorb and oxidize relatively large amounts of SO_2, increasing aerosol acidity. When atmospheric aerosols undergo such acidification, very low pH levels may occur in the early stages of droplet formation (Esman and Fergus 1976). These conditions can enhance metal dissolution by conversion of an increasing proportion of the aerosol-associated metals to relatively soluble sulfate and nitrate compounds.

4.4.2 Dry Deposition

With few exceptions (e.g., mercury and selenium), the dry deposition of trace metals involves particles, with little if any additional input by gaseous deposition. Hence, the study of trace metal dry deposition has been severely limited by a lack of widely accepted sampling strategies. This is reflected in the limited amount of published data on this phenomenon, particularly for complex vegetation canopies such as forests. Research at Walker Branch Watershed on trace metal dry deposition began with aerosol chemistry studies in the mid-1970s during which particle input was approximated by using measured airborne concentrations and literature values of dry deposition velocities (Andren and Lindberg 1977). Beginning in 1976, we developed surface accumulation and leaf-washing methods, as described earlier, which were employed in the actual plant canopy to estimate metal input by depositing particles (Lindberg et al. 1979, 1982b; Lindberg and Harriss 1981).

The measured dry deposition rates to inert surface deposition plates in the forest canopy for several experimental periods are summarized in Table 4.7. Dry deposition rates were generally in the same order as the airborne concentrations of these metals, with the exception of manganese, which exhibited the highest flux: Mn > Pb > Zn > Cd in dry deposition, but Pb > Zn > Mn > Cd in aerosols. The importance of manganese in dry deposition is a result of its larger particle size in the atmosphere relative to the other metals (Table 4.4 and Fig. 4.11). These values at Walker Branch Watershed are comparable to those reported in the literature for various surfaces in generally rural locations (Table 4.8).

The temporal variability in these rates suggested the possible influence of meteorologic and air quality factors. The measured dry deposition rates of Cd, Pb, and Zn to inert surfaces were positively correlated with air stagnation frequency and concentrations of each element in the total aerosol fraction (manganese did not exhibit similar correlations), but were

Table 4.7. Dry deposition rates of acid-leachable components of particles deposited on flat plates situated in the upper canopy at Walker Branch Watershed

Period[a,b]	Dry deposition rate ($\mu g \cdot m^{-2} \cdot d^{-1}$)			
	Cd	Mn	Pb	Zn
W1	0.034 ± 0.007	15 ± 2.5	3.7 ± 0.3	2.0 ± 1.0
W2	0.18	29	15	2.3
W3	0.006	46	2.9	0.78
W6	0.019	21	5.7	0.77

[a]W1 = 5/9–16/77, W2 = 5/16–20/77, W3 = 5/30–6/6/77, W6 = 7/12–18/77.
[b]During W1, the four individual deposition plates located around the perimeter of the canopy area were analyzed separately; given are mean values and standard errors of the estimate. During W2–W6, the four plates were composited before analysis.

not correlated with the frequency of winds from directions influenced by the three local power plants. Thus, dry deposition rates, as well as air concentrations, were not detectably enhanced by local emissions. We investigated other possible sources of the deposited material by calculating upper-level air mass backward trajectories for time intervals comparable to tropospheric residence times for particles in the size range captured by these surfaces. These trajectories traversed two major regional (within 300 km) urban centers and five other nonlocal coal-fired power plants with sufficiently rapid air mass travel times to account for the deposited particles (Fig. 4.13).

Normalizing the measured deposition fluxes (F) to unit air concentrations (C) yields deposition velocities ($V_d = FC^{-1}$) for each element to individual upper canopy surfaces (Table 4.9). The agreement in deposition velocities determined for inert, flat surfaces and estimated from sequential leaf washing during one 7-d period (W1) was good for Cd, Zn (within 30%), and Mn (within a factor of 2) but poor for Pb (a considerably higher value was measured for the inert surface). The poor agreement in the calculated lead fluxes (and hence deposition velocities) resulted from a 1–order of magnitude lower recovery of lead from laboratory-leached vegetation surfaces than from similarly exposed and leached inert surfaces. We do not believe that the lack of agreement in lead fluxes indicates a fundamental difference in the chemistry or particle size distribution of the atmospheric material deposited on each surface. This view is supported by the relatively good agreement seen for the other elements. Sulfate provided a particularly good comparison, because it exhibited a particle size distribution in the atomosphere nearly identical to that exhibited by lead, and the deposition velocities of SO_4^{2-} for each surface agreed to within 30% (Lindberg and Harriss 1981). A possible explanation is loss of the

Table 4.8. Dry deposition to inert surfaces measured at diverse locations

Location	Surface	Airshed type	Dry deposition rate ($\mu g \cdot m^{-2} \cdot d^{-1}$)			
			Cd	Mn	Pb	Zn
Pacific Ocean	[a]	Marine	0.08–0.38	—	4–50	1.2–20
French Congo	[b]	Remote, continental	—	—	0.53	—
San Gabriel Mountains, Calif.	[a]	Remote, continental	0.07–1.0	—	22–40	23–25
Central Wales, U.K.	[c]	Remote, continental	BD[d]	5	30	40
Olympic National Park, Wash.	[a]	Remote, continental	0.008	16	0.024	0.21
Shetland Island, U.K.	[c]	Remote, coastal	BD[d]	BD[d]	40	50
Catalina Island, Calif.	[a]	Remote, coastal	—	—	3–690	0–1700
Ensenada, Mexico	[e]	Remote, coastal	—	160	29	39
Southwest France	[b]	Rural	—	—	7.4	—
Cumberland Plateau, Tenn.	[f]	Rural	0.05 ± 0.01	4.5 ± 0.7	2.4 ± 0.7	13 ± 3

Location	Type				
Oslo, Norway[g]	Rural	0.04 ± 0.005	—	75	—
Cherokee Forest, N.C.[f]	Rural	BD[d]	7.5 ± 1.0	2.3 ± 0.2	15 ± 4
Central South England[c]	Rural	0.06 ± 0.04	8	50	—
Walker Branch Watershed, Tenn.[f]	Rural	—	28 ± 7	6.8 ± 2.8	1.5 ± 0.4
Toulouse, France[b]	Residential	—	—	24	—
La Jolla, Calif.[e]	Suburban	—	27	76	65
Los Angeles basin, Calif.[a]	Suburban	0.7–3.6	—	280–810	73–340
Los Angeles suburbs, Calif.[a]	Suburban	1.1–2.0	—	47–330	23–400
Los Angeles coastal, Calif.[a]	Suburban	0.27–1.6	—	25–140	6–160
Pasadena, Calif.[a]	Suburban	0.7–3.6	—	280–810	73–340
Central Los Angeles, Calif.[a]	Urban	1–2.4	—	1,000–22,000	240–1,500
New York City[h]	Urban	20	270	1,500	2,600

[a] Teflon plates (Davidson 1977; Davidson et al., 1983).
[b] Pluviometer (Servant 1976).
[c] Filter paper (Cawse 1975).
[d] BD = below detection.
[e] Polyethylene buckets (Hodge et al. 1978).
[f] Polycarbonate Petri dish deposition plates (this study: WBW, 1977; Cumberland and Cherokee, 1981–1982). Values given are for the available element fractions only (mean ± SE).
[g] Snow surface (Doveland and Eliasson 1976).
[h] Polypropylene buckets (Eisenbud and Kneip 1975).

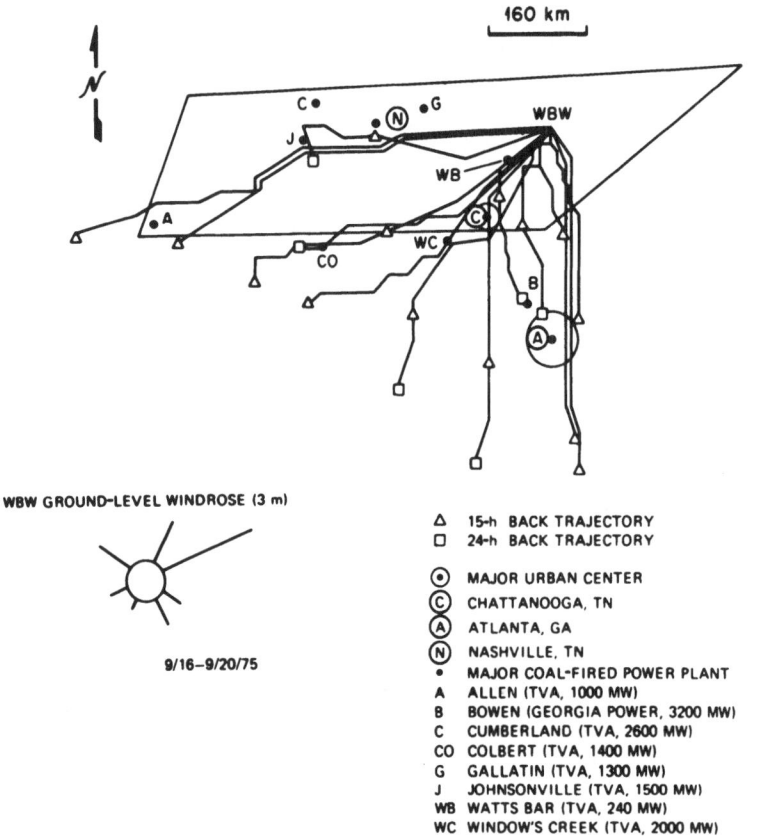

Figure 4.13. Air mass back-trajectories from Walker Branch Watershed (WBW) for the period September 16–20, 1975. Transport layer taken as 100–3000m. Shown are major coal-fired power plants and urban centers in the Tennessee Valley region.

Table 4.9. Dry deposition velocities (V_d) for particle transfer to individual surfaces in the upper canopy

		Deposition velocity (cm/s)			
Period	Surface type	Cd	Mn	Pb	Zn
W1[a]	Inert, flat	0.33	1.7	0.05	0.66
W1	Oak leaves	0.23	0.8	0.005	0.46
X[b]	Inert, flat	0.37 ± 0.18	6.4 ± 3.6	0.06 ± 0.01	0.38 ± 0.10

[a]5/9–16/77.
[b]Mean ± SE of four experimental periods of 4- to 7-d duration each.

water-soluble fraction of the lead initially deposited on the leaves through internal leaf absorption or irreversible surface adsorption processes. The annual mass transfer rates to the forest floor provide further evidence for foliar adsorption of lead and are discussed in the final section of this chapter.

The mean V_d values reflect the general particle size distribution of each element such that elements characterized by larger mass median diameters in the aerosol population exhibited higher deposition velocities. These deposition velocities reflect particle fluxes to individual surfaces in the upper canopy. Because of the considerably larger total surface area of all the leaves in the canopy compared with the soil surface over which they are situated, adjusting the values of V_d for the full canopy effect will result in increased deposition velocities expressed on a ground-area basis. For the forest type at Walker Branch Watershed, the applicable midsummer leaf area index would increase the deposition velocities by a factor of ~3 (see earlier discussions). These data indicate that the often used deposition velocity of 0.1 cm/s for particles is an underestimate for the deposition of Cd, Mn, Zn, and perhaps also Pb to a forest canopy.

Dry-deposited particles exhibited generally lower relative solubilities (<75%) than particles in the air, suggesting a larger particle size than the material in the atmosphere. This was confirmed by scanning electron and light microscopy of biological and inert deposition surfaces, which indicated significant fractions of relatively large (>5 μm) fly ash and dispersed soil particles primarily on upward-facing surfaces. This is illustrated in the scanning electron photomicrograph of a chestnut oak leaf surface from Walker Branch Watershed in the upper left-hand corner of Fig. 4.14. An interesting observation from this and several other (Lindberg et al. 1979) scanning electron photomicrographs was the association of submicron- to micron-size fly ash with considerably larger (5–20 μm) fly ash and scoriaceous ash particles, and the agglomeration of several submicron-size fly ash particles into large aggregates. Fly ash particles were common on leaves collected in regions containing coal-fired electric-power-generating stations (Fig. 4.14).

The dry deposition of large particles from the atmosphere is primarily controlled by sedimentation (Sehmel and Hodgson 1974). Thus, formation of particle aggregates may account for the often high concentrations of small-particle-associated elements (Pb, SO_4^{2-}) in dry-deposited material. The presence of a significant quantity of this material on upward-facing surfaces supports particle sedimentation as an important mechanism of dry deposition to the upper canopy. Davidson (1977) reached a similar conclusion regarding dry deposition to smooth flat surfaces, and a more recent study suggests that sedimentation is also a major process of dry deposition to vegetation surfaces, particularly for canopies with a large leaf area index (Legg and Price 1980).

ORNL-DWG 83-12558

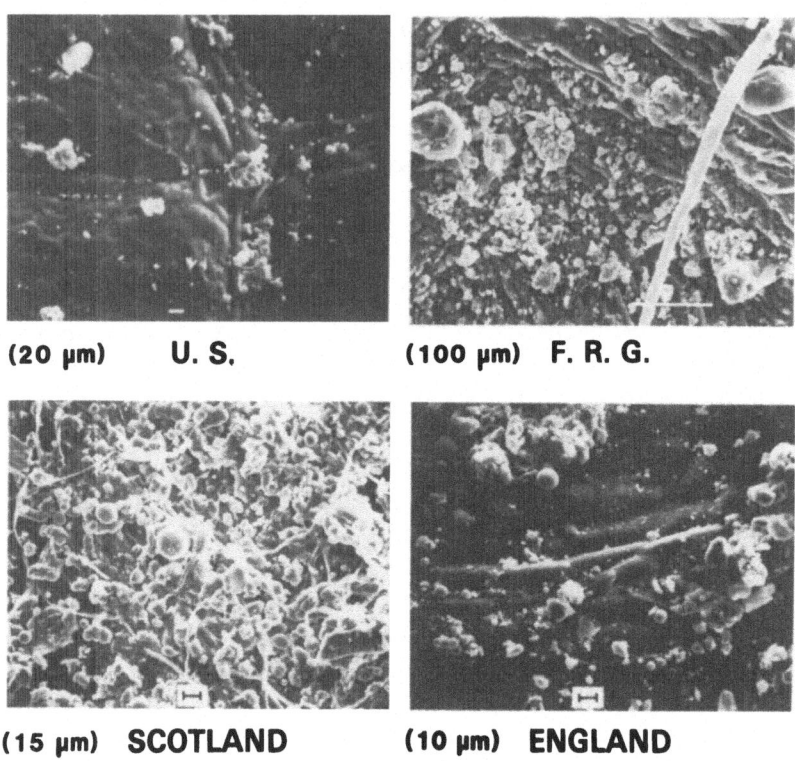

(20 µm)   **U. S.**   (100 µm)   **F. R. G.**

(15 µm)   **SCOTLAND**   (10 µm)   **ENGLAND**

Figure 4.14. Scanning electron microscope photographs of particles deposited on upper surfaces of leaves. At the upper left is a chestnut oak leaf (*Quercus prinus*) collected at Walker Branch Watershed, Tennessee (scale line is 20mm). At the lower left is an English oak (*Quercus robur*) collected near Edinburgh, Scotland (scale line is 15mm). At the upper right is a maple (*Acer* sp.) collected in the Black Forest region of the Federal Republic of Germany [scale line is 100 µm (longest line)]. At the lower right is an English oak collected near Nottingham, England (scale line is 10 µm).

4.4.3 Precipitation Chemistry Above and Below the Canopy

Precipitation is a particularly important component of atmospheric deposition of metals because of its role in delivering material to the canopy in water-soluble forms and because of its ability to dissolve additional material previously deposited to the canopy by dry processes. Material in solution at the leaf surface is available for biological interactions, including surface exchange and direct foliar absorption. Precipitation also provides the mechanism whereby dry-deposited material in the canopy can be transferred to the forest floor.

Research on precipitation chemistry of trace metals at Walker Branch Watershed was begun in 1973 using state-of-the-art methods (Wong wet/dry collectors at ground level; Andren and Lindberg 1977). However, with the development of new collectors and methods (Lindberg et al. 1977) and from the results of an intensive 2-year study of metals in rain collected above the forest canopy (Lindberg 1982), it is clear that some of our earlier data were influenced by sample contamination during collection and analysis. Indeed, much of the early data in the literature on trace metals in precipitation must be regarded with caution because of the well-known susceptibility of such samples to contamination problems (NAS 1980). This can be illustrated by a comparison of historical data collected at Walker Branch Watershed. These data indicate an average decrease of 80% for the metals we measured using the following methods over a 4-year period: rain sampled weekly using polyethylene bottles and funnels in metallic Wong collectors and analyzed by solvent extraction and flame atomic absorption (1974); rain sampled on an event basis using bottles and funnels in plastic-coated Wong collectors and analyzed by graphite furnace atomic absorption (1975); and rain similarly analyzed but sampled on an event basis using bottles and funnels in Teflon-coated positive-sealing HASL collectors (1976–1977). Mean concentrations decreased as follows (1974–1975 to 1976–1977, all in $\mu g/L$): Cd—7.0, 0.50, 0.25; Mn—13, 3.6, 1.9; Pb—13, 7.4, 5.2; and Zn—28, 30, 5.0. This section summarizes the results of this latter study of precipitation chemistry, which included an analysis of rain collected beneath the canopy as well as above (Lindberg 1982).

Precipitation chemistry of these four metals at Walker Branch Watershed is summarized in Table 4.10 along with results for several major ions for comparison. Rainfall at this site was a dilute mineral acid solution of H_2SO_4 and HNO_3, with a pH of ~4.2, containing the metals Cd, Mn, Pb, and Zn at trace levels, as expected for a rural area. These metals comprised only 0.2% of the total cation charge equivalents in average rain, whereas H^+ contributed 75% of the cation charge equivalents and exceeded the next most abundant cations, NH_4^+ and Ca^{2+}, by a factor of ~6 in charge equivalent concentration. Sulfate contributed 70% of the anion charge equivalents and exceeded NO_3^- by a factor of 3.5. The levels of the major ions in rain are comparable to those measured in the northeastern United States (Pack 1980), whereas the levels of the metals are comparable to those reported for other rural areas (Table 4.11), including nearby watersheds. As indicated by the ranges and standard deviations, metal concentrations at Walker Branch Watershed were highly variable. The coefficients of variation $(100 \cdot \sigma / \overline{X})$ approach or exceed 100%, ranging from 80% for lead to 140% for manganese; the values for major constituents are generally less (~60–80%). Frequency histograms of metal concentrations are positively skewed, and probability plots indicate that the distributions of all metal concentrations are log-normal (Fig. 4.15). Such dis-

Table 4.10. Statistical summary of trace and major constituent concentrations in precipitation (wetfall-only) collected at Walker Branch Watershed

Constituent	Weighted mean[a]	Arithmetic mean[b]	SD	Range
Mn, μg/L	1.93	3.62	5.16	0.03–24.0
Zn, μg/L	5.04	5.64	5.80	0.44–29.6
Cd, μg/L	0.25	0.42	0.49	0.003–2.20
Pb, μg/L	5.15	6.83	5.34	0.66–24.0
SO_4^{2-}, mg/L	2.8	3.6	0.94	0.46–16
NO_3^-, mg/L	1.0	1.5	0.94	0.16–4.6
Ca^{2+}, mg/L	0.21	0.26	0.21	0.02–0.96
NH_4^+, mg/L	0.18	0.26	0.21	0.003–0.79
H^+, mmol/L	0.065	0.074	0.038	0.006–0.18
(as pH)	4.19	4.13	—	3.75–5.22

[a]Weighted by rainfall amount.
[b]n = 52 for metals and 53–56 for major ions.

Table 4.11. Concentrations of metals in rain measured at diverse locations

Location	Airshed type	Sample type[a]	Cd	Mn	Pb	Zn	Study
Arctic Ocean	Remote, marine	S	0.004	—	0.013	—	Mart 1983
Corviglia, Switzerland	Remote, continental	B	1.0	3.5	30	—	Muller and Bielke 1975
South Pole	Remote, continental	S	0.0038	0.018	0.040	0.0068	Boutron and Lorius 1979
Northern Minnesota	Rural	W	0.15	3.2	7.1	90	Thornton and Eisenreich 1982
Walker Branch Watershed, Tenn.	Rural	W	0.25	1.9	5.2	5.0	This work[b]
Cumberland Plateau, Tenn.	Rural	W	0.21	2.7	4.8	8.7	This work[b]
Pine Barrens, N.J.	Rural	B	<1	6.0	9.0	8.0	Turner 1983
Cherokee Forest, N.C.	Rural	W	0.04	1.3	3.2	8.6	This work[b]
Chadron, Nebr.	Rural	B	0.26	5.4	4.3	10	Streumpler 1976
Aiken, S.C.	Rural	B	0.49	2.2	6.6	—	Wiener 1979
Sudbury, Ontario, Canada	Rural	W	0.2	—	12	15	Chan et al. 1982
Sudbury, Ontario, Canada	Rural	W	0.5	—	20	—	Beamish and Van Loon 1977
Grassau, F.R.G.	Rural	B	0.31	—	14	25	Thomas 1983
Glen Ellyn, Ill.	Suburban	W	0.15	—	9.5	36	Gatz, et al. 1983
Gainesville, Fla.	Suburban	W	5.7	—	15	28	Hendry and Brezonik 1980
Hof, F.R.G.	Suburban	B	0.68	—	16	93	Thomas 1983
Ochsenkopf, F.R.G.	Urban	B	0.86	—	22	200	Thomas 1983
London, U.K.	Urban	B	18	—	170	—	Harrison et al. 1975
Frankfurt, F.R.G.	Urban	B	3.7	19	250	—	Muller and Bielke 1975

[a]B = bulk precipitation; S = fresh snow; W = wet-only precipitation.
[b]Walker Branch Watershed 1976–1977; Cumberland Plateau and Cherokee Forest 1981–1982.

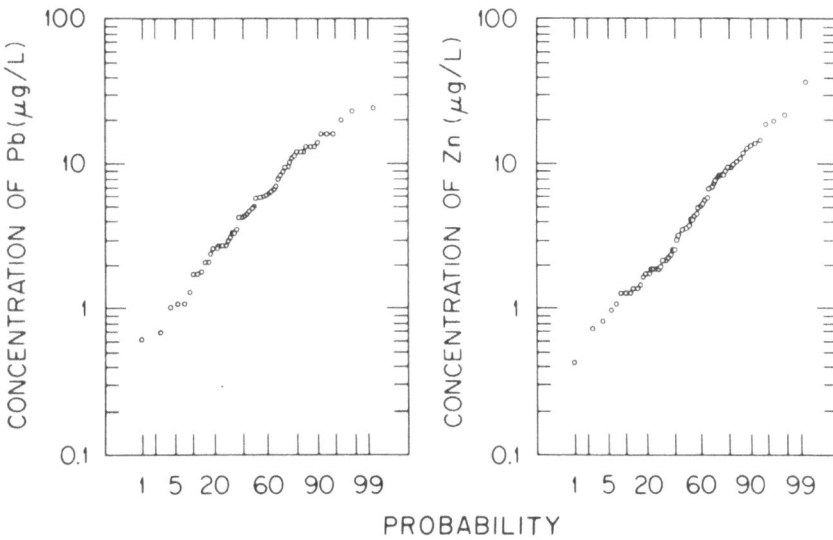

Figure 4.15. Log-normal probability plots of trace metal concentrations in incident precipitation.

tributions have been reported for metals in aerosols but not to our knowledge in precipitation, primarily because of limited data.

The high degree of variability in rain concentrations led us to investigate the controlling factors. There were significant seasonal trends in the level of metals and major ions in Walker Branch Watershed rain, as illustrated in Fig. 4.16 for manganese. With the exception of cadmium, which shows no significant trends, the highest concentrations consistently occur during the warm weather periods. Similar trends were reported elsewhere in the Southeast (Hendry and Brezonik 1980; Wiener 1979), suggesting this to be a regional phenomenon. This trend is particularly apparent for SO_4^{2-}, having occurred at several other eastern U.S. sites, as reflected in monthly weighted mean concentrations for the MAP3S network (Fig. 4.17). The combined effect of elevated concentrations in warm-month rain and the seasonal distribution of precipitation at Walker Branch Watershed resulted in significant wet deposition loading during the forest growing season (Lindberg 1982). Of the total annual wet deposition of these constituents, 35% of the cadmium and 45% of the manganese were deposited in the spring, 40% of the lead, SO_4^{2-}, and H^+ was deposited in summer, and 65% of the zinc was deposited in the fall.

The similarity in temporal patterns for several elements in rain at Walker Branch Watershed and the regional similarity in monthly sulfate concentrations in the eastern United States suggest synoptic conditions as con-

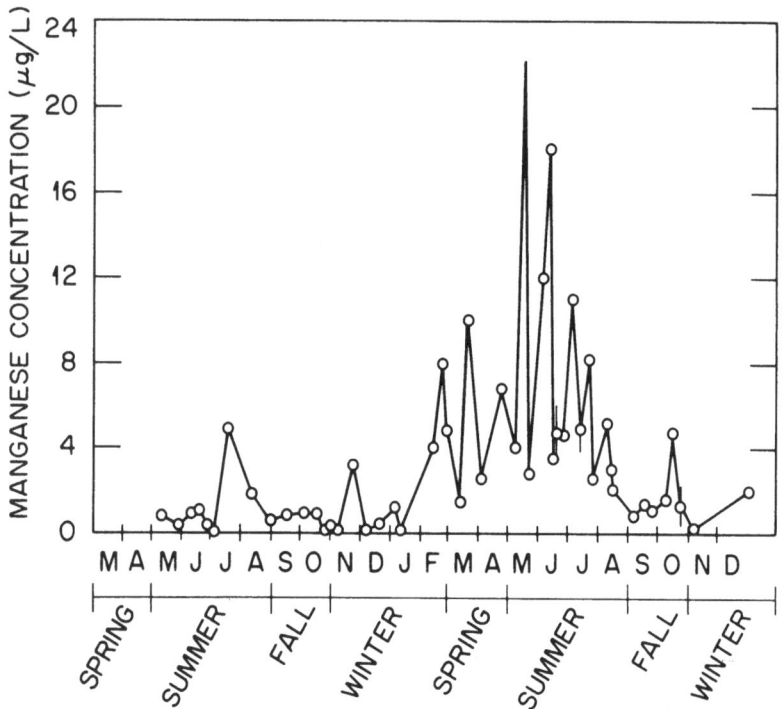

Figure 4.16. Temporal variations in the concentrations of manganese in precipitation over a 20-month period during 1976–1977.

trolling factors. Galloway (1978) speculated that the summer sulfate maximum was related to increased sulfur emission rates combined with higher atmospheric oxidation rates under the elevated temperature and humidity conditions. An alternative, more general explanation is that the summer maxima are related to synoptic meteorologic conditions that result in elevated pollutant concentrations during air stagnation and in generally lower rainfall amounts per event, and hence less dilution of rain-scavenged material.

These dilution phenomena were reflected by significant negative correlation coefficients between element concentrations in rain and the following factors: rainfall amount, storm duration, and intensity. However, rainfall amount appears to exert the most uniform and significant influence, based on correlation analyses. Plots of rainfall amount vs. concentration illustrate the dilution curve as a negative exponential, whereas plots of wet deposition vs. rainfall amount indicate that wetfall input of metals can be predicted using a power law relationship with an exponent of 0.6 (Fig. 4.18) (Lindberg 1982).

To investigate the influence of local power plant emissions on concentrations in rain, we used continuous wind speed and direction data (Lind-

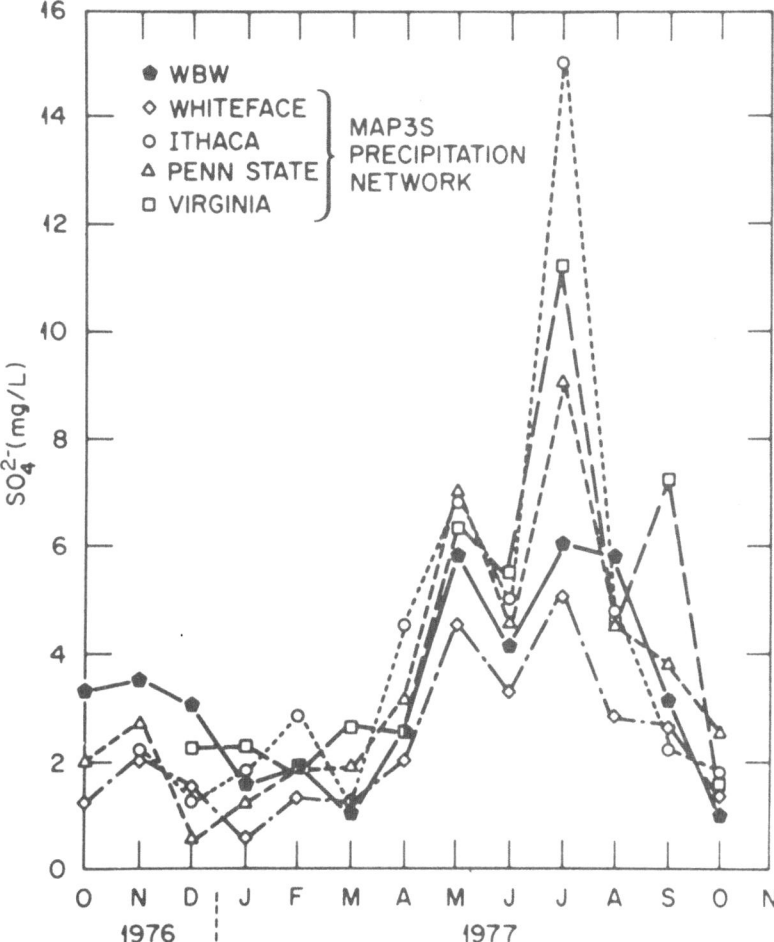

Figure 4.17. Comparison of weighted mean monthly concentrations of sulfate in incident precipitation collected at Walker Branch Watershed (WBW) and at four MAP3S (Multistate Atmospheric Power Production Pollution Study) precipitation chemistry monitoring stations in New York (Whiteface and Ithaca), Pennsylvania, and Virginia. (Source: Adapted from J. N. Galloway. 1978. Sulfur deposition in the eastern United States. Paper presented at the Air Pollution Control Association Middle Atlantic States Regional Meeting, Philadelphia, Pennsylvania, April 13–14, 1978.)

berg et al. 1979) to plot plume trajectories from the three nearby generating stations before and during single-storm events. We selected storms of <1 cm and divided these into several groups similar in rainfall amount but differing in combustion plume presence. The most balanced data set included eight events. There was no significant influence of plume presence

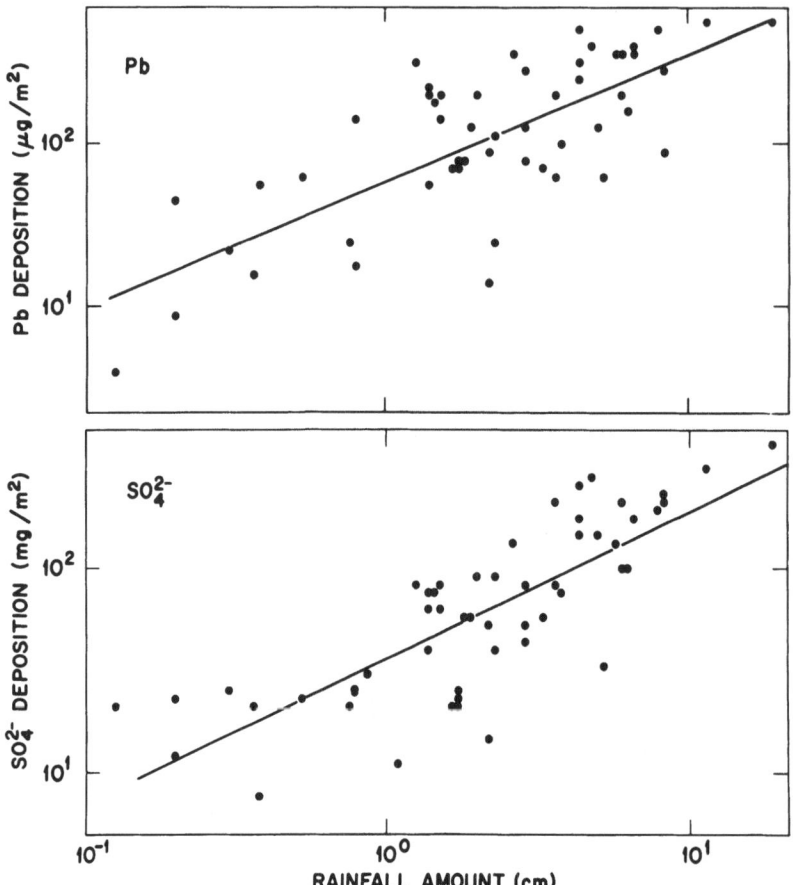

Figure 4.18. The dependence of wet deposition on the rainfall amount per event for lead and SO_4^{2-}, and the best-fit linear regression line for each (determined from log-transformed data).

on precipitation concentrations of any constituent (at $p < 0.05$, t-test of two means), although with this limited data set the differences in mean concentrations would have to be considerable to be significant. The ratios of mean concentrations for plume-influenced events to those not influenced generally fell in the range from 0.7 to 1.2. The ratio for lead (3.9), however, suggested a plume effect, the significance of which was masked by the variability of the data.

The chemistry of rain collected beneath the forest canopy (throughfall) is substantially different from that of the incident precipitation because of ion exchange, foliar leaching, and dry deposition wash-off reactions. To investigate the results of these interactions, we collected throughfall beneath the chestnut oak canopy (*Quercus prinus*) at four sites across Walker Branch Watershed (Fig. 4.1), using the same wet-only samplers described earlier for rain.

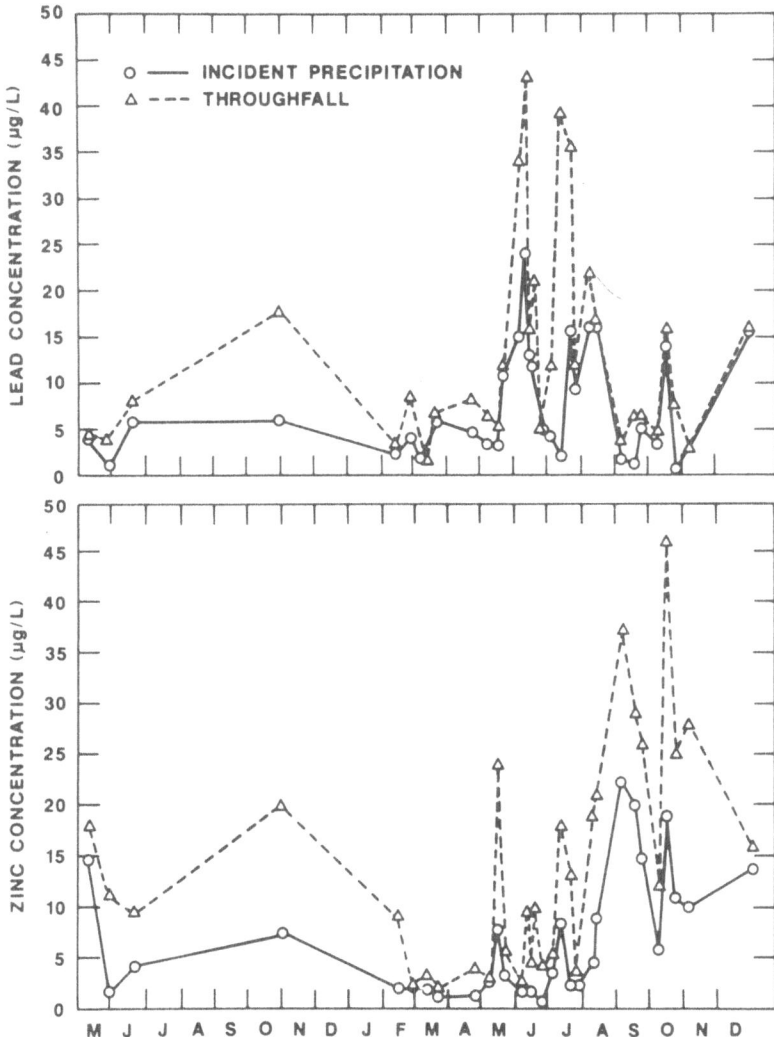

Figure 4.19. Temporal variations in lead and zinc concentrations in rain and throughfall for events sampled simultaneously over a 20-month period. The X axis represents 20 months starting with May (M).

The influence of the canopy on the concentrations of lead and zinc in rain and throughfall is illustrated in Fig. 4.19 for a 19-month period. The interception of incoming rain by the forest canopy resulted in a net increase in the concentrations of Cd, Mn, Pb, and Zn in throughfall.

The mean concentrations of metals in rain and throughfall are summarized in Table 4.12 for 33 well-defined single events sampled simultaneously above and below the canopy. The relative influence of the canopy on the concentrations of each metal is reflected in the through-

Table 4.12. Statistical summary of the relationship between incident precipitation and throughfall concentrations for 33 single-event storms sampled simultaneously above and below the canopy

	Incident precipitation (μg/L)	Combined data, 2 years (mean ± SE)	
		Throughfall (μg/L)	T:I[a]
Cadmium	0.42 ± 0.05	1.57 ± 0.45	4.4 ± 1.2
Manganese	3.40 ± 0.79	211 ± 29	160 ± 40
Lead	7.69 ± 1.14	13.9 ± 2.2	2.7 ± 0.6
Zinc	6.52 ± 6.22	14.2 ± 11.3	2.9 ± 0.3
Precipitation amount, cm	2.59 ± 0.46	2.46 ± 0.43	—

[a]Throughfall-to-incident precipitation ratio.

fall-to-incident precipitation (T:I) ratios. The mean enrichments of concentrations in throughfall relative to incident rainfor 33 events are as follows: Mn, factor of 160; Cd,4; and Zn and Pb, 3.

Enrichment in absolute terms is expressed as the difference in concentration in rain above and below the canopy and is termed net removal or net throughfall. This measure of the influence of the canopy is useful in analyzing factors influencing throughfall chemistry. Analyses of several potential parameters indicated that the most significant were degree of canopy foliation and rainfall intensity (Lindberg et al. 1979). The influence of canopy foliation (leaf status) is illustrated in Fig. 4.20, which shows that during the growing season, the fully foliated canopy releases measurably higher quantities of metals to incoming rain than does the leafless canopy during the dormant season. The leaf area index during the growing season (~5) exceeds the branch area index during winter by a factor of ~3, providing considerably more surface area for the scavenging and subsequent wash-off of dry deposition and the leaching of internal plant pools.

The relationship between rainfall intensity and net throughfall is illustrated in Fig. 4.21 for cadmium and lead. Our data were separated into three classes of rainfall intensity for samples collected during the growing season and the mean net removal calculated for each. Net removal from the canopy decreased significantly ($p < 0.05$) as intensity increased for Cd, Pb, and Mn (not shown), whereas the data for zinc indicated only a general trend in this direction. In each case, for the highest rainfall intensity class, the net removal term was close to zero, indicating little measurable canopy influence on the throughfall chemistry of trace metals for the most intense storms. These observations suggest that the residence time of precipitation on canopy surfaces was the key parameter influencing throughfall

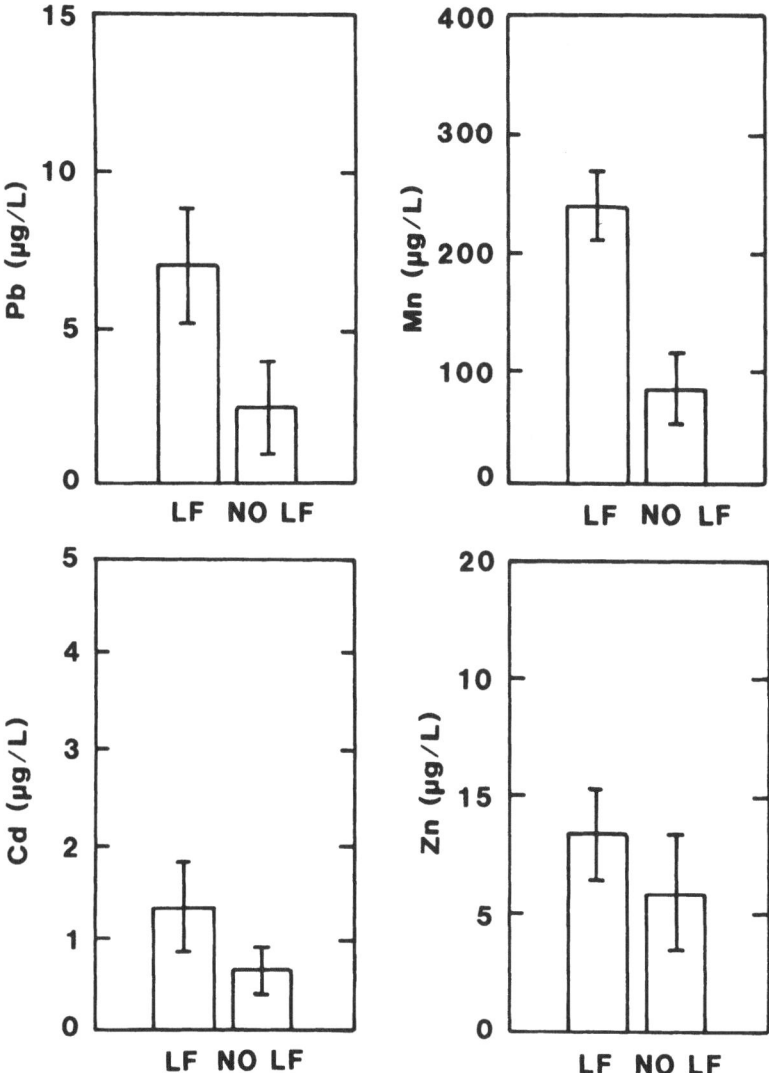

Figure 4.20. Variations in mean (± SE) throughfall net leaching concentrations (throughfall concentration minus incoming rain concentration) of Pb, Mn, Cd, and Zn during two leaf status periods (LF = leaf; NO LF = no leaf).

chemistry. Storms characterized by relatively high intensities would necessarily result in short residence times of the incoming rain on the leaf surface. Given less equilibration time at the leaf surface, the incident rain would be expected to remove less water-soluble material from both internal foliar pools and leaf surfaces, resulting in lower net throughfall concen-

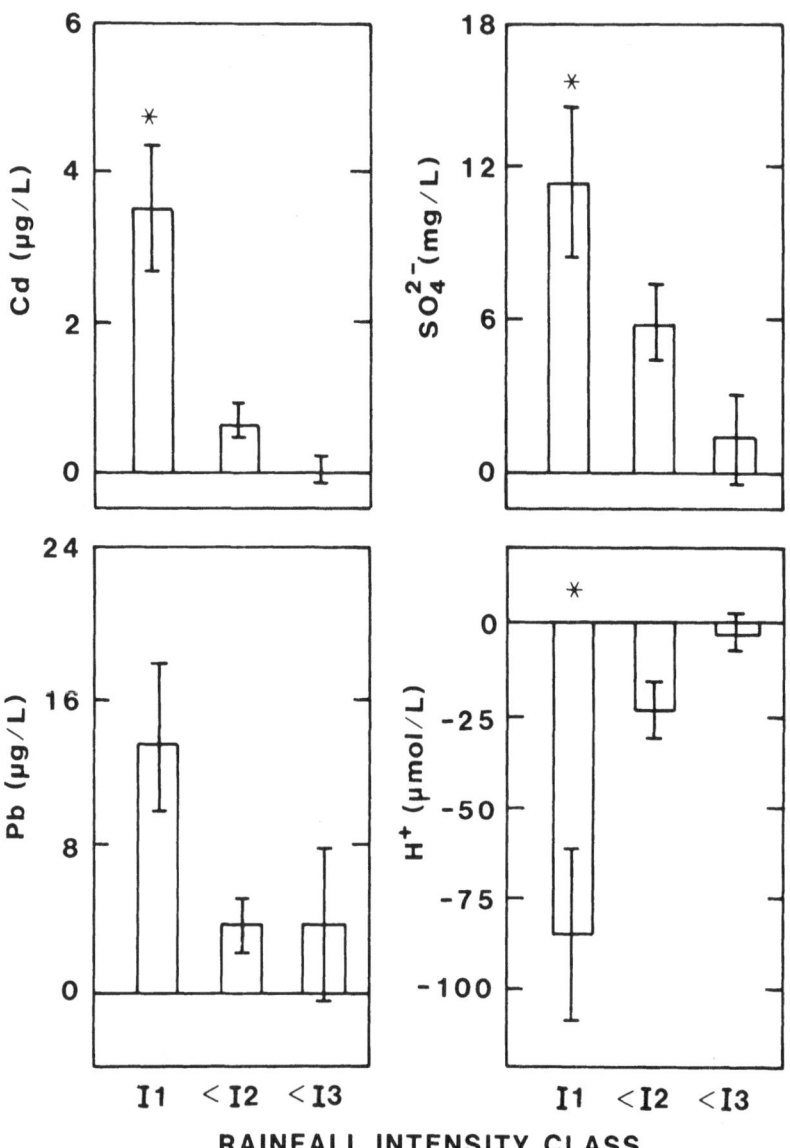

Figure 4.21. Variations in mean (± SE) throughfall net leaching concentrations (throughfall minus rain) of Cd, Pb, SO_4^{2-}, and H^+ for three classes of rainfall intensity (I1 ≤ 1 mm/h, I2 < 5 mm/h, I3 > 5 mm/h). Values marked with an asterisk (*) are significantly different from the associated values ($p < 0.05$).

trations. The importance of each of these sources to net throughfall is discussed below.

4.4.4 Total Atmospheric Input and Implications for Plant Effects

The observations described above were integrated into estimates of the total annual atmospheric input to the forest at Walker Branch Watershed (Lindberg and Harriss 1981). A major question in the interpretation of such data, particularly the estimates of dry deposition, is the most appropriate means of extrapolating point measurements to the full canopy. Because of the spatial extent of the precipitation and throughfall measurement, flux estimates should be considered as representative of ridgetop oak stands at Walker Branch Watershed. Dry deposition measurements for large particles are representative of the same areas but do not reflect the influence of the full canopy, because they were taken using individual inert surfaces in and above the canopy. The extrapolation of such data to the full canopy is widely recognized as a difficult task in dry deposition research (Hicks et al. 1980; Hosker and Lindberg 1982).

In our initial analyses, dry deposition flux was estimated from the mean large-particle dry deposition rates determined from point measurements and extrapolated to the canopy by multiplying these values by suitable leaf area and scavenging indices for several canopy levels during the growing and dormant seasons (Lindberg et al. 1979). Based on subsequent research (Lindberg et al. 1984b; Lovett and Lindberg 1984), we feel that this early approach probably overestimated the large-particle dry deposition flux. Originally, we assumed that nearly all leaves in the canopy collect dry-deposited particles with a similar efficiency; hence, the full-canopy dry deposition rate was ~5 times the individual surface deposition rates (5 is the value for the leaf area index in area of leaves per unit ground area). We now believe that this "extrapolation factor" is on the order of 3 during the growing season (Lovett and Lindberg 1984), as discussed in the methodology section above. The factor for the leafless, dormant season is 1.0. There is another major difference in our revised approach to calculating the dry deposition input of trace metals in this forest. In our earlier estimates we did not include dry deposition of submicron particles in our estimates of total dry deposition. However, from our research on major ions in aerosols and dry deposition, as discussed earlier, we developed appropriate deposition velocity values for estimating small-particle (<2 μm) dry deposition. Hence, the revised metal deposition estimates for Walker Branch Watershed now include this fraction as well. All of these modifications resulted in reductions in total estimated annual input ranging from ~5% (Cd, Zn) to ~30% (Mn). The value for lead is essentially unchanged.

These revised estimates are summarized in Table 4.13 for the growing and dormant seasons and for the full year. The input of these metals at Walker Branch Watershed is strongly influenced by dry deposition, which contributes ~10% (Cd and Zn) to ~50% (Pb) to ~80% (Mn) of the total annual atmospheric flux. The seasonal distribution of atmospheric deposition is such that most (50% for Cd and 70–80% for Pb, Mn, and Zn) of the input occurs during the growing season, when the canopy is fully developed. This is similar to the case for the major ions discussed earlier and is to some extent a result of the high surface area of the canopy when fully foliated.

To put these estimates into the context of an element cycling approach, we determined the relative importance of several pathways of element transport to the forest floor beneath a representative chestnut oak stand for 1year, based on the above data in combination with data on seasonal variations in leaf area index, watershed hydrology, and leaf fall biomass for oak-hickory stands in Walker Branch Watershed (Grizzard et al. 1976; Huff et al. 1977; Lindberg et al. 1979). The components of wet and dry deposition were determined as described above. Wash-off of dry deposition was determined assuming all dry deposition to be ultimately removed from

Table 4.13. Seasonal atmospheric deposition of trace metals to a chestnut oak stand at Walker Branch Watershed by wet and dry processes during 1977

Process	Season[a]	Atmospheric deposition (mg/m^2 per period)			
		Cd	Mn	Pb	Zn
Wet	GS	0.20	2.6	5.0	5.2
Dry (large particle)[b]	GS	0.04	14	3.7	0.81
Dry (small particle)[b]	GS	0.003	0.19	2.8	0.24
Total	GS	0.24	17	12	6.2
Wet	DS	0.23	1.4	2.3	2.4
Dry (large particle)	DS	0.01	3.8	1.0	0.21
Dry (small particle)	DS	0.001	0.04	0.38	0.058
Total	DS	0.24	5.2	3.7	2.7
Wet	Year	0.43	4.0	7.3	7.6
Dry (large particle)	Year	0.05	18	4.7	1.0
Dry (small particle)	Year	0.004	0.23	3.2	0.30
Total	Year	0.48	22	15	8.9

[a]GS = 215-d growing season (April–October), DS = 151-d dormant season.
[b]Large-particle fraction (>2 μm) determined from inert surface samplers, and small-particle fraction (≤2 μm) calculated from airborne particle size and concentration data (see text).

the leaf surface (recall that we are considering the soluble element fractions), whereas foliar leaching (or loss of soluble internal leaf material) was calculated as the difference between the throughfall flux (wet deposition, dry deposition wash-off, and foliar leaching) and the sum of the wet and dry deposition fluxes to the canopy. The total flux to the forest floor is the sum of wet and dry deposition to the upper canopy, foliar leaching, and dry deposition to the soil, but does not include translocation within vegetation to the soil, if this does indeed occur. These values are summarized in Table 4.14.

Wet deposition constituted the largest single fraction of the total annual flux of lead to the forest floor (60%), whereas incident precipitation and foliar leaching contributed nearly equally to the total zinc flux. Foliar leaching was the largest component of both the manganese and the cadmium fluxes (50–60%). The importance of foliar leaching was expected for the micronutrients manganese and zinc, which are mobile in the plant-and-soil system, but was unexpected for cadmium. The ability of the oak trees to rapidly cycle soluble cadmium suggests a mechanism whereby

Table 4.14. Estimated total annual deposition and internal cycling fluxes of metals in an oak stand at Walker Branch Watershed

Deposition process	Annual element flux ($mg \cdot m^{-2} \cdot year^{-1}$)			
	Cd	Mn	Pb	Zn
Dry deposition				
To full canopy	0.04	14	6.5	0.87
To branches and soil[a]	0.01	3.8	1.4	0.45
Wet deposition				
Above canopy	0.43	4.0	7.3	7.6
Below canopy (throughfall)	1.2	150	11	16
Net removal from canopy[b]	0.77	150	3.7	8.4
Foliar leaching or uptake[c]	0.72	130	(−2.8)[d]	7.5
Leaf fall	0.04	90	0.2	5.0
Total flux to canopy[e]	0.47	18	14	8.5
Total flux to forest floor				
Internal flux[f]	0.76	220	0.2	12
External flux[g]	0.48	22	12	8.6

[a] During the dormant season.
[b] (Throughfall) − (wet deposition to the canopy).
[c] (Net removal) − (dry deposition to the canopy).
[d] The negative value suggests possible foliar uptake (see text).
[e] (Wet deposition above canopy) + (dry deposition to full canopy).
[f] (Leaf fall) + (foliar leaching).
[g] (Dry deposition) + (wet deposition) − (foliar uptake, if apparent).

cadmium accumulation may be avoided, and supports earlier data from this site that indicated rapid transport of cadmium through vegetation (VanHook et al. 1977).

The behavior of lead was somewhat different from that of the other elements in that the annual flux of lead to the canopy (wet plus dry deposition, ~ 15 mg·m^{-2}·year^{-1}; Table 4.14) exceeded the flux to the forest floor (~ 12 mg·m^{-2}·year^{-1}). For the other elements, the flux to the canopy represented only 10 to 40% of the flux to the forest floor. If lead in rain passes through the canopy without absorption (the residence time of wet deposition on foliage is much less than that for dry deposition), the remainder of the below-canopy flux, termed net removal (throughfall flux minus precipitation flux), is attributable to dry deposition wash-off plus internal foliar leaching (Heinrichs and Mayer 1980; Nihlgard 1970; Eaton et al. 1973). Net removal of lead was less than the dry deposition flux to the canopy, however, suggesting possible absorption of some fraction of the dry-deposited lead by the leafy canopy. This situation has also been reported for forests in Europe (Heinrichs and Mayer 1980).

Our data indicate that atmospheric deposition plays a significant role in the cycling of metals through the canopy and to the forest floor in a deciduous forest in the southeastern United States. Deposition appears to be of particular importance in the biogeochemical cycle of lead; nearly 100% of the estimated total flux to the forest floor, which includes wet and dry deposition to the canopy, foliar leaching, leaf fall, and dry deposition directly to the forest floor, can be accounted for by atmospheric deposition to the canopy alone. This external flux exceeds the internal flux (leaf fall plus foliar leaching) by a factor of ~ 90. Atmospheric deposition does not dominate the estimated total flux to the forest floor of Cd, Zn, or Mn, but is still significant. The atmosphere contributes $\sim 40\%$ of the cadmium and zinc and $\sim 10\%$ of the manganese flux.

4.4.5 Comparison with Other Data

Total deposition data at Walker Branch Watershed during 1977 is compared with data from a variety of sites characterized according to industrial and population density in Table 4.15. The annual atmospheric deposition rates to Walker Branch Watershed of these elements were similar to, or slightly lower than, those reported in the literature for rural and nonindustrialized locations (Table 4.15). Because many of the studies reported in the literature used samplers that collect both wetfall and dryfall in the same container, there are few directly comparable data on the relative importance of wet deposition and dry deposition to total trace element flux.

Because of the lack of directly comparable trace metal deposition data for other forested sites, we repeated the wet and dry deposition measurements at Walker Branch Watershed during 1981 and 1982, and expanded our studies to include three regional forested watersheds as well

Table 4.15. Total annual atmospheric deposition measured at diverse locations

Location and study	Airshed type	Total atmospheric deposition $(mg \cdot m^{-2} \cdot year^{-1})$			
		Cd	Mn	Pb	Zn
Antarctic (Boutron 1979)	Remote	0.004	—	0.078	0.10
North Atlantic (Buat-Menard and Chesselet 1979)	Remote	—	0.07	3.1	1.3
Greenland (Davidson et al. 1978)	Remote	0.006	0.022	0.06	0.17
Mexico (Hodge et al. 1978)	Isolated (coastal)	0.1	60	17	15
United Kingdom (Cawse 1975)	Isolated	<20	5.9	18	45
Georgia (Wiener 1979)	Rural	0.72	—	8.8	—
Maine (Norton and Kahl 1983)	Rural	—	—	20	15
United Kingdom (Cawse 1974)	Rural	<7	12	35	50
Vermont (Friedland et al. 1984)	Rural	—	—	52	18
Indiana (Peyton et al. 1984)	Rural	0.3	—	15	—
Walker Branch Watershed (this study, 1977)	Rural	0.5	22	15	8.9
Belgium (Navarre et al. 1980)	Rural	1.9	—	26	370
Federal Republic of Germany (Mayer and Ulrich 1982)	Rural	2.0	520	73	170
California (Hodges et al. 1978)	Suburban (coastal)	0.14	10	50	25
United Kingdom (Cawse 1974)	Industrial	4	<10	27	120
Indiana (Peyton et al. 1976)	Industrial	1.5	—	140	—

(Turner et al. 1985). For 12 months we collected wet-only deposition with automatic collectors and dry-deposited particles using inert surfaces, and determined total deposition as described earlier. The watersheds range from 36 to 94 ha in size and from 350 to 750 m in elevation, with similar deciduous vegetation. The locations of these sites relative to Walker Branch Watershed are illustrated in Fig. 4.22.

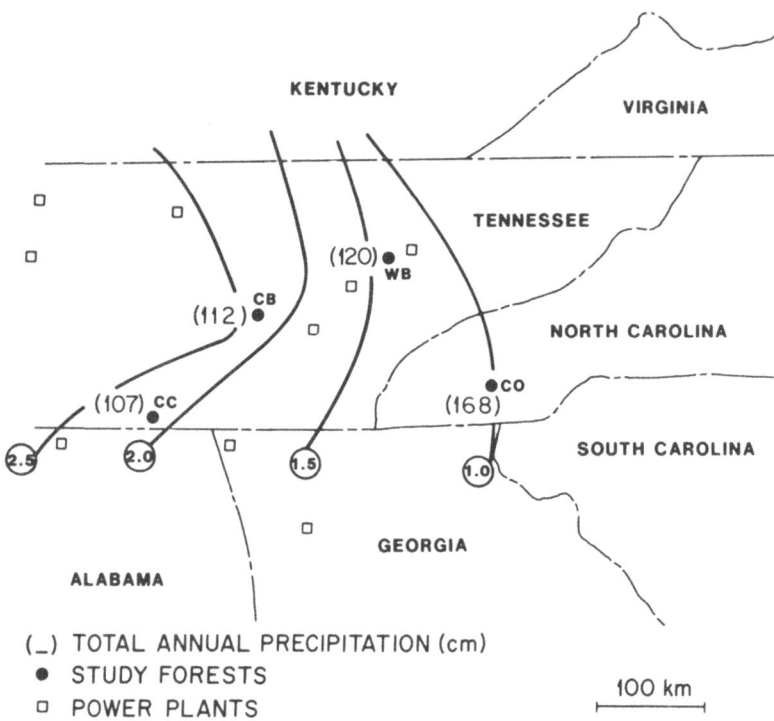

Figure 4.22. Spatial trends in the annual weighted mean concentrations of Cd, Mn, Pb, and Zn in rain. Sites are Walker Branch Watershed (WB); Camp Branch Watershed (CB), Cross Creek watershed (CC) (both operated by the Tennessee Valley Authority); and Coweeta Hydrologic Laboratory (CO) (operated by the U.S. Forest Service).

The general spatial trend in annual weighted mean metal concentrations in rain is shown in Fig. 4.22. Despite the various sources of these metals in rain, there is a notable similarity in the spatial trends, with mean concentrations of each metal generally increasing from east to west. The general path of precipitation systems in this region is from southwest to northeast. The westernmost sites may be more strongly influenced by emissions from the more numerous local power plants in the upwind direction. Trends in overall dry deposition rates (not shown) were different from those for wet input, with the highest mean values occurring at the sites closest to major automobile traffic.

The results of the annual deposition calculations for each watershed are summarized in Table 4.16, which also shows the relative contribution to total input by wet and dry processes. These data indicate that total atmospheric deposition is similar at each site, with deposition varying by

Table 4.16. Total annual atmospheric input (mg/m^2) of metals by wet and dry deposition at four forested watersheds in the southeastern United States during 1981–1982 (see Fig. 4.22)

Site[a]	Cadmium				Manganese				Lead				Zinc			
	Wet	Dry	Total	% Dry	Wet	Dry	Total	% Dry	Wet	Dry	Total	% Dry	Wet	Dry	Total	% Dry
WB	0.099	0.030	0.13	23	2.2	6.2	8.4	74	4.7	2.6	7.3	36	6.3	4.7	11	43
CB	0.14	0.050	0.19	26	3.0	5.8	8.8	65	5.5	2.0	7.5	27	9.5	4.1	14	29
CC	0.19	0.033	0.22	15	2.2	3.7	5.9	63	5.0	2.1	7.1	30	9.8	8.6	18	48
CO	0.079	0.020	0.097	21	1.9	3.6	5.5	65	5.2	1.7	7.5	25	9.1	3.2	12	27

[a] WB = Walker Branch Watershed; CB = Camp Branch watershed; CC = Cross Creek watershed; CO = Coweeta Hydrologic Laboratory.

only 7% for lead and 60% for manganese and zinc, whereas deposition of cadmium varies by a factor of 2. It appears that these fluxes are representative for forests in this region. In contrast, fluxes measured to forests in more industrialized areas, such as those in central Europe, exceed these values by 1 to 2 orders of magnitude (Table 4.15).

Although dry deposition is not commonly measured in forests, we reported that it can contribute significantly to trace metal deposition at Walker Branch Watershed, based on results from our study in 1977 (Table 4.13). The results of this regional study confirm our earlier results. Dry deposition contributes an important fraction of the input of each metal, being most significant for manganese. The mean contribution of particle dry deposition is in the order Mn (67%) > Zn (37%) > Pb (30%) > Cd (21%). This trend is consistent with the size distribution of these elements in atmospheric particles; manganese is dominated by large particles, whereas Pb, Cd, and Zn are primarily associated with submicron aerosols (Fig. 4.11). Large particles are more efficiently removed from the atmosphere by gravitational and inertial forces than are small particles.

4.4.6 Summary and Implications

The potential effects of deposition on the receptor can be illustrated by expressing the wet plus dry atmospheric deposition of soluble elements to an individual leaf relative to the internal pool of the element in the leaf (Lindberg and Harriss 1981). We found that a single dry plus wet event of 2.4-d duration deposited ~150 ng of water-soluble lead to the leaf surface, an amount equal to ~70% of the internal pool of lead in a chestnut oak leaf. The deposited quantity of soluble cadmium represented ~10%, whereas the quantities of manganese and zinc deposited were <1% of the internal pool. Over the growing season, however, the leaf surface was exposed to 1 or 2 orders of magnitude more cadmium and lead, about 4 times as much zinc, but <30% of manganese, relative to the total acid-digestible content of each element in the leaf. This flux is the combined effect of the chronic, cumulative exposure of vegetation to dry deposition and the episodic inundation of the leaf by rain (Lindberg et al. 1979).

The forest canopy is exposed to a wide range of conditions resulting from variable deposition rates and mechanisms. The atmospheric deposition to upper-canopy surfaces in this deciduous forest during wet and dry cycles lasting several days may consist of comparable contributions by wet and dry processes or may be dominated by either. The ratio of wet to dry deposition input, during periods in which we intensively sampled wet and dry deposition, ranged from 0.07 to 26 (mean = 5.0, SD = 7.7). For short-duration, small-amount rain events (storms of <0.5 cm occur with a frequency of ~40%), wet deposition gives rise to a more intense inundation of the vegetation surface with trace metals than does dry deposition. Single-event wet deposition rates exceed dry deposition rates during the intervening periods by factors ranging from 40 to 11,000 (Lindberg et al. 1982b). However, the residence time of dry-deposited material is generally longer than that of precipitation-delivered material. The mean duration of dry periods at this site is ~5 d (range, 0.2–16 d), whereas that for rain events is 0.7 d (range 0.04–2.4 d). This suggests that the leaf contact time for dry deposition is considerably longer than that for wet deposition. The pronounced differences in the temporal character of wet and dry deposition may have important implications for ecological assessment. Exposure dynamics have a strong influence on the effects that gaseous air pollutants may have on plants; however, the data necessary to assess such effects of wet- and dry-deposited metals are not yet available.

Wet and dry deposition interacting with the canopy and with each other over a period of several days also results in a considerable variability in the water-soluble metal concentrations on leaves (Fig. 4.23). Interactions between moisture on vegetation and previously dry-deposited material can result in dissolved metal concentrations at the leaf surface that are considerably higher than those we measured in rain alone. Leaf moisture arising from light showers or intercepted fog is common, has high initial metal concentrations and acidity, wets the leaf but does not necessarily flush it, and is often evaporated before runoff. For example, the events occurring during a 5-d period in the early part of the growing season were used to estimate metal concentrations on leaves due to the interactions of wet and dry deposition. Our measurements resulted in the following estimated metal concentrations (in micrograms per liter) in solution on the leaf surface due to particle dissolution in rain and partial moisture evaporation (Lindberg and McLaughlin 1986): Cd ~13, Zn ~120, Pb ~230, and Mn ~1300. Surface-deposited metals can be absorbed by vegetation, but their physiological effects have not been clearly defined (Krause and Kaiser 1977).

Quantification of these interactions is particularly important in relating precipitation chemistry to vegetation effects observed in the field; the need for both event sampling and segregation of wet and dry deposition is obvious. Figure 4.23 also illustrates the importance of sampling strategy in determining actual field exposure levels during a hypothetical 1-week period (Lindberg et al. 1982a). Shown are the concentrations of lead in solution on a leaf surface in the upper canopy during and following a series of events (A through I), including two rainstorms (D and H) of considerably different rainfall amount (2 and 37 mm), several periods of dry deposition (A, C, and G), one period of intercepted fog (B), and one example of dissolution of dry deposition by rain (E–F). Interactions of moisture with previously deposited particles on the leaf surface can result in extremely high concentrations of soluble material (~ 100–300 μg/L), which exceed concentrations in wetfall alone by factors of approximately 10 to 100. These exposure levels could not be quantified without analysis of dry deposition rates. Another complicating factor involves the choice of collection period

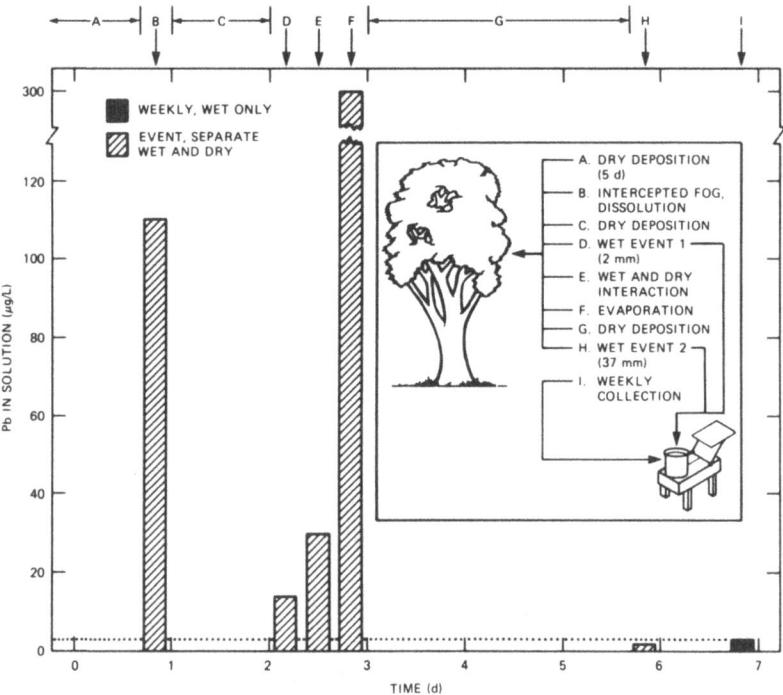

Figure 4.23. The importance of sampling strategy in quantifying pollutant flux to, and concentrations on, canopy surfaces. The ordinate indicates the range in concentrations of lead measured at a leaf surface due to several wet and dry events over the course of a week, as determined by separate collection of wet and dry deposition on an event basis (hatched bars) and by nonevent compositing of wet deposition only over a 7-d period using methods currently employed by several deposition networks [including the National Atmospheric Deposition Program (solid bar)].

for wetfall. If precipitation had been sampled on a weekly basis (I) instead of an event basis during this same period, the occurrence of the larger storm (H) following the 2-mm event (D) would have further diluted the apparent concentrations to which the vegetation was exposed because of the inverse relationship between concentrations of metals in precipitation and rain volume (Lindberg 1982). The effect of weekly wetfall sampling in this situation would be to obliterate the details of the series of events that may or may not have resulted in an observed effect on the vegetation.

The atmospheric deposition of soluble metals results in both long-term and episodic exposure of forest canopy surfaces. Light rain, fog, and dew solubilize particulate trace metals on leaves, enhancing the potential for interaction with internal tissue and possibly increasing the deposition of SO_2 and its associated acidity through the catalytic action of Mn^{2+} (Lindberg 1981). Rainfall removes some fraction of both soluble and particulate metals from the canopy and produces an episodic flux to forest soils. The concentrations and speciation of metals in rain will be modified by interactions in the canopy among rain, particles, and dissolved organic material leached from plant tissue.

The quantification of the interactions between wet and dry deposition and the resulting conditions to which vegetation is exposed require monitoring networks to sample wet and dry deposition separately and on an event basis. Our estimates of the ratios of wet to dry deposition and of atmospheric deposition to the total flux to the forest floor suggest that atmospheric deposition of metals should not be neglected and that dry deposition must be included in future sampling strategies. If the primary sources of atmospheric Pb, Cd, and Zn are anthropogenic (Lantzy and McKenzie 1979), man may be exerting a significant influence on the cycle of these elements in the forest environment.

References

Adams, F., R. Dams, L. Guzman, and J.W. Winchester. 1977. Background aerosol composition on Chacaltaya Mt., Bolivia. Atmos. Environ. 11:629–634.

Albersheim, P. 1965. Biogenesis of the cell wall. pp. 151–188. IN J. Bonner and J. Varner (eds.), Plant Biochemistry. Academic Press, New York.

Andren, A.W., and S.E. Lindberg. 1977. Atmospheric input and origin of selected elements in Walker Branch Watershed, Tennessee. Water Air Soil Pollut. 8:199–215.

Appel, B.R., and Y. Tokiwa. 1981. Atmospheric particulate NO_3 sampling errors due to reactions with particles and gaseous strong acids. Atmos. Environ. 15:1087–1088.

Beamish, R.J., and J.C. Van Loon. 1977. Precipitation loading of acid and heavy metals to a small acid lake near Sudbury. J. Fish. Res. Board Can. 34:649–658.

Boutron, C. 1979. Past and present day fallout fluxes of Pb, Cd, Cn, Zn, and Ag in Antarctica and Greenland. Geophys. Res. Lett. 6:159–162.

Boutron, C., and C. Lorius. 1979. Trace metals in Antarctic snows since 1914. Nature 277:551–554.

Buat-Menard, P., and R. Chesselet. 1979. Variable influence of the atmospheric flux on the trace metal chemistry of area suspended matter. Earth Planet. Sci. Lett. 42:399–411.

Cawse, P.A. 1975. Survey of atmospheric trace elements in the United Kingdom: Results for 1974. R-8038. Atomic Energy Research Establishment, Harwell, England.

Chan, W.H., C.U. Ro, M.A. Lusis, and R.J. Vet. 1982. Impact of INCO smelter emissions on precipitation quality in Sudbury. Atmos. Environ. 16:801–814.

Chow, T., J. Earl, and C. Snyder. 1972. Lead aerosol baseline concentration at White Mountain and Laguna Mountain, California. Science 178:401–402.

Cofer, W.R., D.R. Schnyer, and R.S. Rogswiski. 1981. The oxidation of SO_2 on carbon particles in the presence of O_3, NO_2 and N_2O. Atmos. Environ. 7:1281–1286.

Cronan, C.S., and W.A. Reiners. 1983. Canopy processing of acidic precipitation by coniferous and hardwood forests in New England. Oecologia 59:216–223.

Dams, R., J. Billiet, C. Block, M. Demuynck, and M. Janssens. 1975. Complete chemical analysis of airborne particles. Atmos. Environ. 9:1099–1106.

Davidson, C.I. 1977. Deposition of trace-metal-containing aerosols on smooth, flat surfaces and on wild oat grass. Ph.D. dissertation. California Institute of Technology, Los Angeles, California.

Davidson, C.I., L. Chu, T.C. Grimm, M.A. Nasta, and M.P. Qamoos. 1981. Wet and dry deposition of trace elements onto the Greenland ice sheet. Atmos. Environ. 15:1429–1437.

Davidson, C.I., J.M. Miller, and M.A. Pleskow. 1982. The influence of surface structure on predicted particle dry deposition to natural grass canopies. Water Air Soil Pollut. 18:25–43.

Davidson, C.I., W.D. Goold, and G.B. Wiersma. 1983. Dry deposition of trace elements in Olympic National Park. pp. 871–882. IN H.R. Pruppacher, R.G. Semonin, and W.G.N. Slinn, (eds.), Precipitation Scavenging, Dry Deposition and Resuspension, Vol. 2. Elsevier Science, New York.

Davidson, C.I., S.E. Lindberg, J.A. Schmidt, L.G. Cartwright, and C.R. Landis. 1985. Dry deposition of sulfate onto surrogate surfaces. J. Geophys. Res. 90:2123–2130.

Doveland, H., and A. Eliasson. 1976. Dry deposition on snow surfaces. Atmos. Environ. 10:783–785.

Duce, R.A., G. Hoffman, and W.H. Zoller. 1975. Atmospheric trace metals at remote northern and southern hemisphere sites: Pollution or natural? Science 187:59–61.

Eaton, J.S., G.E. Likens, and F.H. Bormann. 1973. Throughfall and stemflow chemistry in a northern hardwood forest. J. Ecol. 61:495–508.

Eisenbud, M., and T.J. Kneip. 1975. Trace Metals in Urban Aerosols. EPRI-117. Electric Power Research Institute, LaJolla, California.

Elwood, J.W., and G.S. Henderson. 1975. Hydrologic and chemical budgets at Oak Ridge, Tennessee. pp. 31–51. IN A.D. Hasler (ed.), Coupling of Land and Water Systems. Springer-Verlag, New York.

Esman, N.A., and R.B. Fergus 1976. Rain water acidity: pH spectrum of individual drops. Sci. Total Environ. 6:223–226.

Fowler, D. 1980. Removal of S and N compounds from the atmosphere in rain and dry deposition, 1980. pp. 22–32. IN D. Drablos and A.Tollan (eds.), Proc., International Conference on Ecological Impacts of Acid Precipitation. SNSF Project, Sandefjord, Norway.

Friedland, A.J., A.H. Johnson, and T.G. Siccama. 1984. Trace metal content of the forest floor in the Green Mountains of Vermont: Spatial and temporal patterns. Water Air Soil Pollut. 21:14–170.

Galloway, J.N. 1978. Sulfur deposition in the eastern United States. Paper presented at the Air Pollution Control Association Middle Atlantic States Regional Meeting, Philadelphia, Pennsylvania, April 13–14, 1978.

Galloway, J.N., and G.E. Likens. 1978. The collection of precipitation for chemical analysis. Tellus 30:71–82.

Galloway, J.N., and D.M. Whelpdale. 1980. An atmospheric S budget for eastern North America. Atmos. Environ. 14:409–417.

Galloway, J.N., G.E. Likens, and D. Edgerton. 1976. Acid precipitation in the northeastern U.S.—pH and acidity. Science 194:722–723.

Garrells, R., and C.L. Christ. 1965. Solutions, Minerals, and Equilibria. Harper and Row, New York.

Gatz, D.F., B.K. Warner, and L. Chu. 1983. Solubility of metal ions in rain water. pp. 133–151. IN B.B. Hicks (ed.), Deposition, Both Wet and Dry, Vol 4. Acid Precipitation Series. Ann Arbor Science Publishers, Ann Arbor, Michigan.

Gordon, G. 1975. Study of the emissions from major air pollution sources and their atmospheric interactions. Progress Report 75. Department of Chemistry, University of Maryland, College Park.

Grennfelt, P., C. Bengtson, and L. Skarby. 1980. Estimation of the atmospheric input of acidifying substances to a forest ecosystem. pp. 29–40. IN T.C. Hutchinson and M. Havas (eds.), Acid Precipitation Effects on Terrestrial Ecosystems. Plenum Press, New York.

Grizzard, T., G.S. Henderson, E. Clebsch, and D.E. Reichle. 1976, Seasonal nutrient dynamics of foliage and litterfall on Walker Branch Watershed. ORNL/TM-5254. Oak Ridge National Laboratory, OakRidge, Tennessee.

Hammerle, R., and W. Pierson. 1975. Sources and elemental composition of aerosols in Pasadena by energy dispersive X-ray fluorescence. Environ. Sci. Technol. 9:1058–1068.

Harrison, P.R., K. Rahn, R. Dams, J. Robbins, J. Winchester, S. Brar, and D. Nelson. 1971. Area-wide trace metal concentrations measured by multielement neutron activation analysis: A one-day study in northwest Indiana. J. Air Pollut. Control Assoc. 21:563–568.

Harrison, P.R., R. Perry, and R. Wellings. 1975. Lead and cadmium in precipitation—their contribution to pollution. J. Air Pollut. Control Assoc. 25:627–630.

Heinrichs, H., and R. Mayer. 1980. The role of forest vegetation in the biogeochemical cycling of metals. J. Environ. Qual. 9:111–118.

Henderson, G.S., A. Hunley, and W. Selvidge. 1977. Nutrient discharge from Walker Branch Watershed. pp. 307–320. IN D.L. Connell (ed.), Watershed Research in Eastern North America. Smithsonian Institution Press, Washington, D.C.

Hendry, C.D., and P.L. Brezonik. 1980. Chemistry of precipitation at Gainesville, Florida. Environ. Sci. Technol. 14:843–849.

Henry, W.M., and K.T. Knapp. 1980. Compound forms of fossil fuel fly ash emissions. Environ. Sci. Technol. 14:250–256.

Hicks, B.B., M.L. Weseley, and J.L. Durham 1980. Critique of methods to measure dry deposition: Workshop summary. EPA 600/9-80-050. U.S. Environmental Protection Agency, Environmental Sciences Research Laboratory, Research Triangle Park, North Carolina.

Hicks, B.B., D.R. Matt, R.T. McMillan, J.D. Womack, and R.H. Sketler. 1983. Eddy fluxes of nitrogen oxides to a deciduous forest in complex terrain. Paper presented at the Air Pollution Control Association Acid Deposition Conference, Hartford, Connecticut, October 1983. National Oceanic and Atmospheric Administration, Atmospheric Testing and Development Laboratory, Oak Ridge, Tennessee.

Hodge, V., S.R. Johnson, and E.D. Goldberg. 1978. Influence of atmospherically transported aerosols on surface ocean water composition. Geochem. J. 12:7–20.

Hoffman, W.A., Jr., S.E. Lindberg, and R.R. Turner. 1980a. Precipitation acidity: The role of the forest canopy in acid exchange. J. Environ. Qual. 9:95–100.

Hoffman, W.A., S.E. Lindberg, and R.R. Turner. 1980b. Some observations of organic constituents in rain above and below a forest canopy. Environ. Sci. Technol. 14:999–1002.

Hosker, R.P., and S.E. Lindberg. 1982. Review: Atmospheric deposition and plant assimilation of particles and gases. Atmos. Environ. 16:889–910.

Huebert, B.J. 1983. Measurements of the dry deposition flux of HNO_3 to grasslands and forests. pp. 785–794. IN H.R. Pruppacher, R.G. Semonin, and W.G.N. Slinn (eds.), Precipitation Scavenging, Dry-Deposition, and Resuspension. Elsevier, New York.

Huff, D.D., R.J. Luxmoore, and C. Begovich. 1977. TEHM: A Terrestrial Ecosystem Hydrology Model. ORNL/NSL/EATC-27. Oak Ridge National Laboratory, Oak Ridge, Tennessee.

Johannes, A.H., E.R. Altwicker, and N.L. Clesceri. 1981. Characterization of acidic precipitation in the Adirondack region. EPRIEA-1826. Electric Power Research Institute, Palo Alto, California.

Johnson, A.H., and T.G. Siccama. 1983. Acid deposition and forest decline. Environ. Sci. Technol. 17:294A–305A.

Johnson, D.W., G.A. Henderson, D.D. Huff, S.E. Lindberg, D.D. Richter, D.S. Shriner, D.E. Todd, and J. Turner. 1982. Cycling of organic and inorganic S in a chestnut oak forest. Oecologia 54:141–148.

Kelly, J.M., and J.F. Meagher. 1986. Nitrogen input/output relationships for three sites in eastern Tennessee. RP-1727. Electric Power Research Institute, Palo Alto, California.

Keyser, T.R., F.S. Natusch, C.A. Evans, and R.W. Hinton. 1978. Characterizing the surface of environmental particles. Environ. Sci. Technol. 12:768–773.

King, R., J.S. Fordyce, A.C. Leibeck, H. Neustadter, and G. Sidik. 1976. Extensive one-year survey of trace elements and compounds in airborne suspended particulate matter in Cleveland, Ohio. NASATND-8110. National Aeronautics and Space Administration, Washington, D.C.

Krause, G.H.M., and H. Kaiser. 1977. Plant response to heavy metals and SO_2. Environ. Pollut. 12:63–71.

Kretzschmar, J., I. Delespaul, T. DeRyck, and G. Verdruyn. 1977. Belgian network for determination of heavy metals. Atmos. Environ. 11:263–271.

Lantzy, R.J., and F.T. McKenzie. 1979. Atmospheric trace metals: Global cycles and assessment of man's impact. Geochim. Cosmochim. Acta 43:511–525.

Lawson, D.R., and J.W. Winchester. 1978. Sulfur and crystal reference elements in nonurban aerosols from Squaw Mt., Colorado. Environ. Sci. Technol. 12:716–721.

Legg, B.J., and R.I. Price. 1980. The contribution of sedimentation to aerosol deposition on vegetation with a large surface area. Atmos. Environ. 14:305–309.

Likens, G.E., F.H. Bormann, R.S. Pierce, J.S. Eaton, and N.M. Johnson. 1977. Biogeochemistry of a Forested Ecosystem. Springer-Verlag, New York.

Lindberg, S.E. 1981. The relationship between manganese and sulfate ions in rain. Atmos. Environ. 15:1749–1753.

Lindberg, S.E. 1982. Factors in influencing the concentrations of metals, sulfate, and H$^+$ in precipitation Atmos. Environ. 16:1701–1709.

Lindberg, S.E., and R.C. Harriss. 1980. Trace metal solubility in aerosols produced by coal combustion. IN J. Singh and A. Deepak (eds.), Environmental and Climatic Impact of Coal Utilization. pp. 589–608. Academic Press, New York.

Lindberg, S.E., and R.C. Harriss. 1981. The role of atmospheric deposition in an eastern U.S. deciduous forest. Water Air Soil Pollut. 16:13–31.

Lindberg, S.E., and R.C. Harriss. 1983. Water and acid soluble trace metals in atmospheric particles. J. Geophys. Res. 88:5091–5100.

Lindberg, S.E., and J.M. Kelly. 1984. Atmospheric deposition of sulfur to forested watersheds differing in proximity to local coal combustion emission. pp. A1–A9. IN S.E. Lindberg, G.M. Lovett, and J.M. Coe (eds.), Acid Deposition/Forest Canopy Interactions, Final Report for Project RP-1907-1 to the Electric Power Research Institute. EPRI, Palo Alto, California.

Lindberg, S.E., and G.M. Lovett. 1983. Application of surrogate surface and leaf extraction methods to estimation of dry deposition to plant canopies. pp. 837–848. IN H.R. Pruppacher, R.G. Semonin, and W.G.N. Slinn (eds.), Precipitation Scavenging, Dry Deposition and Resuspension, Vol. 2. Elsevier, New York.

Lindberg, S.E., and G.M. Lovett. 1985. Field measurements of particle dry deposition rates to foliage and inert surfaces in a forest canopy. Environ. Sci. Technol. 19:238–244.

Lindberg, S.E., and S.B. McLaughlin. 1986. Current problems and future research needs in the acquisition, interpretation, and application of data in terrestrial vegetation-air pollutant interaction studies. pp. 449–504. IN A. Legge and S.V. Krupa (eds.), Air Pollutants and Their Effects on the Terrestrial Ecosystem. John Wiley and Sons, New York.

Lindberg, S.E., R.C. Harriss, R.R. Turner, D.S. Shriner, and D.D. Huff. 1979. Mechanisms and rates of atmospheric deposition of selected trace elements and sulfate to a deciduous forest watershed. ORNL/TM-6674. Oak Ridge National Laboratory, Oak Ridge, Tennessee.

Lindberg, S.E., D.S. Shriner, and W.A. Hoffman, Jr. 1982a. The interaction of wet and dry deposition with the forest canopy. pp. 385–409. IN Acid Precipitation: Effects on Ecological Systems. Ann Arbor Science Publishers, Ann Arbor, Michigan.

Lindberg, S.E., R.C. Harriss, and R.R. Turner. 1982b. Atmospheric deposition of metals to forest vegetation. Science 215:1609–1611.

Lindberg, S.E., J.M. Coe, and W.A. Hoffman. 1984a. Dissociation of weak acids during Gran plot free acidity titrations. Tellus 36B:186–191.

Lindberg, S.E., G.M. Lovett, and J.M. Coe. 1984b. Acid Deposition/Forest Canopy Interactions. Final Report for Project RP-1907-1 to the Electric Power Research Institute. EPRI, Palo Alto, California.

Lindberg, S.E., G.M. Lovett, E.A. Bondietti, and C.I. Davidson. 1984c. Recent field studies of dry deposition to surfaces in plant canopies. Proc., Air Pollution Control Association 6(108.5):1–15. APCA, Pittsburgh, Pennsylvania.

Lindberg, S.E., G.M. Lovett, D.D. Richter, and D.W. Johnson. 1986. Atmospheric deposition and canopy interactions of major ions in a forest. Science 231:141–145.

Lovett, G.M., and S.E. Lindberg. 1984. Dry deposition and canopy exchange in a mixed oak forest determined by analysis of throughfall. J. Appl. Ecol. 21:1013–1027.

Lovett, G.M., S.E. Lindberg, D.D. Richter, and D.W. Johnson. 1985. The effect of acidic deposition on cation leaching from a deciduous forest canopy. Can. J. For. Res. 15:1055–1060.

Mart, L. 1983. Seasonal variations of Cd, Pb, Cn, and Ni levels in snow from the eastern Arctic Ocean. Tellus 35:131–141.

Mayer, R., and B. Ulrich. 1982. Calculation of deposition rates from the flux balance. pp. 195–200. IN H.W. Georgii and J. Pankrath (eds.), Deposition of Atmospheric Pollutants. Reidel, New York.

Meszaros, E. 1968. On the size distribution of water soluble particles in the atmosphere. Tellus 20:443–448.

Miller, H.G., and J.D. Miller. 1980. Collection and retention of atmospheric pollutants by vegetation. pp. 33–40. IN D. Drablos and A. Tollan (eds.), Ecological Impact of Acid Precipitation. SNSF Project, Sandefjord, Norway.

Mudd, J.B., and T. Kozlowski. 1975. Response of Plants to Air Pollutants. Academic Press, New York.

Müller, J., and S. Bielke. 1975. Wet removal of heavy metals from the atmosphere. pp. 987–999. IN International Conference on Heavy Metals in the Environment, Symposium Proceedings. University of Toronto, Canada.

Multistate Atmospheric Power Production Pollution Study (1980. Quality Assurance Manual. Battelle Northwest Laboratory, Richland, Washington.

National Academy of Sciences (NAS). 1980. Lead in the human environment. NAS, Washington, D.C.

National Academy of Sciences (NAS). 1983. Acid Deposition–Atmospheric Processes in Eastern North America. National Academy Press, Washington, D.C.

National Atmospheric Deposition Program (NADP). 1982. Site Operation Instruction Manual. NADP, Colorado State University, Fort Collins, Colorado.

National Atmospheric Deposition Program (NADP). 1985. Precipitation Chemistry in the United States—1982. S. Lindberg, D. Bigelow, V. Bowerson, W. Knapp, and T. Olsen, (eds.). NADP, Colorado State University, Fort Collins, Colorado.

Navarre, J.-L., C. Ronnean, and P. Priest. 1980. Deposition of heavy elements on Belgian agricultural soils. Water Air Soil Pollut. 14:207–213.

Nihlgard, B. 1970. Precipitation, its chemical composition and effect on soil water in a beech and spruce forest in Sweden. Oikos 21:208–217.

Office of Technology Assessment (OTA). 1984. Acid rain and transported air pollutants. Congress of the United States OTA Report, Washington, D.C.

Pack, D.H. 1980. Precipitation chemistry patterns: A two-network data set. Science 208:1143–1145.

Parkhurst, W.J., T.L. Crawford, N.T. Lee, and J.D. Lokey. 1984. The effect of large-scale utility SO_2 emissions control program on regional sulfur measurements. pp. 1–17 (Paper 84-21.2). IN Proc., Annual Meeting of the Air Pollution Control Association, San Francisco, California, June 25–29, 1984. APCA, Pittsburgh, Pennsylvania.

Peyton, T., A. McIntosh, V. Anderson, and K. Yost. 1976. Areal input of metals into an aquatic ecosystem. Water Air Soil Pollut. 5:443–451.

Richter, D.D., D.W. Johnson, and D.E. Todd. 1983. Atmospheric sulfur deposition, neutralization, and ion leaching in two forest ecosystems. J. Environ. Qual. 12:263–270.

Sehmel, G.A., and W.H. Hodgson. 1974. Particle and gaseous removal in the atmosphere by dry deposition. BNWL-SA-4941. Battelle Pacific Northwest Laboratory, Richland, Washington.

Shannon, J.D. 1981. A model of regional long-term average S atmospheric pollution, surface removal, and net horizontal flux. Atmos. Environ. 15:1155–1163.

Servant, J. 1976. Deposition of atmospheric lead particles to natural surfaces in field experiments. pp. 87–95. IN R. Engelman and G. Sehmel (eds.), Atmosphere-Surface Exchange of Particulate and Gaseous Pollutants (1974). ERDA Symposium Series 38, CONF-740921. National Technical Information Service, Springfield, Virginia.

Streumpler, A.W. 1975. Trace element composition in atmospheric particles during 1973 and summer of 1974 at Chadron, Nebraska. Environ. Sci. Technol. 9:1164–1168.

Streumpler, A.W. 1976. Trace metals in rain and snow during 1973 at Chadron, Nebraska. Atmos. Environ. 10:33–37.

Swank, W.T., and G.S. Henderson. 1976. Atmospheric input of cations and anions to forest ecosystems in North Carolina and Tennessee. Water Resour. Res. 12:541–546.

Taylor, G.E., and D.T. Tingey. 1983. SO_2 flux into leaves of geranium. Plant Physiol. 72:237–244.

Tennessee Valley Authority (TVA). 1982. Tennessee Valley Authority Monitoring Section, Air Resources Program, Ambient Air Quality Monitoring System Data Summary. TVA/ONR/ARP-82/18. TVA, Muscle Shoals, Alabama.

Thomas, W. 1983. The atmospheric emission of heavy metals at 14 stations in Bavaria, Federal Republic of Germany. pp. 144–147. IN Proc., International Conference on Heavy Metals in the Environment. CEP Consultants, Edinburgh, Scotland.

Thornton, J.D., and S.J. Eisenreich. 1982. Impact of land-use on the acid and trace metal composition of precipitation in the north-central U.S. Atmos. Environ. 16:1945–1955.

Tjepkema, J.D., R.J. Cartica, and H.F. Hemond. 1981. Atmospheric concentration of NH_3 in Massachusetts and deposition on vegetation. Nature 294:445–446.

Turner, R.S., 1983. Biogeochemistry of trace elements in the McDonalds Branch watershed, New Jersey Pine Barrens. Ph.D. Thesis. University of Pennsylvania, Philadelphia.

Turner, R.R., S.E. Lindberg, and J.M. Coe. 1985. Comparative analysis of trace metals in forest ecosystems. pp. 356–358. IN Proc., International Conference on Heavy Metals in the Environment. CEP Consultants, Edinburgh, Scotland.

Van Hook, R.I., W.F. Harris, and G.S. Henderson. 1977. Cd, Pb and Zn distributions and cycling in a forested watershed. Ambio 6:281–286.

Weseley, M.L., D.R. Cook, and L.R. Hart. 1983. Fluxes of gases and particles above a deciduous forest in winter. Boundary Layer Meteorol. 27:237–255.

Wiener, J.G. 1979. Aerial inputs of Cd, Cu, Pb and Mn into a freshwater pond in the vicinity of a coal-fired power plant. Water Air Soil Pollut. 12:343–353.

Winchester, J.W., D.L. Meinert, J.W. Nelson, T.B. Johansson, R.E. Van Grieken, C. Orsini, and R. Akselsson. 1974. Trace metals in the St. Louis aerosol. IN Proc., Second International Conference on Nuclear Methods in Environmental Research. ERDA Symposium Series. CONF-740701. National Technical Information Service, Springfield, Virginia.

Winkler, P. 1980. Observations on acidity in continental and in marine atmospheric aerosols and in precipitation. J. Geophys. Res. 85:4481–4486.

Zoller, S.H., E.S. Gladney, and R.H. Duce. 1974. Atmospheric concentrations and sources of trace metals at the South Pole. Science 183:198–200.

Chapter 5
Water

R.J. Luxmoore and D.D. Huff

5.1 Introduction

Water movement in Walker Branch Watershed is the major mass movement each year, and this flux has contributed to the formation of the landscape through erosion of solid materials over millenia. Dissolution and transport of solutes are a major component of weathering and mineral cycling processes. Besides these hydrologic processes, water is an essential component of biological systems, and the movement of water through vegetation (transpiration) seems to be a natural consequence of plant morphology. A wide range of physical, chemical, and biological phenomena are dependent on water, and the investigation of these processes on Walker Branch Watershed has been an important focus of research since the inception of the project.

5.2 Hydrologic Features

Walker Branch is a first- and second-order stream that drains a 97.5-ha forested landscape consisting of two subwatersheds, the West Fork being 38.4 ha and the East Fork 59.1 ha (Curlin and Nelson 1968). The catchment basin is bounded on the north side by Chestnut Ridge, which reaches an elevation of 350 m. The surface slopes southward and drops to an elevation of 265 m at the confluence of the two subwatersheds. A high proportion of the slopes exceed 0.3 m/m (Table 5.1), and the steepest topography is found on the middle to lower slopes adjacent to the stream channels (Fig. 2.2.).

Table 5.1. Area distribution of land slope classes on Walker Branch Watershed

Slope class (m/m)	Area (ha)		Percent area	
	West	East	West	East
0.00–0.05	0.2	0.9	0.6	1.5
0.05–0.12	4.1	6.3	10.8	10.7
0.12–0.2	11.0	11.7	28.7	19.8
0.2–0.3	6.8	11.9	17.7	20.2
0.3 +	16.2	28.3	42.2	47.8

The profiles for the West Fork (Fig. 5.1) and East Fork (Fig. 5.2) and their tributaries show the West Fork to have a shorter and steeper stream system. The stream bed of the East Fork extends over 940 m, with the limit of perennial flow being 750 m upstream from the weir. When the East Fork streamflow is <0.017 m^3/s, flow is subterranean along two reaches. The stream goes underground ~435m from the weir and emerges 265 m from the weir at spring S1E (Fig. 5.3). A second underground reach occurs between 220 and 30 m above the weir. In contrast, the West Fork is a perennial stream that flows for 365 m on exposed bedrock.

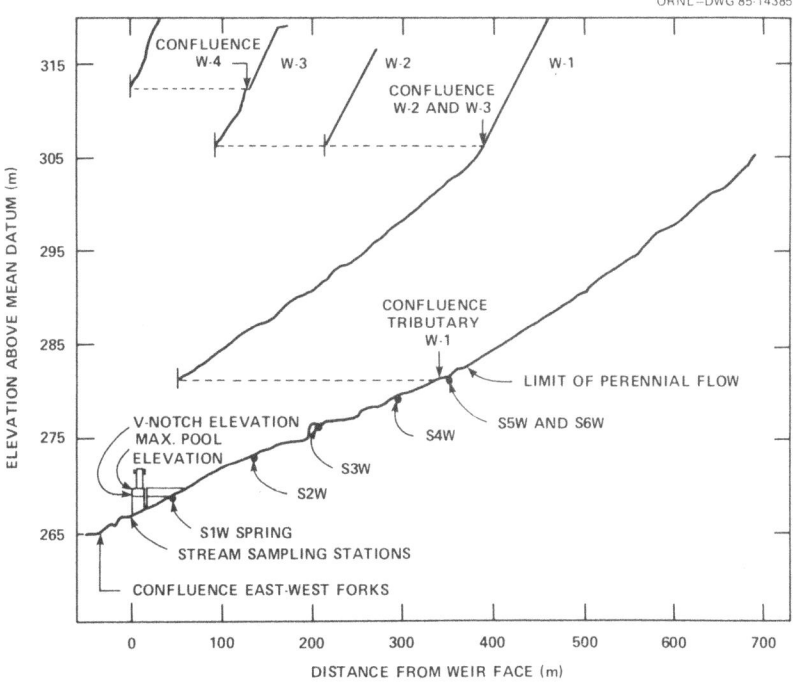

Figure 5.1. Profile of West Fork stream channel and tributary system.

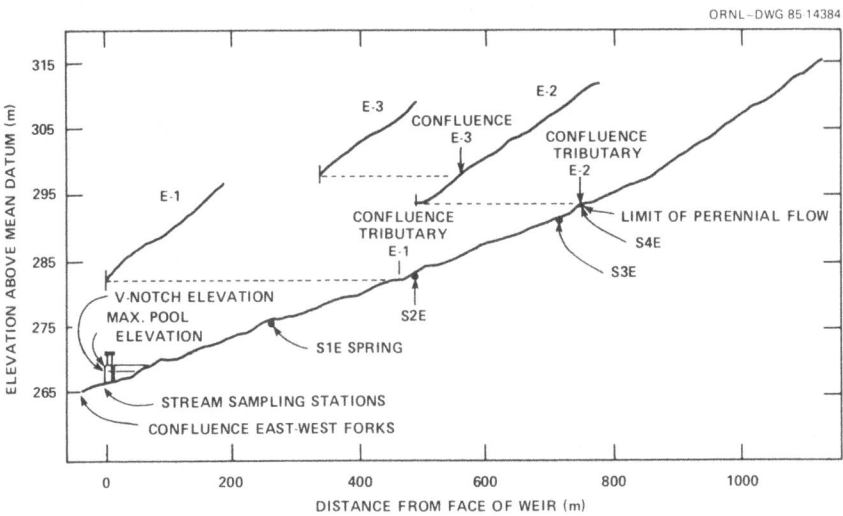

Figure 5.2. Profile of East Fork stream channel and tributary system.

The base flow for both forks is sustained by springs. Two primary springs (S2W, S3W) supply continuous flow into the West Fork, which is supplemented during wet periods by four other intermittent springs and seeps. The East Fork is also fed by two primary springs (S4E, S5E) and two intermittent ones. Numerous minor seeps occur along the stream channels during periods of high rainfall.

5.3 Hydrologic Monitoring

Precipitation was measured at five locations around and within Walker Branch Watershed (Fig. 2.2), using Fischer and Porter model 1548 punched-tape weighing recorders (Fig. 5.4). The data were collected at 5-min intervals with a recorder sensitivity of 2.5 mm. The punched-tape system was operational from 1969 to 1981. Subsequent monitoring has been conducted only at gage stations 1 and 3 with an 8-d weighing-bucket strip chart recorder. The strip charts are read on a digitizing board to a sensitivity of 0.25 mm, giving a 1–order of magnitude increase in sensitivity over the earlier punched-tape system. The digitized data are stored in a data base management system (Jackson 1982) with a convenient report generation capability. All precipitation records for Walker Branch Watershed are available from Oak Ridge National Laboratory.

The two subcatchments of Walker Branch Watershed are monitored for streamflow with 120° V-notch weirs. Streamflow up to 1.18 m³/s can be measured with the V-notch weirs, and higher flows up to 1.86 m³/s,

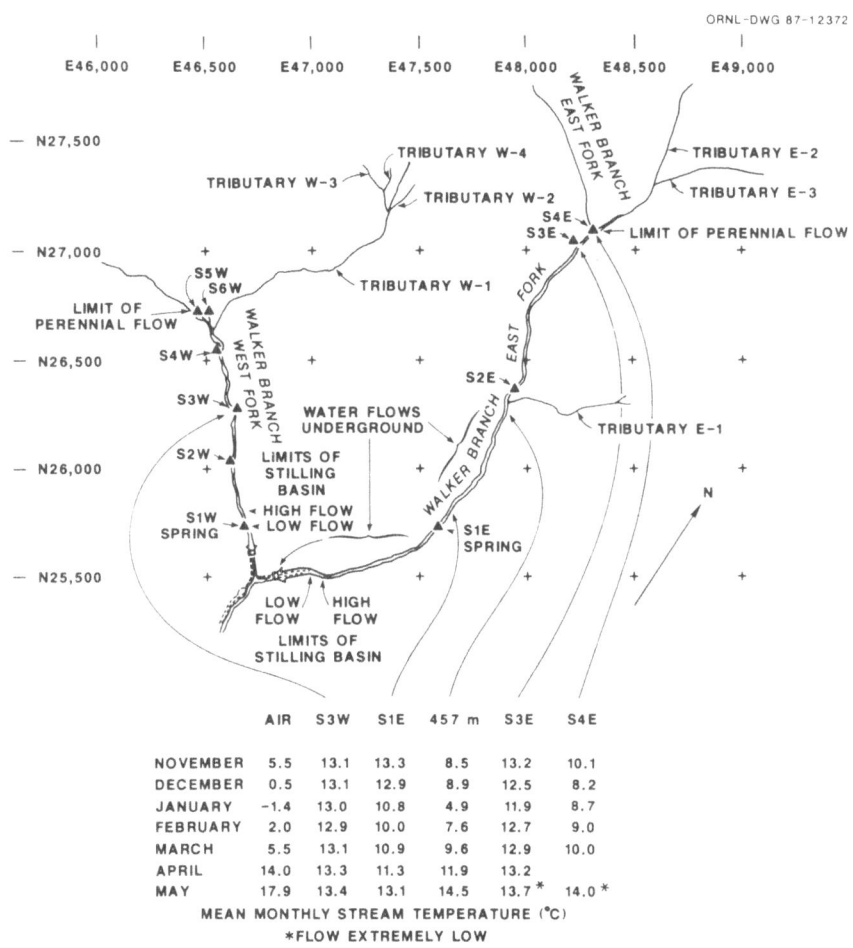

Figure 5.3. Stream channel characteristics and spring locations on Walker Branch.

are monitored with a sharp-crested, rectangular cross section above the V-notch weir (Fig. 5.5). The stream stage heights are monitored with Fisher and Porter model 1542 punched-tape water-level recorders at 5-min intervals with a resolution of 0.3 mm. The conversion of stage height to streamflow is calculated by means of regression equations, and the measurement uncertainty is $\pm5\%$ at low flows (<0.004 m^3/s) and $\pm0.5\%$ at flows >0.125 m^3/s. Flow rates seldom exceed 0.7 m^3/s. The streamflow-monitoring system has recently been changed to a computer-controlled data logger system.

There are a number of springs on the watershed, most having small flows, which have not been permanently monitored. There is one primary spring on the West Fork, which was instrumented for several years with

Figure 5.4. Typical precipitation gage installation on a ridgetop clearing on Walker Branch Watershed.

Figure 5.5. Weir and weir-house on the west subcatchment of Walker Branch Watershed.

Figure 5.6. Parshall flume and water-level recorder installed at spring S3W on the West Fork of Walker Branch.

a 15-cm Parshall flume and Fischer and Porter model 1542 punched-tape water-level recorder similar to those used at the weir sites (Fig. 5.6). Springs contribute a high proportion to the base flow of the West Fork.

5.4 Annual Water Budget

The annual rain gage and stream gage records for the 15-year period 1969–1983 provide an overview of the water budget for the watershed (Fig. 5.7) and illustrate some of the variability in annual hydrologic cycles. The 15-

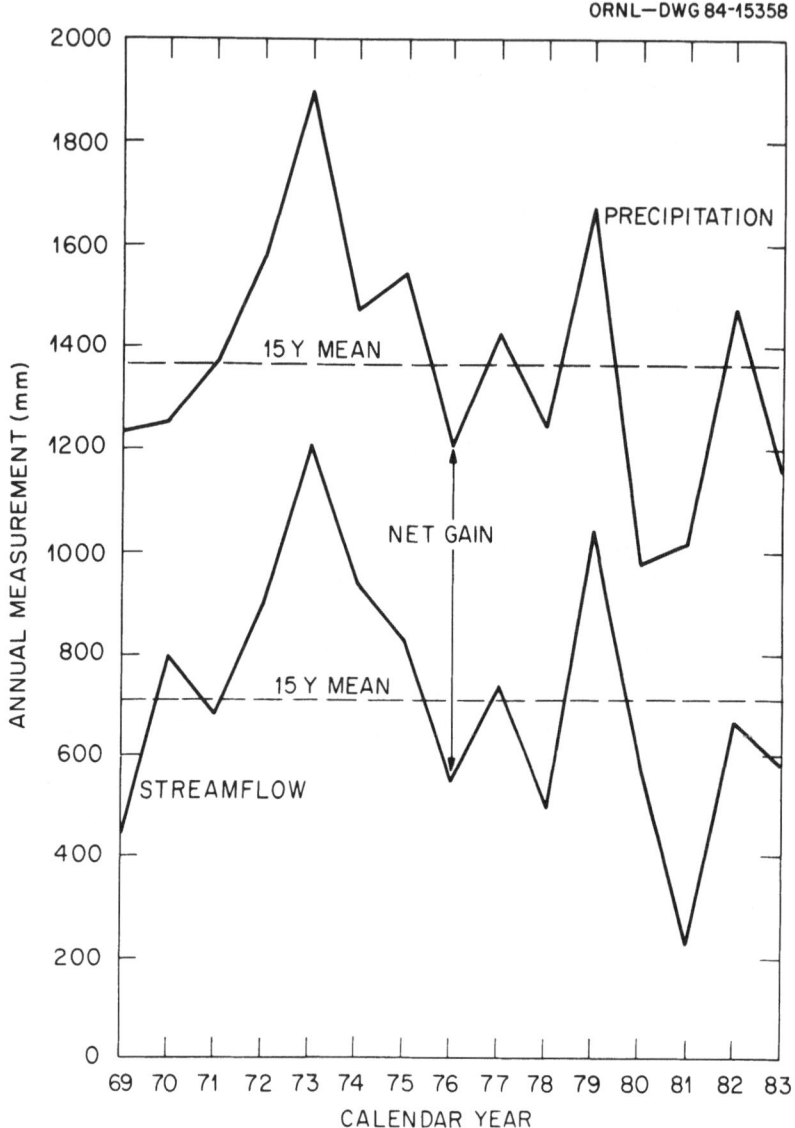

Figure 5.7. Annual precipitation and streamflow for Walker Branch Watershed.

year precipitation mean is 1368mm with a coefficient of determination of 18.3%. The equivalent values for streamflow are 713 mm and 35%, respectively. The difference between precipitation and streamflow is called net gain, which has a 15-year mean value of 655 mm/year and a coefficient of determination of 19%. On an annual basis, net gain is closely related to the loss of water vapor from the landscape through evapotranspiration. This is based on the assumption that the annual net groundwater exchange for the watershed is zero and that initial soil water and groundwater storage levels are unchanged at the end of the annual cycle. The mean estimate of net gain of 655 mm/year is 10% lower than the 730 mm/year value estimated by McMaster (1967) for the Oak Ridge area. This result suggests that the groundwater divide for Walker Branch Watershed could encompass a larger area than the surface water divide on which areal precipitation is determined.

There is a close correlation between streamflow and precipitation (r^2 = 0.78) but a much weaker relationship between net gain and precipitation (r^2 = 0.06) (Fig. 5.8). These relationships show the same pattern as those obtained at Hubbard Brook watershed (Likens et al. 1977); however, there is greater variability in the Walker Branch Watershed data. This variability can be attributed to the large water-storage capacity of the Walker Branch Watershed soils. The weathered zone on Walker Branch is up to 30m deep on the ridges, whereas the Hubbard Brook soils are shallow (<0.6 m) and have a low water-storage capacity (Likens et al. 1977). The average annual streamflow for larger watersheds in the vicinity of Walker Branch was 44% of precipitation for the 24-year period 1936–1960 (McMaster 1967). This is significantly lower than the 52% value determined from Walker Branch Watershed records. A further discussion of this aspect of the hydrologic budget is given in Sect. 5.6.5.

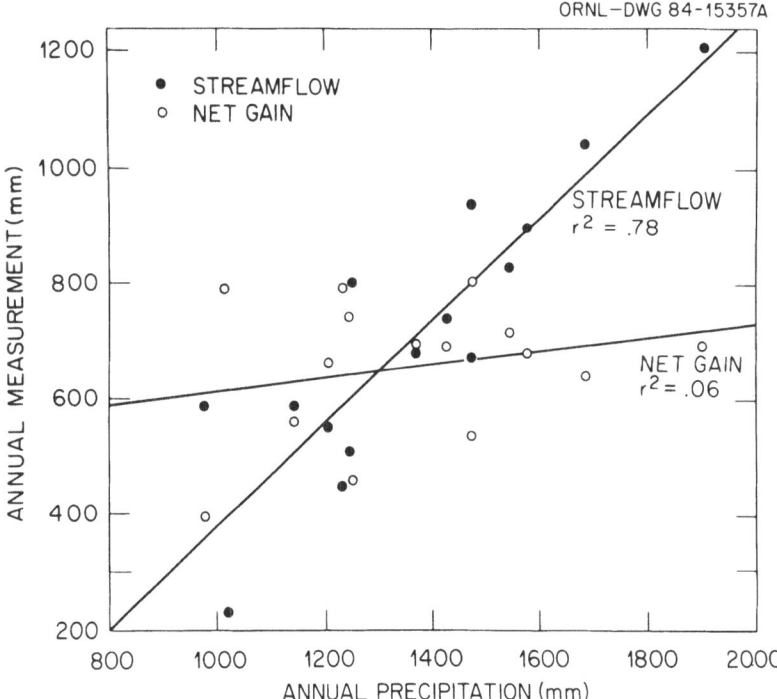

ORNL–DWG 84-15357A

Figure 5.8. Relationship of streamflow and net gain vs. precipitation for the period 1969–1983.

5.5 Precipitation

The heaviest rainfall occurs during the winter and early spring, with the monthly maximum occurring in the period January to March. A secondary maximum usually occurs in July due to convective thunderstorms in the afternoon and evening. The driest months are usually September and October; however, precipitation is highly variable. According to the 20-year record (1951–1971) for Oak Ridge, Tennessee (ATDL 1972), a few of the larger amounts of rainfall have occurred during the normally drier autumnal period. The same record shows that periods of 5 consecutive days without measurable precipitation occur about four or five times per year, with 10 consecutive dry-day periods occurring about once or twice a year. Snow can fall during the period November through March. The snowfalls are usually light (<15 cm accumulation), with the snow melting within 1 or 2 d.

The watershed precipitation was estimated by area-weighting the gage measurements from the five monitoring stations on Walker Branch using the Thiessen triangulation procedure (Thiessen 1911). The contributing station records show small differences in annual totals for the period 1969–1980 (Table 5.2). There is much greater variability in precipitation between years than there is between monitoring stations within a given year. The greatest variation between the five monitoring stations was obtained in

Table 5.2. Annual precipitation measured at monitoring stations 1–5 on Walker Branch Watershed

Year	Annual precipitation (mm) at monitoring stations 1–5					Mean (mm)	SD
	1	2	3	4	5		
1969	1201	1232	1237	1242	1252	1232.8	19.3
1970	1227	1237	1265	1252	1250	1246.2	14.6
1971	1382	1389	1356	1392	1356	1375.0	17.7
1972	1552	1623	1565	1603	1567	1582.0	29.7
1973	1859	1940	1887	1951	1900	1907.4	38.0
1974	1400	1547	1488	1481	1417	1466.6	59.2
1975	1547	1557	1537	1557	1534	1546.4	10.8
1976	1123	1245	1212	1245	1229	1210.8	50.9
1977	1323	1473	1412	1460	1509	1435.4	71.8
1978	1242	1283	1217	1262	1285	1257.8	28.8
1979	1687	1704	1666	1717	1664	1687.6	23.2
1980	996	983	965	978	988	982.0	11.6
Mean (mm)	1378.3	1434.4	1400.6	1428.3	1412.6		
SD	247.3	259.0	246.2	259.1	240.1		

1976 with a coefficient of determination of 5%. Monitoring stations 1 and 2 had the least and greatest mean precipitation, respectively.

Analysis of Walker Branch Watershed precipitation records for a 6-year period (1970–1976) showed that some precipitation occurred, on average, 1 d in 3. A frequency analysis of these rainy days (Henderson et al. 1977) showed precipitation to be <10 mm/d on 55% of the rainy days and <30 mm/d on 90% of the rainy days (Fig. 5.9). In contrast, there was almost 140 mm of rainfall on one particular day. The precipitation pattern is characterized by many days with low amounts of rain and fewer days

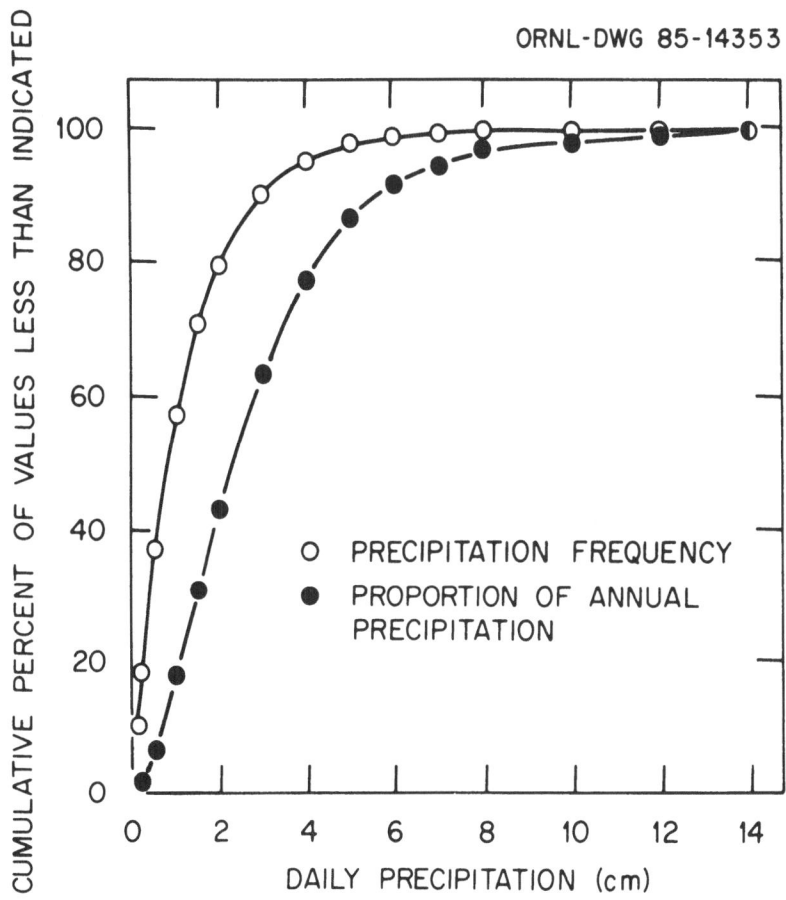

Figure 5.9. Cumulative distributions of frequency and proportion of annual precipitation as a function of daily precipitation. Mean data for the period September 1, 1970—August 31, 1976. Source: G.S. Henderson, D.D. Huff, and T. Grizzard. 1977. Hydrologic characteristics of Walker Branch Watershed. pp. 195–209. IN: D.L. Correll (ed.), Watershed Research in Eastern North America. Chesapeake Bay Center for Environmental Studies, Smithsonian Institution, Edgewater, Maryland.

with high amounts. The 55% of rainy days with <10 mm of precipitation accounted for 20% of the annual total, whereas the 10% of rainy days with >30 mm of precipitation accounted for 35% of annual precipitation.

The monthly distribution of precipitation for the five rain gages (data not presented) shows that the greatest spatial variability occurs during the summer. Convective thunderstorms occur frequently during July and August (ATDL 1972).

5.6 Soil-Plant-Atmosphere Processes

Water movement in the soil-plant-atmosphere system has been investigated, using field and computer modeling approaches. Simulation methods are useful in terrestrial water flow research, since they provide a common mathematical basis for the soil physics, plant physiology, and micrometeorology methods of investigation. A schematic diagram of the structure and components of the Terrestrial Ecosystem Hydrology Model (TEHM)

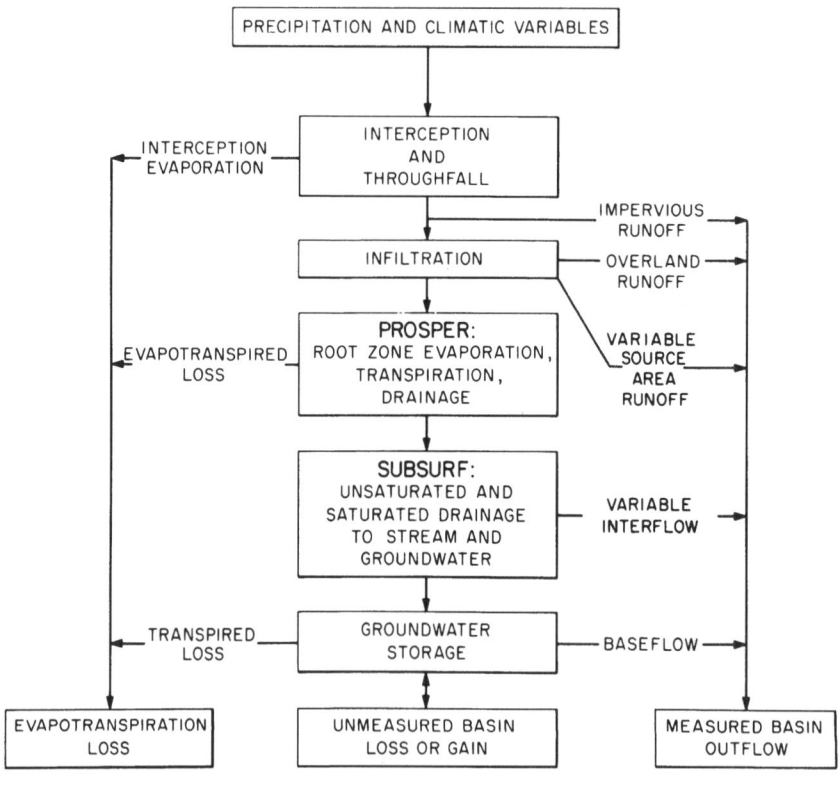

Figure 5.10. Schematic diagram of the structure and components of the Terrestrial Ecosystem Hydrology Model (TEHM).

used in Walker Branch Watershed studies (Fig. 5.10) illustrates the linkages in water movement through the landscape.

The soil, plant, and meteorological data obtained from field measurements are used to obtain parameter values required by the PROSPER subroutines (Fig. 5.11). Hourly precipitation, daily average dewpoint temperature and wind speed, daily total solar radiation, and daily maximum and minimum air temperature are provided as inputs to the model. Algorithms for deriving hourly values of air temperature and irradiance, adjusted to slope and aspect, are used in the calculation of evapotranspiration for the changing conditions during diurnal cycles.

The PROSPER module may be viewed in its simplest form as a set of four equations and four unknowns that are solved for hourly time steps between rainfall events and 15-min time steps during rainfall (Fig. 5.11).

ORNL WS 40335

Location	Properties	Structural Equations
Atmosphere	Environmental conditions Solar radiation Precipitation Dew point temperature Max. and min. air temperature Average wind speed	Vapor flux from surface $F_v = f(R_x)$ Calculation uses combined energy balance-aerodynamic method
Boundary Layer	Resistance to vapor and heat flow	
Evaporating Surface	Resistance to vapor flow (R_x) Surface water potential (ψ)	Surface characteristic $R_x = f(\psi)$
Plant and Soil System	Plant and root resistances Root distribution in upper two soil layers Soil water characteristic for each soil layer Hydraulic conductivity vs water content for each soil layer Soil layer thicknesses	Liquid flux to surface $F_w = f(\psi)$ Calculation uses electrical network equations
Whole System	Steady state	Vapor flux = Liquid flux $F_v = F_w$

Figure 5.11. Data requirements and basic equations used in the soil-plant-atmosphere routines (PROSPER) of the Terrestrial Ecosystem Hydrology Model (TEHM). Source: R.J. Luxmoore, D.D. Huff, R.K. McConathy, and B.E. Dinger. 1978. Some measured and simulated plant water relations of yellow-poplar. For. Sci. 24:327–341.

In the first equation, the flux of water vapor (F_v) from the evapotranspiration surface (leaves and litter) is a function of surface resistance (R_x) and is calculated with the Monteith (1965) form of the combined energy-aerodynamic equation. The second equation relates the foliar surface water potential (ψ_x) and surface resistance, as determined from empirical measurements of leaf water potential and stomatal resistance [$R_x = f(\psi_x)$]. A third equation for the flux of liquid through the soil, roots, and trunk to the leaves is based on electrical network equations of the potential-resistance form. This liquid flux (F_w) to the evapotranspiration surface depends on the surface water potential [$F_w = f(\psi_x)$]. The last equation satisfies the steady-state assumption that $F_v = F_w$ at each time step. Meteorological input data determine the evaporative demand of the atmosphere, whereas the soil water status, as modified by infiltration and drainage, determines the liquid supply. The model matches the demand with the supply through selection of the appropriate combination of surface water potential and surface resistance for the conditions applying to a given time step. More complete descriptions of the model are given by Swift et al. (1975) and Huff et al. (1977b).

In the next four sections, the experimental and modeling studies of water transport in the various components of the soil-plant-atmosphere system are outlined.

5.6.1 Interception and Throughfall

Throughfall was measured during 1971–1973 and compared with the regression equation predictions for eastern deciduous forests developed by Helvey and Patric (1965). Their regression equations were developed from experimental measurements of precipitation, throughfall, and rainfall frequency at many hardwood forest sites in the eastern United States. They separated the growing season response from that of the dormant season, and a comparison of these with the field measurements showed close agreement (Table 5.3) (Henderson et al. 1977).

The precipitation that does not pass through the litter is called interception. The surfaces of leaves, branches, trunks, and litter retain water which can be evaporated directly back to the atmosphere. The deciduous forests on Walker Branch Watershed have an estimated interception storage volume of 0.76 mm during the dormant winter period and 1.27 mm during the summer. These values were determined by means of computer simulation (Fig. 5.12) applied to empirical regression relationships (Huff et al. 1977b). A ramp function is used in the model to represent the changes in leaf area in the spring and autumn. According to simulations of Luxmoore (1983) for the 1975–1977 period, interception evaporation of 145 mm/year was 19.5% of annual evapotranspiration. Henderson et al. (1977) estimated the average interception evaporation to be 187 mm/year for the 6-year period 1970–1976.

Table 5.3. Average measured and calculated throughfall in a deciduous forest on Walker Branch Watershed for the growing and dormant seasons during the 1971–1973 period.

	Precipitation (mm)	Throughfall (mm)	
		Measured	Calculated
Growing season (May–November)	721	610	603
Dormant season (December–April)	807	695	710

Source: G.S. Henderson, D.D. Huff, and T. Grizzard. 1977. Hydrologic characteristics of Walker Branch Watershed. pp. 195–209. IN D.L. Correll (ed.), Watershed Research in Eastern North America. Chesapeake Bay Center for Environmental Studies, Smithsonian Institution, Edgewater, Maryland.

5.6.2 Infiltration

Throughfall usually infiltrates into the litter and mineral soil because of the highly permeable properties of the surface soil horizons of Walker Branch Watershed (Peters et al. 1970). Essentially no rainfall becomes overland flow on unsaturated soil. Infiltration measurements at two sites in a white oak stand on Fullerton cherty silt loam soil showed an average

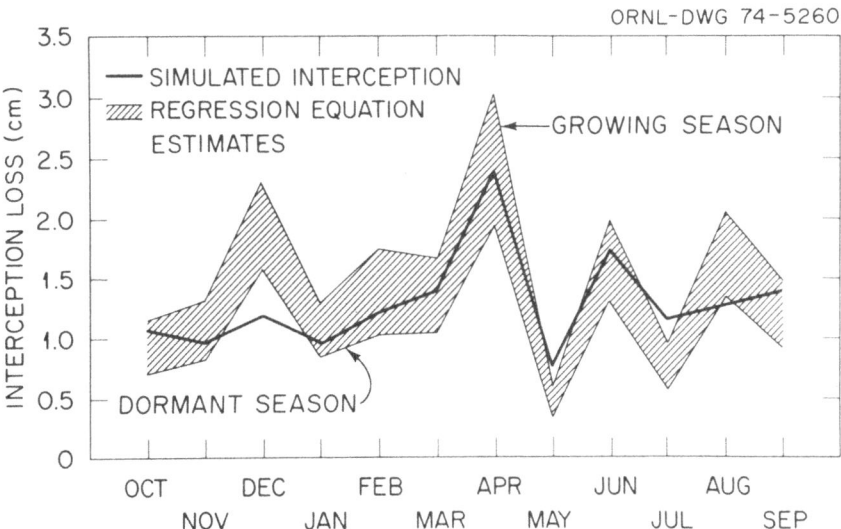

Figure 5.12. Comparison of the Terrestrial Ecosystem Hydrology Model (TEHM) interception simulation results with those of the Helvey and Patric (1965) regression equations.

ınfiltration rate of 26 m/d during 200 min of ponding (Luxmoore 1983). The rainfall rate would have to be >1 m/h to exceed the infiltrability of this Walker Branch Watershed forest soil.

5.6.3 Soil Water

The physical characteristics that determine the status of soil water and the flux of water through the soil have been determined at several sites on Walker Branch Watershed. Soil cores were taken from 13 pits excavated at locations within each of the major soil series, and the water-retention characteristics were determined (Fig. 5.13) by laboratory procedures (Peters et al. 1970). The range of retention characteristics for the surface soil (A2 horizon) and the subsoil (B21 horizon) is shown by the white areas on Fig. 5.13. A mean water content vs. tension function for one site is also shown. The data indicate that there is a wide range in water content at a particular tension. For example, at 10 kPa (the "field capacity" tension of water in a well-drained soil), the water content can range from 0.15 to 0.30 g/g.

The hydraulic characteristics were determined in situ for a soil block of Fullerton cherty silt loam that was hydrologically isolated from the surrounding soil (Luxmoore et al. 1981a). A 2 × 2 m area was excavated, an impervious liner was installed to a 2-m depth on three sides, and a concrete face was constructed on the fourth side to a 3-m depth (Fig. 5.14). Tensiometers were installed laterally through the concrete to measure the soil water matric potential, and vertically installed access tubes were used for inserting a neutron probe to determine water content. Soil

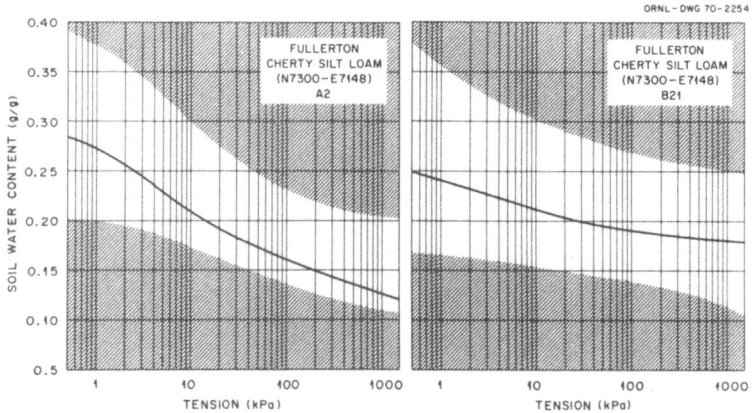

Figure 5.13. Range of water-retention characteristics for Walker Branch Watershed soils, with curves for a specific site on Fullerton cherty silt loam.

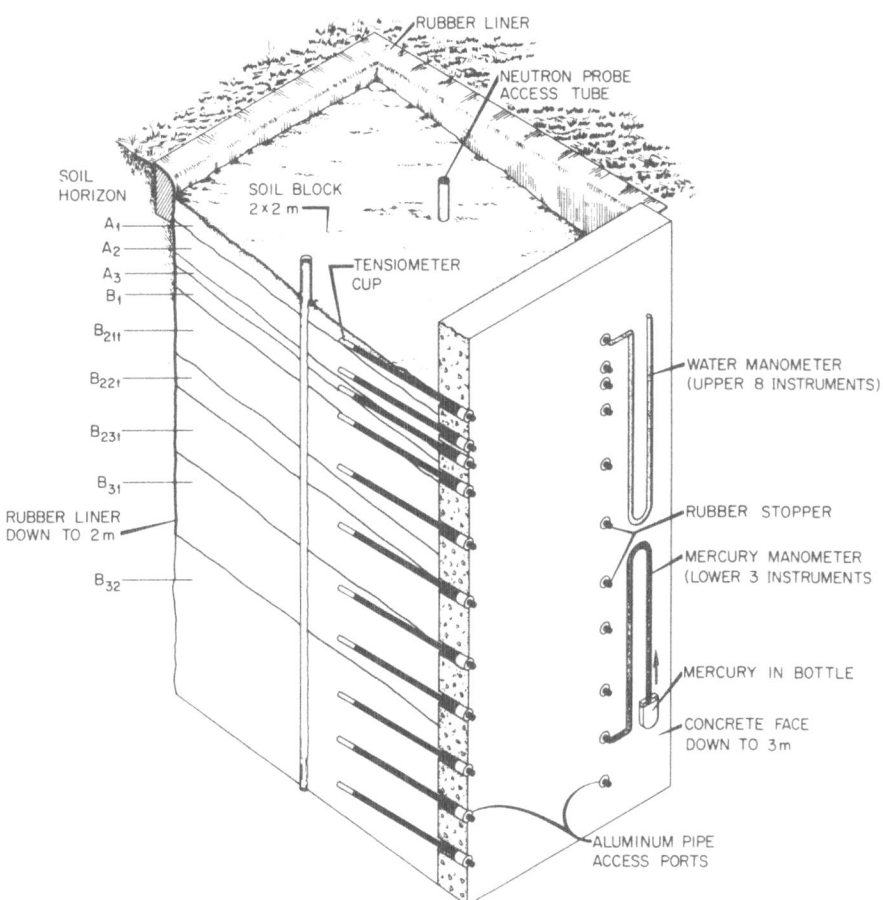

ORNL-DWG 76-4921RA

Figure 5.14. Design of the soil block facility constructed at Walker Branch Watershed for in situ determination of the hydraulic characteristics of the soil. Source: R.J. Luxmoore, T. Grizzard, and M.R. Patterson. 1981. Hydraulic properties of Fullerton cherty silt loam. Soil Sci. Soc. Am. J. 45:692–698.

water content and potential were measured at selected profile positions during drainage of the soil from an initially flooded condition. Evidence of spatial variability in the soil's hydraulic behavior is shown by the soil water drainage profiles (Fig. 5.15) for two positions 1m apart in the soil block. The replicate 1 position shows much less drainage during the 15-min to 3-h period than the replicate 2 position. The soil's hydraulic characteristics provide essential input parameters for use in hydrologic models.

Determinations of water content of the surface soil at several locations on Walker Branch Watershed provide a revealing comparison with sim-

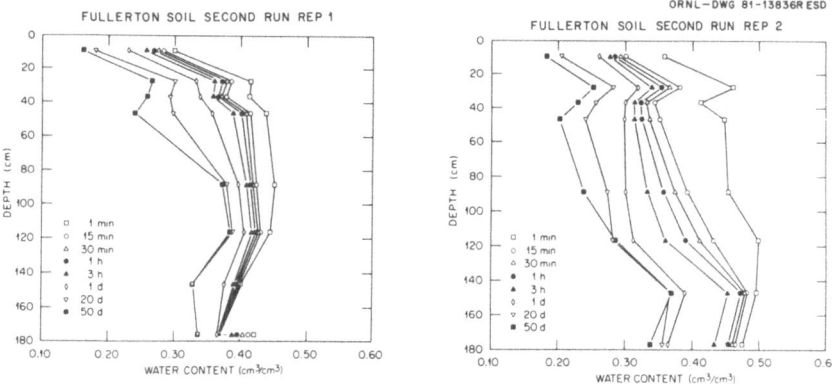

Figure 5.15. Water content profiles of Fullerton soil during a 50-d drainage period for two positions (replicates 1 and 2) in the soil block facility.

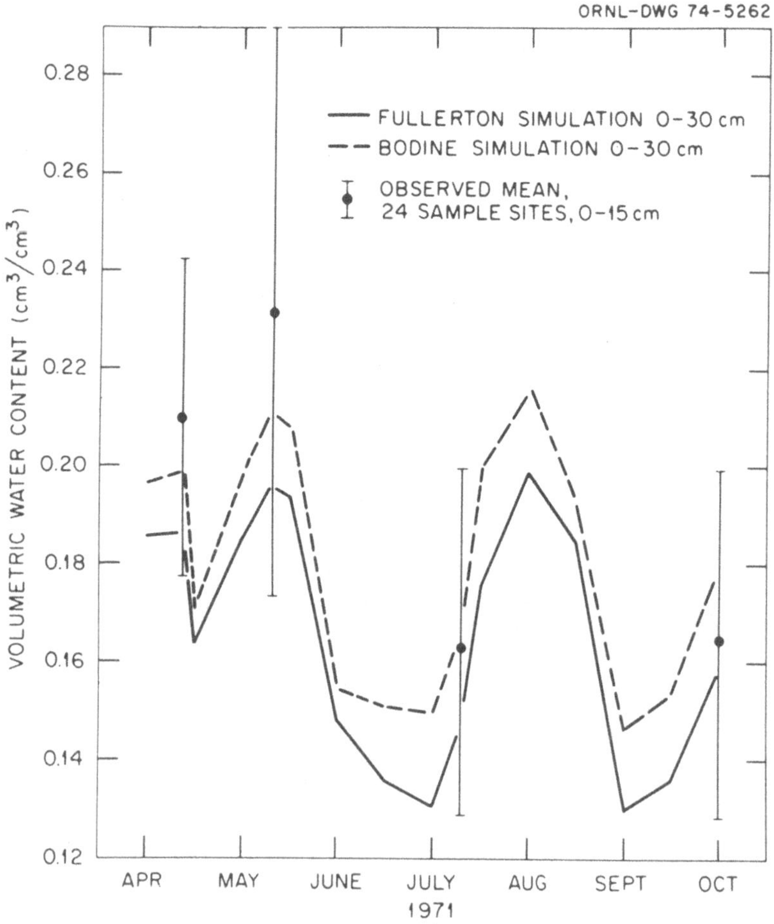

FULLERTON SIMULATION 0-30 cm

BODINE SIMULATION 0-30 cm

OBSERVED MEAN,
24 SAMPLE SITES, 0-15 cm

VOLUMETRIC WATER CONTENT (cm³/cm³)

APR   MAY   JUNE   JULY   AUG   SEPT   OCT

1971

Figure 5.16. Comparison of measured and simulated water content of two dominant Walker Branch Watershed soils.

ulation values (Fig. 5.16) obtained with the TEHM code. There is wide variability in the field measurements, and it is readily seen that the simulation values pass through the range of the field measurements. The wide range in soil water measurements in the field is understandable, given the wide range in water-retention characteristics of the surface soils of Walker Branch Watershed (Fig. 5.13).

The soil matric potential was measured over a 2-year period with tensiometers installed 10m apart at several depths at two locations (Tower and Midway) in a white oak (*Quercus alba*) stand (Luxmoore 1983). The data for tensiometers installed below 1-m depth indicated the hydraulic gradient for soil water drainage. In the autumn, soil water drained downward, whereas during the summer, the gradient of flow was upward (Fig. 5.17). The seasonality of soil water drainage is closely dependent on the water uptake by vegetation. The water-retention characteristics of the soil at the two study sites were determined from the field monitoring records (Fig. 5.18). It is useful to note that the Tower site has a lower water capacity (change in soil water content per unit change in matric potential) than the Midway site at matric potentials below -20 kPa. This has a significant influence on differences in soil water dynamics and in water uptake by plants at the two sites, as discussed in the section below on evapotranspiration.

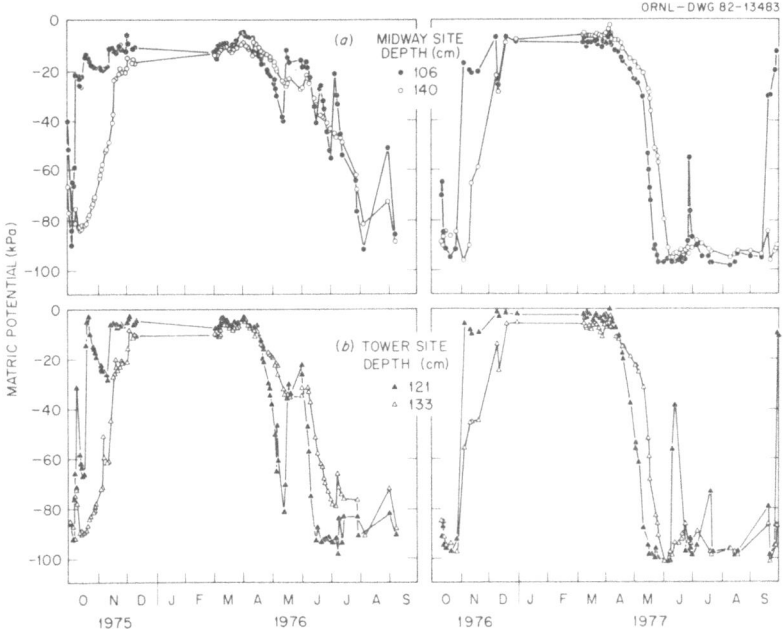

Figure 5.17. Temporal changes in soil matric potential below 1-m depth at two sites on Walker Branch Watershed. Source: R.J. Luxmoore, T. Grizzard, and M.R. Patterson. 1981. Hydraulic properties of Fullerton cherty silt loam. Soil Sci. Soc. Am. J. 45:692–698.

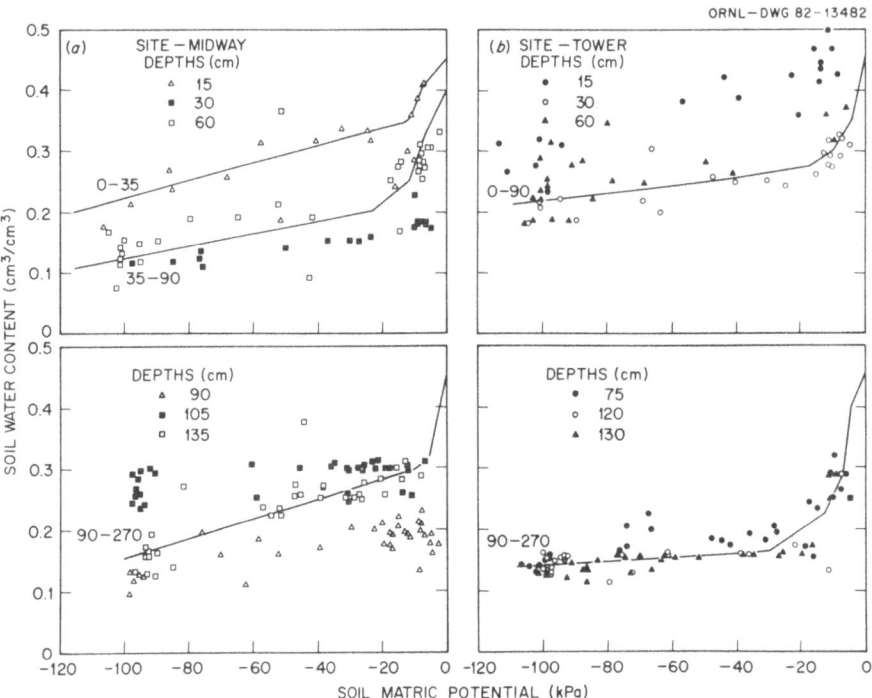

Figure 5.18. Soil-water-retention characteristics of Fullerton cherty silt loam at two sites. Source: R.J. Luxmoore, T. Grizzard, and M.R. Patterson. 1981. Hydraulic properties of Fullerton cherty silt loam. Soil Sci. Soc. Am. J. 45:692–698.

5.6.4 Groundwater

Two wells were drilled to a depth of 26 m at two ridge locations. One well was often dry and occasionally showed a rise in the water table of 60 cm or less. The other well showed seasonal fluctuations in water table elevation, with a general rise of 3 m or more in the saturated zone during the winter and spring and a decline during the summer and autumn (Fig. 5.19).

5.6.5 Evapotranspiration and Plant Water

On a whole-watershed basis, the net gain is largely due to evapotranspiration (see Sect. 5.4). In soil water monitoring studies of a forest stand, the components of the water budget (Eq. 5.1) are either measured or calculated, except for evapotranspiration.

$$\text{Precipitation} = \text{drainage} + \text{runoff}$$
$$+ \text{ change in soil water storage} + \text{evapotranspiration.} \qquad (5.1)$$

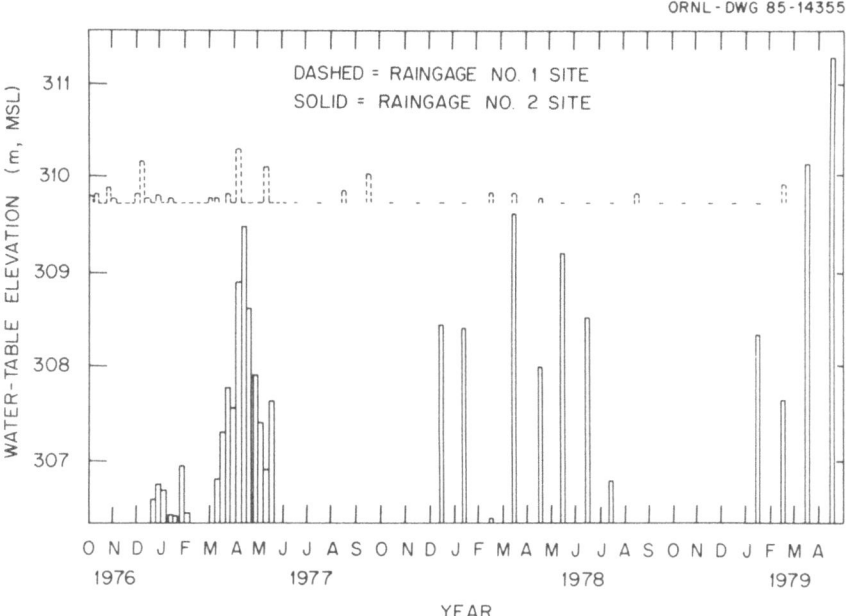

Figure 5.19. Water table elevations for the period October 1976–April 1979 at two observation wells.

The evapotranspiration term is determined by difference and thus includes all the estimation errors of the other terms.

Other empirical methods for estimating evapotranspiration have been used. Henderson et al. (1977) calculated an evapotranspiration estimate of 767 mm/year for the Oak Ridge, Tennessee, area, using the long-term average monthly temperature records in the Thornthwaite equation. McMaster (1967) cited an average water loss for the Oak Ridge area of 760 mm/year without identifying the source of this estimate. The net gain for the watersheds in the Oak Ridge area averaged 730 mm/year (McMaster 1967), and this is an experimental estimate of evapotranspiration for the area. Grigal and Hubbard (1971), using an empirical computer simulation model, calculated evapotranspiration for the Walker Branch Watershed in 1969 to be 729 mm/year.

Simulation with the TEHM provided an annual evapotranspiration estimate of 745 mm/year for ridgetop forest stands on Walker Branch Watershed for the period October 1975 to September 1977 (Luxmoore 1983). It is probable that evapotranspiration is higher at the warmer ridgetop sites than at cooler locations along the stream and in coves. A sensitivity analysis of the TEHM parameters showed that an 11°C decrease in air temperature could reduce evapotranspiration by 60% (Luxmoore et al. 1981b). The annual water use of a yellow-poplar (*Liriodendron tulipifera*) stand in a sinkhole near Walker Branch Watershed was simulated with TEHM to be 665 mm/year (Luxmoore et al. 1978). This estimate could be viewed as a lower bound for evapotranspiration at Walker Branch Watershed; however, it is higher than the net gain value of 655 mm/year for Walker Branch Watershed. The Tennessee Valley Authority's regional map of the net gain (precipitation minus runoff) for watersheds in eastern Tennessee gave an estimate of ~735 mm/year for the Walker Branch vicinity (TVA 1972). From the above discussion of the various evapotranspiration estimates, it seems that the net gain value of McMaster (1967) of 730 mm/year is the best current estimate of evapotranspiration for Walker Branch Watershed.

If Walker Branch Watershed has an evapotranspiration rate of 730 mm/year, then it follows that the groundwater drainage area contributing to streamflow must be larger than the surface area of the watershed. The average annual precipitation of 1368 mm/year, adjusted for evapotranspiration loss, shows that expected streamflow should be 638 mm/year. In fact, the average measured streamflow of 713 mm/year needs to be increased to 733 mm/year to account for an estimated weir leakage of 20 mm/year (Sheppard et al. 1973). An additional groundwater drainage area of 14.5 ha could contribute the additional 95 mm/year base flow (the difference between 733 and 638 mm/year) to Walker Branch, and this represents a 15% increase in the groundwater divide over the surface divide. This is not an unreasonable assessment for groundwater movement in areas dominated by carbonate rocks. Hollyday and Goddard (1979) reported groundwater movement, in terrain similar to Walker Branch Watershed, to occur in solution openings that follow bedding planes and strike joints of the Knox Group of carbonate rocks. These solution openings are intercepted by groundwater, which collects in high-permeability formations in the Knox Group and is routed along strike to discharge as springs.

The TEHM simulations of water loss from two forest stands with contrasting soil-water-retention characteristics (Fig. 5.18) illustrate the significance of the upward flux of soil water back into the root zone (0- to 90-cm depth in the simulations) during the summer. The soil at the Tower site had a low water-storage capacity, and 26% of the annual soil water evaporation plus transpiration was supplied by the upward flux of water stored at depths below 90 cm. The Midway site had soil with a high water-storage capacity, and 9% of the annual soil water evaporation plus transpiration was supplied by the upward flux of soil water. The two sites had similar annual rates of evapotranspiration (Luxmoore 1983). Compensating mechanisms of soil water storage and transport can lead to similar rates of evapotranspiration from sites having spatially varying soil properties. Peck et al. (1977) showed a similar result in simulation studies of soil variability using the TEHM. In contrast, in another simulation study, Luxmoore and Sharma (1984) showed that the evapotranspiration from a prairie grassland was greater in fine-textured soil than in coarse-textured soil. In some soil-plant-weather combinations, variability in the soil's hydraulic characteristics may significantly influence evapotranspiration.

The seasonality of evapotranspiration can be seen in the TEHM simulation results for a ridgetop oak stand. The monthly values for interception, evaporation from soil, and transpiration (Fig. 5.20) show the large contribution of transpiration (74%) to water loss during the summer months (Luxmoore 1983).

ORNL–DWG 85-10964

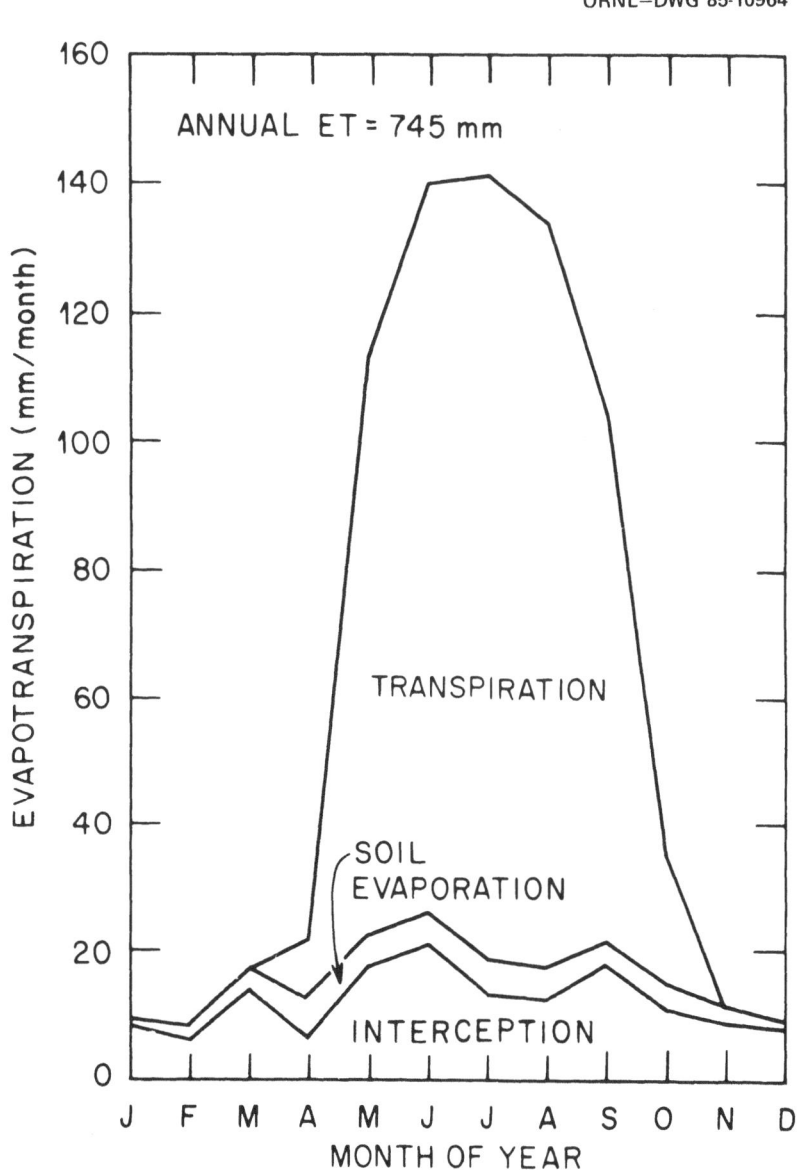

Figure 5.20. Simulated evapotranspiration (ET) components for a deciduous forest stand on Walker Branch Watershed.

The TEHM simulations of plant water potential were compared with field measurements on a yellow-poplar stand (Luxmoore et al. 1978), and reasonable agreement was obtained, except for some morning values (Fig. 5.21). It was concluded that deep roots contributed to the high water potential of foliage in the morning by facilitating nocturnal recovery in plant water status. The hourly patterns of TEHM simulation results for the yellow-poplar stand (Fig. 5.22) for a 5-d drying period following a rainfall agree with the expected patterns described by Slatyer (1967). It is interesting to note that the upward flux of water back into the root zone has a diurnal cycle, with a lag response to evapotranspiration.

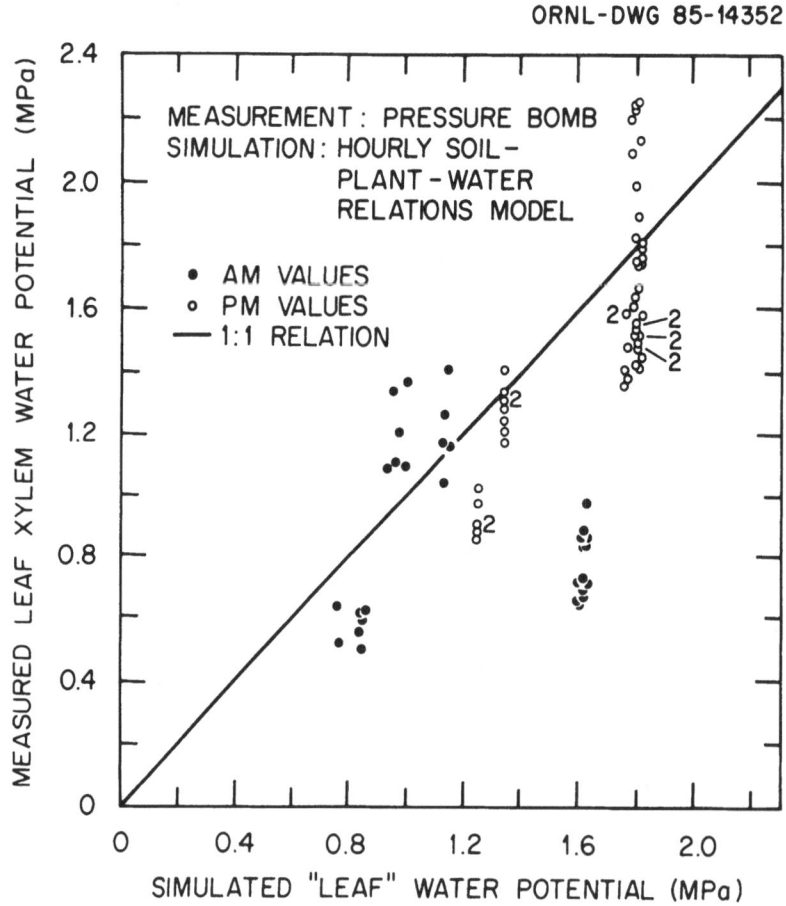

Figure 5.21. Comparison of measured and simulated values of leaf water potential for a yellow-poplar stand. Source: R.J. Luxmoore, D.D. Huff, R.K. McConathy, and B.E. Dinger. 1978. Some measured and simulated plant water relations of yellow-poplar. For. Sci. 24:327–341.

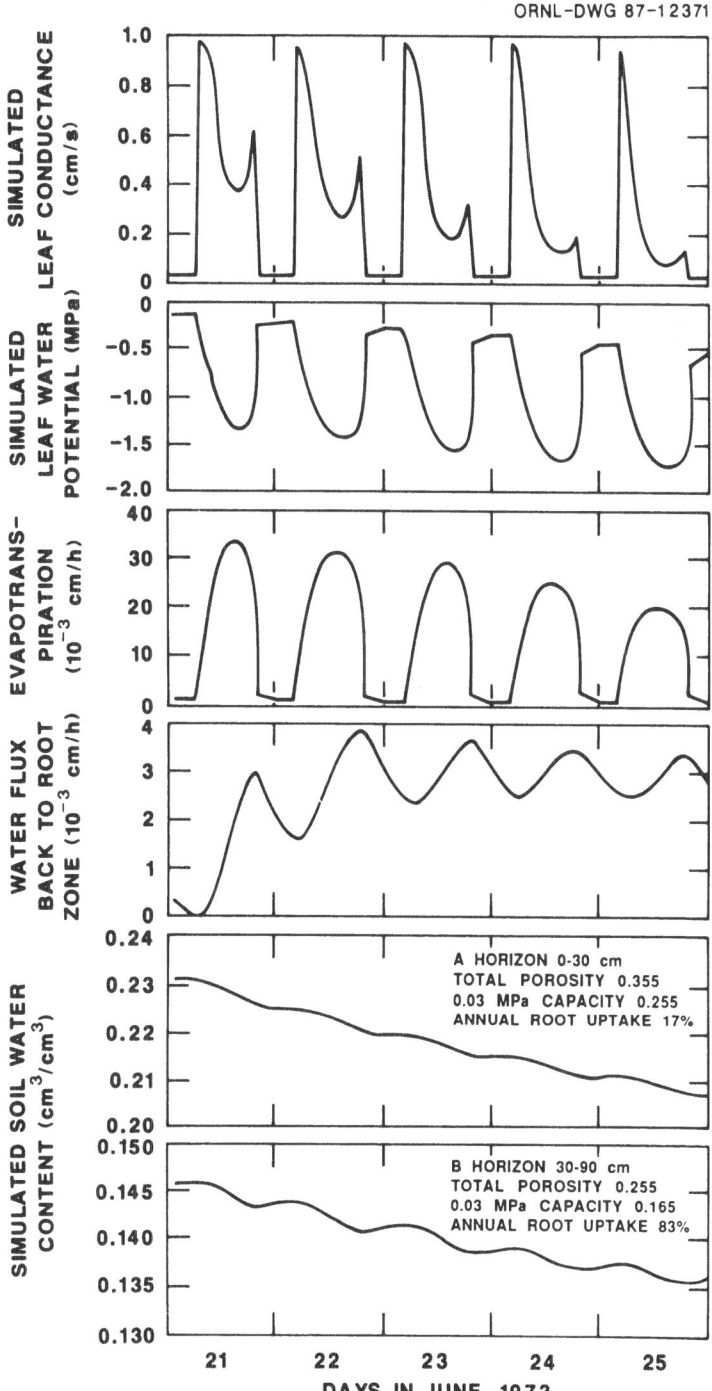

Figure 5.22. Simulated plant and soil water dynamics of a yellow-poplar stand during a 5-d drying period. Source: R.J. Luxmoore, D.D. Huff, R.K. McConathy, and B.E. Dinger. 1978. Some measured and simulated plant water relations of yellow-poplar. For. Sci. 24:327–341.

5.7 Streamflow

Streamflow and precipitation showed a direct relationship on an annual basis, as shown earlier (Fig. 5.8), and a close correspondence can also be seen on a seasonal basis. The precipitation and streamflow summarized on a 4-week basis for a 6-year period (Fig. 5.23) show high streamflows in late winter and spring corresponding with the precipitation during the December–March period (Henderson et al. 1977). Lower streamflows occur during the summer and early autumn.

The daily weighted average streamflow ranged from 0.005 m^3/s to nearly 1.1 m^3/s for the 1970–1976 period. Low daily flows with average flow rates <0.05 m^3/s occur ~90% of the time and account for 50% of annual discharge (Fig. 5.24). Daily mean streamflows >0.2 m^3/s occur <1% of the time but account for 17% of the annual runoff. These infrequent flows carry the major chemical and particulate discharges from Walker Branch (Ch. 8). In both forks, high flows are more frequent in spring and winter than in the summer and autumn. Very high flow rates (>0.7 m^3/s) occur during the autumn at a slightly higher frequency than in the spring and winter (Huff et al. 1977a).

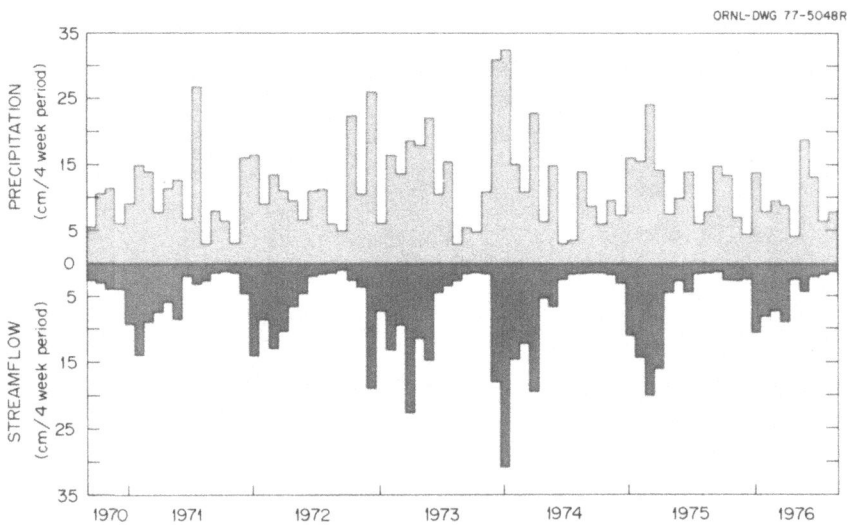

Figure 5.23. Seasonal distributions by 4-week intervals of precipitation and streamflow for Walker Branch Watershed for the period September 1970—August 1976. Source: G.S. Henderson, D.D. Huff, and T. Grizzard. 1977. IN D.L. Correll (ed.), Watershed Research in Eastern North America. Chesapeake Bay Center for Environmental Studies, Smithsonian Institution, Edgewater, Maryland.

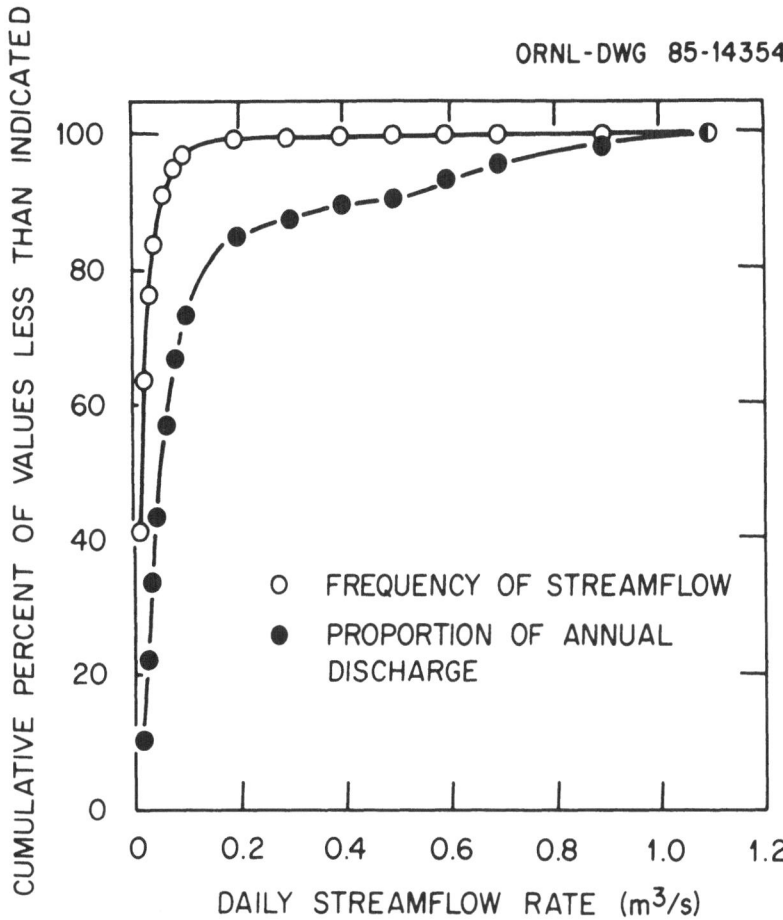

Figure 5.24. Cumulative frequency distribution of daily streamflow and the proportional contribution to annual stream discharge for Walker Branch for the period September 1970—August 1976. Source: G.S. Henderson, D.D. Huff, and T. Grizzard. 1977. Hydrologic characteristics of Walker Branch Watershed. pp. 195–209. IN D.L. Correll (ed.), Watershed Research in Eastern North America. Chesapeake Bay Center for Environmental Studies, Smithsonian Institution, Edgewater, Maryland.

5.7.1 Base Flow

Direct comparison of flow rates observed at the East Fork and West Fork during low-flow conditions showed that the West Fork yields a higher rate of base flow per unit area than the East Fork. This suggested that some groundwater is transferred between the two basins, making it desirable to quantify the extent of exchange and to draw inferences about whether or not the combined base flow of both basins constitutes all the groundwater lost from the combined basin.

ORNL–DWG 85-10965

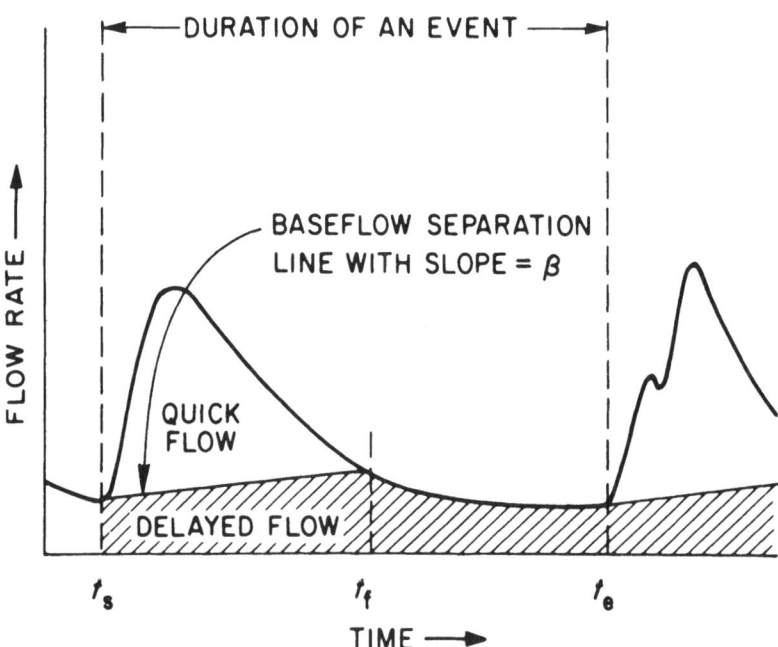

Figure 5.25. Diagram of the method used to separate a hydrograph into quick flow and delayed flow.

In order to differentiate between stormflow and base flow, a hydrograph separation technique of Hewlett and Hibbert (1967) was used because of its simplicity and applicability to small forested watersheds in the Appalachian-Piedmont region. A computer program that implements their procedure has been documented by Huff and Begovich (1976). Briefly, the method separates a hydrograph into quick flow and delayed flow by a straight line of arbitrary slope (Fig. 5.25). The beginning of a storm event is defined by an increase in the stream discharge rate. When the slope of the hydrograph exceeds that of the base flow separation line, a storm begins. The storm ends when the hydrograph drops below the extension of the separation line. Minor events (stormflow volume of <0.025 mm) are not counted as storms, and the resulting runoff is considered to be base flow. Further details of the method are given in a study by Huff et al. (1977a).

The hydrograph separation method was applied to the 5-min flow data from each fork of Walker Branch. The calculated monthly base flow for both the East Fork and West Fork of Walker Branch for the period October

1971 to October 1975 showed that the West Fork had about double the base flow of the East Fork on an areal basis.

A quantitative analysis of the base flow data for the 4-year period supports the thesis that ~30% of the base flow originating on the East Fork is annually transferred to the West Fork. Streamflow simulation results with TEHM for each fork showed much closer agreement with measured flows when algorithms representing a base flow transfer were included (Huff et al. 1977a). There are several springs and seeps that could account for this transfer, as described in Sect. 5.7.3.

5.7.2 Stormflow

The storm runoff determined by hydrograph separation (Fig. 5.25) for East Fork and West Fork showed very close areal agreement in the analysis of data for a 4-year period from October 1971 to October 1975 (Huff et al. 1977a). The mean annual stormflow for both forks was 220 mm, and the monthly flows were very similar. Consequently, there is no evidence for stormflow transfer between subwatersheds. Stormflows occur primarily from November through May. When monthly stormflow is expressed as a percentage of precipitation, a strong seasonal pattern is obtained. During June to September, the monthly stormflow is $(1 \pm 1.5)\%$ of precipitation, whereas during December to March, stormflow averages $(18 \pm 13)\%$ of monthly precipitation. Precipitation directly into the stream channel could account for most of the stormflow during summer; however, a much larger watershed area contributes to stormflow during the winter. The TEHM simulations showed close agreement between the simulated and calculated (hydrograph separation) stormflows for each month during the period October 1971 to October 1975.

5.7.3 Spring and Seepage Flow

A chloride tracer was released at the top of the perennial stream channel of the East Fork during a low-flow period. It was detected in the water reemerging downstream from the subsurface flow sections of the stream channel (Fig. 5.26), illustrating the hydrologic continuity of the whole stream channel. A chloride budget analysis showed that 65% of the tracer was in the S1E spring water, but only 8% was discharged into the stilling basin of the East Fork weir. The chloride was widely dispersed, and some of it may have been diverted to the West Fork; however, chloride was not detected in the S3W spring or in the stilling basin of the West Fork weir.

In another tracer study, tritium was released into the West Fork of Walker Branch as a means of identifying channel sections receiving groundwater inflows. The dilution of tritium between selected water-sampling positions indicates an increase in streamflow. The concentration of

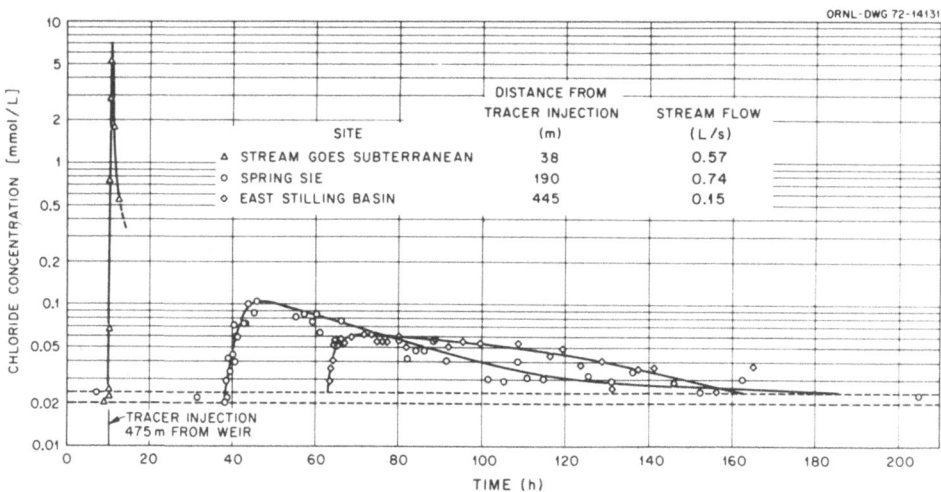

Figure 5.26. Changes in chloride concentration with time at various sites on the East Fork following a calcium chloride release. The time scale begins at 0.00 h on August 29, 1972.

tritium at 11 stations along a 120-m reach of the West Fork was monitored at time intervals of 3 to 6 min during a 30-min period following the release of tritium into the stream. The stream was at base flow. The streamflow rate was determined at each station from the temporal changes in tritium concentration, and the change in flow was calculated for each stream segment (Huff et al. 1982). The flow changes were related to the adjacent topography and the dip and strike of the parent rocks. The study showed that hollows (small coves) were more significant contributors to subsurface flow into the stream than spurs or uniform slopes. In addition, the data suggest that subsurface drainage collected at bedding planes and moved laterally along strike.

5.8 Watershed Comparisons

Knowledge of the status and fluxes of water in forested landscapes is an essential component of the basic research on biogeochemical cycles. This has been the primary benefit of hydrologic research at Walker Branch Watershed. A secondary benefit has been the use of the research data for extrapolation to ungaged areas. Such extrapolation can sometimes be justified on the basis that the experimental watershed is representative of the region. Hubbard Brook watershed is representative of northern New England watersheds, according to comparisons made by Likens et al. (1977). Walker Branch Watershed provides a contrast. This ~100-ha area on Chestnut Ridge in Roane County, Tennessee, generates more streamflow than most of the 100-ha landscape units on the same ridge. The dolomite parent material of this ridge is probably more permeable than that at Walker Branch Watershed, since the surface flow of several neighboring watersheds is ephemeral. Although the groundwater divide is estimated to be 15% larger than the surface divide (see Sec. 5.6.5), the unmeasured inflow of groundwater appears equal to the unmeasured outflow at Walker Branch Watershed.

An adjacent ridge, southeast of and parallel to Chestnut Ridge, has shallow soil on a shale formation. The watersheds on this ridge have a more rapid hydrologic response to storm events than does Walker Branch Watershed because of their shallow soil profiles. Other watersheds with shallow soil profiles developed on a sandstone formation occur extensively on the Cumberland Plateau. This plateau is a major physiographic region in eastern and central Tennessee and lies to the northwest of the Ridge and Valley Province that includes Walker Branch Watershed. Kelly and Meagher (1986) reported that the Cross Creek and Camp Branch Watersheds on the Cumberland Plateau had greater monthly streamflows in winter and spring and lower flows in summer and autumn than Walker Branch Watershed. About 70% of the precipitation contributed to streamflow at the two Cumberland Plateau sites, whereas at Walker Branch Watershed only ~52% of the precipitation becomes surface runoff.

The Coweeta watersheds in the southern Appalachian region of North Carolina have hydrologic features that differ from those of Walker Branch Watershed. The 500-m rise in elevation at Coweeta from the valley floor to the ridgetop is accompanied by a 600-mm increase in precipitation (Dils 1957). Runoff from watershed 18 at Coweeta was 52% of precipitation, a value similar to that measured at Walker Branch Watershed; however, the Walker Branch value is distorted because the groundwater divide is about 15% larger than the surface divide.

These examples illustrate that the hydrology of Walker Branch Watershed cannot be viewed as representative of neighboring areas. However, this does not extend to the water budgets and other investigations conducted on a forest-stand basis. Plot studies are usually representative of areas with similar soil, landscape, vegetation, and climate characteristics. The subsurface characteristics determining the convergent or divergent soil water flow processes and groundwater recharge appear to be unique to Walker Branch Watershed.

5.9 Summary

—Hydrologic monitoring of precipitation and streamflow at Walker Branch Watershed has been completed for a 15-year period (1969–1983).
—The mean precipitation of 1368 mm/year is associated with a mean annual streamflow of 713 mm.

—Precipitation occurred on an average of 1d in 3 with 55% of the rainy days having <10 mm precipitation, and these accounted for 20% of the annual total.

—Stormflow occurs primarily from November through May, contributing 23% of the annual runoff. Large flows >0.2 m³/s occur <1% of the time but account for 17% of the annual runoff.

—Base flow is greatly influenced by springs, and it is estimated that the groundwater divide is 15% larger than the surface divide.

—The forest soils have very high infiltration rates, and surface runoff is generated by subsurface flow processes.

—The seasonality of soil water drainage is inversely related to water uptake by vegetation. Upward soil water fluxes are expected during the summer, particularly in soil profiles with low water-storage capacity.

—Evapotranspiration is estimated to be 730 mm/year, and 20% of this is due to interception evaporation.

—Ridgetop soils are deeply weathered (up to 30-m-deep unsaturated zone). These deep soils have a large water-storage capacity, which buffers the relationship between precipitation and streamflow.

—The streamflow characteristics of Walker Branch Watershed cannot be viewed as representative of the region.

—Simulation models provide an appropriate means of extending knowledge of hydrologic processes at Walker Branch Watershed to other locations.

References

Atmospheric Turbulence and Diffusion Laboratory (ATDL). 1972. Daily, monthly, and annual climatological data for Oak Ridge, Tennessee, townsite and area stations. January 1951 through December 1971. ATDL Report 61. Atmospheric Turbulence and Diffusion Laboratory, Oak Ridge, Tennessee.

Curlin, J.W., and D.J. Nelson. 1968. Walker Branch Watershed project: Objectives, facilities and ecological characteristics. ORNL/TM-2271. Oak Ridge National Laboratory, Oak Ridge,Tennessee.

Dils, R.E. 1957. A guide to the Coweeta Hydrologic Laboratory. Southeastern Forest Experiment Station, Asheville, North Carolina.

Grigal, D.F., and J.E. Hubbard. 1971. SOGGY: An empirical evapotranspiration model for forest soils. pp. 795–800. IN Proceedings of the Summer Computer Simulation Conference, Boston. AFIPS, Montvale, New Jersey.

Helvey, J.D., and J.H. Patric. 1965. Canopy and litter interception of rainfall by hardwoods of eastern United States. Water Resour. Res. 1:193–206.

Henderson, G.S., D.D. Huff, and T. Grizzard. 1977. Hydrologic characteristics of Walker Branch Watershed. pp. 195–209. IN D.L. Correll (ed.), Watershed Research in Eastern North America. Chesapeake Bay Center for Environmental Studies, Smithsonian Institution, Edgewater, Maryland.

Hewlett, J.D., and A.R. Hibbert. 1967. Factors affecting the response of small watersheds to precipitation in humid areas. pp. 275–290. IN W.E. Sopper and

H.W. Lull (eds.), Forest Hydrology. Symposium Publications Division, Pergamon Press, New York.

Hollyday, E.F., and P.L. Goddard. 1979. Groundwater availability in carbonate rocks of the Dandridge area, Jefferson County, Tennessee. U.S. Geological Survey Water-Resources Investigations Open-File Report 79-1263. U.S. Government Printing Office, Washington, D.C.

Huff, D.D., and C.L. Begovich. 1976. An evaluation of two hydrograph separation methods of potential use in regional water quality assessment. ORNL/TM-5258. Oak Ridge National Laboratory, Oak Ridge, Tennessee.

Huff, D.D., G.S. Henderson, C.L. Begovich, R.J. Luxmoore, and J.R. Jones. 1977a. The application of analytic and mechanistic hydrologic models to the study of Walker Branch Watershed. pp. 741–763. IN D.L. Correll (ed.), Watershed Research in Eastern North America. Chesapeake Bay Center for Environmental Studies, Smithsonian Institution, Edgewater, Maryland.

Huff, D.D., R.J. Luxmoore, J.B. Mankin, and C.L. Begovich. 1977b. TEHM: A terrestrial ecosystem hydrology model. ORNL/NSF/EATC-27, Oak Ridge National Laboratory, Oak Ridge, Tennessee.

Huff, D.D., R.V. O'Neill, W.R. Emanuel, J.W. Elwood, and J.D. Newbold. 1982. Flow variability and hillslope hydrology. Earth Surf. Processes Landforms 7:91–94.

Jackson, K. 1982. Primer System 1022, Data base management system. Software House, Cambridge, Massachusetts.

Kelly, J.M., and J.F. Meagher. 1986. Nitrogen input/output relationships for three forested watersheds in eastern Tennessee. pp. 360–391. IN D.L. Correll (ed.), Watershed Research Perspectives. Smithsonian Press, Washington, D.C.

Likens, G.E., F.H. Gormann, R.S. Pierce, J.S. Eaton, and N.M. Johnson. 1977. Biogeochemistry of a forest ecosystem. Springer-Verlag. New York.

Luxmoore, R.J. 1983. Water budget of an eastern deciduous forest stand. Soil Sci. Soc. Am. J. 47:785–791.

Luxmoore, R.J., and M.L. Sharma. 1984. Evapotranspiration and soil heterogeneity. Agric. Water Manage. 8:279–289.

Luxmoore, R.J., D.D. Huff, R.K. McConathy, and B.E. Dinger. 1978. Some measured and simulated plant water relations of yellow-poplar. For. Sci. 24:327–341.

Luxmoore, R.J., T. Grizzard, and M.R. Patterson. 1981a. Hydraulic properties of Fullerton cherty silt loam. Soil Sci. Soc. Am. J. 45:692–698.

Luxmoore, R.J., J.L. Stolzy, and J.T. Holdeman. 1981b. Sensitivity of a soil-plant-atmosphere model to changes in air temperature, dew point temperature and solar radiation. Agric. Meteorol. 23:115–129.

McMaster, W.M. 1967. Hydrologic data for the Oak Ridge area, Tennessee. U.S. Geological Survey Water Supply Paper 1839-N, U.S. Government Printing Office, Washington, D.C.

Monteith J.L. 1965. Evaporation and environment. pp. 205–234. IN C.E. Fogg (ed.), The State and Movement of Water in Living Organisms. Academic Press, New York.

Peck, A.J., R.J. Luxmoore, and J.L. Stolzy. 1977. Effects of spatial variability of soil hydraulic properties in water budget modeling. Water Resour. Res. 13:348–354.

Peters, L.N., D.F. Grigal, J.W. Curlin, and W.J. Selvidge. 1970. Walker Branch Watershed project. Chemical, physical and morphological properties of the soils of Walker Branch Watershed. ORNL/TM-2968. Oak Ridge National Laboratory, Oak Ridge, Tennessee.

Sheppard, J.D., G.S. Henderson, T. Grizzard, and M.T. Heath. 1973. Hydrology of a forested catchment: 1. Water balance from 1969 to 1972 on Walker Branch Watershed. EDFB Memo Report 73-55. Oak Ridge National Laboratory, Oak Ridge, Tennessee.

Slatyer, R.O. 1967. Plant Water Relationships. Academic Press, London.

Swift, L.W., W.T. Swank, J.B. Mankin, R.J. Luxmoore, and R.A. Goldstein. 1975. Simulation of evapotranspiration and drainage from mature and clear cut deciduous forests and young pine plantation. Water Resour. Res. 11:667–673.

Tennessee Valley Authority (TVA). 1972. A continuous daily streamflow model. Research Paper No. 8. TVA, Knoxville, Tennessee.

Thiessen, A.H. 1911. Precipitation for large areas. Mon. Weather Rev. 39:1082–1084.

Chapter 6
Carbon Dynamics and Productivity

N.T. Edwards, D.W. Johnson, S.B. McLaughlin,
and W.F. Harris

6.1 Introduction

Photosynthetically fixed carbon compounds are the building blocks, the energy sources, and the chemical regulators of biological processes involved in the growth, development, and decay of biological systems. These compounds are not only used internally by the autotrophs that synthesize them but are also released by exudation and decay to the heterotrophs that use these compounds for their own growth and development. In forests, a myriad of organisms living in the soil and litter are thus supported. In turn, nutrients released slowly during decomposition are recycled, thus minimizing leaching losses of nutrients essential to the growth and maintenance of the whole system.

The carbon research on Walker Branch Watershed has focused on a better understanding of carbon allocation strategies of forests and on recycling processes and how they relate to the growth and development of different forest types. Because of the complexity of research involving ^{14}C tree tagging and measurements of respiration in different forest floor components, these experiments included only single species or single forest types. The growth and development of forest stands have been followed through periodic biomass inventories of permanent plots established in 1967 in representative forest types. Those inventories have quantified biomass increments and identified shifts in composition among forests due to mortality and subsequent ingrowth of new species. Much of this chapter addresses results from studies performed in four major forest types (chestnut oak, yellow-poplar, oak-hickory, pine-hardwood), whereas the inventory data also include a fifth forest type (loblolly pine).

6.2 Forest Biomass and Productivity

6.2.1 Forest Biomass Inventory

An intensive forest biomass inventory has been kept since 1967 (Harris et al. 1973). A total of 298 permanent plots were originally established, with 129 core plots given a forest type designation (Grigal and Goldstein 1971). Remeasurement of all plots was undertaken in 1970 and 1973, and remeasurement of only the core plots in 1979 and 1983.

The 1983 biomass inventory by forest type is summarized in Table 6.1. Root biomass estimates are not given in the table, but Harris et al. (1973) estimated from root cores that root biomass (including belowground stump) averaged ~26% of aboveground biomass. At this point in time, the chestnut oak and yellow-poplar forest types had the largest biomass, followed by the loblolly pine, oak-hickory, and pine-hardwood types, respectively. There have been some major changes in biomass accumulation rates over the years, however.

6.2.2 Tree Mortality: Shifts in Forest Types and Productivity

Two major bark beetle outbreaks have altered the steady increment of biomass in some forest types. An outbreak of the southern pine beetle in the early 1970s and an outbreak of the hickory borer in the mid- to late

Table 6.1. Biomass (in Mg/ha) of five forest types on Walker Branch Watershed

	Forest type				
Component	Pine-hardwood	Yellow-poplar	Loblolly pine	Oak-hickory	Chestnut oak
No. of plots	36	20	5	32	36
Foliage	4.7 ± 0.3	3.5 ± 0.2		3.4 ± 0.2	4.2 ± 0.1
Branch	30.0 ± 1.9	35.7 ± 2.0		31.4 ± 1.4	37.7 ± 1.4
Bole	105.4 ± 8.1	135.6 ± 8.4		119.1 ± 5.6	143.8 ± 6.1
Stump	10.6 ± 1.1	11.7 ± 0.6		10.4 ± 0.5	12.2 ± 0.4
Σ Tree[a]	143.8 ± 7.3	186.4 ± 11.1	170.8 ± 13.9	163.8 ± 7.7	197.7 ± 7.9
Litter[b]	33	18		32	33
Soil[c]	116	189		116	116
Σ Ecosystem	293	393		312	347

[a]Total tree biomass estimates differ slightly from the sum of biomass components because of averaging procedures (by plot).
[b]Harris et al. (1973).
[c]Cole and Rapp (1981).
Source: D.W. Johnson, G.S. Henderson, and W.F. Harris. Changes in aboveground biomass and nutrient content on Walker Branch Watershed from 1967 to 1983. Paper presented at Sixth Central Hardwood Forest Conference, Knoxville, Tennessee, February 24–26, 1987 (in press).

1970s caused considerable declines in biomass increment and increases in mortality of shortleaf pine and hickory, respectively (Figs. 6.1, 6.2) (Johnson et al. 1987). These two bark beetle infestations combined caused a considerable increase in total mortality (Fig. 6.2) in all forest types from 1967 to 1983 and substantial declines in live pine and hickory biomass (Fig. 6.3). On the other hand, the bark beetle attacks in the loblolly pine stand were relatively mild, and the increased mortality is thought to be due primarily to self-thinning. The basal area of the loblolly stand in 1973 (33.5 m²/ha) was 60% greater than that considered optimal for loblolly pine in the region (21 m²/ha) and far greater than that of any other forest stand (23.4, 23.3, 22.2, and 23.0 m²/ha for the pine, yellow-poplar, oak-hickory, and chestnut oak stands, respectively, in 1973). By 1983, the loblolly stand had thinned considerably (to 28.3 m²/ha), but it still remained somewhat overstocked relative to standard commercial prescriptions for this area.

Although live pine and hickory biomass decreased, yellow-poplar biomass increased, particularly in the yellow-poplar and pine stands (Fig. 6.3). There was an inverse relationship between pine and yellow-poplar biomass in the pine stands (Fig. 6.3), indicating that yellow-poplar growth increased in response to pine mortality. The net effect was a fairly stable biomass and nutrient content (Fig. 6.1a) but a significant change in species composition in the pine stands. From 1967 to 1983, yellow-poplar increased from 14 to 37% of total biomass, whereas pines decreased from 74 to 42% of total biomass in the pine stands. Associated with this species change was a reduction in foliage biomass in pine stands (Fig. 6.1b). Foliage biomass in the other forest stands remained fairly constant, showing relatively less decline following the hickory bark beetle outbreak than did total biomass (Fig. 6.1b).

In even-aged, monoculture forests, foliage biomass normally stabilizes after crown closure (Switzer and Nelson 1972; Turner 1981; Miller 1981). In uneven-aged, mixed species forests, however, the situation can vary as succession proceeds from coniferous to deciduous species, as has occurred in the pine stands. As the pines dropped out and were replaced by yellow-poplar, foliage biomass declined in pine stands (Fig. 6.1b) because of inherent differences in foliage biomass of these species (e.g., the pines retain 2–3 years' needles, whereas the yellow-poplar has only one set of leaves per year). The stability of foliage biomass in the oak-hickory stands, despite the larger biomass loss due to hickory mortality, is due to the death of primarily large trees with low foliage:wood biomass ratios. It is interesting to note the convergence of foliage biomass in the pine stands with that in the other deciduous forest stands as deciduous species become increasingly dominant in the pine stands (Fig. 6.1b).

There is reason to believe that pine beetle and hickory bark beetle outbreaks have occurred in the past also, especially following droughts. Craighead (1949) states in reference to eastern U.S. bark beetles that "almost without exception outbreaks of these insects occur during a period

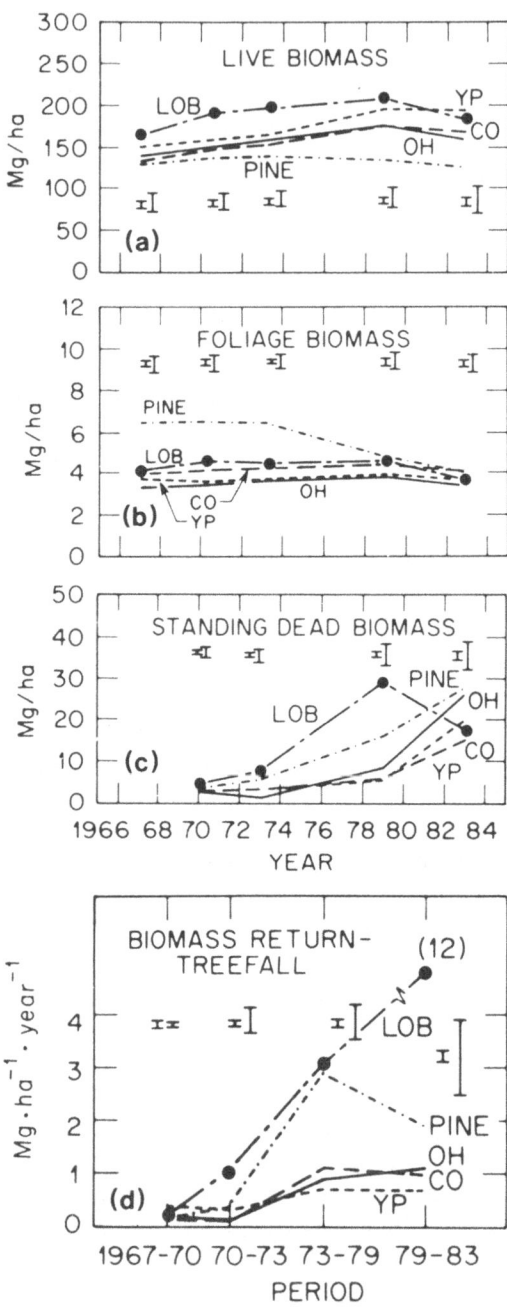

Figure 6.1. Changes in biomass in five forest types on Walker Branch Watershed over 16 years. (a) Total tree biomass; (b) foliage biomass; (c) standing dead biomass; (d) treefall. LOB = loblolly pine type; CO = chestnut oak type; YP = yellow-poplar type; OH = oak-hickory type; PINE = mixed pine-hardwood type. Source: D.W. Johnson, G.S. Henderson, and W.F. Harris. Changes in aboveground biomass and nutrient content on Walker Branch Watershed from 1967 to 1983. IN Proceedings of the Sixth Central Hardwood Forest Conference, Knoxville, Tennessee, February 24–26, 1987. Society of American Foresters, Washington, D.C. (in press).

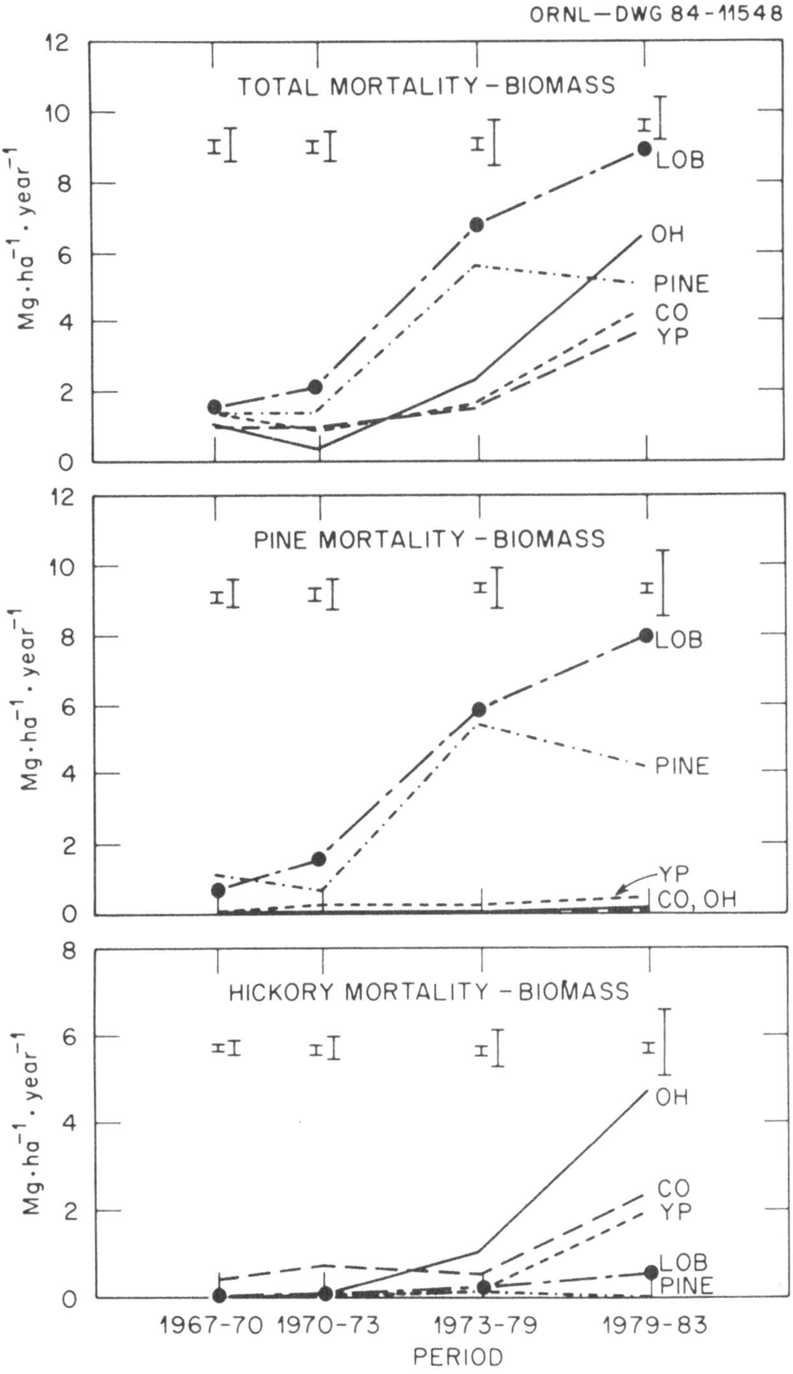

Figure 6.2. Total mortality, pine mortality, and hickory mortality in five forest types on Walker Branch Watershed. See Fig. 6.1 for key. Source: D.W. Johnson, G.S. Henderson, and W.F. Harris. Changes in aboveground biomass and nutrient content on Walker Branch Watershed from 1967 to 1983. IN Proceedings of the Sixth Central Hardwood Forest Conference, Knoxville, Tennessee, February 24–26,1987. Society of American Foresters, Washington, D.C. (in press).

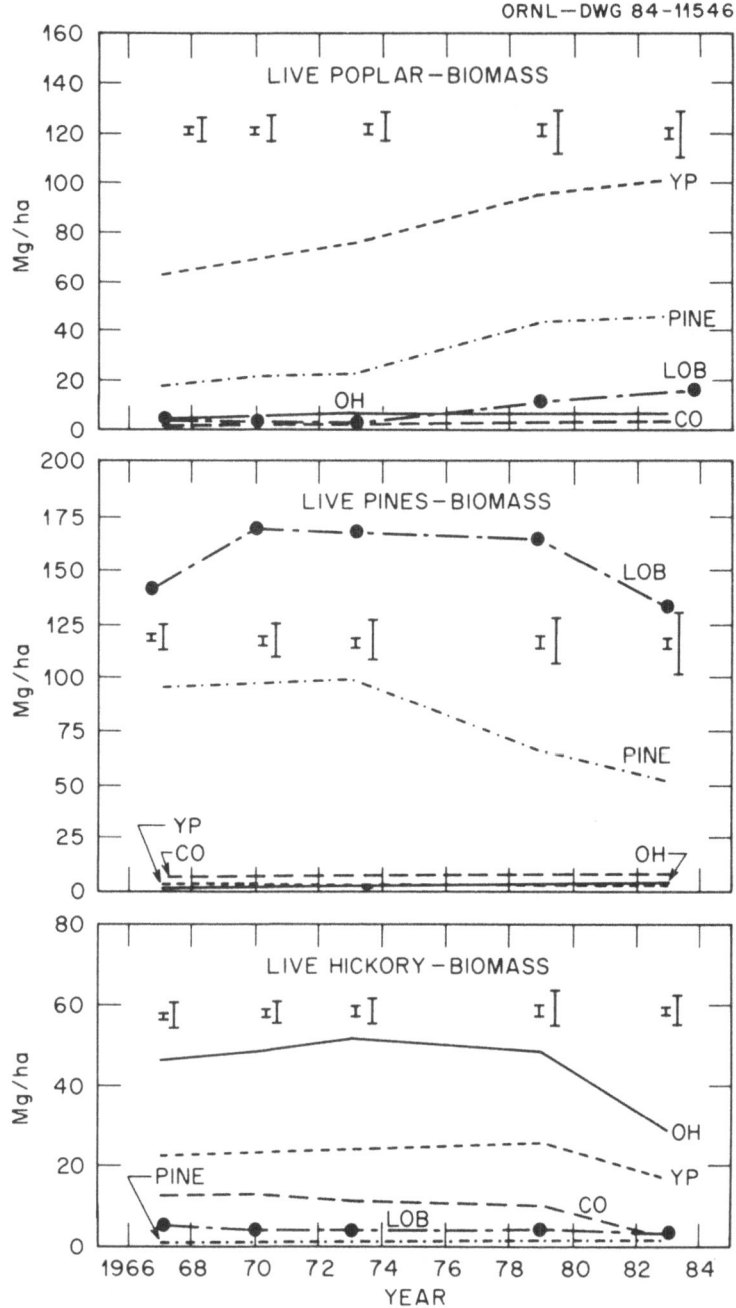

Figure 6.3. Live yellow-poplar, pine, and hickory biomass in five forest types on Walker Branch Watershed. See Fig. 6.1 for key. Source: D.W. Johnson, G.S. Henderson, and W.F. Harris. Changes in aboveground biomass and nutrient content on Walker Branch Watershed from 1967 to 1983. PP 487–496 IN R.L. Hay, F.W. Woods, and H. DeSelm (eds). Proceedings of the Central Hardwood Forest Conference VI., Knoxville, TN, February 24–26, 1987. University of Tennessee Press, Knoxville, TN.

of deficient rainfall. This is particularly true of the southern pine beetle, the hickory bark beetle, and species of *Ips*" (p. 48). Moderate droughts did occur during the spring of 1969 [Palmer Drought Index (PDI) was -2 until June] and in the late summer of 1980 through early spring of 1981 (PDI < -2 from August 1980 until April 1981), which may have contributed to the beetle outbreaks. Other stresses such as air pollution may also have contributed, but the data are insufficient to indicate whether this is the case. Studies of acid deposition effects in a chestnut oak and yellow-poplar stand indicate that no chronic nutritional problems have developed or will develop as a result of acid deposition in the near future (Richter et al. 1983; Johnson and Richter 1984), but other direct effects of air pollution (e.g., Ozone and/or SO_2 damage) cannot be ruled out because studies have not as yet been made. At this time, drought seems the most likely triggering mechanism, especially for the hickory bark beetle whose "outbreaks seem to occur during years in which precipitation is deficient during the summer months" (Craighead 1949, p. 310), as was the case in 1980.

Pine and hickory mortality has also caused a substantial accumulation of biomass in standing dead trees (Fig. 6.1c). Thus, total net aboveground biomass increment declined from 1979 to 1983 as a result of hickory mortality, with a corresponding increase in standing dead biomass (Table 6.2). Estimates in Fig. 6.1c are too large, in that that decay and loss of limbs from standing dead trees are not accounted for. Even so, it is clear that a substantial return of woody litter will occur in localized areas as these trees fall during the next few years. On an area basis, the 1983 estimates of standing dead biomass in the oak-hickory stands rival biomass in the entire forest floor (wood, 01 and 02 horizons). Henderson et al. (1978) reported forest floor biomass of 29.5 Mg/ha for the entire watershed. Harris et al. (1973) estimated a mean residence time of 2 to 3 years for standing dead timber.

Table 6.2. Net annual woody increment and mortality (in $Mg \cdot ha^{-1} \cdot year^{-1}$) in five forest types on Walker Branch Watershed from 1967 to 1983

	Forest type				
	Pine-hardwood	Yellow-poplar	Loblolly pine	Oak-hickory	Chestnut oak
Net woody increment, 1967–1983	2.1	4.1	1.3	3.1	3.5
Mortality, 1967–1983	4.1	1.9	4.2	2.7	1.8

Source: D.W. Johnson, G.S. Henderson, and W.F. Harris. Changes in aboveground biomass and nutrient content on Walker Branch Watershed from 1967 to 1983. IN Proceedings of the Sixth Central Hardwood Forest Conference, Knoxville, Tennessee, February 24–26, 1987. Society of American Foresters, Washington, D.C.

6.2.3 Litter and Soil

Litter mass on the watershed ranged from 17.8 Mg/ha in the yellow-poplar stands to 33.2 Mg/ha in the pine stands in 1970–1971, when it was last measured (Table 6.3). Yellow-poplar total litter mass was significantly ($p < 0.001$) less than in any of the other forest types. There were no significant differences in total litter mass among the other stands. The 01 litter ranged from 44% of the total litter mass in chestnut oak and oak-hickory to 58% in yellow-poplar. Yellow-poplar 01 litter mass was significantly ($p < 0.001$) less than 01 litter in the other stands.

Wood litter (including twigs <2.5 cm diam) ranged from 35% of 01 litter in yellow-poplar stands to 52% in chestnut oak stands. Wood litter (>2.5 cm diam) was significantly ($p < 0.05$) less in yellow-poplar stands than in any of the other stands. There were no significant differences in wood litter among the other forest types. The range of twig litter was narrow among stands but was significantly ($p < 0.05$) greater in pine than in oak-hickory and yellow-poplar.

The 02 litter mass was greatest in the oak-hickory and chestnut oak stands, with no significant difference between the two stands. Pine 02 litter mass was significantly ($p < 0.05$) less than that in oak-hickory and chestnut oak but twice that of yellow-poplar. The lesser amount of yellow-poplar 02 litter was highly significant ($p < 0.0001$) when compared with that from each of the other forest types.

The weighted average total litter mass (01 + 02) for the entire watershed was 30 Mg/ha, assuming an average of 32.8 Mg/ha on 79 ha occupied by pine, oak-hickory, and chestnut oak and 17.8Mg/ha on 18.5 ha of yellow-poplar. Yellow-poplar consistently has less 01 leaf and wood litter and

Table 6.3. Litter biomass estimates (Mg/ha) for Walker Branch Watershed [Values are annual means ± 1 SE based on six seasonal collections of a subsample from six core plots (three replicates/plot) in each stand type ($n = 108$)]

Stand	01 Litter			02 Litter	Total litter	
	Leaves	Twigs[a]	Wood[b]		01	01 + 02
Pine	10.9 ± 0.3	2.1 ± 0.1	4.5 ± 0.7	15.7 ± 0.8	17.5 ± 1.0	33.2 ± 1.0
Oak-hickory	8.7 ± 0.3	1.8 ± 0.1	3.8 ± 0.9	18.1 ± 0.9	14.3 ± 1.1	32.4 ± 1.3
Chestnut oak	7.0 ± 0.2	2.0 ± 0.1	5.5 ± 1.6	18.4 ± 1.1	14.5 ± 1.8	32.9 ± 2.2
Yellow-poplar	6.6 ± 0.2	1.8 ± 0.1	1.9 ± 0.1	7.5 ± 0.5	10.3 ± 0.2	17.8 ± 0.3

[a]Twigs <2.5 cm diam.
[b]Wood >2.5 cm diam.
Source: W.F. Harris, R.A. Goldstein, and G.S. Henderson. 1973. Analysis of forest biomass pools, annual primary production and turnover of biomass for a mixed deciduous forest watershed. pp. 41–64. IN H.E. Young (ed.), Proc., IUFRO Symposium Working Party on Forest Biomass. University of Maine Press, Orono, Maine.

less 02 litter than the other forest types. By contrast, pine has the greatest mass of 01 leaf litter.

Soil organic matter constitutes roughly 35 to 50% of total ecosystem organic matter and is greatest in the more mesic yellow-poplar stands (Table 6.1).

6.2.4 Litter Mass and Fine Root Development

A relationship between 02 litter mass and the strategies of fine-root development among the four forest types appears likely, especially for the deciduous species. Field observations indicate that fine root development in 02 litter and between 02 litter and the mineral soil surface in chestnut oak and oak-hickory stands is much greater than that in yellow-poplar. The 02 litter, besides being rich in available nutrients, has good moisture-holding capacity and is mulched by the 01litter. Trees on dry sites with nutrient-poor mineral soil produce litter that decomposes slowly. The result is thick 01 and 02 litter layers and development of fine roots in the 02 litter. These trees can outcompete species with more rapidly decomposing litter. Yellow-poplar occupies the more mesic sites, produces litter that decomposes rapidly, and develops most of its fine roots below the litter layers in the relatively moist and nutrient-rich upper 10cm of mineral soil.

6.3 Carbon Fluxes

In the previous section we discussed productivity in terms of net wood production and net biomass accumulation, as determined by direct measurements of mass pools and increments. In this section we will emphasize biotic mechanisms that regulate productivity and that, in turn, are regulated by environmental factors. It is through a better understanding of mechanisms at both the individual tree level and the ecosystem level that we hope to improve our capability to evaluate the effects of various kinds of stress (e.g., atmospheric pollution) on individual components of the forest and ultimately on total forest ecosystems. We will discuss these mechanisms in terms of carbon transfers within individual trees and among ecosystem components.

Carbon compounds provide the energy to run the system. Thus, the efficiency of carbon utilization is critical to net ecosystem production (NEP). For example, in a yellow-poplar forest adjacent to Walker Branch Watershed, we found a relatively broad margin between carbon fixed by photosynthesis (gross primary production—GPP) and carbon evolved as a result of energy use for maintenance and growth (autotrophic respiration—R_A) (Harris et al. 1975). However, we also found a very narrow margin between GPP and total ecosystem respiration (R_E). Thus, while net primary production (NPP) was relatively high, NEP was extremely low.

Although studies on Walker Branch Watershed have not been broad enough to quantify all the parameters necessary to estimate NPP and NEP, ^{14}C radiolabeling experiments have provided insight into carbon allocation and utilization within mature white oak trees. In addition, carbon fluxes to the forest floor and forest floor respiration rates (both key parameters in estimating NEP) have been measured.

6.3.1 Carbon Allocation in White Oak

Seasonal patterns of temperate zone climate produce alternating cycles of rapid tree growth and relative inactivity accompanied by corresponding periods of intense utilization or storage of the tree's food reserves. Canopies of late successional deciduous species, such as white oak, typically grow as a single rapid flush from terminal buds after a period of dormancy, whereas evergreens typically grow from repeated flushes of buds. These flushes of growth in evergreens are seasonal, but the year-round maintenance of photosynthetic leaves eliminates the necessity to develop a new canopy at the beginning of each growing period. Thus, deciduous trees show greater fluctuation in their energy reserves than do evergreen species. The storage and mobilization of these reserves play a significant role in competitive strategies of temperate deciduous forest species. A number of carbon allocation experiments have been performed on a representative oak species (*Quercus alba* L.—white oak) on Walker Branch to help us to understand these strategies.

Two types of ^{14}C-tagging experiments were performed on eight mature white oaks on Walker Branch. The first used $^{14}CO_2$ and in situ inflated polyethylene bags around branchlets and leaves to incorporate ^{14}C as photosynthate into two trees (McLaughlin and McConathy 1979; McLaughlin et al. 1979). The $^{14}CO_2$ experiment was designed to relate seasonal changes in allocation of ^{14}C-labeled photosynthate to seasonally changing demands on tree growth and development. Trees were tagged at intervals throughout the growing season. Foliage and branch (<5 years old) samples were collected 7 d after each tag and analyzed for ^{14}C content. Results are summarized in Figs. 6.4 and 6.5. Note the large influx of carbon from larger branches into young branches and leaves in April (Fig. 6.4). Forty-one percent of the ^{14}C-photosynthate remained in the new foliage 7d after the $^{14}CO_2$ was taken up, but an estimated 94% of the ^{14}C-photosynthate was required for respiration during canopy development. To fill this requirement, carbon was translocated into the leaves and twigs from adjacent branch and bole tissues. By May, the canopy was fully developed, and respiratory demands were less (because of decreased leaf growth and expansions), resulting in availability of photosynthate for translocation from leaves and young branches basipetally (Fig. 6.5). In fact, translocation of

DATE   ALLOCATION OF PHOTOSYNTHATE IN WHITE OAK

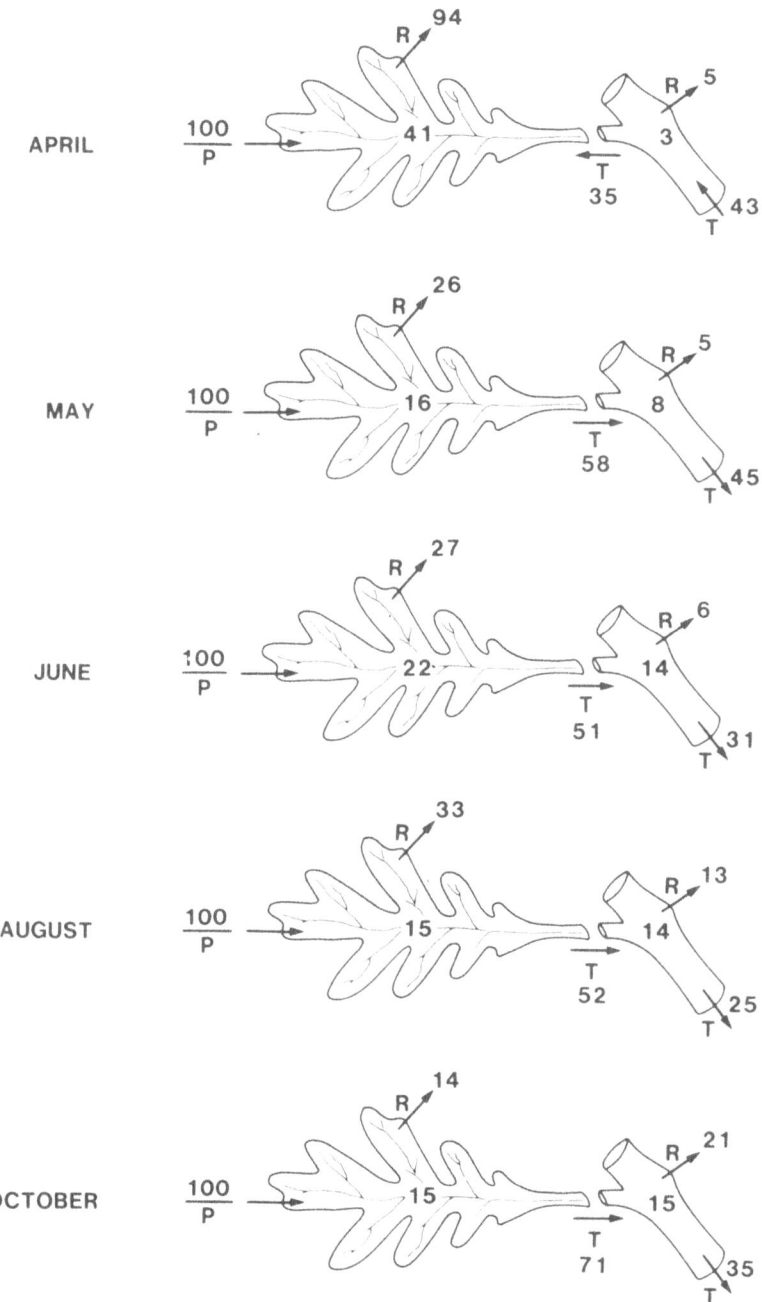

Figure 6.4. Seasonal distribution of ^{14}C-labeled photosynthate in white oak as determined 7d after uptake of $^{14}CO_2$ by foliage and branches of two trees (data from McLaughlin and McConathy 1979). Values are percentages of initial ^{14}C photosynthate. P = photosynthate; R = respiratory losses; and T = translocation. Numbers within figures represent percentages of photosynthate incorporated into tissues. Branches also included buds and acorns when present.

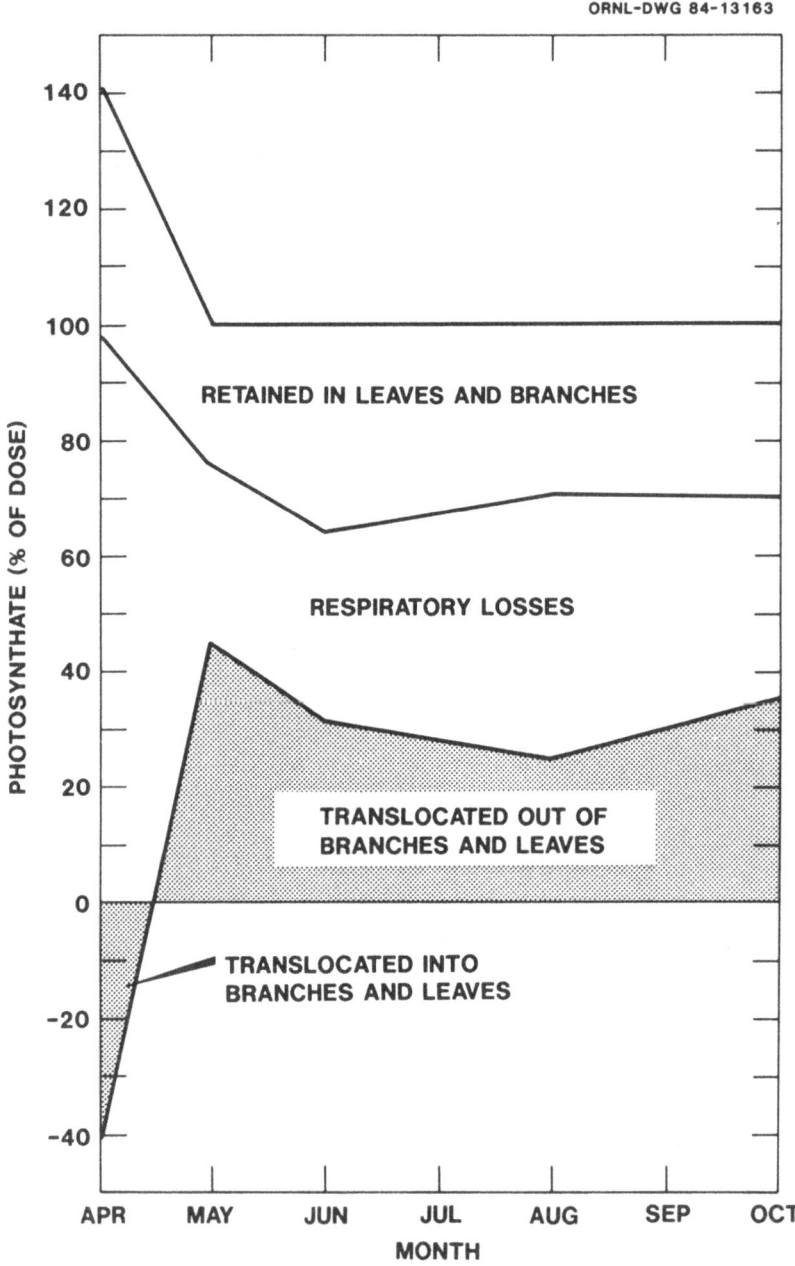

ORNL-DWG 84-13163

Figure 6.5. Seasonal distribution of ^{14}C-labeled photosynthate in white oak determined 7d after uptake of $^{14}CO_2$ by foliage and branches (<5 years old) of two trees. Values for leaf and branch respiratory losses and retention from Fig. 6.4 are summed. Translocation values from branches are shown to represent the amount of photosynthate translocated basipetally. (Data from McLaughlin and McConathy 1979.)

photosynthate from the canopy peaked in May. This peak translocation was in synchrony with peak energy demands by the roots of three tree species, including oak located in a nearby second-growth mixed deciduous forest stand (Edwards et al. 1977). Replenishment of storage reserves in older tree tissues continued from May through October, but with slightly depressed translocation rates from young branches in August as respiratory demands increased in these tissues. As a result of high respiration rates and retention in leaves and young branches, availability of photosynthate for export to tissues basipetal to the crown remained below 50% of gross photosynthesis during the entire growing season.

Young branches, which comprised 12% of total tree biomass, retained an average 13% of gross photosynthate, whereas translocation to tissues below the tree crown, which contained 85% of the total tree biomass, averaged only 34% of gross photosynthate. We suspect that a high percentage of the translocated photosynthate went to the roots. This is based on findings by Harris et al. (1977) that ~40% of NPP in yellow-poplar is translocated to the roots. Intuitively, this energy retention in relatively small fractions of tree biomass makes sense because of high energy demands for new growth of fine roots in late winter and for rapid canopy development in early spring. Translocation from leaves into young branches remained relatively constant from May through August (52–58% of gross photosynthesis) but increased dramatically to 71% in October just prior to leaf fall. This increase coincided with decreased respiration rates by leaves. Thus, energy that would have been lost with leaf fall was retained by the trees. A similar strategy of element conservation, especially nitrogen, has been observed in deciduous tree species.

The second experiment, which utilized ^{14}C-sucrose as a tracer instead of $^{14}CO_2$, was designed to characterize changes in whole-tree biochemical constituents and to relate these changes to the phenology of tree growth and development. The ^{14}C-sucrose was introduced into six white oak tree boles by the trough-uptake technique described by McLaughlin et al. (1977). Two trees were radiolabeled in February, two in May, and two in September. One week after labeling and at three subsequent 1-month intervals, tissues were collected from leaves, twigs, branches, boles, and roots for biochemical and radiochemical analysis. At the end of each study period, two trees were harvested for final chemical analysis and determination of biomass distribution among tissue types.

Biochemical data and weight data averaged over all sample periods are illustrated in Fig. 6.6. The largest chemical component was holocellulose, which ranged from 36% of tissue dry weight in leaves to 77% in boles. The second largest component was lignin, ranging from 14.7% in the bole to 32.7% in the leaves. A labile component (which includes amino acids, organic acids, low-molecular-weight phenols, polar pigments, and oxygenated isoprenols) was the third largest component and ranged from 3.6% in large roots to 17% in the leaves. Sugars are labile but were analyzed separately because of their important role as quick energy sources. Sugars

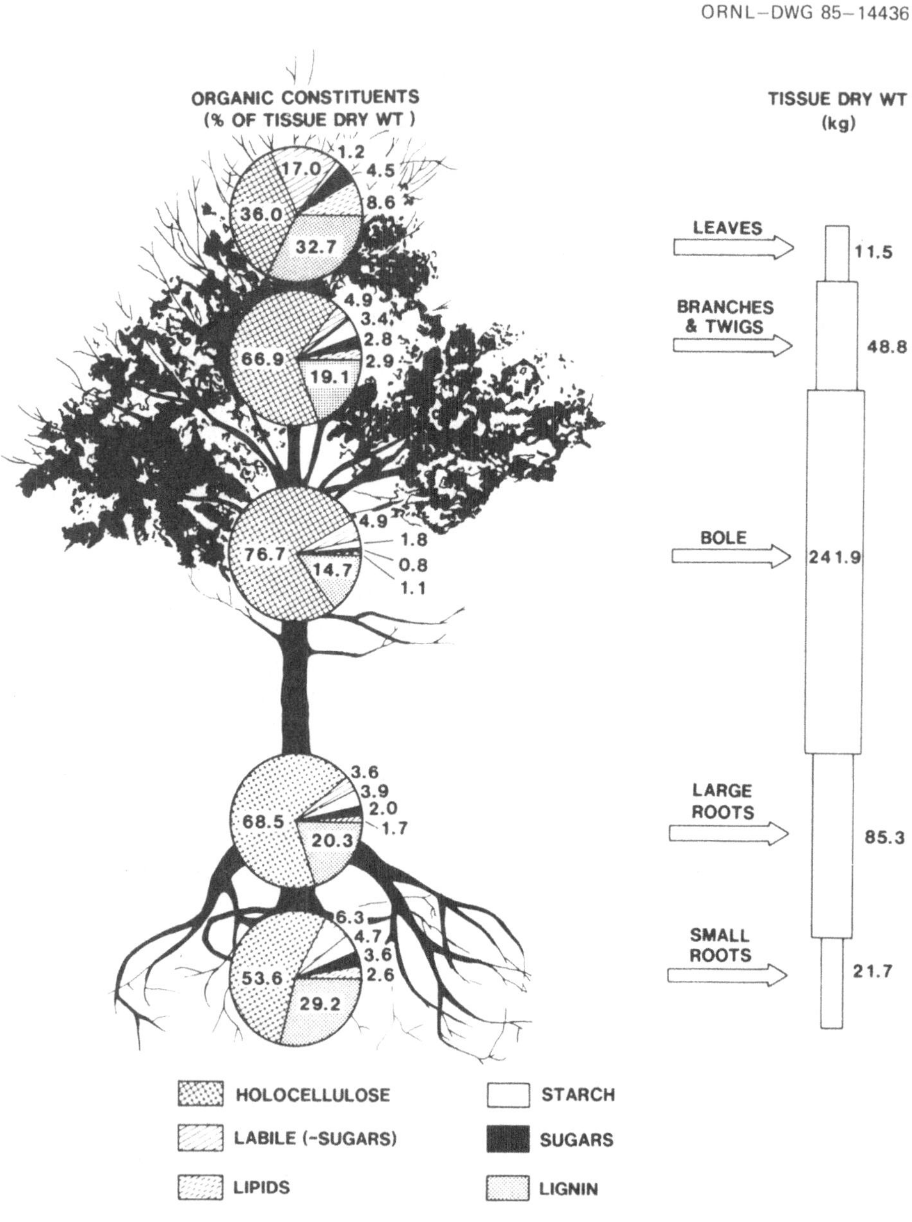

Figure 6.6. Biomass and biochemical constituents of a typical mature white oak. Values are averages from six trees collected over a period of 11 months. The area of rectangular boxes depicts relative weights of individual tree tissues. (Data from McLaughlin et al. 1980.)

were the smallest component overall and ranged from 0.84% in boles to 4.5% in leaves. Lipids and starches averaged across all tissue types were about equal in concentration and were present in only slightly greater concentrations than sugars. Lipids ranged from 1.1% in boles to 8.6% in leaves, and starches ranged from 1.2% in leaves to 4.7% in small roots. Leaves had the greatest concentrations of lipids, sugars, labile components (minus sugars), and lignin. Boles contained the greatest concentrations of holocellulose, but these concentrations were only slightly greater than those in branches and large roots. Roots contained the greatest concentrations of starch, with the concentration in small roots slightly greater than that in large roots. Large roots contained the lowest concentrations of the labile-sugar component, and leaves contained the lowest concentration of holocellulose. Boles contained the lowest concentrations of lipids, sugars, starches, and lignin, but, because of the relatively large bole biomass, all components were greatest in the bole in absolute amounts.

Temporal changes in whole-tree concentrations of the biochemical constituents indicated chemical conversions from one chemical form to another. The greatest fluctuations occurred in holocellulose and in labile and starch fractions (Fig. 6.7). Note that the spring decline in the labile

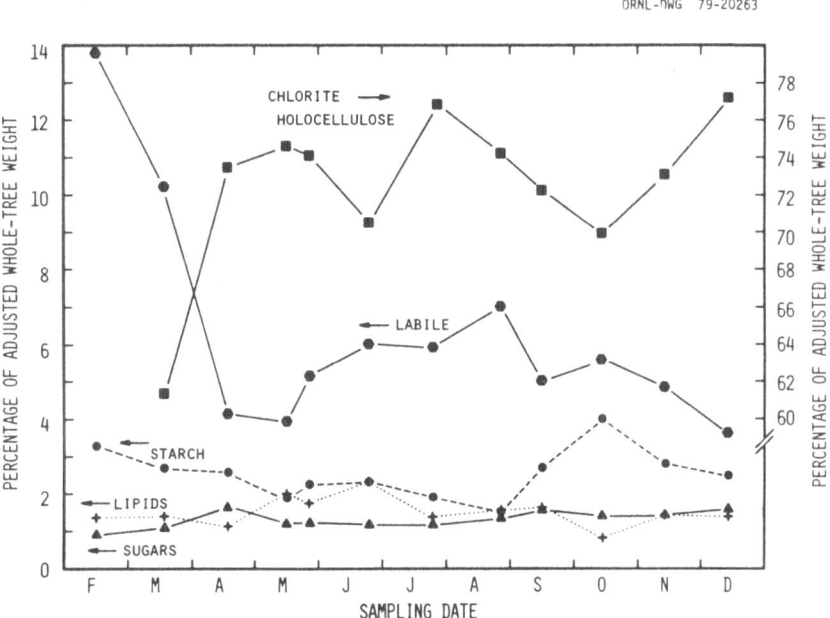

Figure 6.7. Seasonal changes in whole-tree levels of energy reserves in white oak. Source: S.B. McLaughlin, R.K. McConathy, R.L. Barnes, and N.T. Edwards. 1980. Seasonal changes in energy allocation by white oak (*Quercus alba* L.). Can. J. For. Res. 10(3):379–388.

constituents corresponds well, in magnitude and temporally, with an increase in chlorite holocellulose. Other studies (Kimura 1969; Meyer and Splittstoesser 1971) indicate that hemicelluloses, which include the holocellulose fraction, may serve as food reserves in trees. The buildup of holocellulose that we found in white oak parallels the buildup of hemicellulose shown by Kimura (1969) in *Abies*. The spring holocellulose increase also corresponds well to the shift in ^{14}C activity from the labile to the nonlabile compartment shown in Fig. 6.8. However, even with a significant decrease in the labile component between February and March,

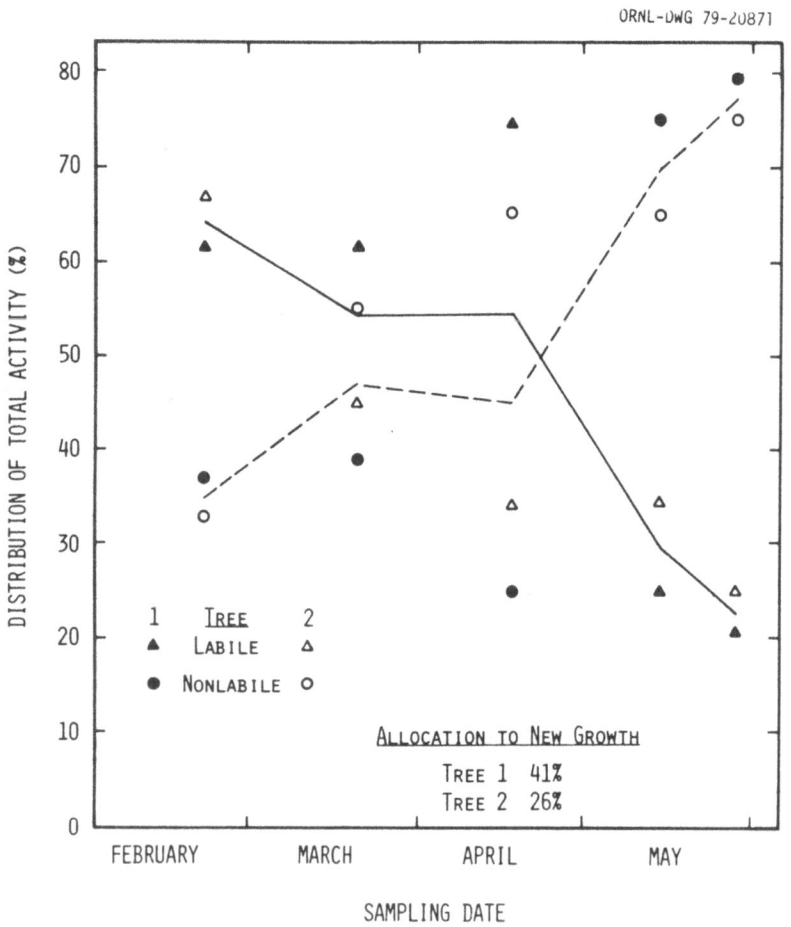

Figure 6.8. Seasonal changes in apportionment of ^{14}C activity between labile (solid line) and nonlabile (dashed line) tissue constituents of two white oak trees labeled with ^{14}C-sucrose. Source: S.B. McLaughlin, R.K. McConathy, R.L. Barnes, and N.T. Edwards. 1980. Seasonal changes in energy allocation by white oak (*Quercus alba* L.). Can. J. For. Res. 10(3):379–388.

the labile concentration remained high throughout the growing season relative to other constituents, except for holocellulose.

Seasonal fluctuations of biochemical constituents were more pronounced in individual tissues than in whole-tree contents. This was expected because of the added effect of translocation between individual tissues. For example, small roots, which contained the highest concentrations of starches, when averaged for all sample dates, fluctuated from two peaks (February–March and September–November) of ~9% of tissue dry weight to a fairly consistent ~2 to 4% during the rest of the year. These two peaks corresponded well temporally to rooting intensity in white oak (Teskey and Hinckley 1981) and to fine root development in yellow-poplar (Edwards and Harris 1977). Thus, the two peaks may reflect translocation of carbohydrates into small roots and subsequent storage as starch to be used as an energy source for growth of new tissues.

Starch concentration in large roots followed a similar but less pronounced seasonal pattern. Starch content in the bole showed little seasonal variability. Canopy starch concentrations were relatively low May through August (~2%) and ranged from ~3 to 6% during the rest of the year. A decline in branch and twig starch content beginning in April and a subsequent increase in the fall reflect mobilization of energy reserves for canopy development, followed by recovery of these reserves during the growing season.

Labile constituents declined rather abruptly in the spring in all tissues (Fig. 6.9), indicating a conversion to structural tissues as discussed earlier. There is also a suggestion of translocation of labile constituents from the lower bole to the upper bole between February and March. This would correspond temporally to a demand for energy and structural growth material for canopy development.

McLaughlin et al. (1980) used the biochemical energy equivalence relationships of Chung and Barnes (1977) to calculate energy requirements for canopy development in a 410-kg white oak (the average size of the six white oaks sampled). Based on their calculations, canopy synthesis would cost the amount of energy found in 17.7 kg of glucose. Only 13k g of this energy came from within the canopy. The $^{14}CO_2$ experiments discussed earlier support this requirement for energy from other parts of the tree as well as a rapid recovery of the energy invested during early spring canopy development.

6.3.2 Litterfall

These whole-tree carbon allocation studies provide a better understanding of the strategies that trees have evolved for efficient use of carbon acquired through photosynthesis. We think similar carbon allocation strategies have evolved in total ecosystems. For example, deciduous trees have evolved the strategy of dropping their leaves each year, which enables them to survive the winter with less energy loss than if the leaves were retained.

ORNL-DWG 79-20869

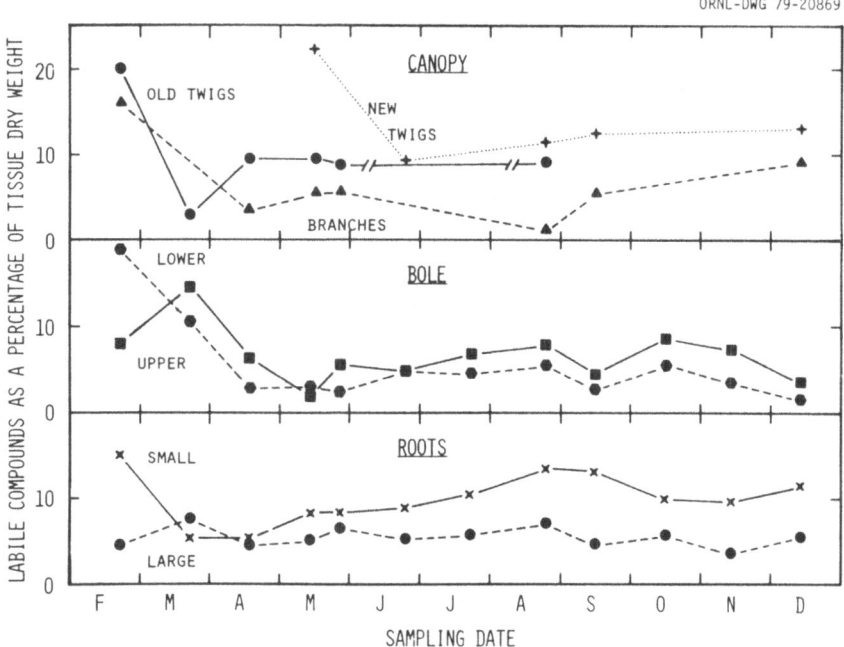

Figure 6.9. Seasonal changes in tissue contents of labile compounds. Source: S.B. McLaughlin, R.K. McConathy, R.L. Barnes, and N.T. Edwards. 1980. Seasonal changes in energy allocation by white oak (*Quercus alba* L.). Can. J. For. Res. 10(3):379–388.

Litterfall on Walker Branch Watershed in 1971 ranged from 4.13 Mg/ha in pine stands to 4.8 Mg/ha in the oak-hickory stands (Table 6.4). Leaves accounted for most of the litterfall, ranging from 82% of the total in pine to 88% in chestnut oak. The only significant difference in leaf fall among the four major forest types was between pine and oak-hickory, with oak-hickory leaf fall significantly ($p < 0.05$) greater than pine. Grizzard et al. (1976) reported similar amounts of litterfall for 1969 on Walker Branch: a range of 4.4 to 4.9 Mg/ha, with leaves contributing from 77 to 87% of total litterfall. No significant differences in litterfall were found between any of the forest types.

Woody litterfall (including twigs <2.5 cm and branches >2.5 cm) was greatest in pine (18% of total litterfall) and least in chestnut oak (12% of the total). Grizzard et al. (1976) also reported pine as having the greatest amounts of branchfall (11% of the total) in 1969, with branchfall in all stands contributing less than that reported by Harris et al. (1973) for 1971.

Peak deciduous leaf fall occurred in November, with an average of 72% of total leaf fall occurring between mid-October and early December (Grizzard et al. 1976). Pine leaf fall peaked in December and was more

Table 6.4. Litterfall estimates (Mg·ha^{-1}·year^{-1}) for Walker Branch Watershed [Values are annual means ±1 SE (n = 24)]

Stand	Leaf litter	Twigs (<2.5 cm)	Wood (>2.5 cm)	Total
		Litterfall		
Pine	3.4 ± 0.2	0.2 ± 0.04	0.53	4.13
Oak-hickory	4.1 ± 0.2	0.4 ± 0.09	0.30	4.80
Chestnut oak	3.9 ± 0.2	0.4 ± 0.09	0.15	4.45
Yellow-poplar	3.7 ± 0.2	0.3 ± 0.11	0.33	4.33

Source: W.F. Harris, R.A. Goldstein, and G.S. Henderson. 1973. Analysis of forest biomass pools, annual primary production and turnover of biomass for a mixed deciduous forest watershed. pp. 41–64. IN H.E. Young (ed.), Proc., IUFRO Symposium Working Party on Forest Biomass. University of Maine Press, Orono, Maine.

spread out over the year, with only 52% falling between mid-October and early December. There was no distinct seasonal pattern of branchfall in 1969 except for pine, where 25% of the annual branchfall occurred in early December.

6.3.3 Litter Decomposition and Forest Floor Respiration

Energy lost from trees in litterfall is utilized by decomposer organisms. Through the growth, reproduction, and death of these decomposers, nutrients in litter are mineralized, immobilized, and remineralized. The synchrony of mineralization by the decomposers and uptake by the autotrophs is the primary ecosystem strategy for preventing the leaching of nutrients from the rooting zone. Thus, litter decay rates and the temporal pattern of decay are extremely important in terms of the nutrient stability of forests. However, the amount of carbon allocated to the decomposers must be small enough to result in a net carbon gain (a positive NEP) or at least to maintain an equilibrium to prevent a decline in total ecosystem biomass.

On Walker Branch, differences in litter mass on the forest floor are apparently due to differences in decay rates, because litterfall mass does not differ significantly among forest types. The reasons for the different decay rates can be attributed to variation in microclimate among the various stands and to differences in litter quality (e.g., nutrient content). For example, yellow-poplar stands, with the most rapidly decaying litter, occupy the most mesic sites. By contrast, pine stands, with the slowest decaying 01 litter, occupy the driest sites on poor soils and have the most nutrient-poor 01 litter (Table 6.5). Note that nitrogen and phosphorus concentrations of pine 02 litter, which decomposes relatively fast, are about equal to concentrations in 02 litter of the other stands. Oak-hickory litter consistently has the highest concentrations of N, P, K, and Ca, but never-

Table 6.5. Nutrient concentrations (%) of litter on Walker Branch Watershed[a]
(Values were calculated from dry weight and nutrient mass data)

| Stand type | 01 Litter | | | 02 Litter |
	Leaves	Twigs	Wood	
		Nitrogen		
Pine	1.07	0.39	0.22	1.11
Chestnut oak	1.17	0.44	0.23	1.15
Oak-hickory	1.24	0.56	0.28	1.30
Yellow-poplar	1.21	0.56	0.29	1.14
		Phosphorus		
Pine	0.065	0.021	0.011	0.070
Chestnut oak	0.065	0.026	0.010	0.072
Oak-hickory	0.077	0.035	0.015	0.084
Yellow-poplar	0.071	0.031	0.018	0.073
		Potassium		
Pine	0.076	0.039	0.036	0.072
Chestnut oak	0.086	0.067	0.037	0.112
Oak-hickory	0.093	0.088	0.055	0.101
Yellow-poplar	0.083	0.075	0.197	0.101
		Calcium		
Pine	1.08	0.46	0.36	0.86
Chestnut oak	1.67	1.19	0.44	1.12
Oak-hickory	2.11	2.25	1.36	1.82
Yellow-poplar	2.11	1.69	0.72	1.83

[a]Data from G.S. Henderson (1972).

theless decays more slowly than litter in the more mesic yellow-poplar stands. Also, the relatively high lignin content of oak and hickory leaves as compared to yellow-poplar leaves results in slower decay of the former.

Rates of litter decay were estimated for a single chestnut oak stand, using CO_2 efflux rates measured by soda-lime adsorption (Edwards 1982). Carbon dioxide efflux from litter was relatively small (23% of the total) when compared with that from other forest floor components. However, this estimate does not include CO_2 efflux from woody litter >0.5 cm in diameter. Annual CO_2 efflux from surface litter (01 + 02) totaled 5073 kg CO_2/ha in 1980–1981 (Table 6.6). This was equivalent to 3450 kg/ha annual weight loss, which is 80% of average annual litterfall in the chestnut oak stands. The weight loss estimate represents an upper limit because of the method used to calculate weight loss from litter CO_2 efflux. In all conversions from CO_2 efflux to weight loss (Table 6.6), the standard conversion factor is 0.68, based on the relative amount of CO_2 produced when carbohydrates are respired. Other compounds, such as amino acids and

Table 6.6. Estimates (kg/ha) of annual CO_2 efflux and weight loss from forest floor components of chestnut oak stands

Component	CO_2 efflux	Weight loss[a]
Litter	5,073[b]	3,450
Roots (<0.5 cm diam)	8,946[c]	6,083
All other roots	7,190[d]	4,890
Soil organic matter	1,171[e]	796
Total	22,380[f]	15,219

[a]Calculated from estimates of CO_2 efflux, using 0.68 as the multiplier to convert CO_2 to mass respired, assuming respiration of glucose.
[b]Annual CO_2 efflux computed from average rates measured over 24-h periods at quarterly intervals over 2 years. Does not include woody litter.
[c]Annual CO_2 efflux computed from quarterly respiration measurements of fine roots in the upper 10 cm of soil (includes live and dead roots).
[d]Assumes respiration rate of large roots to be 25% of fine-root respiration (Edwards and Harris 1977). Rates applied to all lateral and stump root mass (Harris et al. 1973).
[e]Calculated by mass balance.
[f]Based on 24-h field measurements taken at least once a month. Does not include large woody dead material.

fatty acids found in living and dead tissues, respire with varying amounts of CO_2 released per unit of compound respired, but the amount of CO_2 released per unit weight loss of such compounds is greater than that for carbohydrates.

Based on our calculations, the chestnut oak stands (during the period of measurement) are accumulating a minimum of 850 kg/ha of litter detritus (4300 kg/ha litterfall minus 3450 kg/ha litter weight loss) on an annual basis, excluding woody litter. This would amount to 25.5 Mg of litter accumulation over a period of 35 years (approximately the length of time between the last major disturbance in this forest and the time of litter biomass measurements). In fact, the value is in relatively close agreement with the nonwoody litter biomass (27.4 Mg) in the chestnut oak stand (Table 6.3).

Annual respiratory weight loss of soil organic matter was estimated to be 1171 kg/ha. Much of this soil organic matter is derived from an undetermined but theoretically large amount of dead roots.

The temporal pattern of litter decay is highly dependent on two abiotic regulators (temperature and moisture) that vary seasonally and from year to year. Both the autotrophs and the heterotrophs respond quickly to changes in these regulators. In fact, it is this phenomenon that is primarily responsible for the synchrony of mineralization and uptake discussed above.

In this study litter decay was underestimated by a temperature-based regression equation from April through June and October through December, and overestimated in August and September (Fig. 6.10a). This was due in part to moist conditions in the earlier parts of the growing season and dryer conditions in late July through September. However, the major factor in determining CO_2 efflux from litter per unit area is the amount of litter available for decay (Fig. 6.11). Respiration rates per gram of litter were less in November than in September. This decrease in rates per gram of litter is due to cooler temperatures and a lag in microbial colonization of fresh leaf litter. Nutrients leaching from senescent canopy leaves and from fresh surface litter would perhaps be immobilized during the warmer season by rapidly growing microbial populations. Instead, the nutrients are available for increased tree root growth during October and November (Edwards and Harris 1977). Thus, the peak in total CO_2 efflux in October (Fig. 6.10a) probably reflects increased metabolism of growing roots and not increased decay rates.

Year-to-year variations in CO_2 efflux from the forest floor are demonstrated in Fig. 6.10b. Note that the reduction in total forest floor CO_2 efflux that occurred in July through September in 1981 had not occurred in 1980. The respiration differences during this period were probably due to moisture differences. Rainfall amounts in July, August, and September of 1980 were 7.1, 12.2, and 17.5 cm greater, respectively, than the amounts of rainfall in the same months in 1981. The seasonal patterns of total forest floor CO_2 efflux are in close agreement with those observed in two other nearby forest stands, as reported by Edwards and Ross-Todd (1983) and Edwards and Harris (1977).

Annual total CO_2 efflux from this chestnut oak forest floor (Table 6.6), representing respiration losses from litter, soil organic matter, and roots, was 22,380 kg CO_2/ha. The forest floor of an oak-hickory stand at a nearby site on comparable soil produced 19,400 kg $CO_2 \cdot ha^{-1} \cdot year^{-1}$ (Edwards and Ross-Todd 1983), whereas a more mesic yellow-poplar forest floor evolved 39,053 kg $CO_2 \cdot ha \cdot year^{-1}$ (Edwards and Harris 1977). In all three of these forests, roots contributed more than twice as much CO_2 as any other component.

6.3.4 Root Respiration

Annual CO_2 efflux from roots (live and intact dead) accounted for 72% of the total CO_2 evolved from the forest floor of the chestnut oak stand. This estimate is probably a maximum, because the roots were removed from the soil for measurement, resulting in a stimulation of respiration rates. This disturbance effect was reduced, however, by recording measurements only after the initial peak respiration rate had subsided and

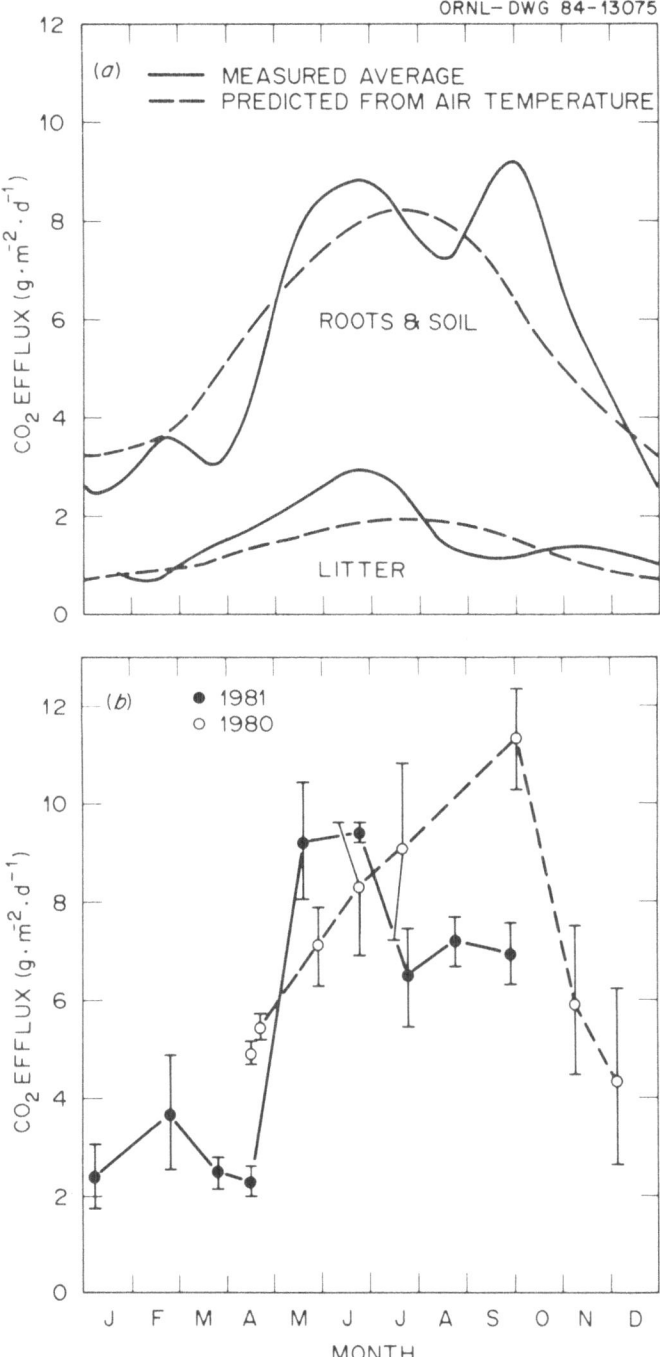

Figure 6.10. (a) Average measured and predicted rates of CO_2 efflux from litter and forest floor of a chestnut oak stand on Walker Branch Watershed in 1980 and 1981. (b) A comparison of forest floor CO_2 efflux in 1980 and 1981 in the same forest stand. Values are means and standard errors ($n = 3$).

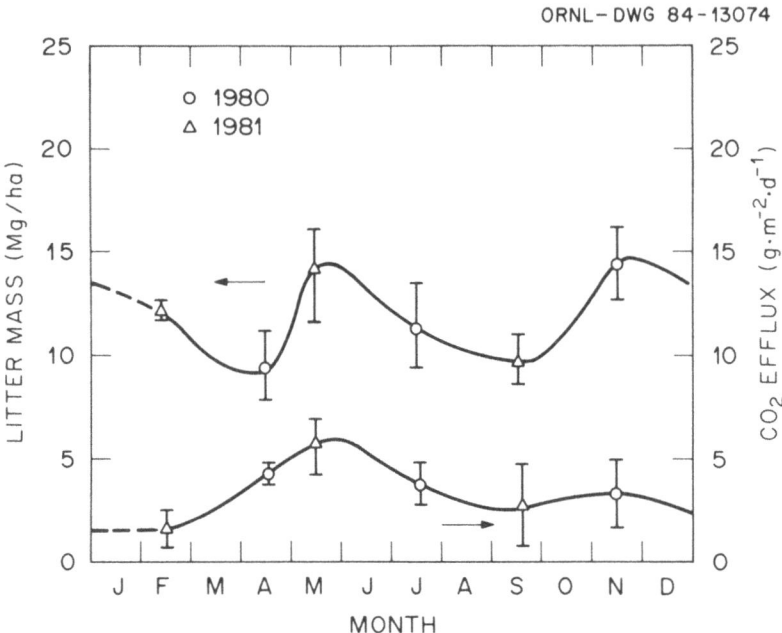

ORNL–DWG 84–13074

Figure 6.11. Seasonal pattern of leaf litter mass and CO_2 efflux from leaf litter in a chestnut oak stand on Walker Branch Watershed. Values are means and standard errors ($n = 3$).

after the rates had stabilized at a lower level. Respiration rates of fine roots (<0.5 cm diam) in the upper 10 cm of soil ranged from 0.18 $mg \cdot g^{-1} \cdot d^{-1}$ in February to 7.69 $mg \cdot g^{-1} \cdot d^{-1}$ in April, with an annual average of 3.09 $mg \cdot g^{-1} \cdot d^{-1}$. These rates are very similar to those reported for yellow-poplar roots (Edwards and Harris 1977). Respiration rates of large roots (0.5 cm diam) were not measured on Walker Branch. Therefore, rates for large roots were estimated by assuming large root respiration to be 25% of fine root respiration (Edwards and Harris 1977). Applying the annual average root respiration rates to the average root biomass (Harris et al. 1973), total annual CO_2 efflux from roots in the chestnut oak stand was calculated to be 16 Mg/ha. Fifty-five percent of that total was from fine roots and 45% from larger roots. Using the 0.68 conversion factor discussed earlier, the annual root mass loss (including live root respiration) was calculated. Fine-root mass loss was 6.1 $Mg \cdot ha^{-1} \cdot year^{-1}$, and large root mass loss was 4.9 $Mg \cdot ha^{-1} \cdot year^{-1}$, for a total of 11 $Mg \cdot ha^{-1} \cdot year^{-1}$. These estimates should not be confused with root turnover (decay), because much (perhaps most) of the CO_2 efflux is from maintenance and growth respiration of live roots in which carbohydrates stored in the living root cells are catabolized. Note, for example, the small peak in total CO_2 efflux in February and again in September (Fig. 6.10a). These peaks correspond with peaks in fine-root biomass in the yellow-poplar stand (Harris etal. 1973). The peak total CO_2 efflux in May and June (in excess of temperature response) may represent CO_2 efflux resulting from the subsequent dieback and decay of fine roots during April, May, and June. This hypothesis is supported by $^{14}CO_2$ efflux data collected from roots of three tree species (yellow-poplar, shortleaf pine, and red oak) injected with ^{14}C-sucrose (Edwards et al. 1977). Their data showed a rapid increase in $^{14}CO_2$ efflux from oak and yellow-poplar roots in May and June, followed by a rapid decline in July and August. A computer simulation of yellow-poplar root processes by Shugart et al. (1977) suggested that much of the mid-summer root mass dies, and, as these roots decompose, their carbohydrates are converted rapidly to CO_2 and H_2O. The rapid decline in respiration in July and August may reflect depletion of the decomposing root carbohydrates. No measurements of temporal patterns of live and dead root mass have been made in any of the forest stands on Walker Branch. However, based on the temporal pattern of total forest floor CO_2 efflux and the very large contribution of root respiration to this total, the existence of a temporal pattern of root growth and turnover in chestnut oak similar to that in yellow-poplar is probable.

6.4 Comparisons with Other Forests

6.4.1 Productivity

Data sets from 14 International Biological Program (IBP) forest study sites (DeAngelis et al. 1981) were selected for comparison with Walker Branch data. These 14 study sites plus the five forest types on Walker Branch and two oak forest sites (Camp Branch and Cross Creek, Tennessee) were divided into three categories (Table 6.7). The categories were (1) oak-dominated and mixed deciduous, (2) pine-hardwood, and (3) pine plantations.

In every forest of the oak-dominated and mixed deciduous category, current annual woody increments were greater than their average lifetime annual average increments (Table 6.7). For example, the current annual wood increment of the oak-hickory forest at Coweeta, North Carolina, is over twice its lifetime annual average increment. Current woody increments on Walker Branch range from 1.3 to 1.7 times greater than their lifetime average increments.

An uncertainty in our calculations of lifetime annual average increments is the ages of the different forests. Increment corings of Walker Branch trees revealed that most of the area is occupied by overstory trees about 70 years old but that the age range is about 45 to 165 years (T.W. Doyle, Oak Ridge National Laboratory, personal communication, 1985). Thus, for uneven-aged forests, we assumed an average age equal to the age of

Table 6.7. A comparison of wood biomass and annual wood increments across a range of age classes and from forests representative of forest types found on Walker Branch Watershed [Data other than those indicated by footnotes are from DeAngelis et al. (1981)]

Forest type	Location	Age (years)	Precipitation (mm)	Temperature (°C)	Wood increment $(Mg \cdot ha^{-1} \cdot year^{-1})$ Lifetime annual average[a]	Current annual average
Oak-pine coastal	Brookhaven, N.Y.	43	1240	9.8	1.4	4.5
Oak	Sikokut, Hungary	65	582	9.9	3.1	3.3
Oak-hickory	Walker Branch, Tenn.	70	1400	13.3	2.1	3.1
Chestnut oak	Walker Branch, Tenn.	70	1400	13.3	2.6	3.5
Yellow-poplar	Walker Branch, Tenn.	70	1400	13.3	2.4	4.1
Oak	Cross Creek, Tenn.[b]	70	1400	13.3	2.1	3.1
Oak	Camp Branch, Tenn.[b]	80	1400	13.3	1.8	2.1
Oak-hickory	Coweeta, N.C.	80	1945	12.6	1.7	3.6[c]
Mixed deciduous	Meathopwood, U.K.	80	1115	7.8	1.5	3.1
Quercetum-Pilosi-Caricosum	Belgorod region, U.S.S.R.	80	537	6.0	3.1	5.2
Oak-hornbean	Ispina, Poland	100	729	7.8	2.6	5.3
Oak	Noa Woods, Wis.	130	777	6.9	2.0	3.5
Oak	Linnebjer, Sweden	135	644	7.5	1.4	5.0
Oak	Meerdink, Netherlands	140	780	8.6	2.0	5.0
Tilieto-Quercetum-Aegopodisum	Belgorod region, U.S.S.R.	250	537	6.0	1.2	4.2
Pine-hardwood	Okita, Japan	20	1467	11.3	4.3	1.3
Pine-hardwood	Walker Branch, Tenn.	70	1400	13.3	1.9	2.1
White pine plantation	Coweeta, N.C.	15	1628	13.6	4.3	9.8
Loblolly pine plantation	Triangle site, N.C.	15	1150	15.6	5.8	10.6
Loblolly pine plantation	Duke Forest, N.C.	16	1150	15.6	8.7	7.5
Loblolly pine plantation	Walker Branch, Tenn.	35	1400	13.3	4.6	1.3

[a]Calculated by dividing the current woody biomass (aboveground) by age of stand.
[b]P.A. Mays, Tennessee Valley Authority, Knoxville, Tennessee, personal communication, June 1985.
[c]Day and Monk 1977.

the youngest trees plus 15% of the difference between the oldest and youngest trees. As a result of this assumption, the calculated age of some of the forests may be biased downward, resulting in an overestimation of lifetime average annual increments.

Currently, the annual increment of 4.1 $Mg \cdot ha^{-1} \cdot year^{-1}$ for the yellow-poplar-dominated stand on Walker Branch is slightly greater than the average of 3.9 for all the hardwood stands, and the chestnut oak and the oak-hickory stands are a little below average at 3.5 and 3.1, respectively.

It is interesting to note that although there is no statistically significant correlation in this limited data set between stand ages and current woody growth increments, the older stands do tend to have higher current growth increments than the younger stands. However, these older, faster-growing stands tend also to be in colder climates, where respiratory demands for maintenance of tree tissues are less than in warmer climates. This would result in more photosynthate available for growth. Thus, these faster growth rates may really be a function more of climate than of stand age. For example, the 80- and 250-year-old forests in the cold Belgorod region of Russia both have relatively high growth increments at 5.2 and 4.2 Mg·ha^{-1}·year^{-1}, respectively. The 250-year-old stand may be experiencing some growth decline as a result of age.

Only one pine-hardwood stand (a 20-year-old stand in Japan) was available from IBP data sets for comparison with Walker Branch. The climates of the two forests are quite similar. However, the Japanese forest was much younger than the Walker Branch forest. The lifetime average increment of the Japanese forest was more than twice that of Walker Branch, but the current increment of the Walker Branch stand was nearly twice that of the forest in Japan. The overstory at the Japanese forest was all pine at the time measurements were made, with oak saplings present in the understory. However, the older Walker Branch stand had a mixture of pine and hardwood in the overstory, with an understory typical of hardwood stands. Thus, the low current increment in the Japanese pine-hardwood stand apparently reflects a decline in pine growth as hardwoods invade and begin to compete for water and nutrients. However, increment measurements for the Japanese forest did not include the young oak saplings in the understory and are therefore underestimates. The Walker Branch pine-hardwood stand has already passed this successional stage and is currently experiencing relatively rapid growth.

Data from three pine plantations were available from IBP data sets for comparison with the Walker Branch pine stand. All three were younger (15–16 years) than the Walker Branch pine stand (35 years), and all three had substantially greater current woody increments (7.5–10.6 Mg·ha^{-1}·year^{-1}) than Walker Branch (1.3 Mg·ha^{-1}·year^{-1}). This low current increment of the Walker Branch pine stand is the result of overcrowding and a recent high rate of mortality. However, the lifetime average increment of the Walker Branch stand (4.6 Mg·ha^{-1}·year^{-1}) is comparable to that of the white pine plantation at Coweeta, North Carolina (4.3 Mg·ha^{-1}·year^{-1}), and the loblolly plantation at the Research Triangle Park site, North Carolina (5.8Mg·ha-1·year-1). This suggests exceptionally rapid growth of the Walker Branch pine stand at a younger age. The plantations at Coweeta and at the Research Triangle Park site have current increments about twice their lifetime average increments, indicating that they are in a period of relatively rapid growth. The 16-year-old Duke Forest stand, however, had a very high lifetime average increment (8.7 Mg·ha^{-1}·year^{-1}), with current increments about equal to its lifetime average.

It will be interesting to observe future changes in species composition and growth rates of the Walker Branch pine stand. Currently, hardwood species such as yellow-poplar and dogwood are invading the pine stand. Young loblolly pine seedlings are also present but primarily in peripheral, disturbed areas. Also, as some of the pine trees die, those remaining may grow at a faster rate than before and compete with the hardwoods for canopy positions. Allowing this natural succession to occur will permit us to compare the productivity of a forest stand with species diversity to that of a previous monoculture on the same site.

6.4.2 Litterfall and Litter Biomass

A comparison of litterfall and litter biomass for eight oak-dominated forests is presented in Table 6.8. The average litterfall for the eight forests is 5.1 Mg·ha^{-1}year^{-1}, with Walker Branch slightly below the average. However, litter biomass on Walker Branch was twice the average. Only one other forest, the forest in Andersby, Sweden, had a higher litter biomass than Walker Branch. The litter biomass in that forest was ~6 Mg/ha greater than that at Walker Branch. The mean annual temperature in the Andersby forest, however, was only 5.5°C as compared with 13.3°C at Walker Branch. Thus, its litter decomposition rates would be expected to be much slower than those for Walker Branch.

With the information available to us we cannot explain the higher-than-average litter biomass on Walker Branch. However, it is interesting to note that two other forests similar to Walker Branch in species composition, climate, and geographic location (Cross Creek and Camp Branch) also have higher-than-average litter biomass. In fact, the forests in colder climates (except for the one in Andersby, Sweden) and with comparable or greater litterfall have less litter biomass than the forests of warmer climates. This is the opposite of what we would expect (i.e., litter decomposition rates with similar species composition should decrease with decreasing temperatures).

Table 6.8. A comparison of litterfall, litter, and soil organic matter in oak-dominated forests at different International Biological Program study sites in the United States and Europe
(Data other than those for the Tennessee forests are from DeAngelis et al. 1981)

Study site	Age (years)	Precipitation (mm)	Temperature (°C)	Litterfall (Mg·ha^{-1}·year^{-1})	Litter (Mg/ha)	Soil organic matter (Mg/ha)
Walker Branch, Tenn.	45–100	1400	13.3	4.5	33.0	116
Cross Creek, Tenn.	45–100	1400	13.3	6.4[a]	18.0[b]	—
Camp Creek, Tenn.	45–100	1400	13.3	4.7[a]	13.0[b]	—
Andersby, Sweden	40–200	566	5.5	3.2	38.9	85
Coweeta, N.C.	60–200	1945	12.6	5.0	8.8	137
Meathopwood, U.K.	80	1115	7.8	5.6	6.7	138
Noa Woods, Wis.	130	777	6.9	6.7	4.7	176
Linnebjer, Sweden	125–190	644	7.5	4.8	6.1	288

[a]Kelly (1984).
[b]Kelly (1979).

Because of the importance of litter decomposition for meeting the nutrient requirements of the trees, the causes of these discrepancies in litter biomass need further study. Initially, the oak-dominated sites listed in Table 6.8, and possibly other oak-dominated forests, should be resampled using identical techniques to be absolutely sure the differences are significant. If the differences are real, we need to design experiments that will determine the causes.

6.4.3 Forest Floor Respiration and Carbon Allocation

Productivity, which is a consequence of carbon fluxes into, within, and out of forest ecosystems, has been estimated for all of the IBP study sites, but data on particular flux pathways are not as readily available. This is especially true for carbon allocation within individual trees and for CO_2 efflux from soil, litter, and vegetation back to the atmosphere. We know of no other study with which to compare the whole-tree carbon allocation studies on Walker Branch.

However, data on carbon fluxes from the forest floor to the atmosphere were available for four of the forest stands (Coweeta, Camp Branch, Cross Creek, and the chestnut oak stand on Walker Branch) for which increment data were available. In addition to these four, we found six additional oak-dominated forests in which forest floor CO_2 efflux had been measured (Table 6.9). Data on factors that could account for respiratory rate differences (e.g., species composition of litter and soil characteristics) were not available for most of the study sites. However, investigators have reported that under similar climatic conditions, forests inhabiting different soil types, and even those with different species composition, tend to have similar rates of total forest floor respiration (Witkamp 1966; Reiners 1968). This suggests that climate may be a predominant factor controlling forest floor respiration rates, at least within a relatively broad range of forest types.

Indeed, climatic conditions, especially temperature, have a strong effect on forest floor respiration rates. For example, seasonal changes in temperature are strongly correlated with seasonal variations in forest floor respiration within individual stands (Witkamp 1966; Reiners 1968; Froment 1972; Edwards 1975). Moisture affects CO_2 efflux primarily under extremely dry or extremely moist conditions, but temperature is the primary influence within a relatively broad range of soil moisture conditions (Reiners 1968; Edwards 1975).

However, differences in CO_2 efflux rates between oak-dominated sites (Table 6.9) do not correlate well with differences in temperature and precipitation rates at the other sites. For example, an oak forest in Minnesota with a mean annual air temperature of 7.1°C and 676 mm precipitation respired 1.3 times more CO_2 than the Walker Branch oak forest, which has a higher mean annual temperature (13.3°C) and greater precipitation (1400 mm/year).

Table 6.9. A comparison of forest floor CO_2 efflux rates among oak-dominated forests.

Location	Stand age (years)	Precipitation (mm)	Temperature (°C)	Forest floor respiration (kg $CO_2 \cdot ha^{-1} \cdot year^{-1}$)				Reference
				Litter	Soil organic matter	Roots	Total	
Coweeta, N.C.	80	1,628	13.6	5,055	1,787	24,603	31,427	J.B. Wade, Coweeta Hydrologic Laboratory, personal communication. 1985
Tennessee[a]	70	1,400	13.3	5,073	1,171	16,136	22,380	Walker Branch—this report
Tennessee[b]	70	1,400	13.3	4,516	916	13,968[c]	19,400	Edwards and Ross-Todd 1983
Tennessee[d]	70	1,400	13.3	6,262	7,184	11,600	25,046[c]	Larkin et al. 1983
Tennessee[f]	80	1,400	13.3	5,563	8,561	10,600	24,724[c]	Larkin et al. 1983
Tennessee[e]	45	1,400	13.3				15,200	Witkamp 1966
Missouri	50	958	12.6				37,150	Garret and Cox 1973
Arnhem, Netherlands	135	780	8.6	5,390	1,043	1,378[h]	13,780	DeBoois 1974
Liege, Belgium	80	952	8.5				6,500	Froment 1972
Minnesota	40	676	7.1				29,120	Reiners 1968

[a] A chestnut-oak-dominated stand on Walker Branch Watershed.
[b] A chestnut-oak-dominated stand near Walker Branch. Litter and root respiration values were apportioned according to values for Walker Branch.
[c] Root values were apportioned according to Walker Branch root respiration data.
[d] Stand in Marion County (Cross Creek Watershed) dominated by chestnut oak and red oak.
[e] Values given by Larkin et al. (1983) were multiplied by 1.4 to account for the loss of water (formed from the chemical reaction between CO_2 and soda-lime) during soda-lime drying following reaction with CO_2.
[f] Stand in Bledsoe County (Camp Branch Watershed) dominated by red oak and white oak.
[g] A white-oak-dominated stand near Walker Branch; litter respiration values are based on measurements of bagged fresh leaves.
[h] Estimated by DeBoois (1974).

Differences in measurement techniques may explain some of the differences in respiration rates given in Table 6.9. The highest rates were measured with infrared gas analysis (IRGA) used in combination with permanently installed chambers inserted, in some cases, several centimeters into the soil. The efficiency of IRGA CO_2 detection, combined with the possible stimulating effects of permanently installed chambers (Edwards 1974), may have resulted in higher apparent CO_2 efflux than techniques employing potassium hydroxide (KOH) as an absorber (only $\sim 60\%$ efficient; Kucera and Kirkman 1971; Edwards and Sollins 1973). For example, Witkamp (1966) reported an efflux of 15,200 kg $CO_2 \cdot ha^{-1} \cdot year^{-1}$ from a white oak forest floor near Walker Branch, using KOH and inverted chambers. This rate is only 68% as high as those at Walker Branch, where a more efficient CO_2 absorber (soda-lime) was used. Using the latter technique, two oak forest floors at Oak Ridge, Tennessee, and two stands near Oak Ridge (Marion and Bledsoe counties) were shown to have similar annual CO_2 efflux rates (Table 6.9). The same technique was used in a northern hardwood forest in New Hampshire, and CO_2 efflux rates were much less (5000 kg $\cdot ha^{-1} \cdot year^{-1}$; J.W.Hornbeck, U.S. Department of Agriculture, Durham, New Hampshire, personal communication, 1985) than those at Walker Branch. This was expected because of the colder, drier climate in New Hampshire.

6.5 Summary and Conclusions

Productivity and carbon dynamics research at Walker Branch Watershed has included (1) monitoring of tree growth and mortality in permanently established plots in each stand type; (2) examination of carbon pools in individual components of each stand and carbon fluxes between vegetation, soil, and atmosphere; and (3) examination of photosynthetically fixed carbon allocation within individual trees.

Tree biomass (excluding roots) ranged from 144 Mg/ha in pine-hardwood stands to 198 Mg/ha in chestnut oak stands, where total ecosystem biomass (excluding roots) ranged from 293 Mg/ha in pine-hardwood to 393 Mg/ha in yellow-poplar. Yellow-poplar, second to chestnut oak in tree biomass, had the greatest ecosystem biomass because of a greater soil organic matter content (189 vs. 116 Mg/ha in each of the other forest types).

Continuing remeasurements of trees in permanently established plots between 1967 and 1983 revealed that net woody increments ranged from 1.3 Mg $\cdot ha^{-1} \cdot year^{-1}$ in loblolly pine stands to 4.1 Mg $\cdot ha^{-1} \cdot year^{-1}$ in yellow-poplar. A relatively small pine increment in both loblolly and pine-hardwood stands reflects (1) a high rate of pine mortality caused by an outbreak of the southern pine beetle and (2) reduced growth rates due to overstocking of the loblolly stand. The dieback of pine has resulted in increased yellow-poplar invasion of the pine stands. Mortality in hickory was also relatively high (2.7 $Mg \cdot ha^{-1} \cdot year^{-1}$) as compared with only 1.9 and 1.8 $Mg \cdot ha^{-1} \cdot year^{-1}$ in yellow-poplar and chestnut oak. Hickory mortality was primarily due to an outbreak of the hickory bark beetle following a drought in 1980.

The current woody increment of the yellow-poplar stand on Walker Branch was slightly higher than the average of 3.9 $Mg \cdot ha^{-1} \cdot year^{-1}$ for 15 selected hardwood stands in the United States and Europe. The chestnut oak and oak-hickory stands, however, were below the average. All stands examined had higher current growth increments than their lifetime average annual increments. This, combined with the observation that generally the older the stand the higher the current growth increment, leads us to cautiously conclude that growth increment in oak-dominated forests increases with age (at least within the age range of 65–250 years). This apparent increase with age, however, may be a reflection of climate and not age, because the older forests were found in colder regions of the world. Forests growing in colder areas have reduced respiration losses of carbon, which may result in greater net carbon gains.

Walker Branch litterfall in the oak-dominated stands was near or below the average for seven other oak stands. However, Walker Branch litter biomass was more than three times the average for the same seven stands. Even the yellow-poplar litter biomass on Walker Branch was more than twice that of most oak stands. With the data available to us, we cannot explain this discrepancy in litter biomass on Walker Branch relative to that of other hardwood forests. It is apparent from the data, however, that litter decomposition rates in Walker Branch are slower than those in other oak forests.

Litter decomposition rates in the Walker Branch chestnut oak stand, based on CO_2 efflux measurements, are about 90% of leaf litterfall. Thus, with a relatively high rate of woody litter input to the forest floor (not included in respiration measurements), enough litter may accumulate each year to account for the relatively large pool of litter biomass. Differences in methodologies used for measuring CO_2 efflux rates among other oak-dominated stands and the absence of partitioned flux rates (i.e., litter vs. roots vs. soil) from most sites made such comparisons inadequate for explaining differences in litter biomass.

Total forest floor CO_2 efflux in the Walker Branch chestnut oak stand was 22.4 $Mg \cdot ha^{-1} \cdot year^{-1}$, with about 23% from litter, 72% from live and dead roots, and 5% from soil. The total value was similar to the rates found in three other oak forests in Tennessee, using identical measurement techniques, and about four times that found in a northern hardwood stand in New Hampshire, also using the same measurement techniques.

Carbon allocation research on mature white oak trees on Walker Branch is unique to this study site. Research of this type, however, is becoming increasingly important as we look for the mechanisms through which stresses (e.g., atmospheric pollution) affect the growth of trees. Photosynthate allocation experiments with $^{14}CO_2$ in white oak demonstrated the complexity of carbon utilization by trees and revealed distinct temporal patterns of carbon translocation. Great energy demands within the canopy resulted in translocation of only ~50% of total photosynthate from the canopy basipetally during the entire growing season. A dramatic increase (71% of gross photosynthesis) in translocation from leaves to young branches occurred just prior to leaf fall.

From experiments using ^{14}C-sucrose injected into boles of mature white oaks, pool sizes of various biochemical constituents, derived from the ^{14}C-sucrose, were estimated, and temporal changes in these constituents were demonstrated. The largest constituent was holocellulose, ranging from 36% of tissue dry weight in leaves to 77% in boles. The next largest constituent was lignin (14–33%), followed by labile components (excluding sugars). Other constituents (i.e., starch, sugars, and lipids) varied greatly between tissues, but on a whole-tree basis, the order was lipids > starches > sugars. Fine roots contained greater concentrations of starches and sugars than lipids, and leaves contained greater concentrations of sugars and lipids than starches. Temporal changes in biochemical constituents were more pronounced in individual tissues than in the whole tree. For example, there were two peaks in fine-root starch content that corresponded with peaks in fine-root growth. Also, there was a decline in branch and twig starch in spring and an increase in the fall, reflecting mobilization of energy reserves for canopy development and subsequent recovery of the trees' reserves during the growing season.

In this chapter we have discussed the growth rates, physiology, mortality, and decay of trees on Walker Branch Watershed. Such information, coupled with concomitant studies of nutrient and water cycles and other regulatory environmental factors (e.g., forest meteorology), is increasing our understanding of interactions between the environment and forests.

References

Chung, H., and R.L. Barnes. 1977. Photosynthate allocation in *Pinus taeda*. I. Substrate requirements for synthesis of shoot biomass. Can. J. For. Res. 7:106–111.

Cole, D.W., and M. Rapp. 1981. Elemental cycling in forest ecosystems. IN D.E. Reichle (ed.), Dynamic Properties of Forest Ecosystems. Cambridge University Press, New York.

Craighead, F.C. 1949. Insect enemies of eastern forests. U.S. Department of Agriculture Misc. Pub. No. 656. U.S. Government Printing Office, Washington, D.C.

Day, F.D., Jr., and C.D. Monk. 1977. Seasonal nutrient dynamics in the vegetation of a southern Appalachian watershed. Am. J. Bot. 64:1126–1139.

DeBoois, H.M. 1974. Measurement of seasonal variations in the oxygen uptake of various litter layers of an oak forest. Plant Soil 40:545–555.

DeAngelis, D.L., R.H., Gardner, and H.H. Shugart. 1981. Productivity of forest ecosystems studied during the IBP: The woodlands data set. IN D.E. Reichle (ed.), Dynamic Properties of Forest Ecosystems. Cambridge University Press, New York.

Edwards, N.T. 1974. A moving chamber design for measuring soil respiration rates. Oikos 25:97–101.

Edwards, N.T. 1975. Effects of temperature and moisture on carbon dioxide evolution in a mixed deciduous forest floor. Soil Sci. Soc. Am. Proc. 39:361–365.

Edwards, N.T. 1982. The use of soda-lime for measuring respiration rates in terrestrial systems. Pedobiologia 23(5):321–330.

Edwards, N.T., and W.F. Harris. 1977. Carbon cycling in a mixed deciduous forest floor. Ecology 58:431–437.

Edwards, N.T., and B.M. Ross-Todd. 1983. Carbon dynamics in a mixed deciduous forest following clear-cutting with and without residue removal. Soil Sci. Soc. Am. J. 47(5):1014–1021.

Edwards, N.T., and P. Sollins. 1973. Continuous measurement of carbon dioxide evolution from partitioned forest floor components. Ecology 54(2):406–412.

Edwards, N.T., W.F. Harris, and H.H. Shugart, Jr. 1977. Carbon cycling in a deciduous forest. pp. 153–158. IN The Belowground Ecosystem: A Synthesis of Plant Associated Processes. Range Science Department Science Series No. 26. Colorado State University, Fort Collins, Colorado.

Froment, A. 1972. Soil respiration in a mixed oak forest. Oikos 23(2):273–277.

Garret, H.E., and G.S. Cox. 1973. Carbon dioxide evolution from the floor of an oak-hickory forest. Soil Sci. Soc. Am. Proc. 37:641–644.

Grigal, D.F., and R.A. Goldstein. 1971. An integrated ordination-classification analysis of an intensively sampled oak-hickory forest. J. Ecol. 59:481–492.

Grizzard, T., G.S. Henderson, E.E.C. Clebsch, and D.E. Reichle. 1976. Seasonal nutrient dynamics of foliage and litterfall on Walker Branch Watershed, a deciduous forest ecosystem. ORNL/TM-5254. Oak Ridge National Laboratory, Oak Ridge, Tennessee.

Harris, W.F., R.A. Goldstein, and G.S. Henderson. 1973. Analysis of forest biomass pools, annual primary production and turnover of biomass for a mixed deciduous forest watershed. pp. 41–64. IN H.E. Young (ed.), Proc., IUFRO Symposium Working Party on Forest Biomass. University of Maine Press, Orono, Maine.

Harris, W.F., P. Sollins, N.T. Edwards, B.E. Dinger, and H.H. Shugart. 1975. Analysis of carbon flow and productivity in a temperate deciduous forest ecosystem. pp. 116–122. IN D.E. Reichle, J.F. Franklin, and D.W. Goodall (eds.), Productivity of World Ecosystems. National Academy of Sciences, Washington, D.C.

Harris, W.F., R.S. Kinerson, Jr., and N.T. Edwards. 1977. Comparison of belowground biomass of natural deciduous forest and loblolly pine plantations. Pedobiologia 17(5):369–381.

Henderson, G.S. 1972. Litter biomass and nutrient contents in four forest stands on Walker Branch Watershed. EDFB-IBP Memo Report No.71–91. Oak Ridge National Laboratory, Oak Ridge, Tennessee.

Henderson, G.S., W.T. Swank, J.B. Waide, and C.C. Grier. 1978. Nutrient budgets of Appalachian and Cascade region watersheds: A comparison. For. Sci. 24:385–397.

Johnson, D.W., and D.D. Richter. 1984. Effects of atmospheric deposition on forest nutrient cycles. Tappi 67:82–85.

Johnson, D.W., G.S. Henderson, and W.F. Harris. Changes in aboveground biomass and nutrient content on Walker Branch Watershed from aboveground biomass and nutrient content on Walker Branch Watershed from 1967 to 1983. IN 1967 to 1983. pp. 487–496 IN R.L. Hay, F.W. Woods, and H. Deselm (eds). Proceedings of the Central Hardwood Forest Conference VI., Knoxville, TN, February 24–26, 1987. University of Tennessee Press, Knoxville, TN.

Kelly, J.M. 1979. Camp Branch and Cross Creek Experimental Watershed Projects: Objectives, Facilities, and Ecological Characteristics. press).

Kelly, J. M. 1979. Camp Branch and Cross Creek Experimental Watershed EPA-600/7-79-053; TVA/ORNL-9/04. U.S. Environmental Protection Agency, Washington, D.C.

Kelly, J.M. 1984. Litterfall sulfur and nitrogen inputs as influenced by power plant proximity. Water Air Soil Pollut. 22:143–152.

Kimura, M. 1969. Ecological and physiological studies on the vegetation of Mt. Shimagre. VII. Analysis of production processes of young *Abies* stand based on carbohydrate economy. Bot. Mag. (Tokyo) 82:6–19.

Kucera, C.L., and D.R. Kirkman. 1971. Soil respiration studies in a tall-grass prairie in Missouri. Ecology 52:912–915.

Larkin, R.P., J.M. Kelly, and F.W. Wood. 1983. Deciduous Forest Litter Decomposition at Two Sulfur Input Rates. TVA/ORNL/AQB-83/11. Tennessee Valley Authority Air Quality Branch, Muscle Shoals, Alabama.

McLaughlin, S.B., and R.K. McConathy. 1979. Temporal and spatial patterns of carbon allocation by white oak canopies. Can. J. Bot. 57:1407–1413.

McLaughlin, S.B., N.T. Edwards, and J.J. Beauchamp. 1977. Spatial and temporal patterns in transport and respiratory allocation of (^{14}C) sucrose by white oak (*Quercus alba*) roots. Can. J. Bot. 55:2971–2980.

McLaughlin, S.B., R.K. McConathy, and B. Beste. 1979. Seasonal changes in within-canopy allocation of ^{14}C-photosynthate by white oak (*Quercus alba* L.). For. Sci. 25:361–370.

McLaughlin, S.B., R.K. McConathy, R.L. Barnes, and N.T. Edwards. 1980. Seasonal changes in energy allocation by white oak (*Quercus alba* L.). Can. J. For. Res. 10(3):379–388.

Meyer, M.M., and W.E. Splittstoesser. 1971. The utilization of carbohydrate and nitrogen reserves by *Taxus* during its spring growth period. Physiol. Plant. 24:306–314.

Miller, H.G. 1981. Forest fertilization: Some guiding concepts. Forestry 54:157–167.

Reiners, W.A. 1968. Carbon dioxide evolution from the floor of three Minnesota forests. Ecology 49(3):471–483.

Richter, D.D., D.W. Johnson, and D.E. Todd. 1983. Atmospheric sulfur deposition, neutralization, and ion leaching in two deciduous forest ecosystems. J. Environ. Qual. 12:263–270.

Shugart, H.H., W.F. Harris, N.T. Edwards, and B.S. Ausmus. 1977. Modeling of belowground processes associated with roots in deciduous forest ecosystems. Pedobiologia 17(5):382–388.

Switzer, G.L., and L.E. Nelson. 1972. Nutrient accumulation and cycling in loblolly pine (*Pinus taeda* L.) plantation ecosystems: The first twenty years. Soil Sci. Soc. Am. Proc. 36:143–147.

Teskey, R.O., and T.M. Hinckley. 1981. Influence of temperature and water potential on root growth of white oak. Physiol. Plant. 52:363–369.

Turner, J. 1981. Nutrient cycling in an age sequence of western Washington Douglas-fir stands. Ann. Bot. 48:159–169.

Witkamp, M. 1966. Rates of carbon dioxide evolution from the forest floor. Ecology 47(3):492–494.

Chapter 7
Terrestrial Nutrient Cycling

D.W. Johnson and G.S. Henderson

7.1 Introduction

Nutrient cycling studies were initiated in the late 1960s at Walker Branch Watershed in eastern Tennessee as part of the International Biological Program's effort to characterize productivity and nutrient cycling in forests throughout the world (summarized by Cole and Rapp 1981). These early studies not only characterized the nutrient cycles and compared them with other forested watersheds, as originally intended (e.g., Henderson et al. 1978), but also served as a prelude to later research on more specific environmental issues such as fertilization (Kelly and Henderson 1978a,b; Johnson et al. 1980a), harvesting (Johnson et al. 1982b), and pollutant deposition, including sulfur (Shriner and Henderson 1978; Johnson et al. 1982a), trace metals (Van Hook et al. 1977), and acid deposition (Richter et al. 1983; Johnson et al. 1985).

The initial characterization of nutrient cycles provided an invaluable base line and thus a strong enticement for subsequent, more specific research projects. The results of each subsequent project, in turn, added to the data base and led to further research questions. Thus, the nutrient cycling program on Walker Branch Watershed has grown considerably in both its scope and sophistication over the past 15 years. This growth continues to this day, but the time is now appropriate for a review and synthesis of results to date in order to convey our current knowledge to others as well as to sharpen our understanding and improve the focus of ongoing research at this site.

7.1.1 The Watershed-Level Vs. Stand-Level Approach

The watershed-level approach to nutrient cycling has been used and advocated for many years as a method of measuring inputs, outputs, and cycling of nutrients in forest ecosystems (Bormann and Likens 1970; Likens et al. 1977). This approach has the advantage of simplicity in that outputs are measured by means of streamflow with little site disturbance and the entire watershed is viewed as a unit (Fig. 7.1). It has decided disadvantages, however, in that fluxes within the soil (e.g., between the rooting zone and bedrock) are not accounted for, nutrient uptake by streams is not accounted for, the heterogeneity within the watershed is ignored, and the exacting requirements for watershed studies (defined watershed with impermeable bedrock) limit their utility to only a few sites such as Hubbard Brook watershed in New Hampshire (Bormann and Likens 1970).

The stand-level approach (Fig. 7.1) predates the watershed-level approach in that it was used by early investigators to examine aboveground nutrient cycling (e.g., Rennie 1955; Ovington 1962; Switzer and Nelson 1972). Measurement of outputs from forest stands was accomplished later by installing lysimeters (e.g., Cole et al. 1968), which have decided disadvantages: soil disturbance, problems with representativeness of samples (e.g., the question of whether the correct soil water fraction has been sampled), and much poorer methods of estimating water flux (Haines et al. 1982). The advantages of the stand-level approach are reduced heterogeneity, estimates of within-soil nutrient flux, and lack of interaction of output water with aquatic ecosystem cycles.

7.1.2 The Walker Branch Watershed Project

The Walker Branch Watershed Project was conceived as a watershed-level nutrient cycling study, and the cycles of most nutrients and trace elements have been characterized on a watershed scale (Henderson and Harris 1975; Shriner and Henderson 1978; Van Hook et al. 1977; Henderson et al. 1978). It was evident from the beginning, however, that the watershed was strongly heterogeneous not only in vegetation associations (Grigal and Goldstein 1971) but also in nutrient content among different forest types (Henderson et al. 1977a). Thus, a combination of stand-level and watershed-level approaches was used to depict the nutrient cycles on Walker Branch Watershed (Cole and Rapp 1981). It was also evident quite early that Walker Branch Watershed did not have impermeable bedrock and that watershed-level outputs of calcium and magnesium were more indicative of the weathering of the dolomite bedrock (which is covered in places by 30 m of weathered soil) than of export from the effective rooting zone of the trees (Curlin and Nelson 1968; Henderson et al. 1978). Since calcium is a nutrient of major interest in the watershed, our current efforts have evolved entirely into a stand-level approach in which certain

ORNL–DWG 87-12370

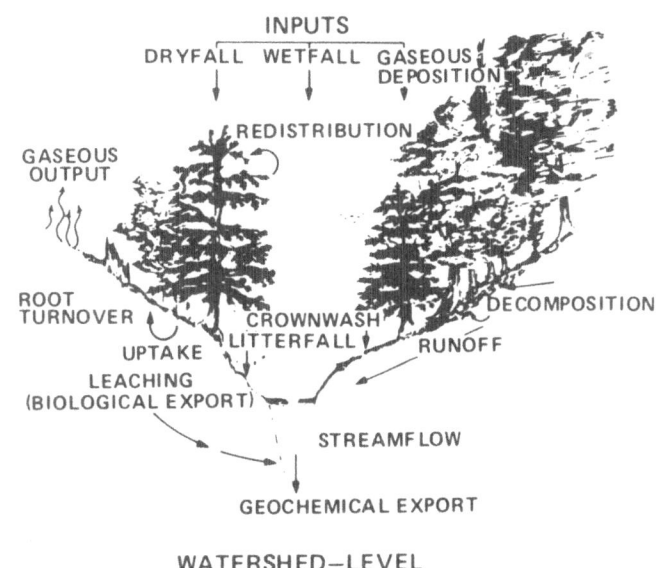

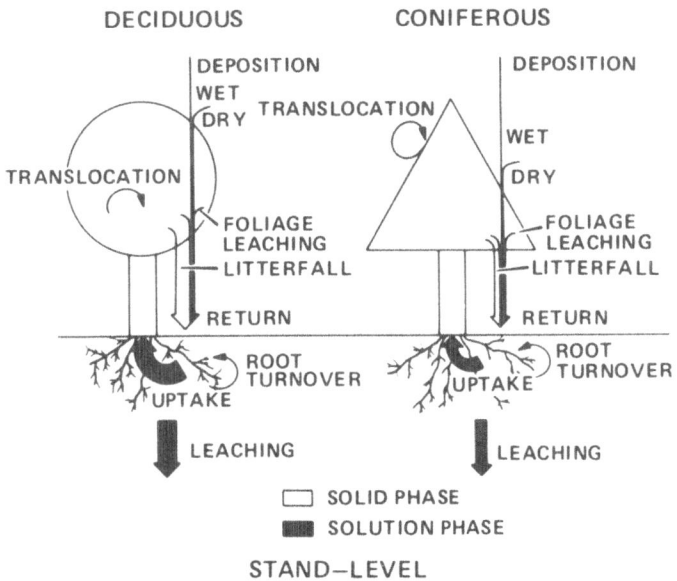

Figure 7.1. Schematic representations of watershed-level vs. stand-level approaches to measuring nutrient cycling in forests.

core plots from the original watershed-level project have been instrumented with tension lysimeters to estimate nutrient export from the top meter of soil (e.g., Johnson et al. 1982a, 1985).

The Walker Branch Watershed Project has evolved in two respects: (1) from a base-line study characterizing nutrient cycles to a series of studies designed to address specific issues, and (2) from a watershed-level to a stand-level approach to evaluating nutrient cycles.

7.2 Nutrient Cycles: Basic Characteristics

7.2.1 Methods

A full description of the methods employed in the individual studies summarized here can be obtained from the references cited for each study. In this section, we briefly outline the methods used in the initial overall nutrient cycling studies on the watershed. Detailed descriptions of the methods used in the original nutrient cycling studies on Walker Branch Watershed are given by Henderson and Harris (1975) and Henderson et al. (1978). Briefly, vegetation biomass was determined from regression equations applied to tree diameter measurements from 298 0.08-ha plots (Ch. 6). Nutrient concentrations were determined for the bole, branch, twig, and leaf components of 40 trees of 15 different species. The incorporation of nutrients in woody growth increment was calculated by applying appropriate concentration values to estimates of net production determined from remeasurement of vegetation on the 298 plots.

Nutrients in organic soil horizons and annual nutrient transfers due to canopy leaching were quantified on 24 of the permanent vegetation plots (Henderson et al. 1977a). Litterfall returns were determined on 80 of the permanent plots (Grizzard et al. 1976). Total and exchangeable (1 N ammonium acetate, pH 7) soil cation analyses were performed for 10 soil profiles representing the two dominant soil series on the watershed (Peters et al. 1970).

Atmospheric inputs were collected at five precipitation gaging sites on the watershed; separate samples of rain-scavenged (wetfall) and dry particulate (dryfall) inputs were collected weekly for cation analysis from September 1, 1970, through August 31, 1974. Cation concentrations were determined by atomic absorption spectroscopy, and ammonium and nitrate concentrations were determined with a Technicon Auto-Analyzer. Concentrations of cations, ammonium, and nitrate in stream water were similarly determined from weekly flow-proportional samples collected from each subcatchment. Precipitation and stream discharge data used to transform concentrations to area loading or discharge values (kilograms per hectare) were collected at 5-min intervals at each gaging site. Streamflow is continuously gaged with a 120° sharp-crested V-notch blade (Henderson et al. 1978).

Four major forest types were originally identified on the watershed: pine-hardwood, yellow-poplar, oak-hickory, and chestnut oak (Grigal and Goldstein 1971). Nutrient cycling studies were conducted in each of these four forest types in the early 1970s as part of the International Biological Program. Subsequently, it was deemed appropriate to identify a fifth major forest type, loblolly pine, which was planted in a 4-ha area in the northwestern corner of the watershed. Nutrient cycling in the loblolly pine forest is being studied at this time (along with an analysis of changes in nutrient cycling patterns in the original four forest types). Some data on nutrient distribution in the loblolly pine forest are available, but a full characterization of nutrient cycles will be reported in later publications.

7.2.2 Distribution of Nutrients on the Watershed

The distribution of N, P, K, Ca, Mg, and S in vegetation, litter, and soil (extractable and nonextractable pools) is depicted in Fig. 7.2. In each case, the amount found in soil-nonextractable pool predominates, but there are large differences in the relative amounts of nutrients in vegetation, litter, and soil-extractable pool; these differences reflect the relative abundance or paucity of the individual nutrients available for plant uptake. The ratio of nitrogen in vegetation to nitrogen in soil-extractable pool (5.1) is greater than that of any of the other five major nutrients. This reflects the fact that nitrogen is the growth-limiting nutrient in Walker Branch Watershed as well as in most forests in the Tennessee Valley region (Farmer et al. 1970). Many reasons might be given for this. Nitrogen is the one nutrient that is not contained in the parent material of the soil (except in trace amounts); thus, its accumulation in the ecosystem must come from the atmosphere, i.e., by fixation or deposition. Once in the ecosystem, nitrogen will not accumulate in inorganic forms, since NH_4^+ is either taken up by plants and soil heterotrophs or nitrified to NO_3^-. Nitrate, in turn, is either taken up or leached from the soil. The amount of mineral nitrogen ($NH_4^+ + NO_3^-$) available for plant uptake at any particular time is determined by the nitrogen demand of soil heterotrophic organisms (which is roughly indexed by the carbon:nitrogen ratio), decomposition rate, and atmospheric nitrogen deposition. If the nitrogen supply exceeds the demands of heterotrophs and trees, nitrification and nitrate leaching usually occur (e.g., Vitousek et al. 1979). Thus, a large soil inorganic nitrogen pool never develops, and the nitrogen cycle is inextricably linked to the carbon cycle.

In contrast to nitrogen, sulfur accumulates as SO_4^{2-} to very high levels on Walker Branch Watershed. For instance, 31.7% of the total ecosystem sulfur is in the form of SO_4^{2-}–S (Fig. 7.2). Furthermore, there are indi-

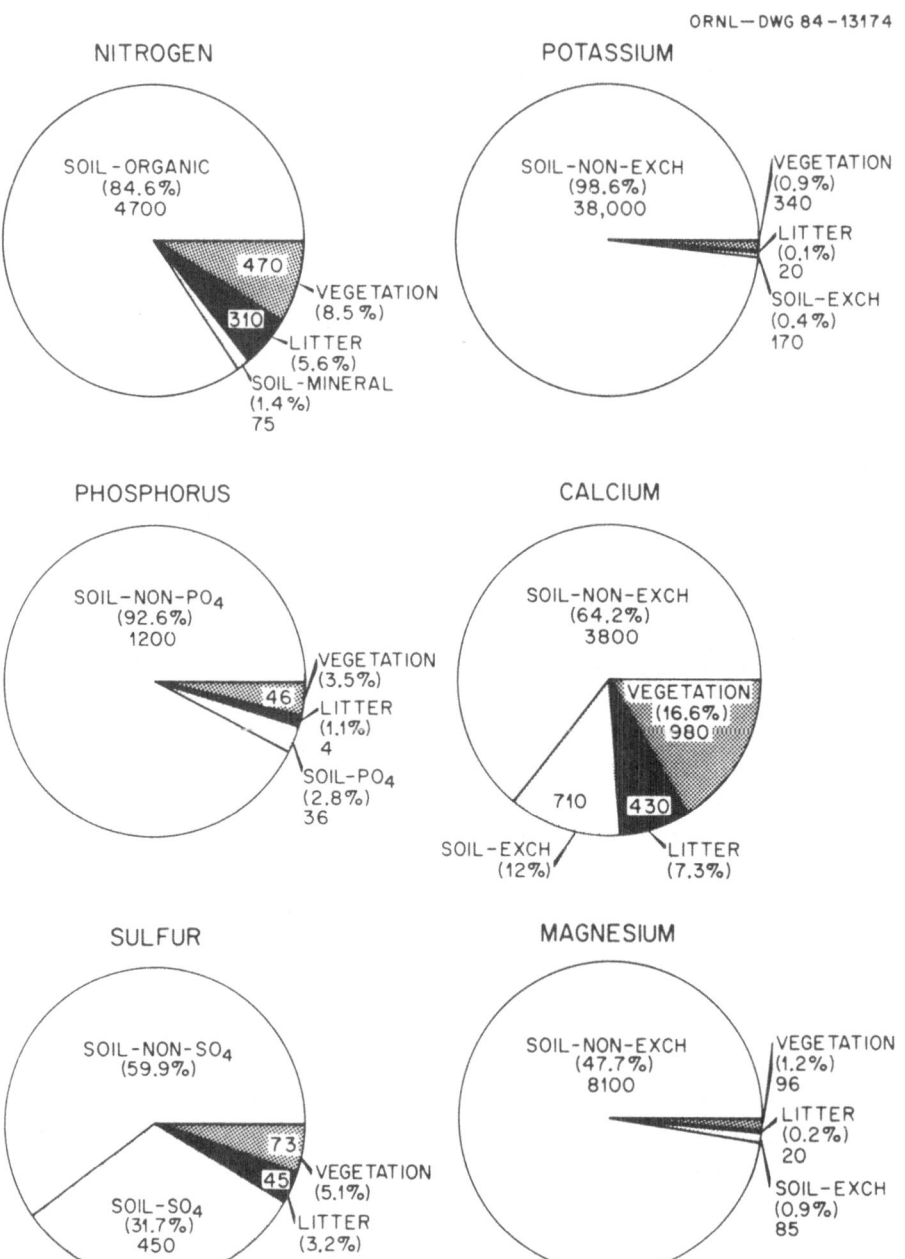

Figure 7.2. Distribution of N, P, K, Ca, Mg, and S on Walker Branch Watershed (kg/ha).

cations that 47% of vegetation sulfur and 50% of litter sulfur are in the SO_4^{2-}–S form as well (see "Sulfur Cycling" in Sect. 7.3.1). Thus, sulfur is unique among the macronutrients on the watershed in that a large portion of the total sulfur in each ecosystem component is in ionic form (SO_4^{2-}). This is due in large measure to very high inputs of sulfur from the atmosphere relative to the biological demands for sulfur, as will be described in detail later.

Phosphorus is intermediate between nitrogen and sulfur in terms of its distribution and availability in the soil. By far the most phosphorus is found in the mineral soil, but only a small fraction of this (2.8%) is in ionic form (orthophosphate as extracted by NH_4F/HCl) (Fig. 7.2). The ratio of vegetation phosphorus to soil orthophosphate (1.25) is less than the ratio of vegetation nitrogen to soil mineral nitrogen (5.1) but much greater than the ratio of vegetation sulfur to soil SO_4^{2-}–S (0.16).

The three major cation nutrients—Ca, K, and Mg—differ considerably in their distribution. Although the soil parent material at this site is dolomite, the calcium and magnesium contents of the soil are relatively low compared with potassium. The calcium and magnesium in the dolomite were originally in readily soluble carbonate forms, which have long since disappeared over millions of years of soil weathering, leaving less soluble secondary mineral components such as chert, iron, and aluminum hydroxy oxides, aluminum hydroxy-interlayered vermiculite, and kaolinite. The high potassium content may be due to fixation in the interlayered vermiculite and/or feldspars.

The percent of total system nutrient content in tree biomass is greatest for Ca (16.6%), followed by N (8.5%), S (5.1%), P (3.5%), Mg(1.2%), and K (0.9%). The unusually high percentage of calcium in tree biomass reflects the high accumulation of calcium in trees, especially in the woody tissues of oaks and hickories (Johnson et al. 1982b).

It is something of a mystery how the trees obtain as much calcium as they do from a soil so low in this nutrient. The current distribution of calcium indicates that the supplies of exchangeable calcium could not support the calcium demands of another forest unless supplemented by supplies obtained by deep rooting or weathering. Similar budgetary considerations concerning a nearby chestnut oak forest on the same soil led to the prescription for calcium fertilization with whole-tree harvesting (Johnson et al. 1982b). Therewas evidence of rooting beyond the 50- to 60-cm depth (up to 150-cm depth) commonly referred to as the "active rooting zone," in the whole-tree harvesting study, but roots were very rare. No evidence of deep rooting has been noticed on Walker Branch Watershed (Harris et al. 1977), although it may well occur in localized areas. In those parts of the watershed where bedrock is near the surface, a nearly inexhaustible supply of calcium and magnesium would be available. On much of the watershed, however, tree roots would have to penetrate 15–30 m of highly weathered soil before reaching bedrock, as evidenced by well-drilling and seismic studies of bedrock depth on the watershed and nearby sites (D.D. Huff, Oak Ridge National Laboratory, personal communication to D.W. Johnson, 1983; Woodward-Clyde 1984). Calcium-accumulating oaks and hickories growing on ridgetops where bedrock is known to be 30 m deep show no evidence of taprooting and every sign of a surface-rooting habit. However, rooting to such depths is not unheard of (Stone 1975). There is also the possibility that dolomite "pedestals" rising nearer the surface may supply the calcium needs of oaks and hickories (D.A. Lietzke, University of Tennessee, personal communication to D.W. Johnson, 1984). However, bedrock depth studies have shown no evidence of such pedestals to date (Woodward-Clyde 1984).

Another mystery is the reason for the high calcium accumulation in woody tissue, especially in oaks and hickories. Is this a form of luxury consumption, or does calcium play a specific role (perhaps a structural role in the deliquescent branching habit of oaks and hickories, requiring high concentrations of this nutrient) in certain species? Luxury consumption of calcium by uptake from dolomite bedrock may require rooting to depths of 15–30 m; therefore, to refer to high rates of calcium uptake from the highly weathered, calcium-poor soil above bedrock as "luxury consumption" seems inappropriate.

Coweeta Watershed in North Carolina has a species composition similar to that of Walker Branch Watershed (oaks and hickories) and a bedrock consisting of granite, mica schist, and gneiss (none of which is especially rich in calcium). In this ecosystem, 24% of total calcium is found in the vegetation (compared with 19% on Walker Branch) (Cole and Rapp 1981). In both systems, the calcium content of vegetation approximately equals the soil-exchangeable calcium content, and the soil-exchangeable calcium content equals a large proportion of the total soil calcium (37.6% at Coweeta and 18.7% at Walker Branch). The weighted average calcium concentration in biomass is 0.44% at Coweeta and 0.63% at Walker Branch. The similarity of the calcium distribution in these two ecosystems suggests that vegetation rather than bedrock influences calcium accumulation, which in turn suggests that accumulating calcium is an inherent property of oaks and hickories.

The potassium and magnesium contents of vegetation constitute much smaller proportions of total system contents (0.9 and 1.2%, respectively) than is the case for calcium (16.6%). Also, the soil-exchangeable potassium and magnesium constitute very small proportions of the total system contents (0.4 and 0.9%, respectively). The litter contains an unusually low proportion of K^+ (0.1%) compared to any of the other nutrients, reflecting the high mobility of K^+ and its association with the hydrologic cycle, as is shown later.

As might be expected, both biomass and nutrient distribution vary among these forest types, especially with respect to calcium (Fig. 7.3). Because of the high calcium concentrations in the woody tissues of oaks

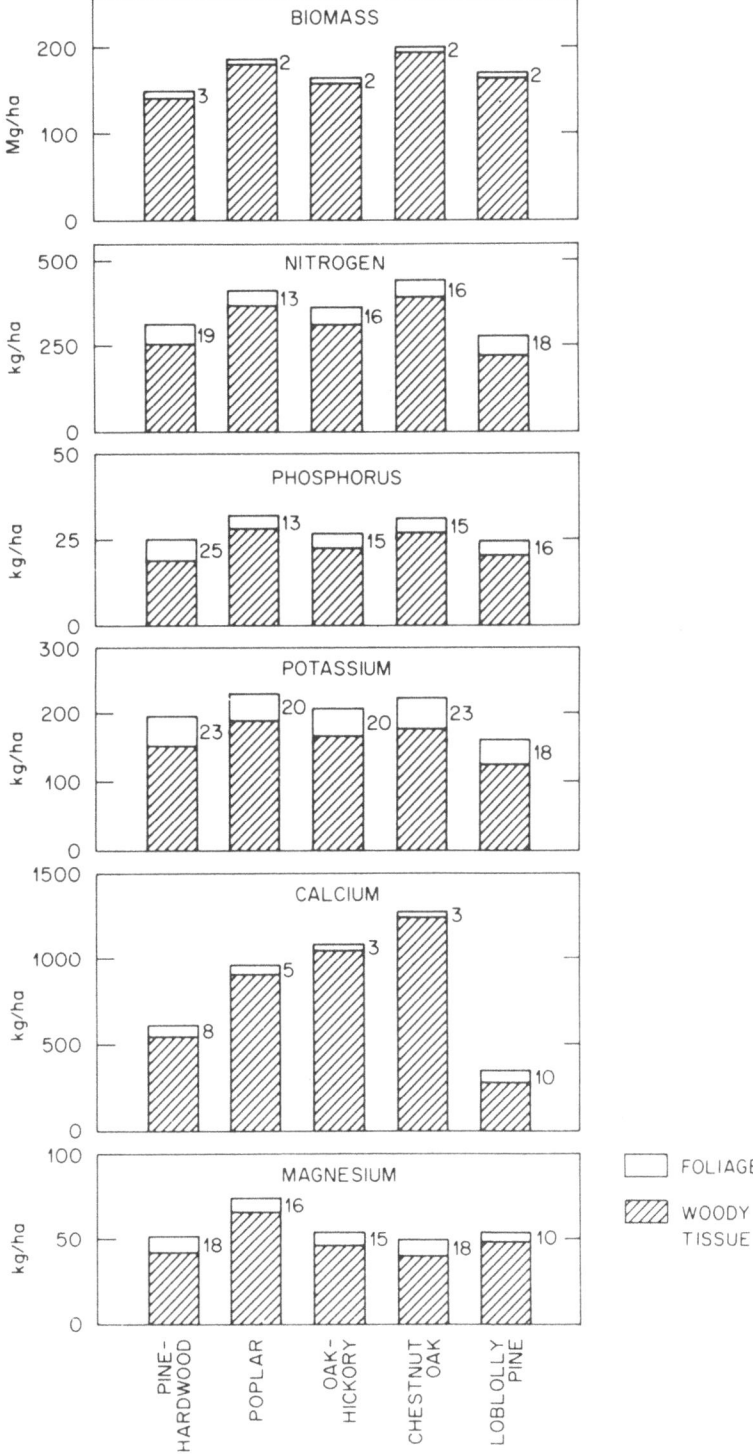

Figure 7.3. Biomass and nutrient distribution in five forest types on Walker Branch Watershed.

Table 7.1. Nutrient concentrations (percent) in tissues of representative tree species on Walker Branch Watershed

Tissue	Black gum	Yellow-poplar	Hickory	White oak	Chestnut oak	Shortleaf pine	Red maple
Nitrogen							
Foliage	1.50 ± 0.00	1.58 ± 0.14	1.40 ± 0.04	1.78 ± 0.28	1.48 ± 0.08	0.84 ± 0.02	1.28 ± 0.08
Branch	0.27 ± 0.02	0.24 ± 0.03	0.28 ± 0.07	0.24 ± 0.05	0.24 ± 0.06	0.26	0.20 ± 0.07
Bole	0.17 ± 0.006	0.18 ± 0.01	0.19 ± 0.04	0.16 ± 0.02	0.18 ± 0.02	0.11 ± 0.01	0.17 ± 0.04
Phosphorus							
Foliage	0.11 ± 0.003	0.11 ± 0.004	0.10 ± 0.006	0.13 ± 0.006	0.11 ± 0.004	0.09 ± 0.03	0.10 ± 0.005
Branch	0.03 ± 0.009	0.02 ± 0.005	0.02 ± 0.008	0.02 ± 0.01	0.02 ± 0.005	0.02	0.02 ± 0.006
Bole	0.018 ± 0.005	0.01 ± 0.0001	0.01 ± 0.003	0.009 ± 0.003	0.01 ± 0.002	0.008 ± 0.007	0.02 ± 0.004
Potassium							
Foliage	1.18 ± 0.02	0.94 ± 0.11	0.92 ± 0.14	1.09 ± 0.05	0.93 ± 0.04	0.68 ± 0.01	0.67 ± 0.07
Branch	0.26 ± 0.17	0.15 ± 0.19	0.30 ± 0.31	0.18 ± 0.18	0.11 ± 0.07	0.15	0.19 ± 0.19
Bole	0.11 ± 0.02	0.06 ± 0.06	0.18 ± 0.08	0.07 ± 0.06	0.06 ± 0.06	0.10 ± 0	0.08 ± 0.007
Calcium							
Foliage	1.09 ± 0.03	1.65 ± 0.11	1.61 ± 0.17	1.10 ± 0.15	0.83 ± 0.05	0.25 ± 0.02	0.79 ± 0.08
Branch	0.71 ± 0.26	0.30 ± 0.20	1.41 ± 0.68	0.89 ± 0.29	0.65 ± 0.32	0.30	0.40 ± 0.28
Bole	0.91 ± 0.35	0.31 ± 0.17	1.08 ± 0.14	0.72 ± 0.26	0.67 ± 0.11	0.19 ± 0.01	0.44 ± 0.19
Magnesium							
Foliage	0.31 ± 0.02	0.41 ± 0.04	0.31 ± 0.02	0.15 ± 0.008	0.16 ± 0.02	0.11 ± 0.02	0.17 ± 0.01
Branch	0.02	0.06	0.05	0.05	0.04	0.04	0.02
Bole	0.05 ± 0.001	0.03 ± 0.01	0.07 ± 0.02	0.02 ± 0.005	0.01 ± 0.002	0.02	0.02 ± 0.003

and hickory (Table 7.1), the calcium content of the chestnut oak and oak-hickory forests is disproportionately higher than that of the other forest types. Patterns of N, P, and K content mirror those in biomass for the most part, although the relatively lower nutrient content of loblolly pine (due primarily to low tissue concentrations) is evident. The magnesium content shows little relation to biomass and is higher in the yellow-poplar forest than in other forest types.

7.2.3 Watershed-Level Nutrient Cycles.

The original watershed-level nutrient cycles are depicted in Fig. 7.4. The watershed as a whole accumulated N, S, and P from atmospheric deposition, whereas K^+, and especially Ca^{2+} and Mg^{2+}, showed a net loss. Dissolution of the dolomite bedrock played an especially important role in the net losses of Ca^{2+} and Mg^{2+}. Johnson et al. (1985) note much lower net losses of Mg^{2+} and a slight net gain of Ca^{2+} at a depth of 80 cm in the soil of a chestnut oak forest on Fullerton soil near the ridgetop on Walker Branch Watershed. Data from other, newly established lysimeter plots show less Ca^{2+} and Mg^{2+} export than watershed-level (streamflow) export. Clearly, watershed-level exports of Ca^{2+} and Mg^{2+} do not represent net losses from the most active rooting zone (although roots may well extend beyond 120 cm in some cases). The nutrient contents of soils from five forest types depicted in Fig. 7.3 represent the 0- to 60-cm depths only, and the highly weathered soils extend for depths of up to 30 m over bedrock. Dividing the total content of Ca^{2+} and Mg^{2+} in the soil by the net export via streamflow gives a value of 32 years for total soil Ca^{2+} supply and 113 years for total soil Mg^{2+} supply. At these rates, the total soil Ca^{2+} would have been depleted by one-third from the time these cycles were first characterized (1970–1973) until 1984, and there is no evidence of such a decline (Johnson, Henderson, and Todd, in press). For the other elements, however, the net losses on a watershed level, as depicted here, are of similar magnitude to the net losses from lysimeter plots (Johnson et al. 1985).

Nitrogen, sulfur, and phosphorus accumulate in the watershed, but the mechanisms of accumulation are very different. The net accumulation of nitrogen and phosphorus (10 and 0.5 $kg \cdot ha^{-1} \cdot year^{-1}$, respectively) could be accounted forby increment in biomass (15 and 2.9 $kg \cdot ha^{-1} \cdot year^{-1}$). However, the net accumulation of sulfur (6$kg \cdot ha^{-1} \cdot year^{-1}$) could not be entirely accounted for by biomass increment (2.9 $kg \cdot ha^{-1} \cdot year^{-1}$) and therefore must have occurred partly in the soil. Also, atmospheric input was probably much greater than 18$kg \cdot ha^{-1} \cdot year^{-1}$ owing to dry deposition to the forest canopy (Ch. 4). In their initial characterization of the sulfur cycle, Shriner and Henderson (1978) noted that a large portion of the very high net removal term (throughfall minus precipitation, or 44 $kg \cdot ha^{-1} \cdot year^{-1}$) must have been due to dry deposition rather than foliar

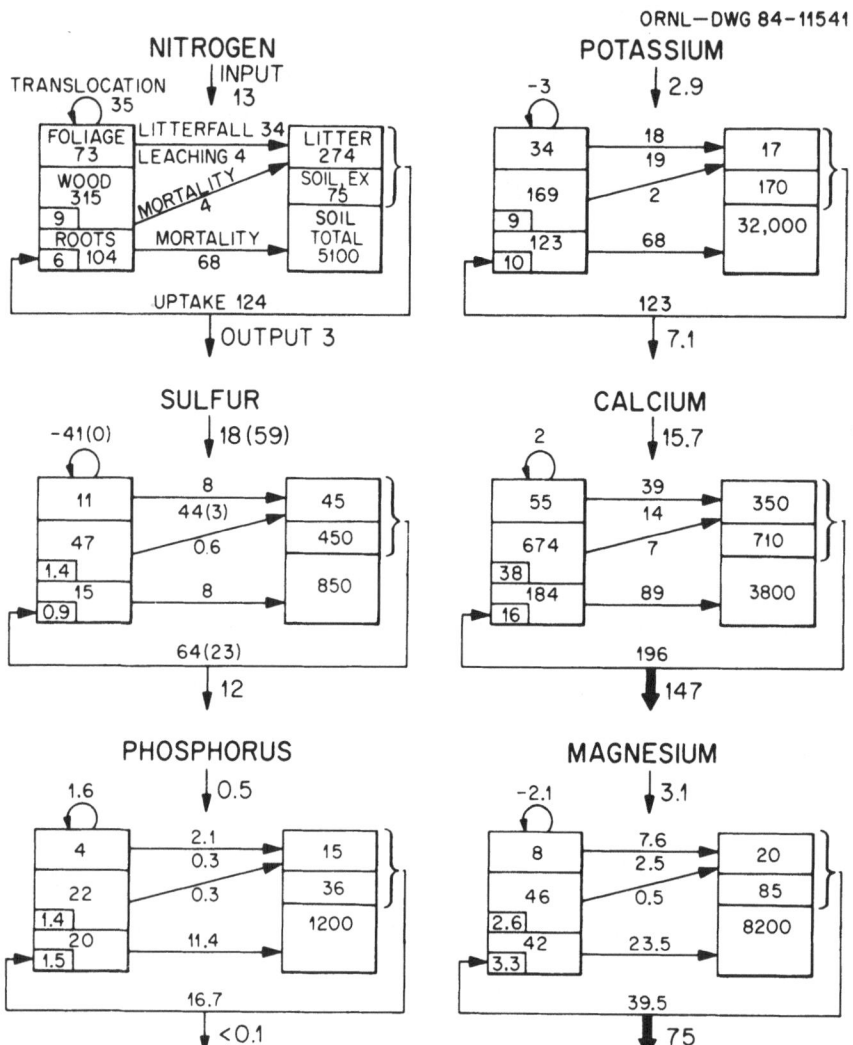

Figure 7.4. Cycles of N, S, P, K, Ca, and Mg on Walker Branch Watershed (kg/ha or kg·ha⁻¹·year⁻¹). The numbers in parentheses in the sulfur cycle represent early values (Shriner and Henderson 1978). The other values in the sulfur cycle are current best estimates.

leaching, since flux alone exceeded foliar sulfur content by four times (Fig. 7.4). Assuming no translocation, the maximum foliar leaching rate would have actually been 3 kg·ha⁻¹·year⁻¹ (shown in parentheses), making the total sulfur input ~59 kg·ha⁻¹·year⁻¹ at that time (in parentheses). This input rate implies a net annual SO_4^{2-}–S accumulation of 47 kg·ha⁻¹·year⁻¹ in the watershed. Clearly, most of this accumulation must have occurred in the soil, and we believe it was primarily by SO_4^{2-} adsorption. Later estimates of sulfur inputs (and therefore accumulation rates) are much lower—perhaps owing to changes in the height of local power plant stacks (a source of SO_2) or changes in collection methodology. Further aspects of the sulfur cycle on Walker Branch Watershed are discussed in Sect. 7.3.1.

7.2.4 Differences in Cycling Among Forest Types

The cycles of N, P, K, Ca, and Mg in the original four forest types are depicted in Figs. 7.5 to 7.9. Despite the differences in species composition, biomass, and nutrient content, the differences in aboveground biogeochemical cycling rates among these four types were not great. The rates of uptake and return of the major nutrients (omitting sulfur) among these four forest types were remarkably constant, especially for nitrogen and phosphorus. (Belowground cycling via root turnover was not characterized for each forest type.) The yellow-poplar forest had appreciably greater soil N, P, and Ca content, but this did not appear to cause appreciably greater rates of accumulation and cycling of these nutrients in the vege-

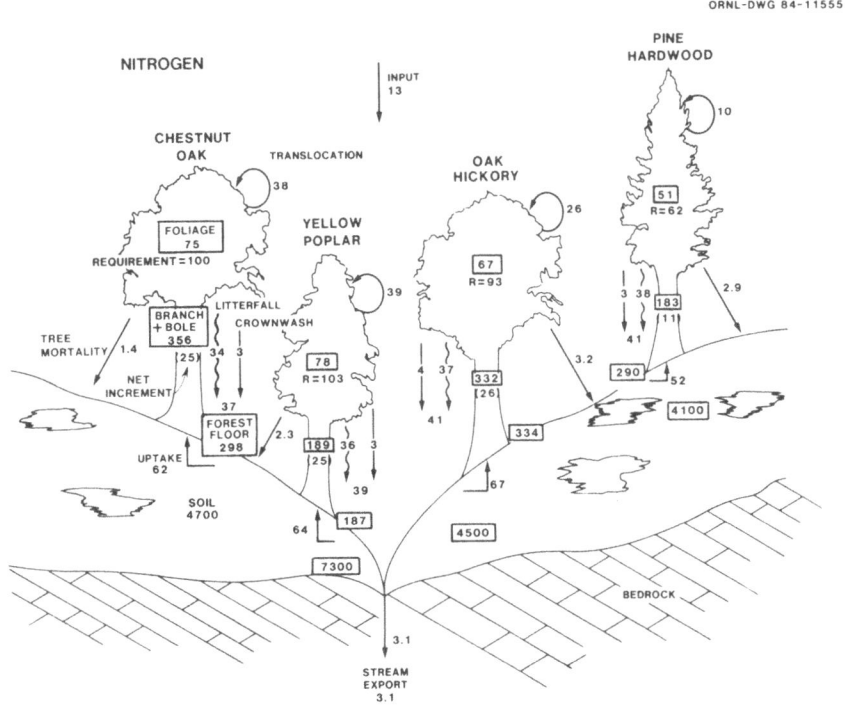

Figure 7.5. Nitrogen cycling in four forest types on Walker Branch Watershed (kg/ha or kg·ha^{-1}·year^{-1}).

ORNL-DWG 84-11554

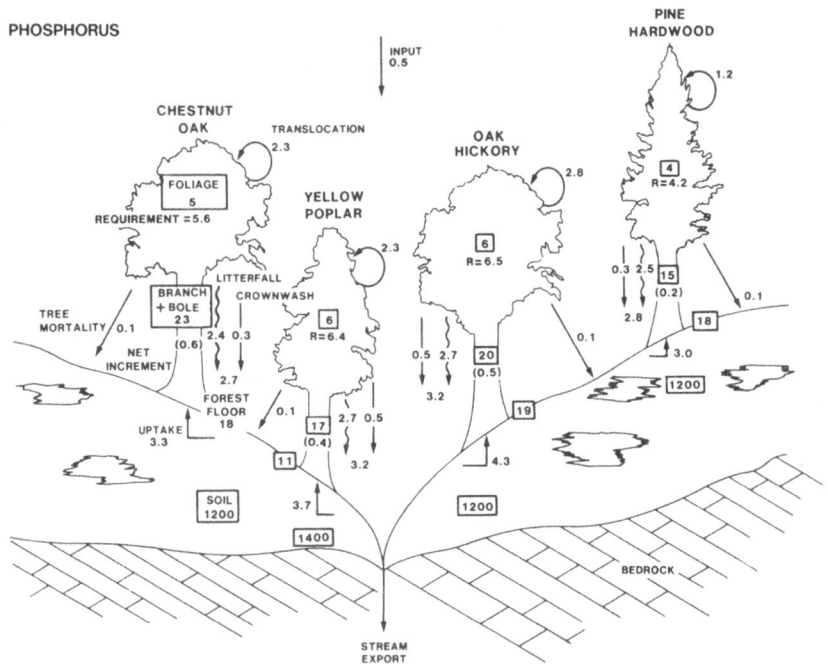

Figure 7.6. Phosphorus cycling in four forest types on Walker Branch Watershed (kg/ha or kg·ha^{-1}·year^{-1}).

tation. The amounts of N, P, and K returned to the forest floor via litterfall and throughfall in the yellow-poplar forest were similar to those in the other forest types. However, the yellow-poplar forest floor had appreciably lower amounts of N, P, and K and therefore greater turnoverrates of these nutrients than the other forest types.

Nitrogen and phosphorus were the only elements that showed significant translocation (transfer from foliage back to woody tissue prior to abscission), as evidenced by a comparison of foliage content with return via litterfall and net removal. This translocation of nitrogen and phosphorus is not surprising in view of the fact that forest growth in this region is frequently limited by nitrogen and sometimes by phosphorus (Farmer et al. 1970). Estimates of translocation derived in this manner are too low, inasmuch as dry deposition (which also contributes to throughfall) is not accounted for; therefore, significant positive translocation of the other elements may occur also.

There were some major differences between these forest types in the rates of biochemical cycling (translocation) and net increment of woody biomass. Translocation was markedly lower in the pine-hardwood forest

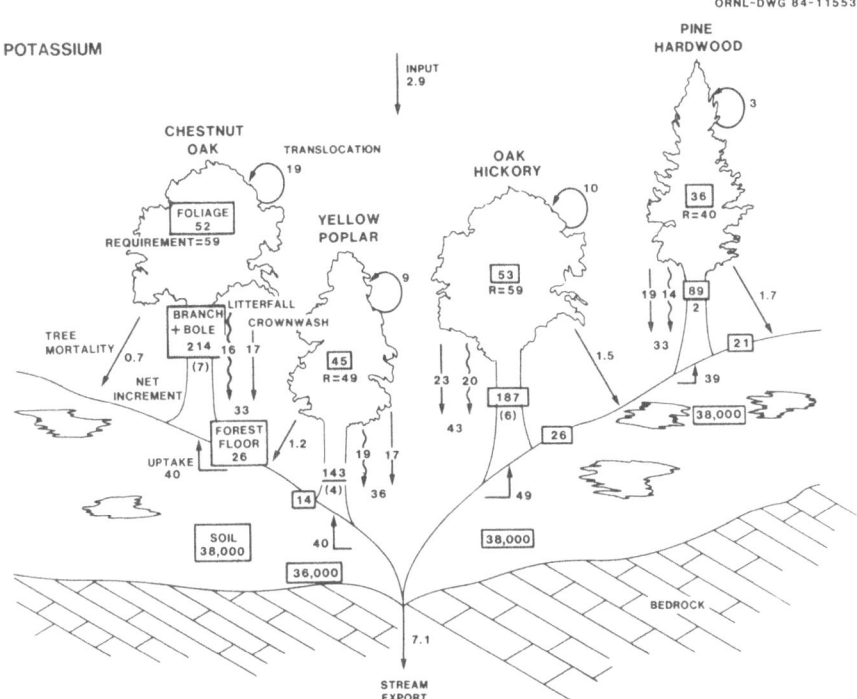

Figure 7.7. Potassium cycling in four forest types on Walker Branch Watershed (kg/ha or $kg \cdot ha^{-1} \cdot year^{-1}$).

than in the other three predominantly deciduous forest types (Figs. 7.5–7.9). This is consistent with the observation of Cole and Rapp (1981) that coniferous species generally show lower internal translocation than do deciduous species, presumably because of the the latter's need to shed all foliage each year.

Luxmoore et al. (1981) conducted detailed studies of nutrient translocation into leaves during spring and out of leaves prior to senescence in autumn. They found that evergreen species had lower translocation rates than deciduous trees, but there were no significant differences in translocation rates between various species of evergreen and deciduous trees (both understory and overstory). The rates of translocation into leaves of deciduous species showed a very rapid increase during spring; however, by late May, foliar phosphorus was being translocated at a slow rate back to stems. A similar trend was established for nitrogen by mid-June (Fig. 7.10). An internal storage pool is suggested as the major source of foliar nitrogen during the spring flush, since a simulation of nitrogen uptake from soil could account for only one-fourth of the quantity of nitrogen transported to the leaves by the end of May. Simulation further

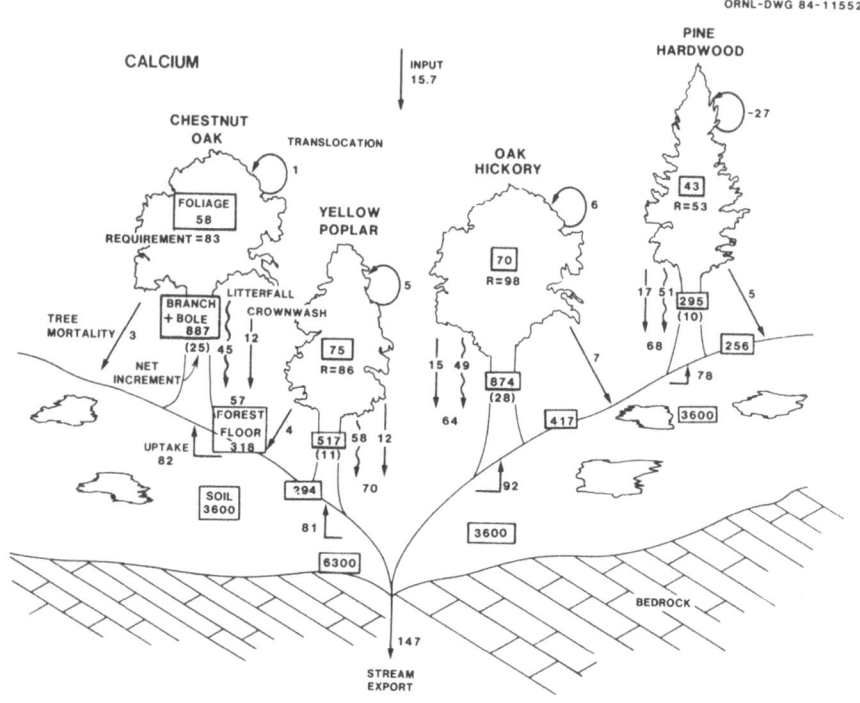

ORNL-DWG 84-11552

Figure 7.8. Calcium cycling in four forest types on Walker Branch Watershed (kg/ha or kg·ha^{-1}·year^{-1}).

showed that trace levels of soluble nitrogen (0.01 mg/kg) in soil were sufficient to supply a deciduous forest with an estimated nitrogen uptake of 100 kg N·ha^{-1}·year^{-1} (Henderson and Harris 1975).

At an early stage in the study, it was recognized that root turnover, which was estimated by multiplying root biomass and nutrient content on Walker Branch by root turnover rates in a nearby *Liriodendron* forest (Harris et al. 1973), played a major role in the cycling of all elements on Walker Branch. Root turnover accounted for ~50% of the total return of N, K, Ca, Mg, and (assuming foliar leaching = 3 kg·ha^{-1}·year^{-1}) S and >80% of the total return of P to the soil on a watershed scale (Fig. 7.4). Since the methods of estimating root turnover are difficult, tedious, and, to some extent, based on untested assumptions (e.g., the assumption that translocation is zero), the values can be questioned to a greater extent than thoseof other fluxes. More recent studies in other ecosystems have confirmed the large role played by roots in the turnover of nutrients in the soil (e.g., Vogt et al. 1983). Thus, although estimates of root turnover are subject to even larger errors than those of litterfall and foliar leaching, it is safe to assume that root turnover plays a major role in nutrient cycling in these forests.

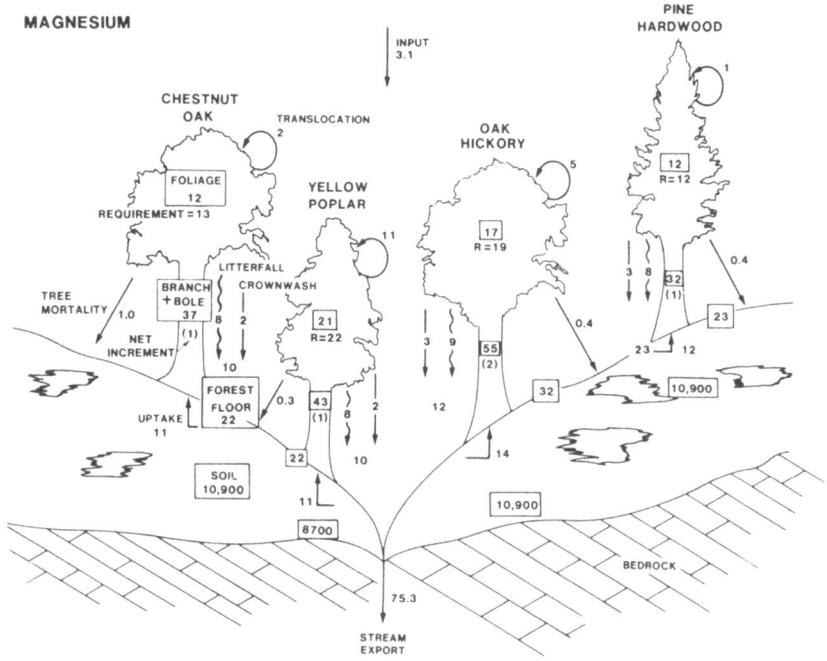

Figure 7.9. Magnesium cycling in four forest types on Walker Branch Watershed (kg/ha or kg·ha^{-1}·year^{-1}).

Forest floor turnover (i.e., nutrient content of forest floor divided by litterfall and throughfall return) varied considerably among nutrients from lows of 0.5, 0.9, and 2.0 years for K, S, and Mg to 6.3, 6.6, and 7.2 years for Ca, P, and N, respectively (Fig. 7.4). These differences reflect both the chemical mobility of (i.e., potassium), and the biological demand for (i.e., nitrogen and phosphorus), the individual nutrients.

Net increment in woody biomass varied considerably among nutrients both in magnitude and as a percentage of total uptake. On a watershed scale, woody increment accounted for 12% of nitrogen uptake, 10% of sulfur uptake, 17% of phosphorus uptake, 15% of potassium uptake, 28% of calcium uptake, and 15% of magnesium uptake (Fig. 7.4). Woody increment exceeded atmospheric input for all elements except sulfur, implying that the soil's supplies of all elements except sulfur are being depleted by vegetation increment even if leaching is ignored. This is especially striking for calcium, where the exchangeable calcium content of soil plus atmospheric input could support current estimates of calcium increment for only 18.5 years [i.e., (soil exchangeable calcium)/(woody increment − atmospheric input) = 18.5 years].

The pine-hardwood forest had considerably lower net increments of all

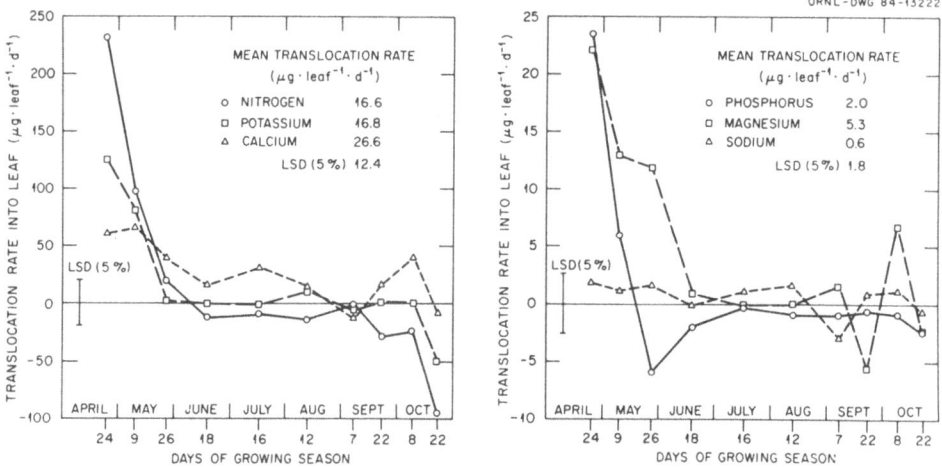

Figure 7.10. Mean translocation rates of N, K, Ca, P, Mg, and Na into and out of tree leaves on Walker Branch Watershed. SD = least significant difference. Source: R.J. Luxmoore, T. Grizzard, and R.H. Strand. 1981. Nutrient translocation in the outer canopy and understory of an eastern deciduous forest. For. Sci. 27:505–518.

nutrients in woody tissue than the deciduous forests (Figs. 7.5–7.9), a pattern due in part to lower biomass increment in the pine forest (as described in Ch. 6) and in part to generally lower nutrient concentrations in woody tissue in pines than in deciduous species (Table 7.1) (Cole and Rapp 1981). In the case of calcium, the yellow-poplar forest also had a low woody increment compared with the oak-hickory and chestnut oak forests (Fig. 7.8). In this case, the differences were entirely due to calcium concentration in wood, since biomass increment in the yellow-poplar forest exceeded that in all other forest types (Ch. 6).

7.2.5 Changes in Nutrient Distribution with Time

The high rates of calcium increment in biomass combined with low calcium supplies in the soil lead to some interesting forecasts of long-term trends in nutrient distribution on Walker Branch Watershed. Although nitrogen is thought to be currently limiting growth, long-term projections based on existing cycling data indicate that calcium will become limiting unless (1) calcium increments decline, (2) atmospheric calcium inputs increase, or (3) drains in calcium supplies through vegetation increment are offset by weathering and deep rooting.

As described in Ch. 6, an intensive inventory of forest vegetation on Walker Branch Watershed has been kept since 1967. The data have been combined with tissue analysis data for the major tree species on and near the watershed (Henderson et al. 1978; Johnson et al. 1982b) to estimate trends in aboveground biomass and nutrient distribution over the entire watershed over a 16-year period (1967–1983) (Johnson, Henderson, and Harris, in press).

Sampling soils for changes is much more time-consuming and difficult than estimating changes in vegetation nutrient content. Consequently, periodic soil sampling was limited to selected core plots within the watershed. The following sections describe watershed-wide changes in vegetation nutrient content and changes in vegetation, litter, and soil nutrient contents in selected plots.

Changes in Aboveground Nutrient Content in Major Forest Types

The southern pine beetle and hickory bark borer attacks described in Ch. 6 have had some striking effects on nutrient distribution and accumulation rates in most forest types. There was a sharp decline from positive to negative biomass and nutrient increment from 1979 to 1983 in all forest types (Fig. 7.11). Most of this can be explained by the southern pine beetle and hickory bark borer outbreaks. Biomass increment and nutrient accumulation rates had begun declining in the early 1970s in the loblolly pine forest as a result of overstocking, and a drop to negative increment was noted from 1979 to 1983 also.

Mortality of shortleaf pine in the early 1970s stimulated increased yellow-poplar growth and resulted in a large change in species composition. In the pine-hardwood forest, yellow-poplar biomass increased from 13 to 37% of total biomass, and pines decreased from 74 to 42% of total biomass from 1967 to 1983 (Ch. 6). Despite this large change in species composition, there was little change in total aboveground biomass and nutrient content in the pine-hardwood forest. It appears that the effects of pine mortality were almost exactly offset by the effects of increased yellow-poplar growth.

There was a major change in foliar biomass and nutrient content associated with the species shift from pine to poplar. As the pines dropped out and were replaced by yellow-poplar in the pine-hardwood forest, foliage biomass and nutrient content declined because of inherent differences in foliage biomass of these species (e.g., the pines retain 2 to 3 years' growth of needles, whereas the yellow-poplars produce only one set of leaves per year; Fig. 7.12). In the other forest types, foliage biomass remained relatively stable. (The stability of foliage biomass in the oak-hickory forest despite the large biomass loss caused by hickory mortality in the late 1970s was due to the death of primarily large trees with low foliage:wood biomass ratios.) It is interesting to note the convergence of foliage biomass in the pine-hardwood forest with that in the other deciduous forest types as deciduous species become increasingly dominant in the pine-hardwood forest (Fig. 7.12).

Since deciduous species usually have higher foliar nutrient concentra-

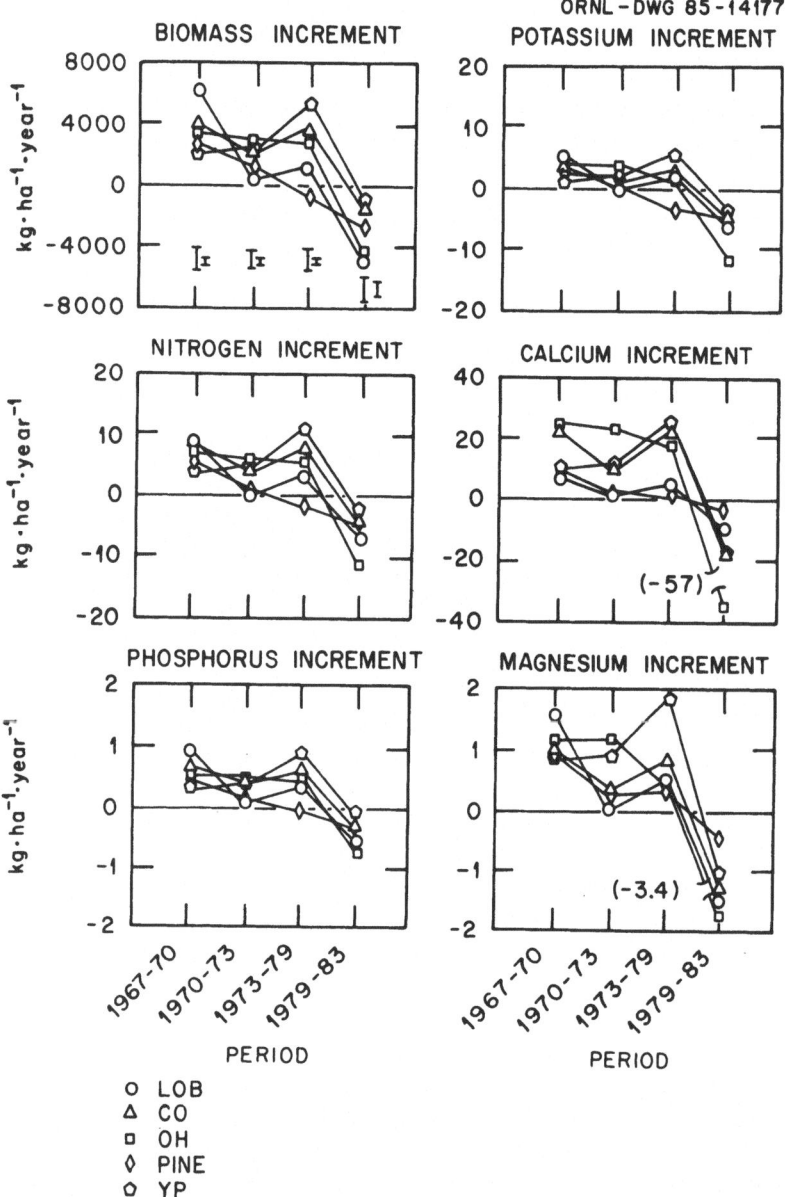

Figure 7.11. Net increments of biomass, N, P, K, Ca, and Mg in live biomass on Walker Branch Watershed (west catchment). (LOB = loblolly pine; CO = chestnut oak; OH = oak-hickory; PINE = pine-hardwood; YP = yellow-poplar.) Source: D.W. Johnson, G.S. Henderson, and W.F. Harris. Changes in aboveground biomass and nutrient content on Walker Branch Watershed from 1967 to 1983. pp. 487–496 IN R.L. Hay, F.W. Woods, and H. DeSelm (eds). Proceedings of the Central Hardwood Forest Conference VI., Knoxville, TN, February 24–26, 1987. University of Tennessee Press, Knoxville, TN.

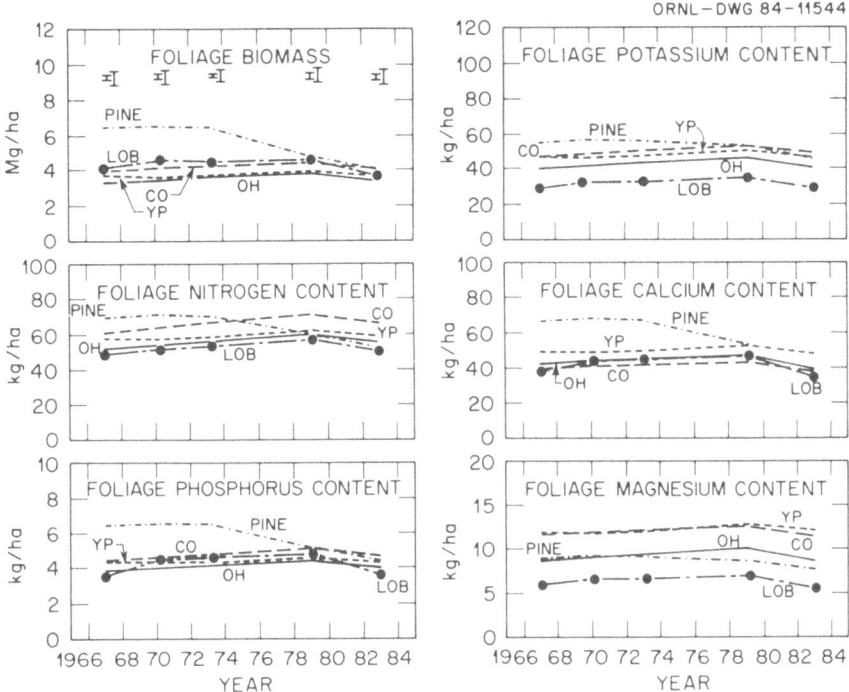

Figure 7.12. Foliage biomass and N, P, K, Ca, and Mg content in five forest types on Walker Branch Watershed. (LOB = loblolly pine; CO = chestnut oak; OH = oak-hickory; PINE = pine-hardwood; YP = yellow-poplar.) Source: D.W. Johnson, G.S. Henderson, and W.F. Harris. Changes in aboveground biomass and nutrient content on Walker Branch Watershed from 1967 to 1983. pp. 487–496 IN R.L. Hay, F.W. Woods, and H. DeSelm (eds). Proceedings of the Central Hardwood Forest Conference VI., Knoxville, TN, February 24–26, 1987. University of Tennessee Press, Knoxville, TN.

tions than coniferous species (Table 7.1), the patterns in foliage biomass are not necessarily reflected in foliage nutrient content. For instance, there were much smaller differences in foliage nutrient content (especially N, K, and Mg) than in biomass between the pine-hardwood and the three other deciduous forest types (Fig. 7.12).

Both pine and hickory mortalities have caused a substantial accumulation of biomass and nutrients in standing dead trees (Fig. 7.13). Thus, the total net aboveground biomass and nutrient increment declined from 1979 to 1983 owing to hickory mortality, and most net increment on an area basis went into standing dead trees. Although the estimates in Fig. 7.13 are too large to the extent that decay and loss of limbs are not accounted for (foliage is deleted, but standing dead biomass and nutrient content are estimated from regressions and nutrient concentrations in the

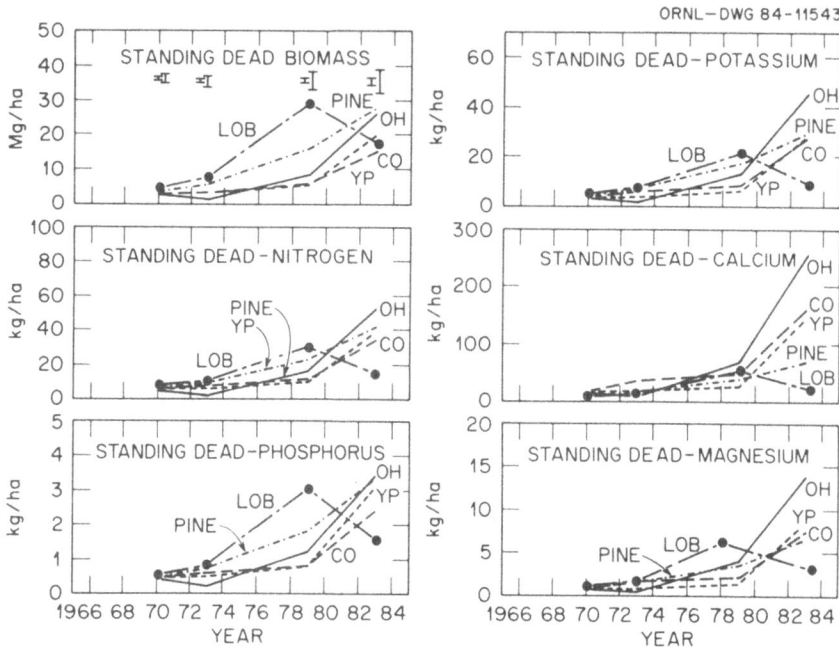

Figure 7.13. Biomass, N, P, K, Ca, and Mg content in standing dead biomass on Walker Branch Watershed. (LOB = loblolly pine; CO = chestnut oak; OH = oak-hickory; PINE = pine-hardwood; YP = yellow-poplar.) Source: D.W. Johnson, G.S. Henderson, and W.F. Harris. Changes in aboveground biomass and nutrient content on Walker Branch Watershed from 1967 to 1983. pp. 487–496 IN R.L. Hay, F.W. Woods, and H. DeSelm (eds). Proceedings of the Central Hardwood Forest Conference, VI., Knoxville, TN, February 24–26, 1987. University of Tennessee Press, Knoxville, TN.

woody tissue of live trees), it is clear that a substantial return of woody litter and nutrients will occur in localized areas as these trees fall over the next few years. On an area basis, the 1983 estimates of standing dead biomass, K, Ca, and Mg contents in the oak-hickory forest rival those in the entire forest floor (wood, 01 and 02 horizons) in the early 1970s. Henderson et al. (1978) report forest floor biomass, potassium, and calcium contents of 29,500, 20, and 430 kg/ha for the entire watershed; Cole and Rapp (1981) report forest floor magnesium values of 22 to 32 kg/ha on Walker Branch Watershed. Nitrogen and phosphorus contents in standing dead timber are much less than in the forest floor (310 and 11–22 kg/ha, respectively) (Henderson et al. 1978; Cole and Rapp 1981). However, the input of woody litter having very wide carbon:nitrogen and carbon:phosphorus ratios may have substantial localized effects on nitrogen and phosphorus mineralization and immobilization in all forest types.

Biomass and estimated nutrient return via treefall increased consid-

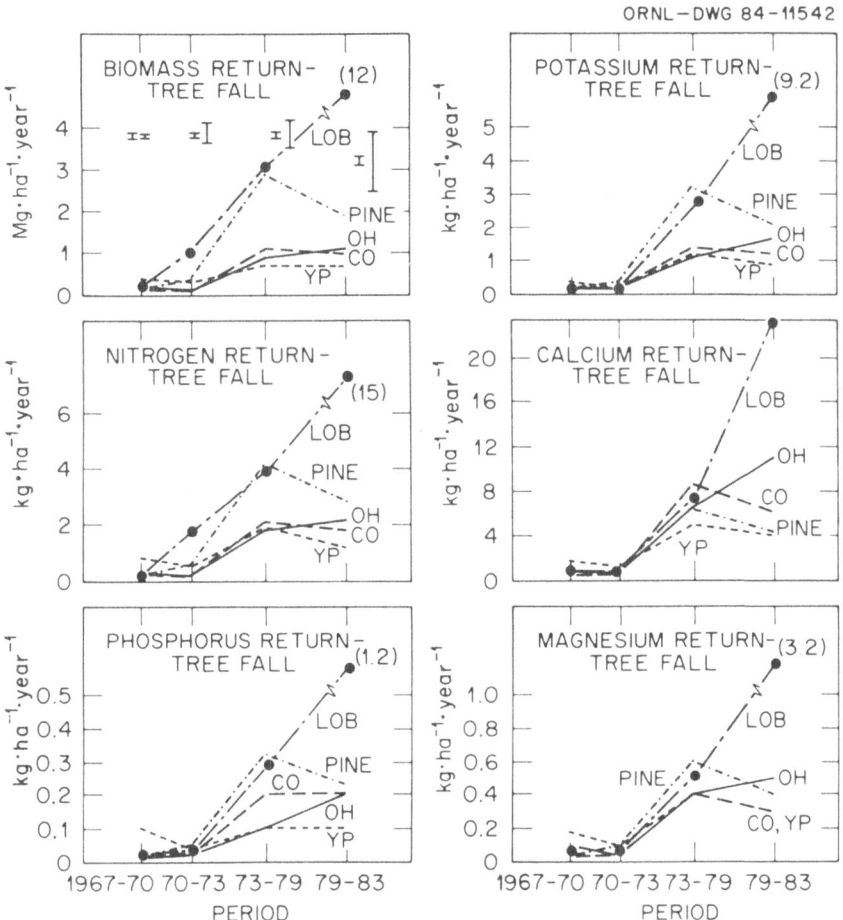

ORNL—DWG 84—11542

Figure 7.14. Return of biomass, N, P, K, Ca, and Mg via treemortality in five forest types on Walker Branch Watershed. (LOB = loblolly pine; CO = chestnut oak; OH = oak-hickory; PINE = pine-hardwood; YP = yellow-poplar.) Source: D.W. Johnson, G.S. Henderson, and W.F. Harris. Changes in aboveground biomass and nutrient content on Walker Branch Watershed from 1967 to 1983. pp. 487–496 IN R.L. Hay, F.W. Woods, and H. DeSelm (eds). Proceedings of the Central Hardwood Forest Conference, VI., Knoxville, TN, February 24–26, 1987. University of Tennessee Press, Knoxville, TN.

erably with the onset of pine and hickory mortality from 1970 to 1983 (Fig. 7.14). Nutrient return was highest in the pine forest, except in the case of calcium: the calcium-rich woody tissues of the hickories (and, to a lesser extent, other assorted hardwood species) cause calcium return via mortality in the oak-hickory and chestnut oak forest to exceed that in the pine forest (Fig. 7.14). Clearly, the bulk of the biomass and nutrients in the hickories that died during 1979–1983 has not yet returned to the forest floor, as discussed previously, although an upward trend in return is already indicated (Fig. 7.14).

Changes in Vegetation, Litter, and Soils in Selected Plots

In addition to the watershed-wide vegetation survey described above, 24 core plots were intensively sampled for forest floor weight and nutrient content and soil nitrogen at 15-cm intervals to a 60-cm depth on a seasonal (quarterly) basis during 1972. Samples were taken randomly from a 10 × 10 m grid subplot (one sample per square meter, 3–7 samples per subplot). Samples were dried, and soils were sieved and stored. Unfortunately, the 15- to 30- and the 30- to 45-cm-depth soil samples were lost. In the spring of 1982, eight of the 24 original core plots were resampled, using the same 10 × 10 m grid and the same sampling procedures as originally established. All soils from both periods were analyzed simultaneously for pH (1:1 soil:solution in H_2O and in 0.01 M $CaCl_2$), total nitrogen (Kjeldahl, using block digester and autoanalyzer analysis for NH_4^+), cation exchange capacity and exchangeable cations (by 1 M NH_4Cl extraction, followed by an ethanol wash and 1 M KCl to displace NH_4^+), and extractable phosphorus (0.5 M HCl plus 1 M NH_4Cl; Olson and Dean 1965). See Johnson, Henderson, and Todd (in press) for full details.

The changes in aboveground tree (live and standing dead) and forest floor biomass and nutrient content from 1972–73 to 1982–83 are given in Fig. 7.15. Because the intermediate (15–45 cm) soil depths were not sampled in 1972, a full accounting of changes in soil nutrient cannot be made, but comparisons between the top (0–15 cm) and the deepest (45–60 cm) soil sampling depths are given later.

All plots had a net increase in aboveground (vegetation plus forest floor) biomass from 1973 to 1982, although in some cases (plots 26, 107), this was due in large measure to an increase in standing dead biomass. (Obviously, in these cases mortality exceeded the growth of live trees.) Outbreaks of the southern pine beetle in the early 1970s and the hickory borer in the late 1970s are largely responsible for the mortality in the pine and oak-hickory forest types.

Despite the overall increase in aboveground biomass, several plots showed either no change or a net decrease in aboveground nutrient content, primarily due to decreases in forest floor nutrient content (Fig. 7.15). In some cases (plots 26 and 91), the decreases in forest floor nutrient content were due largely to decreases in forest floor biomass, but in other cases the decrease in nutrient content was clearly due to a decrease in nutrient concentration: several plots showed a decrease in forest floor nutrient content, with no change in forest floor biomass (nitrogen in plot 42; calcium in plots 281 and 179), and some plots showed either no change

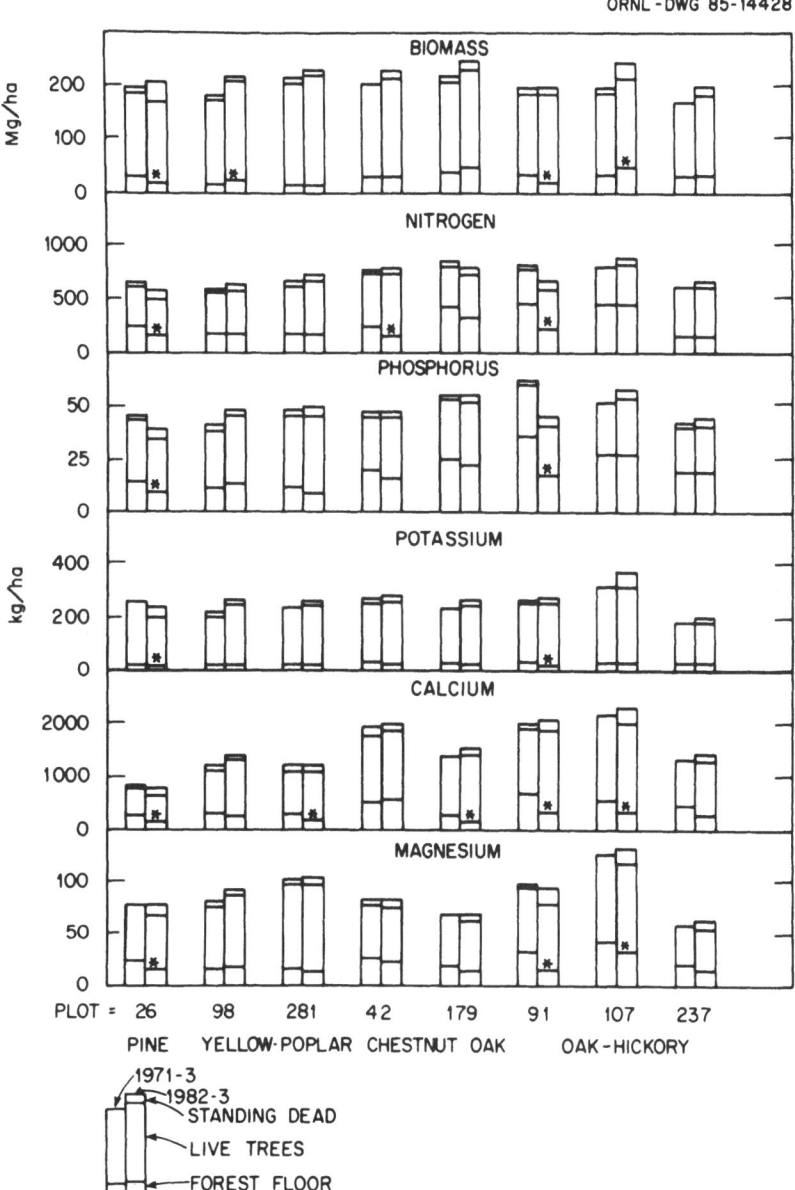

Figure 7.15. Changes in biomass and nutrient content in vegetation and forest floor in selected plots from Walker Branch Watershed from 1971–73 to 1982–83 (* denotes statistically significant difference, 95% confidence level, t-test). Source: D.W. Johnson, G.S. Henderson, and D.E. Todd. Changes in nutrient distribution in forests and soils of Walker Branch Watershed, Tennessee, over an 11-year period. Biogeochemistry (in press).

or a decrease in forest floor nutrient content with increased forest floor biomass (plots 98 and 107).

There were several statistically significant differences in soil nutrient contents and chemical properties between the 1972 and 1982 samplings at both soil depths resampled (0–15 cm and 45–60 cm) (Fig. 7.16). As a general rule, surface (0–15 cm) soils either showed no statistically significant changes or showed increases in concentrations of all nutrients.

Subsoils (45–60 cm) showed either no changes or significant decreases in total nitrogen and exchangeable Ca^{2+} and Mg^{2+}, and both increases and decreases in extractable phosphorus and exchangeable K^+. The most consistent patterns in the subsoils were the decreases in exchangeable Ca^{2+} and Mg^{2+}, which were most notable (60–90% reductions) in subsoils from those cherty, ridgetop plots (91, 107, 179) with the lowest soil contents of these nutrients (Fig. 7.16). There were also marked decreases (70–75%) in exchangeable Mg^{2+} in subsoils from plots 281 (yellow-poplar) and 237, a marked increase (300%) in exchangeable K^+ in the subsoil of plot 107, and a marked decrease in extractable phosphorus in the subsoil of plot 91 (Fig. 7.16).

There are several possible explanations for the apparent changes in forest floor and soil nutrient content from 1972 to 1982. Vegetation increment could have caused the subsoil exchangeable Ca^{2+} decreases in the three cherty, upland chestnut oak, and oak-hickory forests (plots 91, 107, and 179). Vegetation increment per se could not have accounted for the subsoil exchangeable Mg^{2+} decreases in these three plots or in the other two plots (281 and 237) in which they occurred. However, it is possible that total vegetation uptake (which exceeds the net vegetation increment by five- to tenfold) from subsoils, and subsequent return via litterfall to the soil surface, has resulted in a redistribution of nutrients from subsoils to surface soils, as suggested by Thomas (1967) for calcium "pumping" by dogwood trees on Walker Branch Watershed. Without a full accounting of changes in all soil horizons, it is not possible to either confirm or deny this hypothesis.

It is safe to assume that soil leaching, which is thought to have been increased approximately twofold by acid deposition on Walker Branch Watershed (Johnson et al. 1985), has played some role in the observed decline in subsoil exchangeable base cations. This would apply especially to the reductions in subsoil exchangeable Mg^{2+} that cannot be accounted for by net vegetation increment.

The decreases in forest floor nutrient content may have resulted from either a decrease in litterfall nutrient concentration or an increase in forest floor nutrient turnover rate. Given the fact that litter (01) concentrations often decreased whereas litter weights increased, the former (decrease in litterfall nutrient concentration) appears more likely than the latter (increase in forest floor nutrient turnover rate). In short, it appears that a

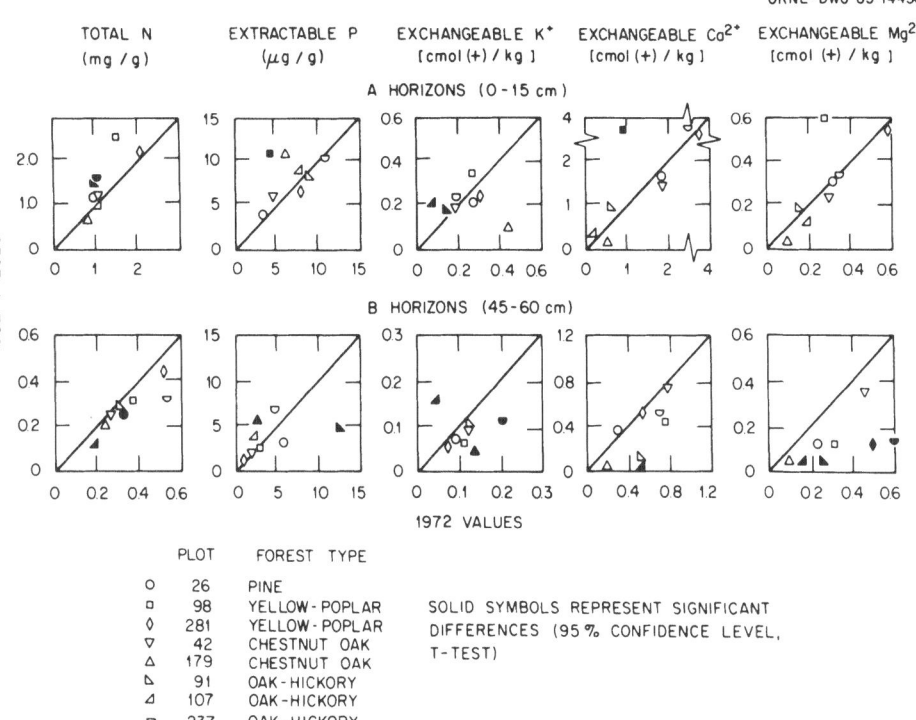

ORNL-DWG 85-14430

Figure 7.16. 1982 vs. 1972 average values for total nitrogen, extractable phosphorus, and exchangeable Ca^{2+}, K^+, and Mg^{2+} in surface (0–15 cm) and subsurface (45–60 cm) soils from selected plots on Walker Branch Watershed. Solid symbols represent significant differences (95% confidence level, t-test) between 1972 and 1982 values. Source: D.W. Johnson, G.S. Henderson, and D.E. Todd. Changes in nutrient distribution in forests and soils of Walker Branch Watershed, Tennessee, over an 11-year period. Biogeochemistry (in press).

general decrease in site fertility is manifested by lower subsoil and litter nutrient concentrations.

7.3 Perturbations

As noted earlier, the Walker Branch Watershed Project was originally set up as a watershed-scale fertilizer experiment. It was decided later that the watershed as a whole would be left relatively undisturbed and that manipulative studies would be done on plots within the watershed or on adjacent, similar sites (Harris 1977). One form of disturbance that could not be avoided, however, was air pollution. A summary of air pollutant effects research on Walker Branch Watershed is given in Sect. 7.3.1.

7.3.1 Pollutant Inputs

Since the early characterization of nutrient cycles on the watershed as part of the International Biological Program, air pollution effects research has been the dominant activity on Walker Branch Watershed. Unfortunately, air pollution effects are especially difficult to study, because there are no comparable control watersheds nearby that are not subject to air pollution. Our approach to the problem has been to make systems-level measurements of the inputs, cycling, and ultimate fate of atmospherically deposited substances, using basically the same methods as those used in the initial characterization of nutrient cycles, and from that information, to deduce the effects of such deposition on the ecosystem.

Sulfur Cycling

Because of the great concern over atmospheric sulfur deposition and its associated effects, sulfur has been the most intensively studied nutrient on the watershed. The initial watershed-scale characterization of the sulfur cycle by Shriner and Henderson (1978) showed some interesting and unique properties of this nutrient as compared with the other macronutrients on the watershed.

One of the most striking aspects of the sulfur cycle was the apparently very rapid turnover of foliar sulfur (Fig. 7.17): the net return of sulfur equaled 4.8 times the foliar sulfur content, mostly because of the very large net removal (i.e., throughfall flux minus precipitation input as measured in the open). Shriner and Henderson (1978) strongly suspected that dry deposition of sulfur was responsible for this anomaly, a suspicion that was later confirmed by detailed deposition studies (Lindberg et al. in press). However, the magnitudes of deposition needed to account for the large net removal term were greater than those measured in subsequent, more detailed deposition studies [e.g., Lindberg et al. (in press), Richter et al. (1983), and Johnson et al. (1982a) obtained sulfur input values of ~26 $kg \cdot ha^{-1} \cdot year^{-1}$]. Questions as to methodology exist [e.g., the type of screen used on throughfall funnels (Shriner and Henderson 1978; Johnson et al. 1982a)], but there is also the possibility that increasing the height of local power plant stacks after the Shriner and Henderson study caused a reduction in dry deposition of sulfur.

A second major feature of the sulfur cycle characterized by Shriner and Henderson (1978) was the accumulation of sulfur in the mineral soil. This was subsequently shown to be due primarily to SO_4^{2-} adsorption onto iron and aluminum hydrous oxides in subsurface Bhorizons (Johnson and Henderson 1979; Johnson et al. 1981). The deep (up to 30 m), highly weathered soils on Walker Branch Watershed provide a very large sink for SO_4^{2-} by adsorption, whereas the surface (A2 horizon) soils contain little adsorbed SO_4^{2-} but large amounts of soluble and readily mineralizable

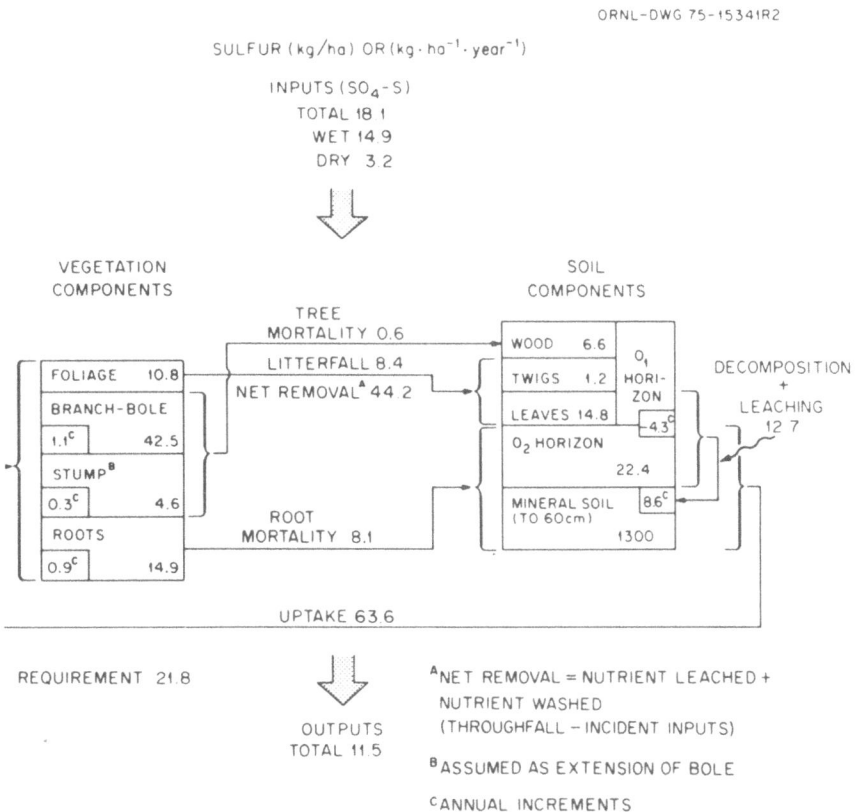

Figure 7.17. Sulfur cycling on Walker Branch Watershed. Source: D.S. Shriner and G.S. Henderson. 1978. Sulfur distribution and cycling in a deciduous forest watershed. J. Environ. Qual. 7:392–397.

SO_4^{2-} (Fig. 7.18a,d). Fully one-half of the total sulfur in the surface soils mineralized during aerobic incubation (Fig. 7.18d). In sharp contrast, only 1% of the soil nitrogen mineralized during the same aerobic incubations in the laboratory (Fig. 7.18d).

The distribution of adsorbed vs soluble SO_4^{2-} in the profile corresponds to the distribution of hydrous iron oxides, as evidenced by the darker color of the B horizon (Fig. 7.18b). Not all water encounters adsorbing Bhorizons before entering the stream, however. During storm events, water may move laterally within the A2 horizon across the top of the argillic B horizon (with its lower hydraulic conductivity) and thereby shunt around the SO_4^{2-}-adsorbing B horizon, carrying SO_4^{2-} directly to the stream. This is thought to account for the fact that the SO_4^{2-} concentration in stream water increases during storm events (Fig. 7.18c), whereas concentrations of other ions generally decrease (Henderson et al. 1977b).

ORNL–DWG 84-13254

(a)

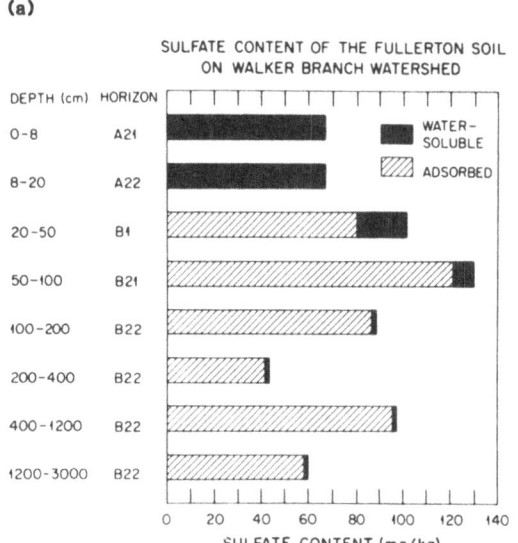

SULFATE CONTENT OF THE FULLERTON SOIL
ON WALKER BRANCH WATERSHED

(b)

FULLERTON PROFILE

ORNL–DWG 87-12335

(c)

(d)

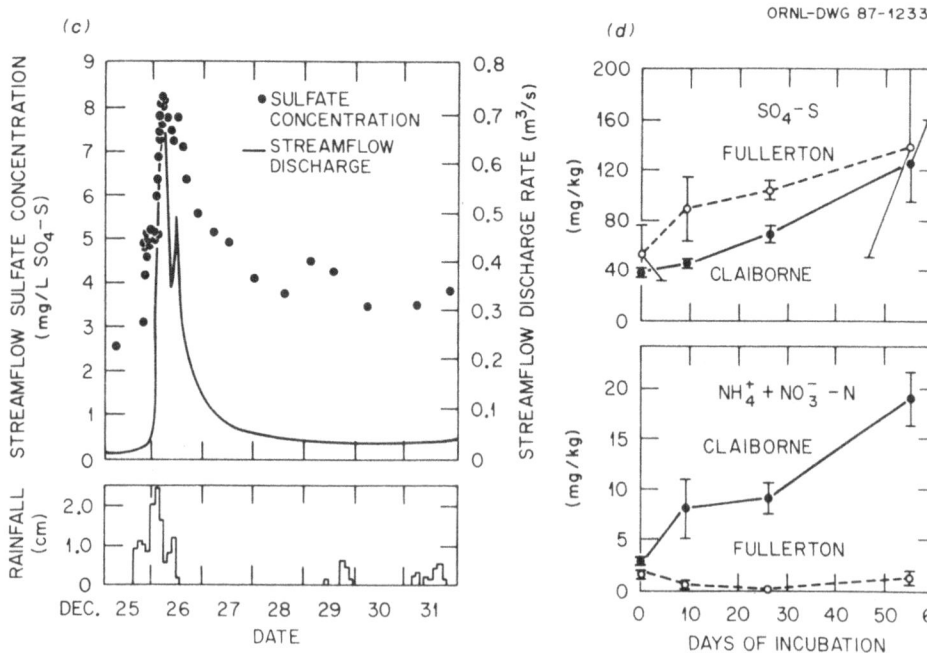

Although a large proportion of soil sulfur is in SO_4^{2-} form, most of it (66%) is in organic sulfur form. Swank et al. (1984) recently suggested that incorporation of sulfur into soil organic matter may account for sulfur retention in ecosystems at Coweeta, North Carolina. Incorporation of sulfur into organic matter may well account for some sulfur retention on Walker Branch Watershed as well, but the patterns of accumulation and export (i.e., considerably lower SO_4^{2-} fluxes through B2 horizons than through A2 horizons as well as the previously mentioned stormflow patterns) suggest that adsorption is the primary sulfur-retention mechanism in Walker Branch.

The imbalance of sulfur and nitrogen in the ecosystem has led to accumulation of sulfur as SO_4^{2-} in vegetation as well as in soils. Several investigators have noted that sulfur in excess of that required for protein synthesis (which is functionally the amount of sulfur present in excess of a sulfur:nitrogen ratio of 0.03 on a molar basis) accumulates in tissues as SO_4^{2-} (Kelly and Lambert 1972; Turner et al. 1977, 1980). We have tested this relationship for vegetation on Walker Branch Watershed and found it to be valid (Johnson et al. 1982a). We also tested for organic sulfates in foliage and found none, thus concluding that all foliar SO_4^{2-} is inorganic (Richter et al. 1983). Sulfate can be an important component of litterfall, and thus SO_4^{2-} can be a major component of sulfur cycling in forests subject to elevated inputs of atmospheric sulfur (Hesse 1957; Turner and Lambert 1980; Turner et al. 1980).

Our current understanding of sulfur cycling in a chestnut oak forest on Walker Branch Watershed is depicted in Fig. 7.19. We find that sulfur cycling is dominated by geochemical processes involving sulfate. In contrast to the results of Shriner and Henderson (1978) for the entire Walker Branch Watershed, sulfur uptake and requirement were equal in the chestnut oak forest described here (Fig. 7.19). The large net removal term in Shriner and Henderson's results accounted for the fact that uptake greatly exceeded requirement. In the chestnut oak forest study, the sulfur utilized in foliar production (12 kg/ha) equaled the net return of sulfur

◁————————————————————————————————————

Figure 7.18. (a) Sulfate content of the Fullerton soil on Walker Branch Watershed. (b) The Fullerton soil profile. (c) Relation between sulfate concentration in streamflow and discharge rate for alarge storm on Walker Branch Watershed. Solid lines are stream discharge rates, and points are sulfate concentrations during thestorm. Rainfall during the storm is also shown. Source: D.W. Johnson and G.S. Henderson. 1979. Sulfate adsorption and sulfur fractions in a highly weathered soil under a mixed deciduous forest. Soil Sci. 128:34–40. (d) Changes in soluble SO_4^{2-}–S and KCl-extractable $NH_4^+ + NO_3^-$ during aerobic incubation of surface soils (A1 and A2 horizons) of the Claiborne and Fullerton series from Walker Branch Watershed. Source: D.W. Johnson, G.S. Henderson, D.D. Huff, S.E. Lindberg, D.D. Richter, D.S. Shriner, D.E. Todd, and J. Turner. 1982. Cycling of organic and inorganic sulphur in a chestnut oak forest. Oecologia 54:141–148.

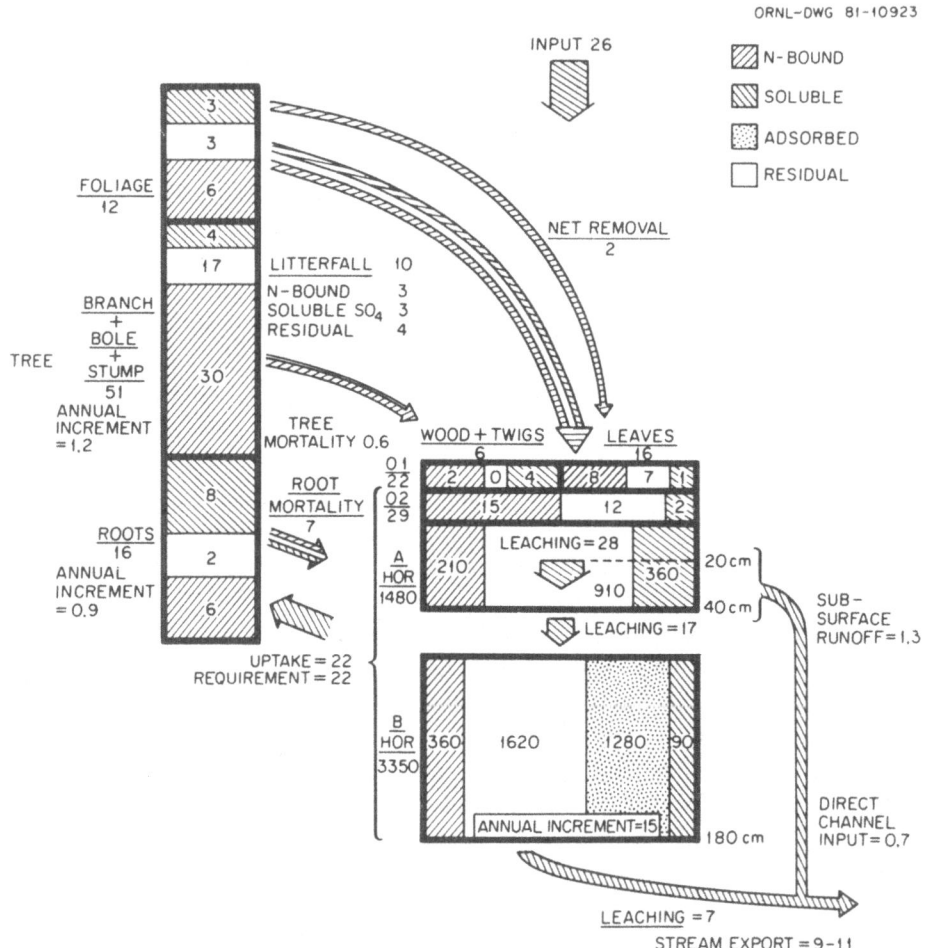

Figure 7.19. Schematic diagram of total sulfur, soluble SO_4^{2-}, nitrogen-bound sulfur, and residual sulfur in a chestnut oak forest on Walker Branch Watershed. Input is given as SO_4^{2-}–S flux in bulk precipitation. Source: D.W. Johnson, D.S. Henderson, D.D. Huff, S.E. Lindberg, D.D. Richter, D.S. Shriner, D.E. Todd, and J. Turner. 1982. Cycling of organic and inorganic sulfur in a chestnut oak forest. Oecologia 54:141–148.

from the foliage to the forest floor (10 kg·ha^{-1}·year^{-1} in litterfall and 2 kg·ha^{-1}·year^{-1} in canopy leaching); thus, uptake and requirement were equal. However, defining sulfur requirement as above is misleading in that it includes the sulfur taken up and cycled as excess SO_4^{2-}. A more realistic value for requirement would be the sulfur required to satisfy a sulfur:nitrogen ratio of 0.03 (shown as nitrogen-bound sulfur in Fig. 7.19), which would give a requirement value of ~11 kg·ha^{-1}·year^{-1} (i.e., about half the total sulfur "requirement").

Sulfur cycling in the chestnut oak forest is clearly dominated by geo-chemical processes. The input of SO_4^{2-}–S via bulk precipitation alone (26 kg·ha^{-1}·year^{-1}) exceeded sulfur return via litterfall (10 kg·ha^{-1}·year^{-1}), crown wash (2 kg·ha^{-1}·year^{-1}), and root mortality (7 kg·ha^{-1}·year^{-1}). Furthermore, atmospheric sulfur input was adequate to fulfill the calculated annual total sulfur requirement (22 kg·ha^{-1}·year^{-1}) even when the re-quirement was defined in such a way that it included the utilization of SO_4^{2-}–S in excess of the nutritional needs of the trees. The soluble SO_4^{2-}–S content of the soil exceeded the total sulfur content by a factor of 5 and the annual tree uptake by a factor of 16. Furthermore, organic sulfates provided a pool of readily mineralizable sulfur in surface horizons that exceeded soluble SO_4^{2-}–S by a factor of 3 (Fig. 7.18d).

Despite the fact that sulfur was available to this forest far in excess of its nutritional needs, ~60% (16-18 kg·ha^{-1}·year^{-1}) of the incoming sulfur (via bulk precipitation) accumulated within the ecosystem. Most (~90%) of this annual accumulation occurred in the mineral soil; only a small amount (~10%) accumulated in vegetation. Of the total annual accumu-lation in the ecosystem, only ~6%, or 1 kg·ha^{-1}·year^{-1}, was sequestered as nitrogen-bound or protein sulfur, whereas most of it (>80%) accu-mulated by means of sulfate adsorption to subsurface soil horizons.

The cycles of sulfur and nitrogen on Walker Branch Watershed were closely coupled and similar to one another in some ways but vastly different in others. Both of these nutrients are involved in protein synthesis (Turner et al. 1980), and both were accumulating in this ecosystem (Henderson and Harris 1975; Shriner and Henderson 1978). However, whereas avail-able nitrogen supplies were extremely limited, available sulfur supplies were excessive. Nitrogen was accumulating bymeans of biological im-mobilization in the ecosystem; sulfur was accumulating mostly by means of inorganic chemical immobilization in the soil. In short, nitrogen cycling was dominated by biological processes, whereas sulfur cycling was dom-inated by geochemical processes.

Acid Deposition

Acid deposition studies were begun on Walker Branch Watershed in 1980 primarily with funding from the Electric Power Research Institute. Two sites—one in a chestnut oak forest and one in a yellow-poplar forest—were chosen for intensive studies of the relative contributions of acid dep-osition and natural acid production to base cation depletion in soils. Much of the previous research on nutrient cycling—especially for sulfur cycling—proved invaluable in facilitating acid deposition studies and later in the interpretation of the results of such studies, which are briefly summarized here. The reader is referred to studies by Richter et al. (1983), Johnson etal. (1985), Lindberg et al. (in press), and Lovett et al. (1986) for further details.

Atmospherically deposited strong acids were rapidly transformed into sulfate-salt solutions by these forest systems. The volume-weighted mean pH for bulk precipitation was 4.3; for bulk throughfall, it was ~4.8; and for leachates from O2, A1, and B21 soil horizons, it was ~6.0 (Fig. 7.20). At both sites, base cations that exchanged with H^+ from atmospheric sources were almost entirely supplied by forest canopies and litter layers rather than directly from exchangeable soil pools. The hydrologic flux of HCO_3^- alkalinity from B21 horizons indicated that the natural H^+ input from the ionization of H_2CO_3 was the same order of magnitude as the H^+ input via bulk precipitation (Table 7.2) but considerably less than the estimated H^+ deposition via wet plus dry deposition (Lovett et al. 1986) (Fig. 7.21). In the soil with a high accumulation of hydrous iron oxides (Fullerton series), hydrologic fluxes of SO_4^{2-} and base cations were reduced by ~50% between the A1 and B21 horizons as a consequence of SO_4^{2-} adsorption by subsoils. In the soil with a low accumulation of hydrous oxides (Tarklin series), fluxes of SO_4^{2-} and cations were not reduced with soil depth.

Coincidentally, each ecosystem received a total estimated H^+ input of ~2.1 kmol$(+)\cdot$ha$^{-1}\cdot$year^{-1}, 70% of which was from atmospheric deposition and 30% from internal carbonic acidformation (Fig. 7.21) (Johnson et al. 1985). Vegetation increment, or the net annual accumulation of base cations in vegetation, effectively added another 0.9 and 1.1 kmol$(+)\cdot$ha$^{-1}\cdot$year^{-1} of internal acidification potential (defined as depletion of exchangeable base cations) in the yellow-poplar and chestnut oak sites, respectively. The extent to which this acidification potential will be realized depends on the size of the exchangeable cation pools, the effectiveness of incoming H^+ in causing the leaching of base cations, and the rate of replenishment of exchangeable base cations by weathering.

Both sites showed large net exports of Na^+, and the yellow-poplar site showed net exports of K^+, Ca^{2+}, and Mg^{2+} as well (Table 7.2). Except possibly in the case of Ca^{2+} at the yellow-poplar site, net exports of cations were not regarded as posing a potential for significantly reducing soil cation reserves. When the net increment of Ca^{2+} in vegetation is added to leaching losses, it would seem that the Ca^{2+} status of the soil at the yellow-poplar site will decline significantly over the next 50 to 100 years, even if weathering rates are high (Fig. 7.22) (Johnson et al. 1985). Several processes might mitigate against such a decline, however. First, only the top 50 cm of soil (down to fragipan) is considered here, whereas the depth of rooting is probably much greater. Second, the Tarklin site is situated in a sinkhole, which receives and entraps unknown amounts of leaf litter from surrounding forests, resulting in a net import of Ca^{2+} as well as other nutrients. Finally, Ca^{2+} leaching rates can be expected to decline if exchangeable Ca^{2+} reserves decline. Declines in Ca^{2+} leaching must be accompanied by greater losses of other cations (including Al^{3+} if the soil becomes extremely acid) if inputs remain constant.

Leaching presents little apparent threat to soil calcium supplies at the

ORNL–DWG 84-13247

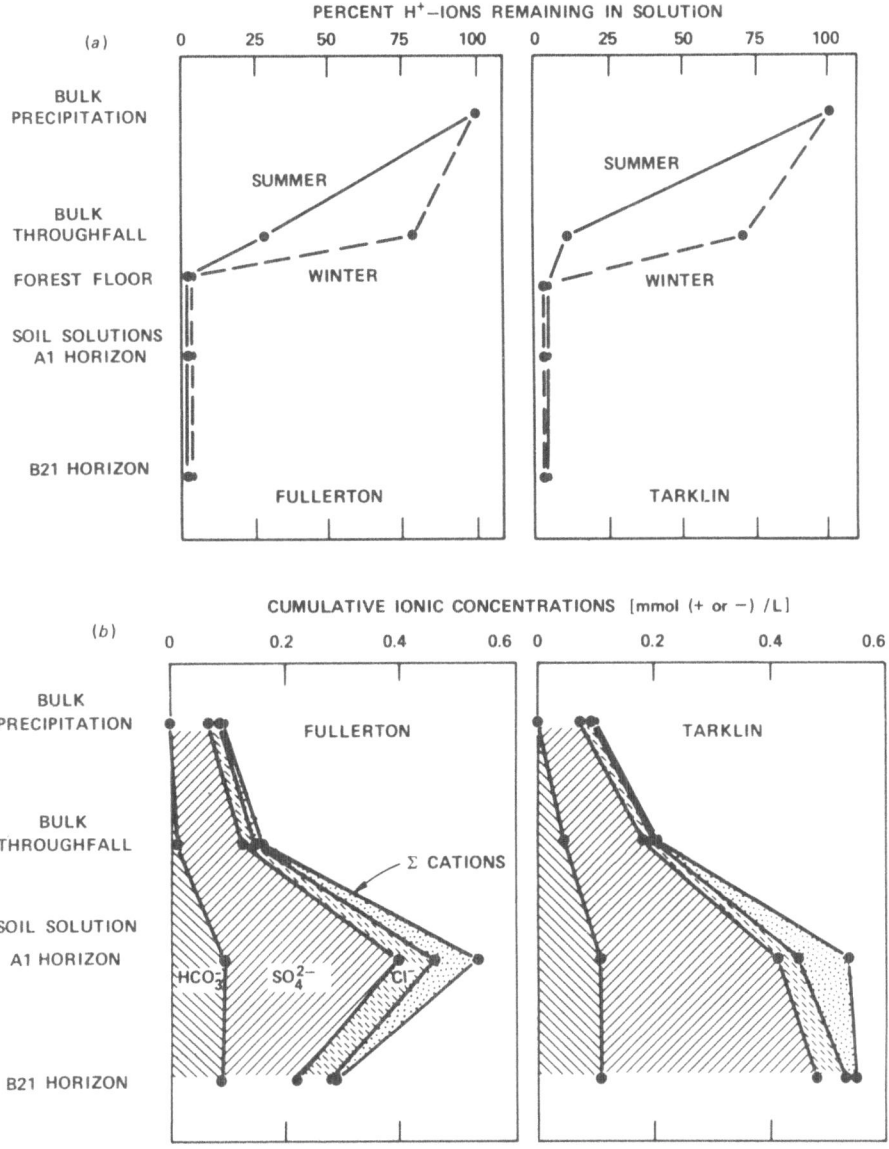

Figure 7.20. (a) Percent of precipitation H^+ ions remaining in throughfall and soil solution in the Fullerton (chestnut oak forest) and Tarklin (yellow-poplar forest) soil series on Walker Branch Watershed. (b) Mean volume-weighted concentrations of major anions and total cations in precipitation, throughfall, and soil solutions at the Fullerton and Tarklin sites. Source: D.D. Richter, D.W. Johnson, and D.E. Todd. 1983. Atmospheric sulfur deposition, neutralization, and ion leaching in two deciduous forest ecosystems. J. Environ. Qual. 12:263–270.

Table 7.2. Annual fluxes of anions and cations in the Fullerton and Tarklin sites (November 1980 to October 1982)

Sampling level	H₂O (cm)	HCO₃⁻	SO₄²⁻	NO₃⁻	H⁺	Ca²⁺	K⁺	Mg²⁺	Na⁺	Σ Cations	Σ Anions
						[kmol(+ or −)·ha⁻¹·year⁻¹]					
Bulk precipitation	126	0	0.78	0.13	0.71	0.27	0.04	0.07	0.05	1.22	1.02
						Fullerton site					
Throughfall	111	0.21	1.28	0.23	0.36	0.66	0.36	0.25	0.05	1.86	1.93
Stemflow	6	0.02	0.40	0.01	0.0006	0.20	0.24	0.05	0.006	0.49	0.46
Soil leaching (B2 horizon, 80 cm)	56	0.59	1.01	0.01	0.002	0.24	0.06	0.22	1.13	1.66	1.92
						Tarklin site					
Throughfall	122	0.69	1.54	0.24	0.28	0.96	0.61	0.41	0.06	2.52	2.72
Stemflow	4	0.001	0.19	0.004	0.02	0.08	0.07	0.03	0.003	0.17	0.22
Soil leaching (B2 horizon, 50 cm)	56	0.48	2.17	0.001	0.02	1.40	0.10	0.44	0.69	2.65	2.84

Source: After D.W. Johnson, D.D. Richter, G.M. Lovett, and S.E. Lindberg. 1985. The effects of atmospheric deposition on potassium, calcium, and magnesium cycling in two deciduous forests. Can. J. For. Res. 15:773–782.

ORNL-DWG 84-13077

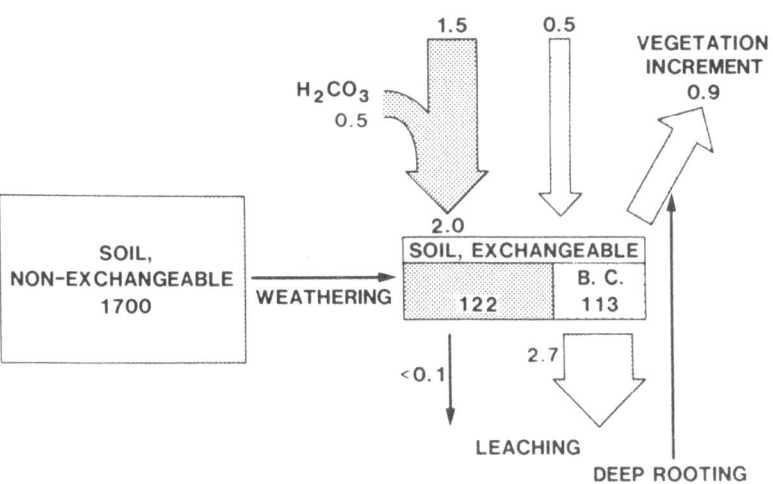

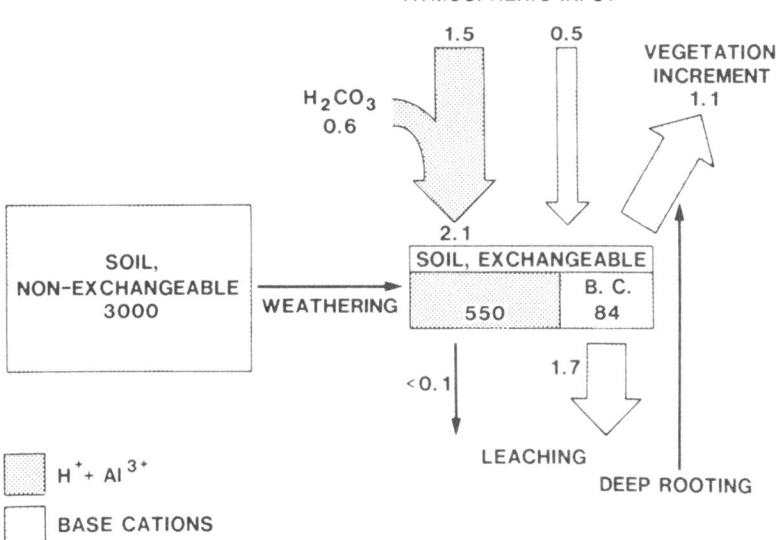

Figure 7.21. Total cation contents and fluxes in soils from the yellow-poplar and chestnut oak sites [kmol(+)/ha or kmol(+)·ha^{-1}·year^{-1}]. Source: D.W. Johnson, D.D. Richter, G.M. Lovett, and S.E. Lindberg. 1985. The effects of atmospheric deposition on potassium, calcium, and magnesium cycling in two deciduous forests. Can. J. For. Res. 15:773–782.

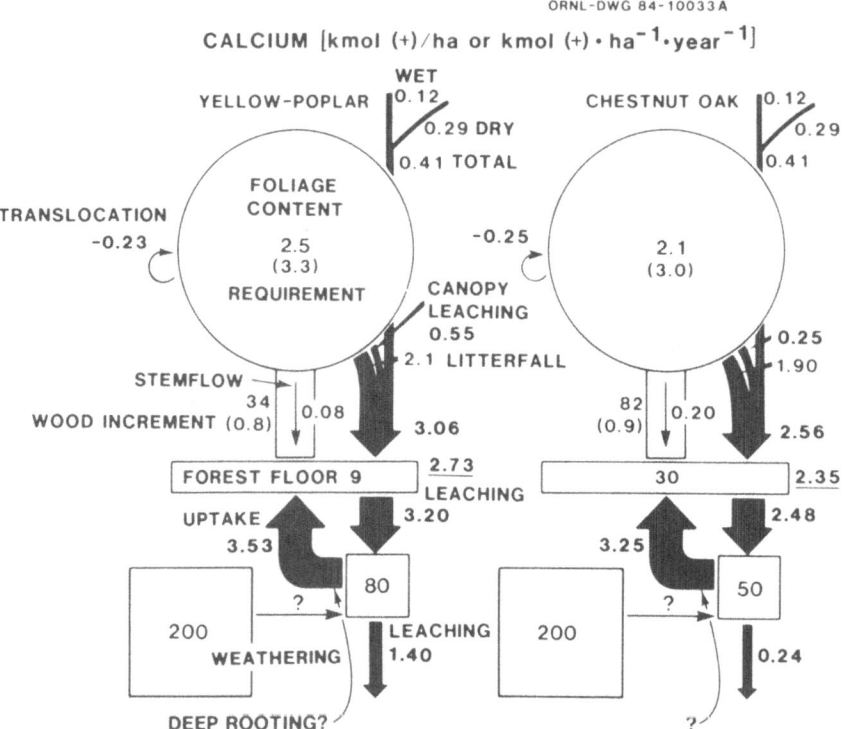

Figure 7.22. Calcium cycles in the yellow-poplar and chestnut oak sites. Source: D.W. Johnson, D.D. Richter, G.M. Lovett, and S.E. Lindberg. 1985. The effects of atmospheric deposition on potassium, calcium, and magnesium cycling in two deciduous forests. Can. J. For. Res. 15:773–782.

chestnut oak site, since calcium input exceeds output (Fig. 7.22). However, increment in woody biomass represents a drain on soil calcium, accounting for 1.8 and 0.5% of exchangeable and total soil calcium supplies, respectively. Johnson et al. (1982b) note that whole-tree harvesting over several rotations in a chestnut oak forest could cause a significant drain of calcium from such sites if not offset by calcium fertilization.

The leaching rates of cations from foliage and through soil have apparently been accelerated by acid deposition. These increased rates of leaching may have altered the cycling of cation nutrients in these forests in one or more of several possible ways, some of which would accelerate the rate of cycling (i.e., accelerate both uptake and return), and one of which would simply change the cycle toward a more hydrologic one (i.e., increasing leaching from foliage while decreasing litterfall and maintaining constant uptake and return). The effects of such changes in nutrient cycling on productivity are unknown, but could be either positive (i.e., increased nutrient availability) or negative (i.e., increased physiological cost for nutrient uptake, decreased nutrient-use efficiency). Manipulative experiments are needed to determine the extent to which various nutrient transfer processes change in response to accelerated cation leaching.

Aside from cation leaching from soils, concern has been expressed that acid deposition might affect the rates of decomposition and nitrogen mineralization (Baath et al. 1980). Abrahamsen (1980) reported that irrigation with sulfuric acid in Norwegian coniferous forests stimulated nitrogen mineralization, nitrification, and nitrate leaching. Since nitrogen is the limiting nutrient on Walker Branch Watershed, we initiated a study to determine whether irrigating with sulfuric and nitric acid would affect the soil's available nitrogen status.

After 1 year of irrigation with H_2SO_4 and HNO_3 at two and 10 times the current inputs on a Tarklin soil, no effect of any treatment was noted on the soil's available N, P, or Al^{3+} status or on CO_2 evolution (Table 7.3). Large seasonal variations in all of the above parameters were noted, however, and the existence of this seasonality must be considered when sampling for long-term status, including Al^{3+} (e.g., Ulrich et al. 1980).

Cadmium, Lead, and Zinc Cycling

Walker Branch Watershed was one of the first ecosystems for which the cycles of Cd, Pb, and Zn were characterized (Van Hook et al. 1977). Although one of these elements (zinc) is an essential plant nutrient, the primary motivation for this research was to obtain background information and data that would aid in the assessment of the effects of these potentially toxic trace elements. Zinc was studied because of its geochemical kinship with cadmium (Lagerwerff 1972), providing a comparison between nutrient and nonnutrient trace metal cycles. The primary objective of this research was to identify those ecosystem components that were accumulating Cd, Pb, and Zn and to at least begin to determine the dominant processes governing the transport and retention of these elements.

All three of these trace elements show a large net annual accumulation in the watershed in that atmospheric inputs exceed streamflow outputs by a substantial margin (Fig. 7.23). [The input and output figures have been revised downward from the orgnal values published by Van Hook et al. (1977), based on subsequent, more accurate measurements by Lindberg et al. (Ch. 4, this book).] For cadmium and lead, annual accumulations in biomass (1.2 and 16 g·ha^{-1}·year^{-1}, respectively) are only fractions of the net annual ecosystem gains (4.9 and 119.3 g·ha^{-1}·year^{-1}, respectively), implying that litter and soils are the primary sinks for these elements (Fig. 7.21). Several other studies of trace metal cycling have shown that both cadmium and lead accumulate by adsorption to both litter and mineral soils (e.g., Heinrichs and Mayer 1977; Zöttl 1985; Turner et al. 1985). All three elements were also significantly concentrated in lateral roots (which contained 11% of total biomass but 28% of the Cd and 40% of the Pb and

Table 7.3. Soil extractable N (before and after aerobic incubation), P, and Al, and CO_2 evolution before and after acid irrigations[a]

Treatment	NH_4^+-N[b]	NO_3^--N[b]	After incubation[c] Mineral N (mmol/kg)	After incubation[c] NO_3^--N (mmol/kg)	Extractable P[d]	Al[b]	CO_2 evolution (mmol·m^{-2}·h^{-1})
			August 31, 1981 (20 d after last irrigation)				
Control	A 0.6	A <0.01	A 5.2	A 2.4	A 0.071	A 4.3	A 12.3
H_2O	A 0.6	A <0.01	A 4.1	A 3.2	A 0.039	A 4.0	A 12.4
H_2SO_4 (50 mmol (–)/m^2)	A 0.3	A <0.01	A 7.1	A 5.5	A 0.048	A 5.7	A 14.4
H_2SO_4 (500 mmol (–)/m^2)	A 0.9	A <0.01	A 5.0	A 3.8	A 0.042	A 4.1	A 13.0
HNO_3 (50 mmol (–)/m^2)	A 0.4	A 0.01	A 7.4	A 4.9	A 0.068	A 4.6	A 11.8
HNO_3 (500 mmol (–)/m^2)	A 0.4	A <0.01	A 6.0	A 4.4	A 0.055	A 4.8	A 12.8
			Effects of season (treatments pooled)				
Date (treatments pooled)							
October 14	A 0.5	A 0.01	A 6.6	A 4.2	A 0.010	A 4.1	A 4.3
March 4	B 0.2	A 0.01	—	—	A 0.023	A 3.3	B 7.3
June 2	C 1.2	A <0.01	—	—	A 0.010	C 2.2	C 12.0
August 31	A 0.6	B 0.03	A 5.8	A 4.1	B 0.097	A,B 3.8	C 11.7

[a] Numbers sharing the same letter are not significantly different at the 95% confidence level.

[b] 1 M KCl.

[c] 1 M KCl after 120-d aerobic incubation.

[d] 1 M NH$_4$F + 0.5 M HCl.

Source: After D.W. Johnson and D.E. Todd. 1984. Short-term effects of sulfuric acid and nitric acid irrigation on CO_2 evolution, extractable N, P, and Al in a deciduous forest soil. Soil Sci. Soc. Am. J. 48:664–666.

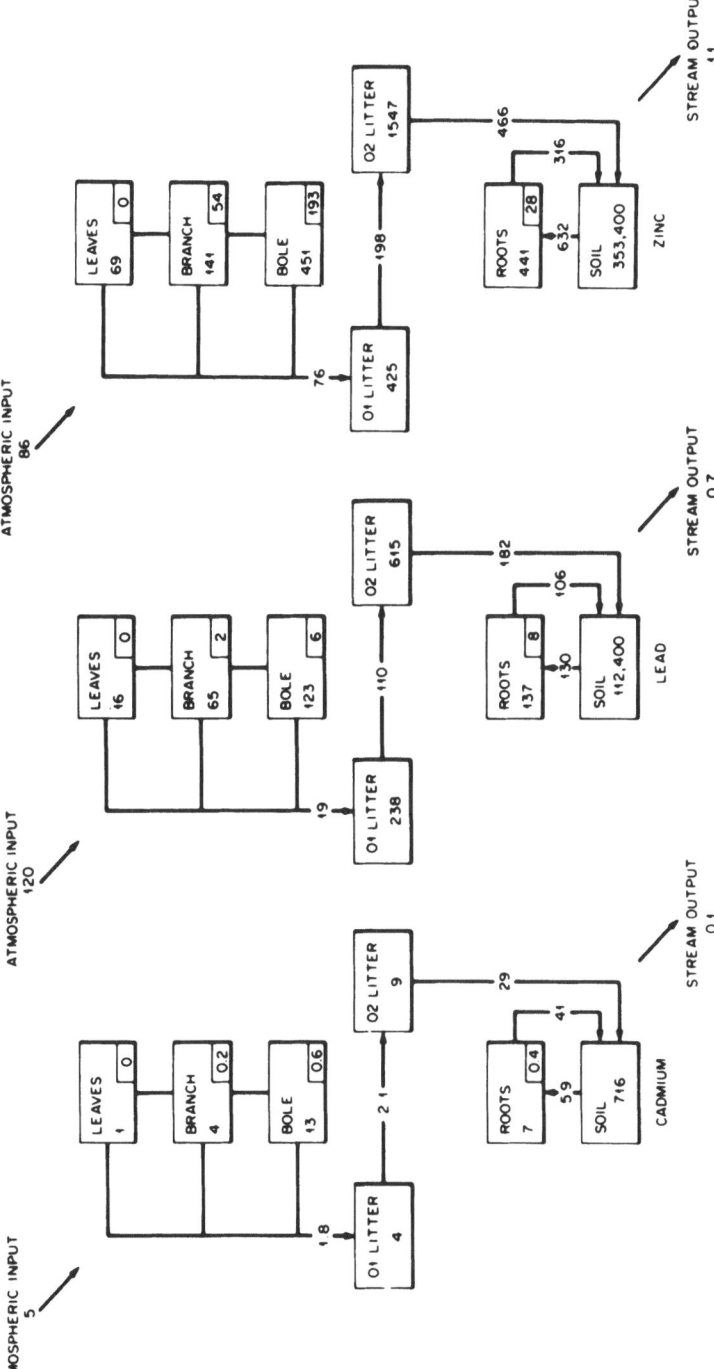

Figure 7.23. Summary of Cd, Pb, and Zn cycles for an eastern Tennessee mixed deciduous forest. Standing pools are in units of g/ha (values in center of boxes), annual fluxes in g·ha^{-1}·year^{-1} (values on arrows), and annual increments in g/ha (values in lower right corner of boxes). Source: Adapted (and updated) from R.I. VanHook, W.F. Harris, and G.S. Henderson. 1977. Cadmium, lead, and zinc distributions and cycling in a mixed deciduous forest. Ambio 6(5):281–286.

Zn in the biomass) and litter (which contained 13% of the organic matter but 34% of the Cd, 71% of the Pb, and 64% of the Zn in organic components of the ecosystem), implying that these are sites of trace metal accumulation in the ecosystem. However, soils are the major long-term sink for these elements.

Of what significance are these accumulations? Logically, continued accumulations of lead or cadmium must ultimately lead to toxic levels in the ecosystem. According to the results of research on highly affected ecosystems near smelters, toxic effects are likely to be manifested through disruption of decomposition and nutrient mineralization processes long before concentration levels become toxic to the trees themselves (e.g., Jackson et al. 1978). Trees are known to have a rather high tolerance for heavy metals (Zöttl 1985). According to the literature [see review by Zöttl (1985)], the levels of cadmium and lead in Walker Branch Watershed are far below the levels thought to produce negative effects on ecosystems. An illustration of this is provided in Table 7.4, where lead concentrations in foliage and soil from Walker Branch Watershed, a relatively unpolluted site in the Federal Republic of Germany (Bärhalde), and abandoned mine sites in the Black Forest in the Federal Republic of Germany are compared. Zöttl (1985) reports that the elevated concentrations of lead and other trace metals in foliage and the extremely high concentrations in the soil at the mine sites are producing no noticeable ill effects on tree growth.

7.3.2 Harvesting

Demands for biomass—both for wood and fiber products and for energy—are increasing. Whole-tree harvesting and removal of logging residue are being considered in many areas, and in some cases these practices are being implemented (Leaf 1979; Van Hook et al. 1981). Concerns have been expressed over the environmental effects of whole-tree harvesting, however, especially in terms of the nutrient drain associated with biomass

Table 7.4. Lead concentrations (µg/g) in foliage and soils from Walker Branch Watershed, a relatively unpolluted site in the Federal Republic of Germany (Bärhalde), and abandoned mine sites in the Black Forest of the Federal Republic of Germany

	Walker Branch[a]	Bärhalde[b]	Abandoned mine sites in Black Forest[b]
Foliage	2.1–10.4	0.56–2.30	12–21
Soil	12–50	36–43	1,900–15,550

[a]Data from Van Hook et al. (1977).
[b]Data from Zöttl (1985).

removal. The base-line data on biomass and nutrient content for Walker Branch Watershed have proved to be invaluable in assessing the effects of whole-tree harvesting in this region, even though no actual harvesting has taken place on the watershed. If the proportions of bole, branch, and foliar tissues removed are known, calculations of nutrient drain by various harvesting intensities can readily be made.

Implementation of an actual harvesting study in a chestnut oak forest near Walker Branch Watershed was greatly aided by Walker Branch data. Cross-checking showed that biomass and nutrient distributions at the harvesting site were comparable to those in the chestnut oak forest on Walker Branch (Johnson et al. 1982b), and biomass equations used on Walker Branch were checked by weighing many trees during the harvesting operation.

In terms of percent removal from the ecosystem, harvesting had the greatest effect on organic matter and lesser effects (in descending order) on Ca, N, P, and K (Mg removal was not determined) (Fig. 7.24). Whole-tree harvesting caused the removal of 60% of total organic matter, 15% of total Ca, 9% of total N, 2% of total P, and 1% of total K in the ecosystem (defined to a soil depth of 50 cm; Johnson et al. 1982b). Harvesting after leaf fall resulted in a 23% reduction in potential K removal compared with harvesting before leaf fall, as opposed to only 7, 7, and 5%reductions for N, P, and Ca, respectively. The rather small differences in N and P removal by harvesting before and after leaf fall were due to translocation, whereas the small difference in Ca was due to a relatively small foliar Ca pool. Since the ecosystem's reserves of potassium were very large, harvesting before leaf fall would not have added significantly to the nutrient drain caused by whole-tree harvesting.

Nutrient removal in harvesting should be viewed not only in terms of percent of ecosystem reserves but also in terms of potential replenishment of nutrients by means of atmospheric deposition. The latter is, of course, difficult to assess, since changes in nutrient inputs and outputs over the full rotation are not known. A rough idea of the *maximum* potential replenishment of nutrients by atmospheric deposition can be obtained by dividing nutrient removal by harvesting (kg/ha) by the current atmospheric deposition rate ($kg \cdot ha^{-1} \cdot year^{-1}$). This gives the number of years of atmospheric input (at current levels) that would be required to replenish nutrient removal in biomass. It is a maximum value, since output is not accounted for (and therefore assumed to be zero by default). Despite the many questionable assumptions, such calculations are useful in giving "ball park" figureswith which to estimate potential future nutritional problems.

These calculations show that N, P, and K removal *could* be offset by atmospheric inputs over the next 50-year rotation, but Ca removal could not (unless deposition rates increases) (Table 7.5). Thus, while nitrogen is currently the limiting element, budget projections suggest that calcium will become limiting if whole-tree harvesting is practiced.

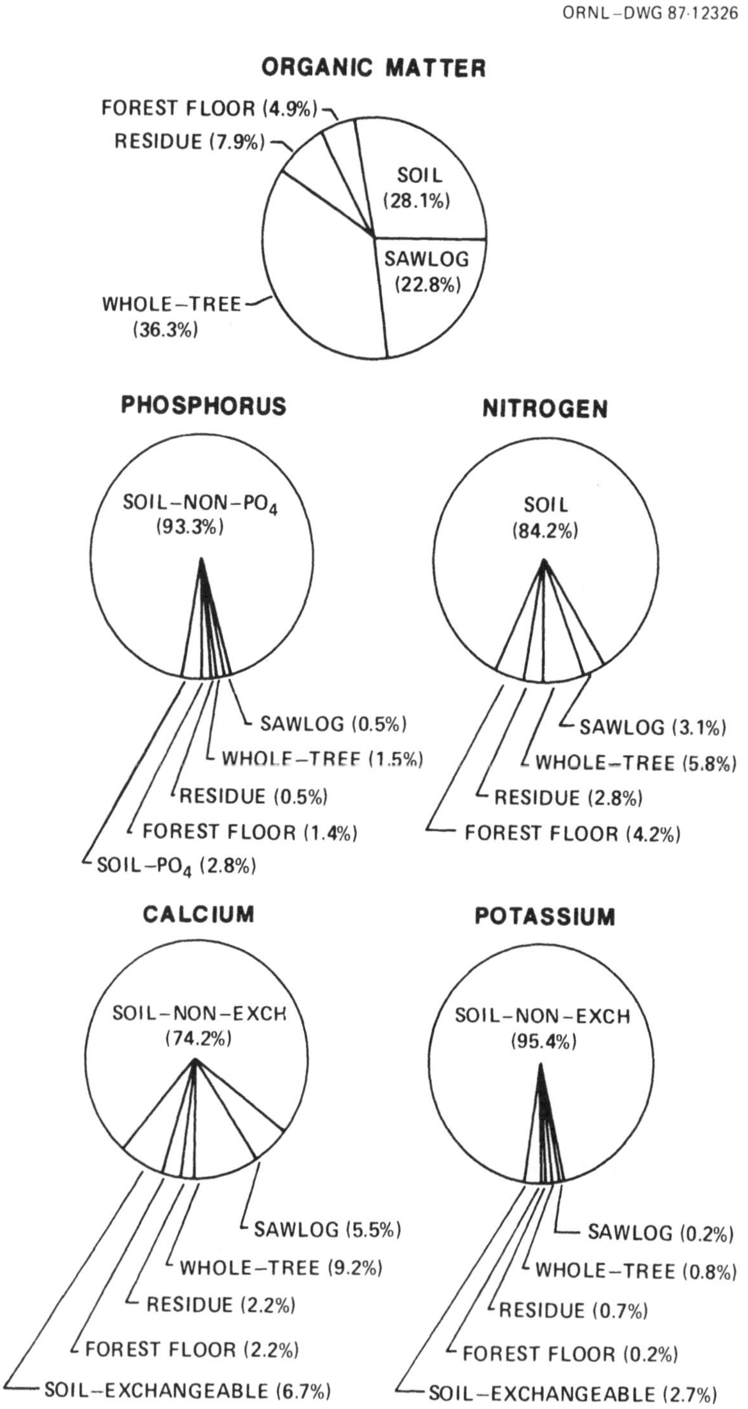

Figure 7.24. Removal of organic matter, N, P, K, and Ca from a chestnut oak site by sawlog and whole-tree harvesting. Data from D.W. Johnson, D.C. West, D.E. Todd, and L.K. Mann. 1982. Effects of sawlog vs whole-tree harvesting on the nitrogen, phosphorus, potassium, and calcium budgets of an upland mixed oak forest. Soil Sci. Soc. Am. J. 47:792–800.

Table 7.5. Atmospheric deposition, leaching, and removal by harvest/atmospheric deposition of N, P, K, and Ca in a chestnut oak forest

	N	P	K^+	Ca^{2+}
Atmospheric input, kg·ha^{-1}·year^{-1}	6.9	0.55	4.2	4.6
Removal by harvest, kg/ha				
Sawlog	110	7	36	410
Whole-tree	315	22	120	1090
Harvest removal/ atmospheric input, years				
Sawlog	15.9	12.7	8.6	89.1
Whole-tree	45.7	40.0	28.6	236.9

Source: After D.W. Johnson, D.C. West, D.E. Todd, and L.K. Mann. 1982. Effects of sawlog vs. whole-tree harvesting on the nitrogen, phosphorus, potassium, and calcium budgets of an upland mixed oak forest. Soil Sci. Soc. Am. J. 46:1304–1309.

Using the experience gained from actual harvesting as a guideline for calculating the amount of biomass removed, we can estimate the effects of sawlog vs. whole-tree harvesting on nutrient export from the various forest types on Walker Branch. Table 7.6 summarizes the potential nutrient removals through sawlog and whole-tree harvesting based on the 1983 inventory. Removal by sawlog harvesting was estimated as 40% of bole content except in loblolly pine, where 100% bole removal is probable. Removal by whole-tree harvesting is estimated as branch plus bole content plus translocation (assuming harvest after leaf fall). For the loblolly pine forest, nutrient losses from harvesting all species vs harvesting pines only (assuming invading hardwoods are cut and left as residue) are calculated separately.

The differences in nutrient content among the various forest types clearly affect potential nutrient removal through harvesting, especially whole-tree harvesting (Table 7.6). Some of the variation is due to variation in biomass, but in most cases it is primarily due to differing nutrient concentrations in tree tissues. For instance, whole-tree harvesting in the oak-hickory forest would remove 5% less biomass than in the loblolly pine forest, but removal of N, P, K, and Ca in the chestnut oak forest would exceed those in the loblolly pine forest by 12, 15, 16, and 200%, respectively, despite the fact that nutrient-rich foliage would be removed in the loblolly pine forest. Only magnesium removal in the loblolly pine forest would exceed that in the chestnut oak forest (by 9%). Within the loblolly pine forest, invading hardwoods constitute ~20% of the biomass but 26, 25, 27, 46, and 27% of the N, P, K, Ca, and Mg, respectively, so that

Table 7.6. Biomass and nutrient content (kg/ha) of trees in five forest types on Walker Branch Watershed and potential removal by sawlog and whole-tree harvesting

Forest type	Biomass	N	P	K	Ca	Mg
			Tree content			
Pine-hardwood	150,600	308	25	191	596	51
Yellow-poplar	186,700	410	32	226	946	73
Oak-hickory	164,300	358	27	207	1078	54
Chestnut oak	197,900	434	31	222	1268	53
Loblolly pine	170,800	279	24	165	363	52
Loblolly pines only[a]	137,800	206	18	121	203	38
		Removal by sawlog harvest[b]				
Pine-hardwood	42,100	64	5	37	151	10
Yellow-poplar	54,300	96	7	44	260	16
Oak-hickory	47,700	82	6	42	303	12
Chestnut oak	57,500	100	6	44	362	10
Loblolly pine	142,400	158	12	94	225	34
Loblolly pines only[a]	108,600	112	9	64	124	25
	Removal by whole-tree harvest[c]					
Pine-hardwood	137,900	276	24	157	300	43
Yellow-poplar	170,700	343	27	169	835	57
Oak-hickory	150,500	296	23	157	976	43
Chestnut oak	181,500	356	28	161	1155	41
Loblolly pine	157,300	263	20	135	324	46
Loblolly pines only[a]	126,700	168	15	114	190	34

[a]Loblolly pines within the loblolly pine forest type.
[b]Assuming 40% bole removal in hardwoods, 100% bole removal in loblolly pine.
[c]Assuming branch and bole removal during the dormant season and corrected for translocation into wood from foliage in hardwoods. Assumes foliage removal in pines.

removing only the pines yields 80% of the biomass but removes 75, 65, 74, 58, and 74% of the aboveground N, P, K, Ca, and Mg, respectively.

Besides the unknowns associated with changing inputs and outputs over a full rotation, the budget approach has one serious limitation: it does not account for potential changes in nutrient availability. The effect of harvesting on soil nutrient availability is the least understood yet perhaps the most important aspect of harvesting effects on the nutrient status and growth of forests. Nitrogen availability is especially difficult to assess and predict (Keeney 1980). Although very few studies of whole-tree harvesting suggest that nitrogen will be a limitation, these conclusions must be viewed with caution because they are drawn from considerations of total nitrogen budgets, not of potential changes in nitrogen availability.

Nitrogen availability is strongly affected by the carbon:nitrogen ratio in forest litter and soils. The carbon:nitrogen ratio of forest residues left after harvesting may therefore have a significant impact on soil nitrogen availability. In general, the carbon:nitrogen ratio of residues will decrease as the percent of woody biomass removal increases, as illustrated by studies of a loblolly pine forest and achestnut oak forest on Walker Branch Watershed (Johnson 1983) (Fig. 7.25). Harvesting 30% of bole biomass results in the removal of 25 and 18% of total aboveground biomass and nitrogen, respectively, inthe loblolly pine stand. Residue left on site has a total carbon:nitrogen ratio of 250 (considering bole, branch, and foliage as one unit) (Fig. 7.25). A 30% bole removal in the chestnut oak forest results in slightly lower percentages of total aboveground biomass (23%) and nitrogen (17%) removal, but residues have a considerably lower carbon:nitrogen ratio because of the higher nitrogen concentrations in woody tissues (Fig. 7.25). In both cases, the percent of total aboveground nitrogen removal increases with increasing bole removal, but the percent of aboveground nitrogen removal increases less rapidly than the percent of above-

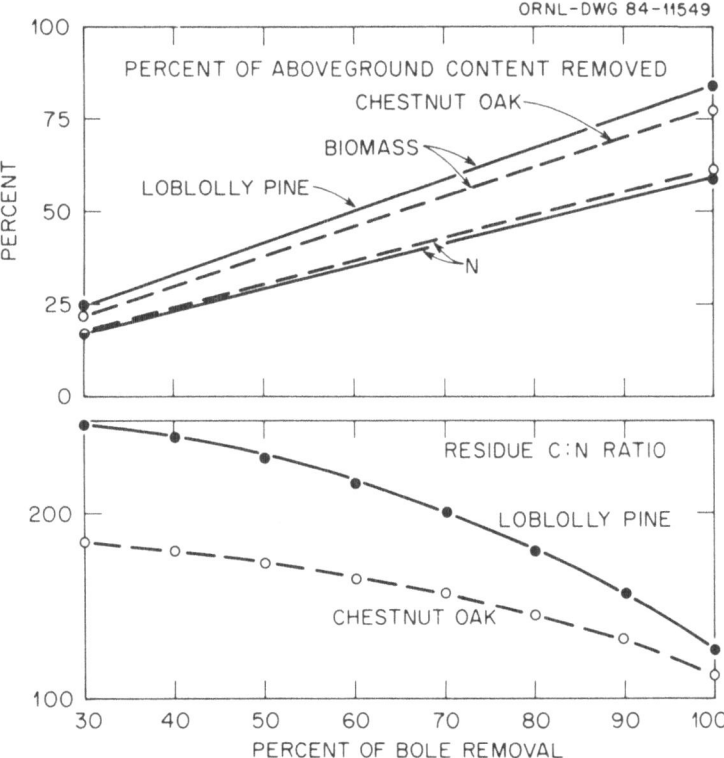

Figure 7.25. Percent of aboveground biomass and nitrogen removed and carbon:nitrogen ratio of residue as a function of percent bole removed in loblolly pine and chestnut oak forests.

ground biomass removal because of the low nitrogen concentration of bole tissue as compared to branch and foliar tissue (i.e., boles constitute a greater proportion of total aboveground biomass than of total aboveground nitrogen.)

The carbon:nitrogen ratio of residues decreases drastically, however, as the percent of bole removal increases because less bole tissue (with its high carbon:nitrogen ratio) is left on site. Thus, although increasing utilization of woody tissues will cause increases in total nitrogen removal from the site, the nitrogen left in nonbole residues is probably more available than that in bole residues. In fact, bole residue, because of its high carbon:nitrogen ratio, will probably cause a net reduction in nitrogen availability as it decomposes. Thus, the nitrogen status and productivity of a forest can be temporarily reduced by leaving too much woody residue after harvesting (Bollen 1974). However, the long-term effects of residues on the soil organic matter and nitrogen status are probably beneficial. The ultimate effects of carbon-nitrogen interactions in forest residues on forest growth will depend on the extent to which the regrowing stand recovers from the short-term nitrogen deficiency (if such a deficiency occurs) caused by leaving too much woody residue. After an early deficiency stage, the regrowing stand may respond to the long-term improvements in the soil organic matter and nitrogen status caused by the woody residues. This, in turn, depends on the decomposition rate, which is affected by temperature, moisture, aeration, pH, nutrient supply, and a variety of complex biological interrelationships. One might expect the period of nitrogen deficiency to be shorter in warm temperate or tropical ecosystems with inherently high decomposition rates. Only long-term studies of forest growth and nitrogen cycling can provide answers to these questions, however, and such answers are bound to be highly sitespecific.

The availability of nutrient cations and phosphorus to the regrowing forest will be strongly affected by soil weathering rates (Clayton 1979). Unfortunately, weathering rates are usually unknown or only poorly known. Furthermore, the rate of weathering can be affected by the trees themselves. Boyle and Voigt (1973) found that organic acids exuded from pine roots (and associated fungi) were effective in weathering silicate minerals. The numerous forecasts of potential calcium deficiency due to intensive harvesting must therefore be regarded as tentative.

7.4 Comparison of Walker Branch Watershed with Other Forest Ecosystems

Having described the nutrient cycles of Walker Branch Watershed, we now consider a question that naturally arises: To what extent is Walker Branch representative of the southern Appalachian region or of decidous forest ecoystems in general? There is every reason to believe that certain aspects of Walker Branch are unique to that site; but to what extent might the data from Walker Branch be extrapolated to other watersheds? To what extent do the processes regulating nutrient flux on Walker Branch Watershed operate in other ecosystems?

7.4.1 Nutrient Pools and Net Annual Accumulations

Henderson et al. (1978) compared the N, K, and Ca cycles at Walker Branch Watershed; Coweeta, North Carolina (Watershed 18); and the H.J. Andrews watershed in Oregon. They found that while nitrogen inputs varied by nearly 10-fold at these sites, the NH_4^+ and NO_3^- discharge through streamflow was uniformly small, resulting in large net nitrogen accumulations in all three ecosystems. Similarly, Likens et al. (1977) and Cole and Rapp (1981) noted net nitrogen accumulations in a variety of forest ecosystems throughout the world. This is thought to be largely due to biological uptake (since nitrogen is not retained in soil mineral form), but the role of denitrification as an unmeasured output is seldom quantified.

Most forest ecosystems show net annual exports of (most) base cations (Likens et al. 1977; Cole and Rapp 1981), but exceptions to this rule (including the aforementioned chestnut oak site on Walker Branch) do occur where soils are extremely low in one or more of the exchangeable base cations. In their comparison of Walker Branch, Coweeta, and H.J. Andrews watersheds, Henderson et al. (1978) noted the strong effect of bedrock type on base cation export in streamflow. The role of the dolomite bedrock in causing large net Ca^{2+} and Mg^{2+} exports from Walker Branch was particularly striking in contrast to the more modest Ca^{2+} and Mg^{2+} exports from a similar deciduous forest at Coweeta, where the bedrock is composed of gneiss and schist. Despite these differences in net export, however, Henderson et al. noted great similarities in the nutrient pool sizes and transfer rates between the Walker Branch and Coweeta sites. One of the most striking features of both sites was the large accumulation of Ca^{2+} in forest biomass even though both exchangeable and total soil calcium were low (Table 7.7). Some argument might be made for luxury consumption of Ca^{2+} from bedrock on Walker Branch (although this would require rooting to 30 m in many cases), but this argument cannot be made for Coweeta. It appears that Ca^{2+} accumulation in woody and bark tissues is characteristic of oak and hickory species, regardless of bedrock type or availability of soil Ca^{2+}. Johnson and Risser (1974) found extremely high calcium concentrations (up to 9% of dry weight) in the bark of post oak (*Quercus stellata*) growing on sandstone-derived soil in Oklahoma. Similarly, Rolfe et al. (1978) reported large accumulations of calcium (1600 kg/ha) in forest biomass (240,500 kg/ha) of an oak-hickory forest in southern Illinois. Much earlier, Duvigneaud and Denaeyer-DeSmet (1970) reported high accumulations of calcium (1248 and 1648 kg/ha at Virelles and Wav-

Table 7.7. Calcium distribution and cycling (kg/ha) in three watershed ecosystems

Item	H. J. Andrews	Coweeta	Walker Branch
Amount in:			
Aboveground vegetation	750	830	980
Forest floor[a]	569	130	430
Mineral soil[b]			
Exchangeable	4450	940	710
Total	[c]	2500	3800
Annual transfers as:			
Litterfall	41	44	55
Canopy leaching	8	8	14
Woody increment	−4	23	31
Uptake	45	75	100

[a] O1 and O2 horizons (organic).
[b] A and B horizons to a 60-cm depth (exchangeable).
[c] Data not available.
Source: After G.S. Henderson, W.T. Swank, J.B. Waide, and C.C. Grier. 1978. Nutrient budgets of Appalachian and Cascade region watersheds: A comparison. For Sci. 24(3):385–397.

reille, respectively) in the biomass (156,000 and 380,000 kg/ha) of oak-beech forests growing on calcareous soils in Belgium, which they interpreted as luxury consumption. Other species known to accumulate calcium are aspen (*Populus tremuloides*), white spruce (*Picea glauca*), and western red cedar (*Thuja plicata*) (Alban 1969, 1982). Alban (1982) presented evidence that high rates of calcium accumulation in aspen and spruce vegetation caused significant increases in soil acidity over a period of 40 years.

Comparisons of nutrient pool sizes can be informative in a descriptive way, but comparisons of fluxes and, if possible, the mechanisms that control them are more meaningful, since pool sizes are merely the result of a number of fluxes integrated through time.

As noted previously, there is a general pattern of net retention of nitrogen and phosphorus and net loss of Ca^+, K^+, and Mg^{2+} from forest ecosystems (Likens et al. 1977; Cole and Rapp 1981). The situation with respect to sulfur is different, however; some ecosystems show a net accumulation of sulfur (Swank et al. 1984; Kelly 1984; Meiwes and Kanna 1981), whereas others show either no net retention or a net loss (Likens et al. 1977; Turner et al. 1980; Stednick 1982).

As discussed in the subsection titled "Sulfur Cycling" in Section 7.3.1, we hypothesized (and presented evidence for our interpretation) that sulfur retention was due to SO_4^{2-} adsorption onto iron and aluminum hydrous oxides on Walker Branch Watershed. In an attempt to extrapolate these results, we also tested the hypothesis that sulfur retention in an ecosystem is related to the soil SO_4^{2-} adsorption capacity, which, in turn, is negatively related to the soil organic matter content (soil organic matter is thought to block SO_4^{2-} adsorption sites) and positively related to the iron and aluminum hydrous oxide content. If this is true, it should be possible to relate the soil SO_4^{2-} adsorption capacity (and the ecosystem sulfur-retention capacity) to the U.S. Department of Agriculture (USDA) soil classification system and use existing soil maps to delineate geographical areas hypothesized to have soils that retain SO_4^{2-} and those that do not. Such a regional extrapolation would be useful not only in explaining ecosystem sulfur retention per se, but also as one of several indices of sensitivity to leaching by acid deposition (since SO_4^{2-} adsorption prevents leaching of base cations or aluminum from the soil by sulfuric acid, as described previously).

We tested these hypotheses by conducting laboratory studies of the SO_4^{2-} adsorption capacities of soils from a variety of forest ecosystems with known sulfur budgets (Johnson et al. 1980b). The SO_4^{2-} adsorption capacities of the soils were also correlated with their organic matter and iron and aluminum hydrous oxide contents (Johnson and Todd 1983).

The overall results of this study (Johnson and Todd 1983) are depicted in Fig. 7.26 and Table 7.8. As a general rule, Spodosols had lower SO_4^{2-}

ORNL–DWG 85-14289

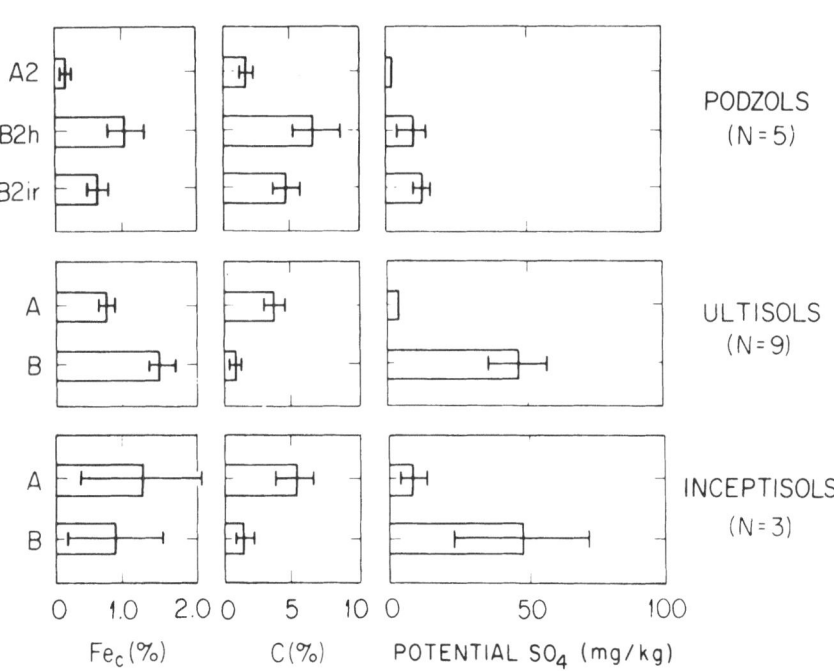

Figure 7.26. Percent crystalline iron hydrous oxides (Fe$_c$), percent carbon (C), and potential sufate (SO$_4$) adsorption in Podzols (Spodosols and heavily podzolized soils), Ultisols, and Inceptisols. Source: Adapted from D.W. Johnson and D.E. Todd. 1983. Relationships among iron, aluminum, carbon, and sulfate in a variety of forest soils. Soil Sci. Soc. Am. J. 47:792–800.

Table 7.8. Net sulfur retention by forests with Ultisols, Spodosols, and Inceptisols

Site	Vegetation	Soil order	Net ecosystem sulfur retention		Reference
			(kg·ha^{-1}·year^{-1})	(% of input)	
Walker Branch, Tenn.[a]	Mixed deciduous	Ultisols	+6.6	+36%	Shriner and Henderson (1978)
Coweeta, N.C.[a] (Watershed 18)	Oak-hickory	Ultisols	+11.2	+88%	Swank and Douglass (1977)
Camp Branch, Tenn.[a]	Mixed oak	Ultisols	+21.7	+70%	Kelly (1984)
Cross Creek, Tenn.	Mixed oak	Ultisols	+23.5	+70%	Kelly (1984)
White Oak Run, Va.	Oak-hickory	Ultisols	+10.7	+67%	Shaffer (1984)
Hubbard Brook, N.H.[a]	Northern hardwood	Spodosols	+1.2	+6%	Likens et al. (1977)
Vilas County, Wis.	Aspen-mixed hardwood	Spodosols	+2	+43%	Pastor and Bockheim (1984)
Chesuncook, Maine[a]	Spruce-fir	Spodosols	−1.2	19%	J.W. Hornbeck[b]
Panther Lake, N.Y.	Spruce-fir	Spodosols	0	0	Goldstein et al. (1984)
Findley Lake, Wash.	Subalpine fir	Spodosols	+4	25%	Johnson (1975)
Thompson, Wash.[a]	Douglas-fir	Inceptisols	−1.9	45%	Cole and Johnson (1977)
Pack Forest, Wash.	Douglas-fir	Inceptisols	+2.2	+25%	Stednick (1982)

[a]Soils from these sites were tested for SO_4^{2-} adsorption (Johnson and Todd 1983).
[b]J.W. Hornbeck, U.S. Forest Service, Durham, North Carolina, personal communication to D.W. Johnson, March 1985.

adsorption capacities than Ultisols (Fig. 7.26), which appeared to be related to the higher organic matter content and lower quantity of crystalline iron hydrous oxides (Fe$_c$) in subsurface horizons. These findings are generally consistent with ecosystem sulfur budgets: in most cases, ecosystems with Spodosols show little or no net sulfur retention, whereas ecosystems with Ultisols usually show marked sulfur retention (Table 7.8; see also Rochelle et al. 1987). More soil samples must be analyzed before any firm conclusions can be reached, but thus far, the hypothesis that the retention of SO$_4^{2-}$ in soil is the major factor in ecosystem sulfur retention is supported. This, in turn, would imply that Spodosols, which occupy much of the northeastern United States, are more susceptible to leaching (of either base cations or Al^{3+}, depending upon how acid they are) than Ultisols, which occupy much of the southeastern United States (Fig. 7.27) (Johnson and Todd 1983).

ORNL-DWG 85-14175

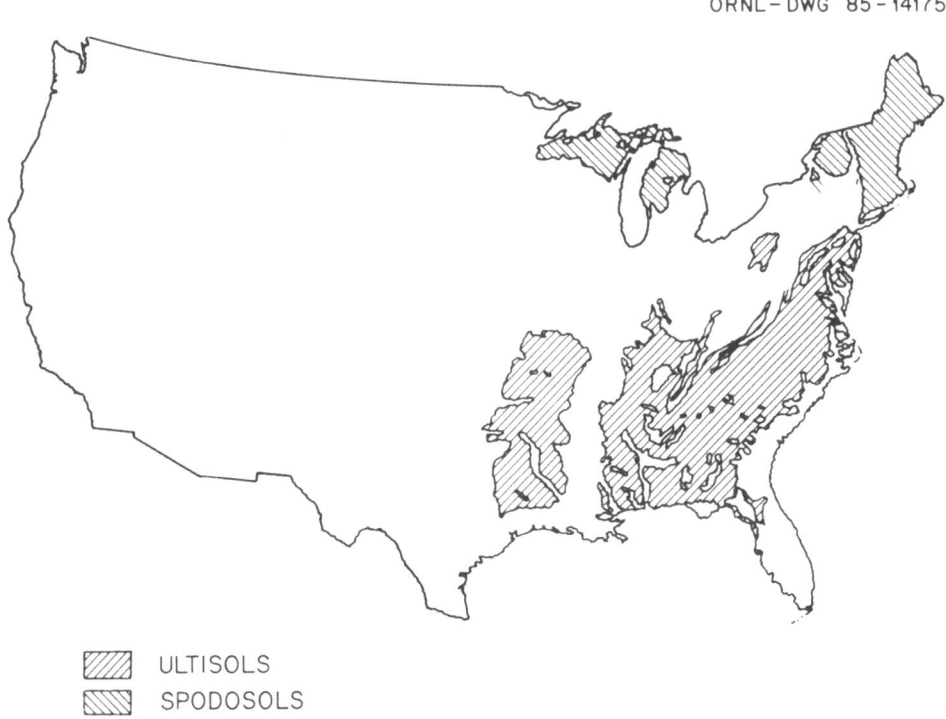

ULTISOLS

SPODOSOLS

Figure 7.27. Major soil orders of the eastern United States. Source: Adapted from U.S. Department of the Interior. 1967. The National Atlas of the United States of America. U.S. Department of the Interior, Geological Survey, Washington, D.C.

A complication arises, however, when the sulfur deposition history of a site (as well as its soil sulfate adsorption capacity) is considered. Given a steady input, all soils will eventually reach a steady-state condition with respect to sulfate input and output. The time required to reach a steady-state condition with respect to SO_4^{2-} flux in a SO_4^{2-}-adsorbing soil is related to the depth of the soil and the slope of the adsorption isotherm (Johnson 1985). This is illustrated in Fig. 7.28, where SO_4^{2-} adsorption isotherms for two fictitious soils are plotted. Soil A (an Ultisol) is richer in iron and aluminum oxides and adsorbs more SO_4^{2-} than soil B (a Spodosol). Let both soils be initially in equilibrium (i.e., no net SO_4^{2-} adsorption) with 40 $\mu mol(-)/L$ SO_4^{2-}. Concentrations are then raised to 300 $\mu mol(-)/L$, and each soil will adsorb additional SO_4^{2-} until a new steady state is reached. For soil A, this requires an additional 2.2 mmol($-$)/100 g of adsorption, and for soil B only 1.0 mmol($-$)/100 g of adsorption. Thus, a given depth of soil A will show a net accumulation of SO_4^{2-} for a longer period than a given depth of soil B under the same input conditions. Conversely, if SO_4^{2-} adsorption is reversible, soil A will show a net export of SO_4^{2-} for a longer period than soilB if SO_4^2s;- inputs are reduced.

7.4.2 Nutrient Cycling by Biotic Components

The requirements for, and annual uptake of, nutrients in the forests of Walker Branch Watershed fall within the range of values reported by Cole and Rapp (1981) for temperate deciduous forests (Table 7.9). However, it appears that uptake of nitrogen and phosphorus in Walker Branch forests is at the low end of the range, whereas calcium uptake is near the high end (Table 7.9). The former may reflect above-average nitrogen and phosphorus limitations in Walker Branch forests, whereas the latter reflects the normally high calcium accumulation in oaks and hickories, as discussed previously. A further useful comparison is the ratio of uptake to requirement, which provides an index of the degree to which the trees rely on external sources (atmospheric input, uptake from soil) for their annual growth requirements as opposed to internal translocation. Once again, the Walker Branch forests fall at the low end of the spectrum for nitrogen, implying a greater than average need for internal translocation rather than external inputs for annual nitrogen requirements. The ratios for P, K, and Mg are near average, whereas the ratio for Ca is below average, implying a minimum of luxury consumption of Ca (Table 7.9).

Decomposition is an extremely important nutrient-transfer pathway, but because measurements of decomposition are inconsistent and use varying methods, it is difficult to compare decomposition at different sites except in the broadest sense. One index of decomposition is forest floor mean residence time (MRT), which is measured in years and is calculated by dividing forest floor content by litterfall input. Forest floor MRTs for biomass, N, P, K, Ca, and Mg for Walker Branch are well within the

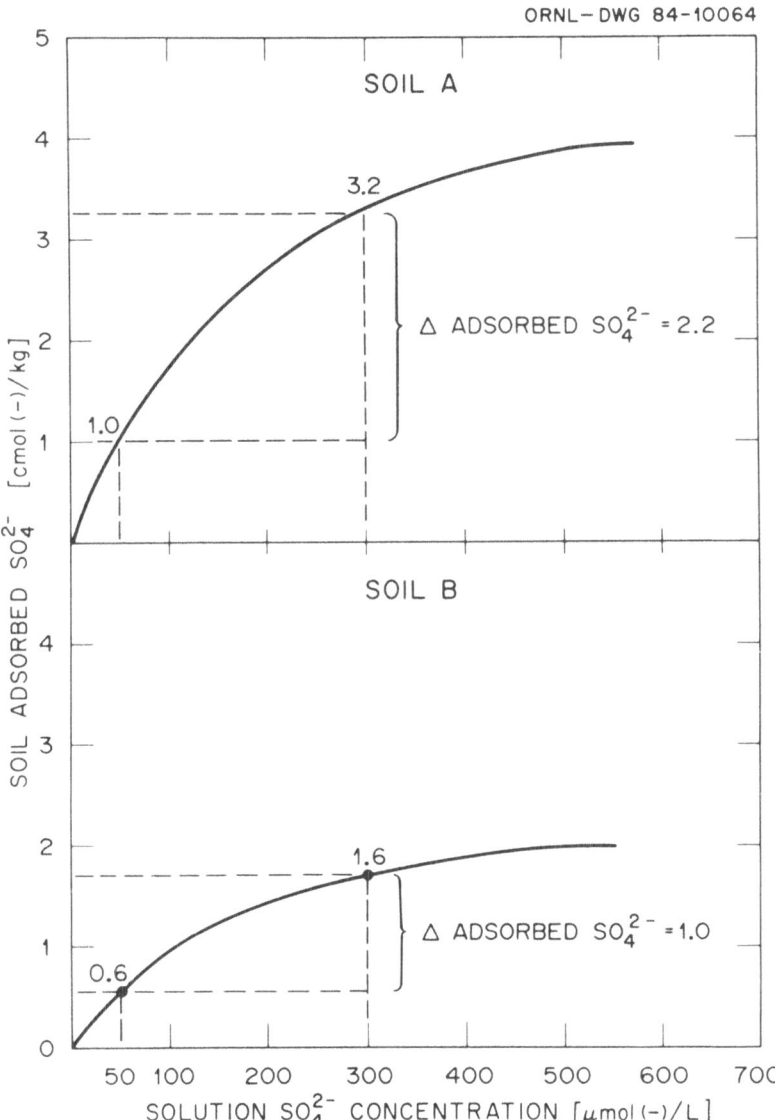

Figure 7.28. Idealized SO_4^{2-} adsorption isotherm for a soil with high adsorption capacity (soil A) and low adsorption capacity (soil B). Amounts of SO_4^{2-} adsorption required to re-equilibrate from 50 to 300 μmol$(-)$/L SO_4^{2-} are shown on the right. Source: D.W. Johnson. 1985. Sulfur cycling in forests. Biogeochemistry 1:29–43.

Table 7.9. Uptake, requirement, and uptake: requirement ratio for 14 temperate deciduous forests[a] as compared with forests on Walker Branch Watershed

	Organic matter	N	P	K	Ca	Mg
		Uptake (kg·ha⁻¹·year⁻¹)				
Average, 14 forests		75	5.6	51	85	13.2
Range, 14 forests		48–115	3.3–10.3	40–86	50–169	6.4–25.0
Walker Branch Watershed		45	3.8	46	91	12.7
		Requirement (kg·ha⁻¹·year⁻¹)				
Average, 14 forests		98	7.2	48	56	10.4
Range, 14 forests		61–129	5.5–10.5	31–77	18–106	3–22
Walker Branch Watershed		84	5.4	43	93	10.6
		Uptake: requirement ratio				
Average, 14 forests		0.77	0.79	1.09	1.70	1.46
Range, 14 forests		0.47–0.97	0.54–0.94	0.69–1.44	0.95–2.67	0.94–2.38
Walker Branch Watershed		0.54	0.70	1.07	0.98	1.20
		Forest floor mean residence time (years)				
Average, 14 forests	4.0	5.5	5.8	1.3	3.0	3.4
Range, 14 forests	0.9–9.6	0.9–19.3	0.8–13	0.8–2.6	1.0–9.3	0.9–24
Walker Branch Watershed	6.6	7.2	6.2	0.5	6.6	2.0

[a]Data from Cole and Rapp (1981).

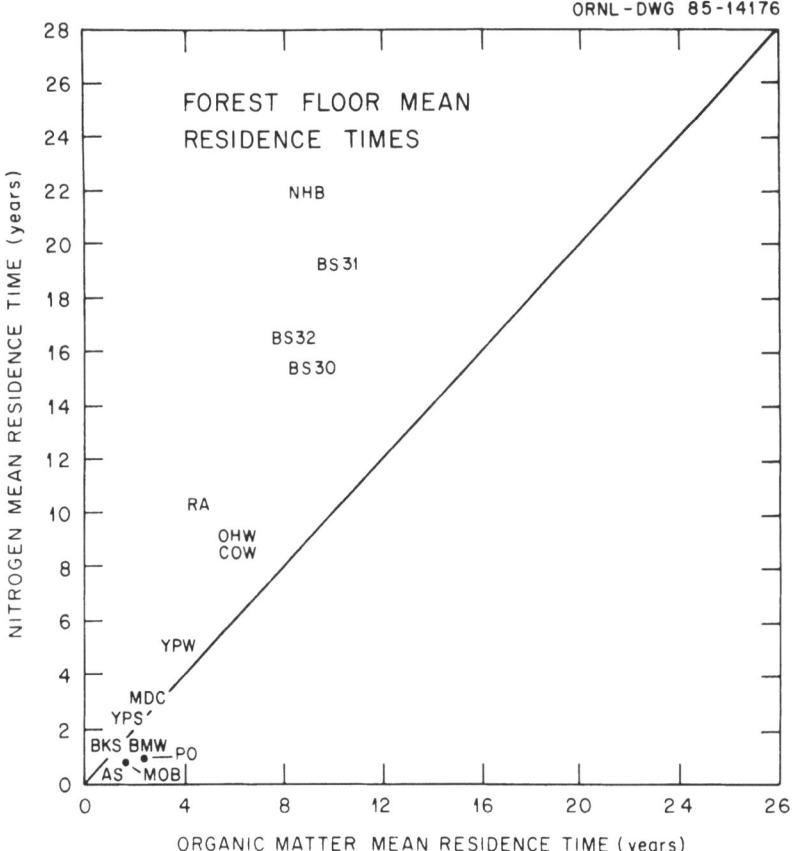

Figure 7.29. Plot of the mean residence time of nitrogen vs. the mean residence time of organic matter in a variety of forest ecosystems. Data from Cole and Rapp (1981): YPS = yellow-poplar, sinkhole, Oak Ridge, Tennessee; OHW, COW, and YPW = oak-hickory, chestnut oak, and yellow-poplar at Walker Branch Watershed, Tennessee; BS30, 31, 32 = beech stands in Solling, Federal Republic of Germany; BMW = beech stand in Meathop Wood, England; MDC = mixed deciduous forest at Coweeta, North Carolina; NHB = northern hardwood, Hubbard Brook, New Hampshire; MOB = mixed oak site in Belgium; BKS = beech site in Sweden; RA = red alder site in Washington. Data from Johnson and Risser (1974): PO = post oak stand in Oklahoma. Data from Pastor and Bockheim (1984): AS = aspen-mixed hardwood stand in Wisconsin.

range of values given for temperate deciduous forests by Cole and Rapp (1981) (Table 7.9).

Johnson et al. (1982c) found poor correlations between foliar MRT (foliage content divided by litterfall) and forest floor MRT for both biomass and nitrogen in coniferous forests of the Pacific Northwest. They interpreted this result as an indication of nitrogen conservation by the trees (low foliar MRT) under conditions where forest floor decomposition and nitrogen mineralization (forest floor MRT) are low. In deciduous forest ecosystems, foliar MRT is merely the inverse of internal translocation, since no more than 1 year's foliage may be retained, and no relationship was found between the percent of nitrogen translocation and the forest floor nitrogen MRT, nor were relationships found between the latter and temperature or precipitation when the data for temperate deciduous forests from Cole and Rapp (1981) were tested. Johnson et al. (1982c) also noted longer MRTs for nitrogen than for biomass in coniferous forests of the Pacific Northwest, as is the case for Walker Branch Watershed (Table 7.9). This is not the case in all deciduous forest ecosystems, however. In cases where forest floor MRT is short, the MRTs of both biomass and nitrogen are near unity, whereas nitrogen MRT increases more rapidly than the biomass MRT in forests with slower decomposition rates [i.e., the plot of nitrogen MRT vs. biomass MRT goes well above the 1:1 line (Fig. 7.29)]. It appears that nitrogen is more tightly conserved in the forest floor, if not in foliage, at sites having slow decomposition rates.

7.5 Summary and Conclusions

Perhaps the best way to summarize and synthesize a large portion (but by no means all) of the element cycling work that has been done on Walker Branch Watershed is to briefly review the water and element budgets, including the dominant forms of various elements entering, accumulating in, and leaving the ecosystem, and to describe what we currently understand about the dominant processes controlling the net gains or losses of these elements. Tables 7.10 to 7.12 represent a summary of the information about the pools and fluxes of water and 11 elements on the watershed. These tables were compiled from the data and other information provided in this book. The information supporting these summaries varies from good data bases (e.g., dissolved Ca^{2+} and Mg^{2+} losses), to speculation based on the literature (e.g., the forms of phosphorus in soils), to unknown (e.g., N_2 and N_2O loss via denitrification). The speculative information is enclosed in parentheses to identify it as such, and the unknowns are identified with question marks.

As is true with all forest ecosystems, gaseous fluxes dominate the carbon budgets for the watershed and are a major component of the water budget. The water budget is almost completely balanced, with only a slight net annual increment in live and dead vegetation (0.004% of annual input; Table 7.10). Eighty-four percent of the CO_2 fixed by photosynthesis is rereleased as CO_2 by autotrophic (60%) and heterotrophic (24%) respiration (Van Hook et al. 1977). Current (1967–1983 average) estimates indicate that only 8% of total CO_2 fixed is stored in live trees (Table 7.10). As a result of self-thinning and bark beetle attacks (Ch. 6), mortality currently

Table 7.10. Water and element budgets for Walker Branch Watershed[a]

	C	N	S	P	K	Ca	Mg	Mn	Cd	Pb	Zn	H$_2$O
			(kg·ha^{-1}·year^{-1})						(g·ha^{-1}·year^{-1})			(mm)
Atmospheric input	23,900	10	26	0.54	0.9	8.6	1.3	220	5	120	86	1,370
Gaseous/vapor loss	20,000	?	?	Trace	0	0	0	0	0	0	0	730[b]
Vegetation uptake[c]	9,600[d]	120	22	14.7	10.9	170	35.4	—	7.7	154	557	—
Live vegetation increment	1,900	4	1.1	0.3	0.9	8	0.5	?	0.4	4	125	0.02
Mortality	3,400	6	2.0	0.5	4	18	1.1	?	0.7	7	224	0.03
Soil retention	?	(−3)	+10.5	−0.4	−11.1	−17.4	−0.3	?	+3.9	+108	−274	0
Output	160	3	12	0.02	7.1	147	75	38	0.1	0.7	11	640

[a]Speculative values (based on the literature) are enclosed in parentheses; question marks indicate unknowns.
[b]Transpiration.
[c]Includes estimates of root uptake and turnover.
[d]Net primary production.

exceeds live tree increment, and some carbon is currently accumulating as woody material on the forest floor. The extent to which this woody material will be decomposed to release CO_2 or stored in soil organic matter is unknown.

The output of carbon via streamflow is almost entirely a function of the dissolution of the dolomite bedrock resulting in large quantities of Ca^{2+}, Mg^{2+}, and HCO_3^- flux via streamflow. Normal HCO_3^-–C flux from a noncalcareous watershed would be less than one-tenth of the value for Walker Branch Watershed given in Table 7.10.

To the best of our knowledge, most nitrogen input to the watershed is via dissolved NO_3^- and NH_4^+ in rainfall and HNO_3 vapor deposition. Aside from a few small scattered plants, there are no nitrogen-fixing species of significance on the watershed, and tests for free-living nitrogen fixers have proved negative (D.S. Shriner, Oak Ridge National Laboratory, personal communication, 1978). Gaseous losses of N_2 and N_2O via denitrification have not been measured. Streamflow export of nitrogen (~ 3 kg·ha^{-1}·year^{-1}), most of which is in the form of dissolved organic matter rather than NH_4 and NO_3^-, is less than one-third of atmospheric input, indicating a net ecosystem accumulation of nitrogen (if gaseous fluxes are ignored). Since the forests are generally deficient in nitrogen, this is to be expected. The calculated net annual accumulation in the watershed (~ 7 kg·ha^{-1}·year^{-1}) is roughly equal to the total net annual increment in live (4 kg·ha^{-1}·year^{-1}) and dead (6 kg·ha^{-1}·year^{-1}) vegetation, implying little net gain or loss of nitrogen from the soil.

Sulfur inputs are primarily in the form of dissolved SO_4^{2-} in rainfall with dry deposition of SO_2 as the second most important form of input.

Table 7.11. Dominant forms of inputs and outputs of elements and dominant sources of sinks of elements on Walker Branch Watershed[a]

| Element | Dominant forms of flux | | Significant sources and sinks | |
	Input	Output	Source(s)	Sink(s)
C	CO_2 gas	CO_2 gas	—	Vegetation, litter
N	Dissolved NH_4^+ + NO_3^-, HNO_3 vapor	Dissolved organic N, NH_4^+, NO_3^- (N_2, N_2O?)	—	Vegetation, litter
S	Dissolved SO_4^{2-}, SO_2 dry deposition	Dissolved SO_4^{2-}	—	Soil adsorption
P	(Particulate)	Dissolved organic P	—	Vegetation, soil organic matter
K	Particulate	Dissolved K^+	Soil, bedrock	—
Ca	Particulate	Dissolved Ca^{2+}	Bedrock	Vegetation
Mg	Particulate	Dissolved Mg^{2+}	Bedrock	Vegetation
Mn	Particulate	Suspended particulate	—	(Soil organic matter)
Cd	Dissolved Cd^{2+}	Dissolved Cd^{2+}	—	(Soil organic matter, adsorption)
Pb	Dissolved Pb^{2+}	Dissolved Pb	—	(Soil organic matter, adsorption)
Zn	Dissolved Zn^{2+}	Suspended particulate	—	Vegetation

[a] Speculative information (based on the literature) is enclosed in parentheses.

Table 7.12. Major ecosystem pools and dominant forms of elements in Walker Branch Watershed[a]

Element	Major ecosystem pool	Dominant form
C	Vegetation, soil	Organic
N	Soil	Organic
S	Soil	Organic, adsorbed SO_4^{2-}
P	Soil	(Organic, secondary minerals)
K	Soil	Secondary minerals
Ca	Soil, bedrock	Secondary minerals in soil, dolomite bedrock
Mn	Soil	Organically bound
Mg	Soil, bedrock	Secondary minerals in soil, dolomite bedrock
Cd	Soil	Organically bound, adsorbed
Pb	Soil	(Organically bound)
Zn	Soil	(Secondary minerals, organic)

[a]Speculative information (based on the literature) is enclosed in parentheses.

Both forms are acidifying to the ecosystem (i.e., SO_4^{2-} is deposited primarily as H_2SO_4, and SO_2 oxidizes and hydrolyzes to H_2SO_4 in the ecosystem). Based on some measurements as well as the literature for Ultisols (Adams et al. 1980), gaseous losses of sulfur are thought to be negligible in comparison to SO_4^{2-} leaching. As in many other watersheds in the southeastern United States, SO_4^{2-} output via streamflow (12 kg·ha^{-1}·year^{-1}) is less than half of total sulfur deposition (26 kg·ha^{-1}·year^{-1}). Our best evidence suggests that this apparent net accumulation is due to SO_4^{2-} adsorption onto iron and aluminum oxides in the deep, highly weathered subsoil of the watershed. Perhaps because of this annual accumulation, soil-adsorbed SO_4^{2-} constitutes nearly 30% of total ecosystem sulfur (Table 7.12). However, soil organic matter contains most of the total sulfur as well as the total nitrogen capital in the ecosystem. Since current sulfur deposition exceeds tree uptake and greatly exceeds tree sulfur increment, and since the soil SO_4^{2-} pool is quite large, we hypothesize that little of the sulfur currently accumulating in the watershed is incorporated into organic matter.

Little research has been done on phosphorus deposition to the watershed, but we speculate that the dominant form of phosphorus input is in particulates. Even though phosphorus input is quite low, the watershed accumulates 96% of incoming phosphorus. The deep subsoils have a much greater phosphate adsorption capacity than a sulfate adsorption capacity (Johnson et al. 1986), but biological uptake appears to play a major role in ecosystem phosphorus retention at current input levels in that (1) the increment of phosphorus in vegetation (0.3 kg·ha^{-1}·year^{-1} in live vegetation and 0.5 kg·ha^{-1}·year^{-1} in dead) approximately equals net ecosystem phosphorus accumulation (0.5 kg·ha^{-1}·year^{-1}) and (2) soluble reactive phosphate rapidly disappears in forest litter layers rather than in subsoils, implying tight tree and microbial conservation of this potentially limiting nutrient (Seagers et al. 1986). Phosphorus application studies have shown that adsorption in subsoils quickly comes into play in immobilizing phosphorus once biological demands are saturated, however (Johnson et al. 1986). The major forms of phosphorus in soil have yet to be determined, but based on the literature (Smeck 1985), we hypothesize that secondary mineral and organic forms predominate.

The watershed shows a net annual loss of K$^+$, Ca^{2+}, and Mg^{2+}, all of which occur primarily in secondary minerals in the soil component. The very large net exports of Ca^{2+} and Mg^{2+} are due to dissolution of the dolomite bedrock lying up to 30 m beneath these ancient, highly weathered soils. Stand-level studies have shown that leaching losses of Ca^{2+} and Mg^{2+} from a 1-m soil depth are <10% of the values for streamflow export. The amount of potassium leaching from the top 1 m of soil in these plot studies is somewhat lower (2-4 kg·ha^{-1}·year^{-1}) but within the same order of magnitude as streamflow export (Johnson et al. 1985), suggesting that bedrock influences are far less significant for K$^+$ than for Ca^{2+} and Mg^{2+}.

While bedrock influences overwhelm any possible effect of acid deposition on watershed-level Ca^{2+} and Mg^{2+} export, process-level plot studies on Walker Branch have shown that acid deposition has caused an approximate twofold increase in total base cation leaching from foliage and soils of the watershed. The potential for soil leaching by sulfur inputs is greater than that currently occurring, but soil SO$_4^{2-}$ adsorption prevents this potential from being fully realized. Despite acid-deposition-accelerated leaching, however, some sites show less Ca^{2+} leached from the top 1 m of soil than is deposited from the atmosphere (Johnson et al. 1985). This apparent accumulation the ecosystem appears to be due in part to very low exchangeable Ca^{2+} in some soils, which in turn is likely to be due to very high rates of calcium uptake and accumulation in oaks and hickories. The accumulation of Ca^{2+} in oaks and hickories is thought to be the primary reason for the observed decrease in exchangeable Ca^{2+} in subsoils from cherty, upland sites. The observed decreases in exchangeable Mg^{2+} in these subsoils may be due primarily to leaching, however.

The watershed shows a net annual accumulation of Mn^{2+}, Cd^{2+}, Pb^{2+}, and Zn^{2+} from atmospheric inputs. The soil is clearly the major sink for Cd^{2+} and Pb^{2+}, since vegetation increment is far less than net watershed accumulation (Table 7.10). From a mass balance perspective, it appears that vegetation is the major sink for Zn^{2+}, since vegetation increment exceeds atmospheric input. Vegetation uptake and increment data for Mn^{2+} are not available.

Whole-tree harvesting removes a substantial fraction of ecosystem calcium content in the deciduous forests studied. Since atmospheric calcium inputs will not replenish this loss, fertilization with calcium may be desirable to offset the removal unless it is compensated for by deep rooting. Calcium removal by whole-tree harvesting does not seem to pose problems in the pine-hardwood forests, nor does the removal of N, P, K, S, or Mg result in a significant depletion of total site capital in the deciduous forests. The effects of harvesting at various intensities on available pools of nutrients are not known, however, and harvesting is likely to be an important factor for nitrogen.

Walker Branch Watershed is comparable to nearly all forest ecosystems in some ways (e.g., net retention of atmospheric nitrogen and phosphorus, net losses of Ca^2, K^+, and Mg^{2+} on a watershed scale), to forest ecosystems of the southeastern United States in other ways (net retention of sulfur), to deciduous forest ecosystems in yet other ways (rapid turnover of forest floor and within-tree nitrogen and phosphorus conservation by internal translocation), and to oak-hickory forests in still other ways (large calcium accumulation in woody biomass). Walker Branch Watershed fits into, and is representative of, each of these broad catagories of forest ecosystems in one or more (but not all) respects, but it is most representative of southeastern oak-hickory forests.

The Walker Branch Watershed Project has evolved and continues to evolve toward more in-depth process-level studies now that the basic framework of element cycling has been established. Descriptive characterizations of element pools and fluxes, while essential as background information, cannot lead to generalizations and extrapolation of results. We look toward the process-level research as a means ofunderstanding the general principles of element cycling and conservation, which can be tested by extrapolation to other systems (e.g., the role of SO_4^2s;- adsorption in sulfur retention in an ecosystem). The element pools and fluxes in ecosystems that we measure at any point in time are in fact a function of element transfer processes integrated over time.

References

Abrahamsen, G. 1980. Impact of atmospheric sulfur deposition on forest ecosystems. pp. 397–416. IN D.S. Shriner, C.R. Richmond, and S.E. Lindberg (eds.), Atmospheric Sulfur: Environmental Impact and Health Effects. Ann Arbor Science, AnnArbor, Michigan.

Adams, D.F., S.O. Farwell, M.R. Pack, and E. Robinson. 1980. Estimates of natural sulfur sources and strengths. pp. 35–46. IN D.S. Shriner, C.R. Richmond, and S.E. Lindberg (eds.), Atmospheric Sulfur Deposition: Environmental Impact and Health Effects. Ann Arbor Science, Ann Arbor, Michigan.

Alban, D.H. 1969. The influence of western hemlock and western red cedar on soil properties. Soil Sci. Soc. Am. Proc. 33:453–457.

Alban, D.H. 1982. Effects of nutrient accumulation by aspen, spruce, and pine on soil properties. Soil Sci. Soc. Am. J. 46:853–861.

Baath, E., B. Berg, U. John, B. Lundgren, H. Lundkvist, T. Rosswall, B. Soderstrom, and A. Wiren 1980. Effects of experimental acidification and liming on soil organisms and decomposition in a Scots pine forest. Pedobiologia 20:85–100.

Bollen, W.B. 1974. Soil microbes. pp. B-1 to B-41. IN O.P. Cramer (ed.), Environmental Effects of Forest Residues Management. U.S. Department of Agriculture Forest Service Gen. Tech. Rep. PNW-24. Pacific Northwest Forest and Range Experiment Station, Portland, Oregon.

Bormann, F.E., and G.E. Likens. 1970. The nutrient cycles of an ecosystem. Sci. Am. 223:92–101.

Boyle, J.R., and G.K. Voigt. 1973. Biological weathering of silicate minerals: Implication of tree nutrition and soil genesis. Plant Soil 38:191–201.

Clayton, J.L. 1979. Nutrient supply by rock weathering. pp. 75–96. IN A.L. Leaf (ed.), Effects of Intensive Harvesting on Forest Nutrient Cycling. State University of New York, Syracuse.

Cole, D.W., and D.W. Johnson. 1977. Atmospheric sulphate additions and cation leaching in a Douglas-fir ecosystem. Water Resour. Res. 13:313–317.

Cole, D.W., and M. Rapp. 1981. Elemental cycling in forest ecosystems. pp. 341–409. IN D.E. Reichle (ed.), Dynamic Properties of Forest Ecosystems. Cambridge University Press, London.

Cole, D.W., S.P. Gessel, and S.F. Dice. 1968. Distribution and cycling of nitrogen, phosphorus, potassium, and calcium in a second-growth Douglas-fir forest. pp. 197–213. IN H.E. Young (ed.), Primary Production and Mineral Cycling in Natural Ecosystems. University of Maine Press, Orono.

Curlin, J.W., and D.J. Nelson. 1968. Walker Branch Watershed Project: Objectives, facilities, and ecological characteristics. ORNL/TM-2271. Oak Ridge National Laboratory, Oak Ridge, Tennessee.

Duvigneaud, P., and S. Denaeyer-DeSmet. 1970. Biological cycling of minerals in temperate deciduous forests. pp. 199–255. IN D.E. Reichle (ed.), Analysis of Forest Ecosystems. Springer-Verlag, New York.

Farmer, R.E., G.W. Bengston, and J.W. Curlin. 1970. Response of pine and mixed hardwood stands in the Tennessee Valley to nitrogen and phosphorus fertilization. For. Sci. 16:130–136.

Goldstein, R.A., S.A. Gherini, C.W. Chen, L. Mok, and R.J.M. Hudson. 1984. Integrated Acidification Study (ILWAS): A mechanistic ecosystem analysis. Phil. Trans. R. Soc. Lond. Ser. B 305:151–165.

Grigal, D.F., and R.A. Goldstein. 1971. An integrated ordination-classification analysis of an intensively sampled oak-hickory forest. J. Ecol. 59:481–492.

Grizzard, T., G.S. Henderson, E.E.C. Clebsch, and D.E. Reichle. 1976. Seasonal nutrient dynamics of foliage and litterfall on Walker Branch Watershed, a deciduous forest ecosystem. ORNL/TM-5254. Oak Ridge National Laboratory, Oak Ridge, Tennessee.

Haines, B.L., J.B. Waide, and R.L. Todd. 1982. Soil solution concentrations sampled with tension and zero-tension lysimeters: Report of discrepancies. Soil Sci. Soc. Am. J. 46:658–661.

Harris, W.F. 1977. Walker Branch Watershed: Site description and research scope. pp. 4–17. IN D.L. Correll (ed.), Watershed Research in Eastern North America—A Workshop to Compare Results. Chesapeake Bay Center for Environmental Studies, Smithsonian Institution, Edgewater, Maryland.

Harris, W.F., R.A. Goldstein, and G.S. Henderson. 1973. Analysis of forest biomass pools, annual primary production, and turnover of biomass for a mixed deciduous forest watershed. pp. 41–64. IN Harold Young (ed.), IUFRO Biomass Studies, Mensuration, Growth and Yield. University of Maine Press, Orono.

Harris, W.F., R.S. Kinerson, and N.T. Edwards. 1977. Comparison of belowground biomass of natural deciduous forest and loblolly pine plantations. Pedobiologia 17:369–381.

Heinrichs, H., and R. Mayer. 1977. Distribution and cycling of major and trace elements in two central European forest ecosystems. J. Environ. Qual. 6:402–407.

Henderson, G.S., and W.F. Harris. 1975. An ecosystem approach to characterization of the nitrogen cycle in a deciduous forest watershed. pp. 179–193. IN B. Bernier and C.H. Winget (eds.), Forest Soils and Land Management. Les Presses de l'Universite Laval, Quebec.

Henderson, G.S., W.F. Harris, D.E. Todd, Jr., and T. Grizzard. 1977a. Quantity and chemistry of throughfall as influenced by forest type and season. J. Ecol. 65:364–374.

Henderson, G.S., A.E. Hunley, and W.J. Selvidge. 1977b. Nutrient discharge from Walker Branch Watershed. pp. 307–320. IN D.L. Correll (ed.), Watershed Research in Eastern North America—A Workshop to Compare Results. Chesapeake Bay Center for Environmental Studies, Smithsonian Institution, Edgewater, Maryland.

Henderson, G.S., W.T. Swank, J.B. Waide, and C.C. Grier. 1978. Nutrient budgets of Appalachian and Cascade region watersheds: A comparison. For. Sci. 24(3):385–397.

Hesse, P.R. 1957. Sulphur and nitrogen changes in forest soils of East Africa. Plant Soil 19:86–96.

Jackson, D.R., W.J. Selvidge, and B.S. Ausmus. 1978. Behavior of heavy metals in forest microcosms: II. Effects on nutrient cycling processes. Water Air Soil Pollut. 10:13–18.

Johnson, D.W. 1975. Processes of elemental transfer in some tropical, temperate, alpine, and northern forest soils: Factors influencing the availability and mobility of major leaching agents. Ph.D. thesis, University of Washington, Seattle.

Johnson, D.W. 1983. The effects of harvesting intensity on nutrient depletion in forests. pp. 157–166. IN R. Ballard and S.P. Gessel (eds.), IUFRO Symposium on Forest Site and Continuous Productivity, Seattle, Washington, August 23–25, 1982. General Technical Report PNW-163. U.S. Forest Service, Portland, Oregon.

Johnson, D.W. 1985. Sulfur cycling in forests. Biogeochemistry 1:29–43.

Johnson, D.W., and G.S. Henderson. 1979. Sulfate adsorption and sulfur fractions in a highly weathered soil under a mixed deciduous forest. Soil Sci. 128:34–40.

Johnson, D.W., and D.E. Todd. 1983. Relationships among iron, aluminum, carbon, and sulfate in a variety of forest soils. Soil Sci. Soc. Am. J. 47:792–800.

Johnson, D.W., and D.E. Todd. 1984. Short-term effects of sulfuric acid and nitric acid irrigation on CO_2 evolution, extractable nitrogen, phosphorus, and aluminum in a deciduous forest soil. Soil Sci. Soc. Am. J. 48:664–666.

Johnson, D.W., N.T. Edwards, and D.E. Todd. 1980a. Nitrogen mineralization, immobilization, and nitrification following urea fertilization of a forest soil under field and laboratory conditions. Soil Sci. Soc. Am. J. 44:610–616.

Johnson, D.W., J.W. Hornbeck, J.M. Kelly, W.T. Swank, and D.E. Todd. 1980b. Regional patterns of soil sulfate accumulation: Relevance to ecosystem sulfur budgets. pp. 507–520. IN D.S. Shriner, C.R. Richmond, and S.E. Lindberg (eds.), Atmospheric Sulfur Deposition: Environmental Impact and Health Effects. Ann Arbor Science, Ann Arbor, Michigan.

Johnson, D.W., G.S. Henderson, and D.E. Todd. 1981. Evidence of modern accumulation of sulfate in an east Tennessee forested Ultisol. Soil Sci. 132:422–426.

Johnson, D.W., G.S. Henderson, D.D. Huff, S.E. Lindberg, D.D. Richter, D.S. Shriner, D.E. Todd, and J. Turner. 1982a. Cycling of organic and inorganic sulphur in a chestnut oak forest. Oecologia 54:141–148.

Johnson, D.W., D.C. West, D.E. Todd, and L.K. Mann. 1982b. Effects of sawlog vs. whole-tree harvesting on the nitrogen, phosphorus, potassium, and calcium budgets of an upland mixed oak forest. Soil Sci. Soc. Am. J. 46:1304–1309.

Johnson, D.W., D.W. Cole, C.S. Bledsoe, K. Cromack, R.L. Edmonds, S.P. Gessel, C.C. Grier, B.N. Richards, and K.A. Vogt. 1982c. IN R.L. Edmonds (ed.), Analysis of Coniferous Forest Ecosystems in the Western United States. Hutchinson Ross, Stroudsburg, Pennsylvania.

Johnson, D.W., D.D. Richter, G.M. Lovett, and S.E. Lindberg. 1985. The effects of atmospheric deposition on potassium, calcium, and magnesium cycling in two deciduous forests. Can. J. For. Res. 15:773–782.

Johnson, D.W., D.W. Cole, H. Van Miegroet, and F.W. Horng. 1986. Factors affecting anion movement and retention in four forest soils. Soil Sci. Soc. Am. J. 50:776–783.

Johnson, D.W., G.S. Henderson, and D.E. Todd. Changes in nutrient distribution in forests and soils of Walker Branch Watershed, Tennessee, over an eleven-year period. Biogeochemistry (in press).

Johnson, D.W., G.S. Henderson, and W.F. Harris. Changes in aboveground biomass and nutrient content on Walker Branch Watershed from 1967 to 1983. pp. 487–496 IN R.L. Hay, F.W. Woods, and H. DeSelm (eds). Proceedings of the Central Hardwood Forest Conference VI., Knoxville, TN, February 24–26, 1987. University of Tennessee Press, Knoxville, TN.

Johnson, F.L., and P.G. Risser. 1974. Biomass, annual net primary production, and dynamics of six mineral elements in a post oak blackjack oak forest. Ecology 55:1246–1258.

Keeney, D.R. 1980. Prediction of soil nitrogen availability in forest ecosystems: A literature review. Forest Sci. 26:159–171.

Kelly, J.M. 1984. Power plant influences on bulk precipitation, throughfall, and stemflow nutrient inputs. J. Environ. Qual. 13:405–409.

Kelly, J.M., and G.S. Henderson. 1978a. Nutrient flux in litter and surface soil after nitrogen and phosphorus fertilization. Soil Sci. Soc. Am. J. 42:963–966.

Kelly, J.M., and G.S. Henderson. 1978b. Effects of nitrogen and phosphorus additions on deciduous litter decomposition. Soil Sci. Soc. Am. J. 42:972–976.

Kelly, J., and M.J. Lambert. 1972. The relationship between sulfur and nitrogen in the foliage of *Pinus radiata*. Plant Soil 37:395–408.

Lagerwerff, J.V. 1972. Lead, mercury, and cadmium as environmental contaminants. pp. 593–636. IN J.J. Mortvedt, P.M. Giordano, and W.L. Lindsay (eds.), Micronutrients in Agriculture. Soil Science Society of America, Madison, Wisconsin.

Leaf, A.L. (ed.) 1979. Impact of Intensive Harvesting on Forest Nutrient Cycling. State University of New York, Syracuse, New York.

Likens, G.E., F.H. Bormann, R.S. Pierce, J.S. Eaton, and N.M. Johnson. 1977. Biogeochemistry of a Forested Ecosystem. Springer-Verlag, New York.

Lindberg, S.E., R.R. Turner, and G.M. Lovett. 1983. Mechanisms of the flux of acidic compounds and heavy metals onto receptors in the environment. *In* Löbel, J. and W.R. Thiel (eds.) *Acid Precipitation: Origin and Effects*, p. 165–172. Lindau, FRG, June 1983. Verlag des Vereins Deutscher Ingenieure, Düsseldorf, FRG.

Lovett, G.M., S.E. Lindberg, D.D. Richter, and D.W. Johnson. 1986. The effect of acidic deposition on cation leaching from a deciduous forest canopy. Can. J. For. Res. 15:1055–1060.

Luxmoore, R.J., T. Grizzard, and R.H. Strand. 1981. Nutrient translocation in the outer canopy and understory of an eastern deciduous forest. For. Sci. 27:505–518.

Meiwes, K.J., and P.K. Khanna. 1981. Distribution and cycling of sulphur in the vegetation of two forest ecosystems in an acid rain environment. Plant Soil 60:369–375.

Olson, S.R., and L.A. Dean. 1965. Phosphorus. pp. 1035–1049. IN C.A. Black, D.D. Evans, J.L. White, L.E. Engsminger, and F.E. Clark (eds.), Methods of Soil Analysis, Part 2: Chemical and Microbiological Properties. American Society of Agronomy, Madison, Wisconsin.

Ovington, J.D. 1962. Quantitative ecology and the woodland ecosystem concept. Adv. Ecol. Res. 1:103–192.

Pastor, J.J., and J.G. Bockheim. 1984. Distribution and cycling of nutrients in an aspen-mixed-hardwood-Spodosol ecosystem in northern Wisconsin. Ecology 65:339–353.

Peters, L.N., D.F. Grigal, J.W. Curlin, and W.J. Selvidge. 1970. Walker Branch Watershed Project: Chemical, Physical, and Morphological Properties of the Soils of Walker Branch Watershed. ORNL/TM-2968. Oak Ridge National Laboratory, Oak Ridge, Tennessee.

Rennie, P.J. 1955. The uptake of nutrients by mature forest growth. Plant Soil 7:49–95.

Richter, D.D., D.W. Johnson, and D.E. Todd. 1983. Atmospheric sulfur deposition, neutralization, and ion leaching in two deciduous forest ecosystems. J. Environ. Qual. 12:263–270.

Rochelle, B.P., M.R. Church, and M.B. David. 1987. Sulfur retention at intensively studied sites in the U.S. and Canada. Water Air Soil Pollut. 33:78–83.

Rolfe, G.L., M.A. Akhtar, and L.E. Arnold. 1978. Nutrient distribution and flux in a mature oak-hickory forest. For. Sci. 24:122–130.

Seagers, J.E., R.A. Minear, J.W. Elwood, and P.J. Mulholland. 1986. Chemical Characterization of Soluble Phosphorus Forms Along a Hydrologic Flowpath of a Forested Stream Ecosystem. ORNL/TM-9737. Oak Ridge National Laboratory, Oak Ridge, Tennessee.

Shaffer, P.W. 1984. Acid precipitation: Sulfate dynamics and the role of sulfate in cation mobility in White Oak Run watershed, Shenandoah National Park, Virginia. Ph.D. thesis, Department of Environmental Sciences, University of Virginia, Charlottesville.

Shriner, D.S., and G.S. Henderson. 1978. Sulfur distribution and cycling in a deciduous forest watershed. J. Environ. Qual. 7:392–397.

Smeck, N.E. 1985. Phosphorus dynamics in soils and landscapes. Geoderma 36:185–199.

Stednick, J.D. 1982. Sulfur cycling in Douglas-fir on a glacial outwash terrace. J. Environ. Qual. 11:43–45.

Stone, E.L. 1975. Effects of species on nutrient cycles and soil change. Phil. Trans. R. Soc. Lond. Ser. B 271:149–162.

Swank, W.T., and J.E. Douglass. 1977. Nutrient budgets for undisturbed and manipulated hardwood forests in the mountains of North Carolina. pp. 343–363. IN D.L. Correll (ed.), Watershed Research in North America, Vol. 1. Smithsonian Institution, Washington, D.C.

Swank, W.T., and G.S. Henderson. 1976. Atmospheric inputs of some cations and anions in North Carolina and Tennessee. Water Resour. Res. 12:541–546.

Swank, W.T., J.W. Fitzgerald, and J.T. Ash. 1984. Microbial transformation of sulfate in forest soils. Science 223:182–184.

Switzer, G.L., and L.E. Nelson. 1972. Nutrient accumulation and cycling in loblolly pine (*Pinus taeda* L.) plantation ecosystems: The first twenty years. Soil Sci. Soc. Am. Proc. 36:143–147.

Thomas, W.A. 1967. Accumulation and cycling of calcium by dogwood trees. Ecol. Monogr. 39:101–120.

Turner, J., and M.J. Lambert. 1980. Sulfur nutrition of forests. pp. 321–334. IN D.S. Shriner, C.R. Richmond, and S.E. Lindberg (eds.), Atmospheric Sulfur Deposition: Environmental Impact and Health Effects. Ann Arbor Science, Ann Arbor, Michigan.

Turner, J., M.J. Lambert, and S.P. Gessel. 1977. Use of foliage sulphate concentrations to predict response to area application by Douglas-fir. Can. J. For. Res. 7:476–480.

Turner, J., D.W. Johnson, and M.J. Lambert. 1980. Sulphur cycling in a Douglas-fir forest and its modification by nitrogen applications. Oecol. Plant. 15:27–35.

Turner, R.S., A.H. Johnson, and D. Wang. 1985. Biogeochemistry of lead in McDonalds Branch, New Jersey, Pine Barrens. J. Environ. Qual. 14:305–314.

Ulrich, B., R. Mayer, and R.K. Khanna. 1980. Chemical changes due to acid precipitation in a loess-derived soil in central Europe. Soil Sci. 130:193–199.

U.S. Department of the Interior (USDI). 1967. The National Atlas of the United States of America. U.S. Department of the Interior, Geologic Survey, Washington, D.C.

Van Hook, R.I., W.F. Harris, and G.S. Henderson. 1977. Cadmium, lead, and zinc distributions and cycling in a mixed deciduous forest. Ambio 6(5):281–286.

Van Hook, R.I., D.W. Johnson, D.C. West, and L.K. Mann. 1981. Environmental effects of harvesting forests for energy. For. Ecol. Manage. 4:79–94.

Vitousek, P.M., J.R. Gosz, C.C. Grier, J.M. Melillo, W.A. Reiners, and R.L. Todd. 1979. Nitrate losses from disturbed ecosystems Science 204:469–474.

Vogt, K.A., E.E. Moore, D.J. Vogt, M.J. Redlin, and R.L. Edmonds. 1983. Conifer fine root and mycorrhizal root biomass within the forest floors of Douglas-fir stands of different ages and site productivities. Can. J. For. Res. 13:429–437.

Woodward-Clyde, Inc. 1984. Subsurface Characterization and Geohydrologic Site Evaluation of West Chestnut Ridge—Report to Oak Ridge National Laboratory, Vol. I. ORNL Sub/83-64764. Woodward-Clyde, Inc., Wayne, New Jersey.

Zöttl, H.W. 1985. Heavy metal levels and cycling in forest ecosystems. Experimentia 41:1104–1113.

Chapter 8
Streams: Water Chemistry and Ecology

J.W. Elwood and R.R. Turner

8.1 Introduction

When viewing watersheds as ecological units of the landscape, geochemists and geomorphologists have historically considered streams as "sculptors" of the landscape and as conduits, providing the primary mode of egress of nutrients and other products of weathering, mineralization, and erosion from the watershed through hydrologic transport. A stream, however, is also a subsystem of a watershed, with biotic structure and function that can influence the timing, magnitude, and form of material fluxes carried in streamflow. In addition, the structure and function of streams are embodied in the physical, chemical, and biological processes of their associated watershed. Thus, considering streams only as conduits or agents of geomorphology is, as Hynes (1963) pointed out, a failure to perceive the terrestrial-aquatic linkage that exists between streams and their surrounding watershed.

Much of the early research on the streams draining Walker Branch Watershed focused on the spatial and temporal patterns in stream chemistry, quantifying the flux of nutrients and trace metals from the watershed in streamflow and characterizing the ecology of the streams. Subsequent work addressed the dynamics of organic carbon inputs and transport in the streams, the role of nutrient limitation in stream productivity, and the spatially dependent cycling of nutrients. In this chapter, the results of these studies are reviewed and discussed in terms of how they compare with those for similar studies in other watersheds. Because of subterranean flow in two reaches of the East Fork of Walker Branch when discharge rates are low, stream studies (other than the hydrologic and chemical mass balances) have largely focused on the West Fork and on a reach of stream below the confluence of the East Fork and West Fork.

8.2 Site Description

The East Fork and West Fork of Walker Branch Watershed are drained by first- and second-order streams (Fig. 8.1). The perennial reaches of the streams above the weirs on both forks, however, are first-order under base flow but become second-order during major storms as the channel expands into intermittent first-order tributaries. The channel length of West Fork in the reach above the weir with perennial flow is 365 m, and the drainage density is 9.5 m of perennial stream length per hectare of watershed area. The average gradient of the West Fork is ~5.6%; the average width of the perennial stream channel is ~2.8 m, whereas the average width of the wetted perimeter is ~2 m. The total length of the stream channels (i.e., perennial plus intermittent) in the West Fork is 1300 m.

The East Fork has a perennial stream length of 750 m upstream of the weir and a drainage density of 12.7 m/ha. The catchment drained by the East Fork is thus more deeply incised than the West Fork in terms of the drainage density of perennial streams and should therefore have a lower capacity for groundwater storage. Base flow per unit area is higher on the West Fork than on the East Fork (Luxmoore and Huff, Ch. 5, this book), which is consistent with lower groundwater storage on the East Fork. However, transfer of groundwater from the east to the west basin and/or deep leakage of water around the weir on the East Fork could also explain the lower base flow on an areal basis (Ch. 5).

The stream system on the East Fork is not as steep as that on the West Fork, having an average gradient of ~4.2%. At low discharge rates (<0.017 m^3/s), flow in two reaches (a 178-m segment and a 140-m segment) of the East Fork upstream of the weir is subterranean (Fig. 8.1). The total length of stream channels in the East Fork upstream of the weirs is 1795 m. Profiles of the two streams are shown in Figs. 5.1 and 5.2 (Ch. 5).

The two forks of Walker Branch converge ~25m below the weirs and flow for a distance of ~1 km before emptying into Melton Hill Reservoir. At low discharge rates, flow is subterranean over approximately the last 300 m upstream of Melton Hill Reservoir. A profile for the reach of Walker Branch downstream of the weirs has not been determined, but the gradient is less than that for the two forks above the weirs.

Both the East Fork and West Fork are fed by perennial and intermittent springs and seeps. The hydrology of the streams, including the location of primary and intermittent springs and seeps, is described in Ch. 5.

The streams in both subcatchments are characterized by intervening riffles and pools, with small cascades in areas where the stream flows across or through a more resistant rock type. The dominant substratum on the stream bottom consists of residual chert, a cryptocrystalline sedimentary rock. Depth to bedrock in riffles is generally <30 cm; there are reaches of stream on both forks flowing over exposed dolomite bedrock.

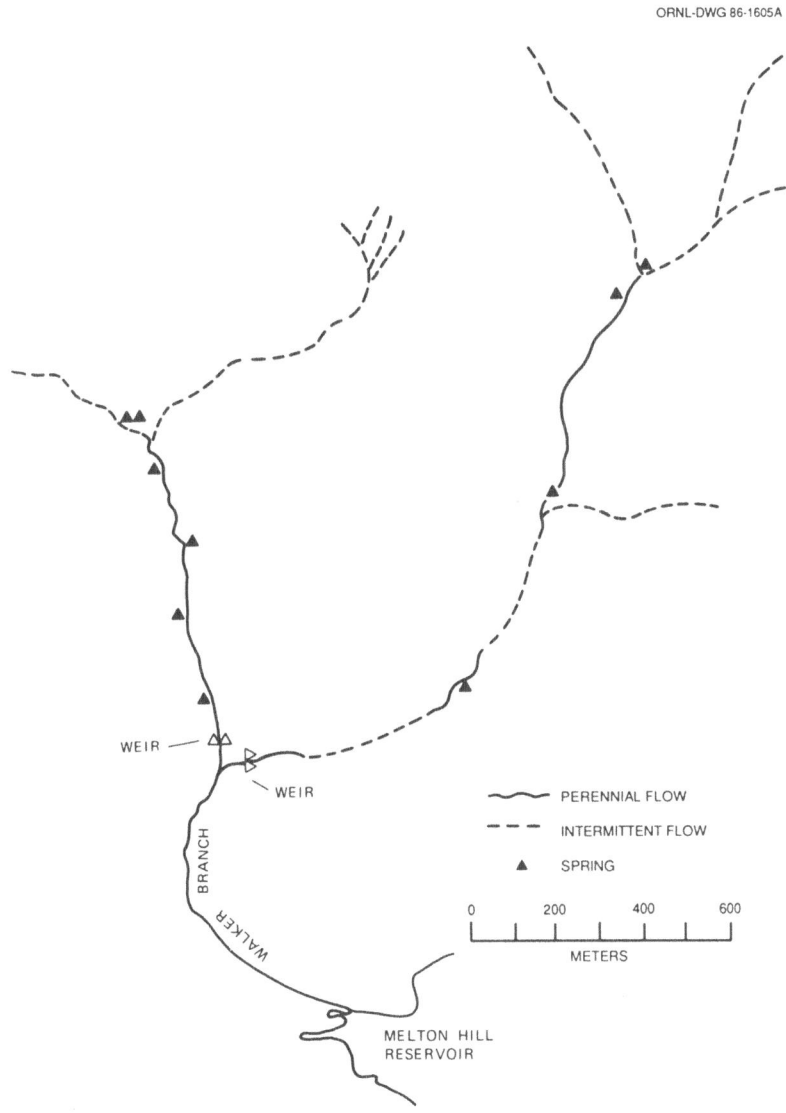

Figure 8.1. Intermittent and perennial reaches of streams draining Walker Branch Watershed. Also shown are the major springs feeding the streams on each fork. Source: Adapted from J.W. Curlin and D.J. Nelson. 1968. Walker Branch Watershed Project: Objectives, Facilities, and Ecological Characteristics. ORNL/TM-2271. Oak Ridge National Laboratory, Oak Ridge, Tennessee.

8.3 Stream Water Chemistry

A detailed description of the sample collection and analysis methods for stream water chemistry is given by Elwood and Henderson (1975) and Henderson et al. (1977) for major ions and nutrients and by Turner et al. (1977) for trace metals. Briefly, samples of streamflow for analysis of major ions (Ca^{2+}, Mg^{2+}, K^+, Na^+, SO_4^{2-}) and nutrients (NO_3^-, NH_4^+, P) from each subcatchment were composited (in a refrigerated container) proportional to discharge and collected weekly under base flow conditions. During storm events, separate samples were collected at 15- or 30-min intervals with a fraction collector. These samples were either analyzed individually or combined over short time intervals to determine concentration changes associated with rapidly increasing and decreasing flow rates. Samples were also collected manually for trace metal analyses and for certain specific studies.

Base cations (Ca^{2+}, Mg^{2+}, K^+, and Na^+) were analyzed by atomic absorption spectrophotometry (AAS), with lanthanum added to samples for calcium analysis to eliminate interferences. Ammonium was determined by the indophenol method, nitrate by reduction to nitrite and reaction with sulfanilimide, and phosphorus by the molybdate blue method (Technicon Industrial Systems 1971). Prior to 1985, sulfate was determined by the methylthymol blue method (McSwain and Watrous 1974), and after that by ion chromatography. Total nitrogen (ammonium-nitrogen plus organic nitrogen) was determined by Kjeldahl digestion and distillation and analysis of the distillate for ammonium-nitrogen as above. In most cases, samples were not filtered prior to analysis. Thus, results for digested, unfiltered samples (e.g., total nitrogen) include both the soluble and particulate fractions. Analyses of undigested unfiltered samples (e.g., base cations, ammonium, nitrate, sulfate, molybdate reactive phosphorus), however, include only, or predominantly, the soluble fraction, depending on whether the analytical method used causes elements associated with particulates to be released into solution.

Total soluble phosphorus (TSP) and soluble reactive phosphorus (SRP) were also determined on manually collected samples of stream water. These samples were filtered immediately after collection through preleached (in 100 ml of deionized water), 0.4-μm pore size Nuclepore membrane filters. SRP was measured using the method of Murphy and Riley (1962), and TSP was measured using the same method after persulfate oxidation of the sample by the procedure of Menzel and Corwin (1965). Soluble unreactive phosphorus (SUP), which is assumed to be a measure of soluble organic phosphorus, was calculated as the difference between TSP and SRP. Hydrolysis of organic phosphorus during the analysis for SRP with the molybdate reactive method, however, will result in an underestimate of organic phosphorus when calculated from the difference between TSP and SRP (Stainton 1980; Tarapchak 1983). Concentrations of SRP determined with this method are thus probably overestimates of orthophosphorus, which, if so, means that SUP concentrations are underestimates of soluble organic phosphorus.

For trace metals, grab samples were collected at least weekly; during and following major storm events, sampling was done more frequently (hourly to daily). Methods of sample collection, processing, and analysis for trace metals are described in detail by Turner et al. (1977). Briefly, trace metal concentrations were measured in both unfiltered (so-called "bulk" concentration) and filtered (so-called "dissolved" concentration) samples. Unfiltered samples were acidified immediately after collection to a pH of ≤ 1.5. For analysis of "dissolved" trace metals, samples were filtered through prerinsed 0.4-μm Nuclepore filters and then acidified with Ultrex HNO_3 to a pH of ≤ 1.5. All metal concentrations were determined by the method of standard additions, using a graphite furnace AAS.

8.3.1 Classification of Stream Water Chemistry

Streams draining both forks of Walker Branch have moderately hard water, with calcium and magnesium being the major cations (Table 8.1) and bicarbonate being the major anion in solution. The ratio of the average flow-weighted concentration of calcium to magnesium in stream water is approximately 2 to 1, reflecting the influence of dolomite weathering on the relative abundance of these cations in stream water. The average concentration ratio of calcium to magnesium in samples of dolomite collected from both the East Fork and West Fork of Walker Branch was 2.2 ± 0.1 (± 1 SD; $n = 6$).

Table 8.1. Mean annual flow-weighted concentration of Ca, Mg, Na, K, sulfate, nitrate, ammonium, total nitrogen ($NO_3 - N + NH_4^+ - N$ + organic nitrogen), and soluble reactive phosphorus in stream water draining the East Fork and West Fork of Walker Branch combined (based on weekly, flow-proportional composite samples)

	Period of record	Concentration (± 1 SD)
Calcium, mg/L	1970–74	16 ± 2
Magnesium, mg/L	1970–74	8.4 ± 1.2
Sodium, mg/L	1970–74	0.48 ± 0.05
Potassium, mg/L	1970–74	$0.73 + 0.02$
Sulfate, mg/L	1973–76	4.0 ± 0.2
Nitrate, μg/L	1972–74	57.3 ± 26.6
Ammonium, μg/L	1972–74	27.8 ± 8.1
Total nitrogen, μg/L	1972–74	156 ± 38
Soluble reactive phosphorus, μg/L	1970–74	2.3 ± 0.7

The average sulfate concentration in the streams draining Walker Branch Watershed is relatively high (Table 8.1) compared with that in many streams in the southeastern United States that are not affected by acid mine drainage or other internal sources of sulfate. This watershed is located within 22 km of three coal-fired power plants, one of which is often immediately upwind of Walker Branch, suggesting that atmospheric deposition is the primary source of sulfate in the stream water. As previously discussed by Lindberg et al. (Ch. 4, this volume), sulfur deposition rates on Walker Branch are higher than on other forested watersheds in the Southeast located much farther from local sources of sulfur emissions. While Walker Branch Watershed exhibits a net retention of sulfur, Johnson et al. (1986) found that the sulfate adsorption capacity of two soils from this watershed was lower than that for other highly weathered soils from forested watersheds in the southeastern United States that have a relatively high content of hydrous oxides of iron and aluminum. They suggested that the observed variation in retention of sulfate among watersheds may be due to differences in both their soil properties (e.g., pH, base saturation, organic matter content, hydrous oxide content) and their sulfur deposition histories. The sulfate adsorption capacity of soils on Walker Branch Watershed, for example, may be closer to equilibrium (i.e., the sulfate concentration at which there is no net adsorption or desorption) with respect to current atmospheric sulfur inputs owing to the historically higher deposition rates from local point sources, resulting in less net adsorption of sulfate.

In contrast, other watersheds in the Southeast with similar soils but with lower deposition rates may be further from equilibrium, resulting in lower sulfate concentrations in their drainage waters due to greater adsorption by their soils. While the relative importance of soil properties vs the history of sulfur deposition in controlling sulfate concentrations in stream water remains unresolved, the high sulfate levels in stream water draining Walker Branch Watershed indicate that sulfate is relatively more mobile in the soils of this watershed than in other watersheds in the Southeast with deep, highly weathered soils.

Despite the higher sulfate levels in streams draining Walker Branch Watershed, the pH and alkalinity of the stream water remain relatively high (pH >7.0), except during high flows. These conditions are due to the generation of alkalinity by weathering of the underlying dolomitic bedrock. Walker Branch may thus be somewhat unique among forested watersheds in the Southeast in terms of its lower retention of sulfate and the lack of a significant effect of acidic deposition on surface water chemistry. As discussed later, however, the alkalinity and pH decrease, and sulfate levels increase in stream water during storms, the former most likely because of both dilution and titration (i.e., real acidification).

In contrast to sulfate, the concentrations of nitrate, ammonium, and soluble reactive phosphorus in the stream water are relatively low (Table

8.1) and typical of streams draining second-growth forested watersheds in the southeastern United States (Messer et al. 1987; Omernik 1977). Nitrate is the dominant form of inorganic nitrogen in stream water, a pattern consistent with both the preferential uptake of ammonium over nitrate by benthic algae and bacteria and the nitrification of ammonium by bacteria in stream sediments.

Concentrations of trace metals (Cd, Cu, Cr, Mn, Fe, Pb, and Zn) in the streams draining Walker Branch are low (Table 8.2), which is consistent with the relatively undisturbed condition of this catchment and the low level of suspended solids (typically <5 mg/L) in the streams. In addition, the relatively high pH (geometric mean of 7.9 for the West Fork) and well-oxygenated condition of the stream waters imply low solubility of most naturally occurring solid phases (oxides, hydroxides, carbonates) of these metals if only theoretical mineral equilibria are considered. The mechanisms of trace metal transport in the streams draining Walker Branch vary, depending on the element and stream stage, with Cd, Cu, and Cr transported primarily in the "dissolved" form during nonstorm flows and Fe and Mn transported primarily in the particulate form.

Concentrations of the divalent cations, as well as other ions and trace metals in stream water, exhibit considerable spatial and temporal variation within the watershed. This variation can be attributed to (1) variation in elemental inputs from atmospheric deposition in precipitation and dryfall, (2) variation in the duration and extent of contact of rainfall with weatherable minerals in the soil and bedrock, (3) differences in hydrologic flow paths and source areas that contribute water to the stream, and (4) variation in biological processes that remove solutes from, or contribute solutes to, drainage water. Since water is a major carrier that mediates nutrient and

Table 8.2. Flow-weighted mean concentrations (μg/L) of trace metals in the West Fork of Walker Branch between September 1975 and May 1977, excluding large storm of April 4, 1977 (based on grab samples collected at least weekly)

	Bulk[a]			Dissolved[b]		
	Mean	1 SD	n	Mean	1 SD	n
Cadmium	0.02	0.01	142	0.02	0.01	79
Chromium	0.10	0.04	67	0.09	0.11	46
Copper	0.12	0.06	69	0.10	0.07	29
Iron	33.0	15.0	141	1.1	0.5	139
Lead	0.35	0.08	143	0.20	0.10	85
Manganese	28.0	14.0	142	3.2	1.4	141
Zinc	1.7	1.1	142	0.35	0.17	114

[a]Unfiltered sample.
[b]Filtered through $0.4-\mu m$ pore size Nuclepore membrane filter.

trace metal transfers between components of a watershed, information on stream water chemistry can be used to assess the basic factors controlling the biogeochemical cycling and export of these materials within a landscape. It is in this context that the chemistry of the streams draining Walker Branch Watershed is discussed.

8.3.2 Spatial Patterns in Stream Water Chemistry

In general, variation in the concentration of major ions between the streams and springs feeding into Walker Branch is greater on the East Fork than on the West Fork. Calcium concentrations of some springs feeding the East Fork during nonstorm (base flow) periods, for example, are relatively low compared with the concentrations in adjacent springs and in stream water at the weir (Fig. 8.2). As a consequence, calcium concentrations in stream water at the weir on the East Fork during base flow are lower than those at the weir on the West Fork. The greater spatial variations in calcium concentrations on the East Fork and for the lower average calcium level in stream water are attributed to differences between the two subcatchments in the proportion of flow emanating from groundwater that has been in contact with the dolomite bedrock. Tarklin and Linside soils are distributed along certain portions of the East Fork (Fig. 8.3). These two soils, which may be of both colluvial and alluvial origin, have a compact, relatively impermeable fragipan zone between 45 and 150 cm below the surface. This fragipan is known to impede vertical drainage and to foster lateral drainage, as reflected in the much lower hydraulic conductivity of these two soil types compared with that of other soils on Walker Branch (Peters et al. 1970). As a consequence, water in these two soil types moves laterally to the stream channel rather than vertically into the groundwater. Springs whose drainage areas are occupied by the Tarklin and Linside soils (Fig. 8.3) are thus perched groundwater sources, the water from which appears to have had minimal contact with the underlying dolomite bedrock, as reflected in the lower concentrations of calcium and magnesium compared with those in springs emanating in areas where these soil types do not occur (Fig. 8.2). Calcium concentrations in springs feeding the West Fork, where the Tarklin and Linside soils do not occur, are all relatively uniform and greater than those in the major springs feeding the East Fork (Fig. 8.2).

Other solutes in stream water also reflect this difference in the flow path of water, and the resulting difference in the extent of bedrock weathering, between the East Fork and West Fork of Walker Branch. The conductivity and pH of water in the East Fork, for example, are consistently less than those in the West Fork (Fig. 8.4) owing to less weathering of the dolomite, which releases alkalinity and base cations (Ca^{2+}, Mg^{2+}) to solution.

Within short stream reaches of Walker Branch, especially on the East

ORNL—DWG 86—1605

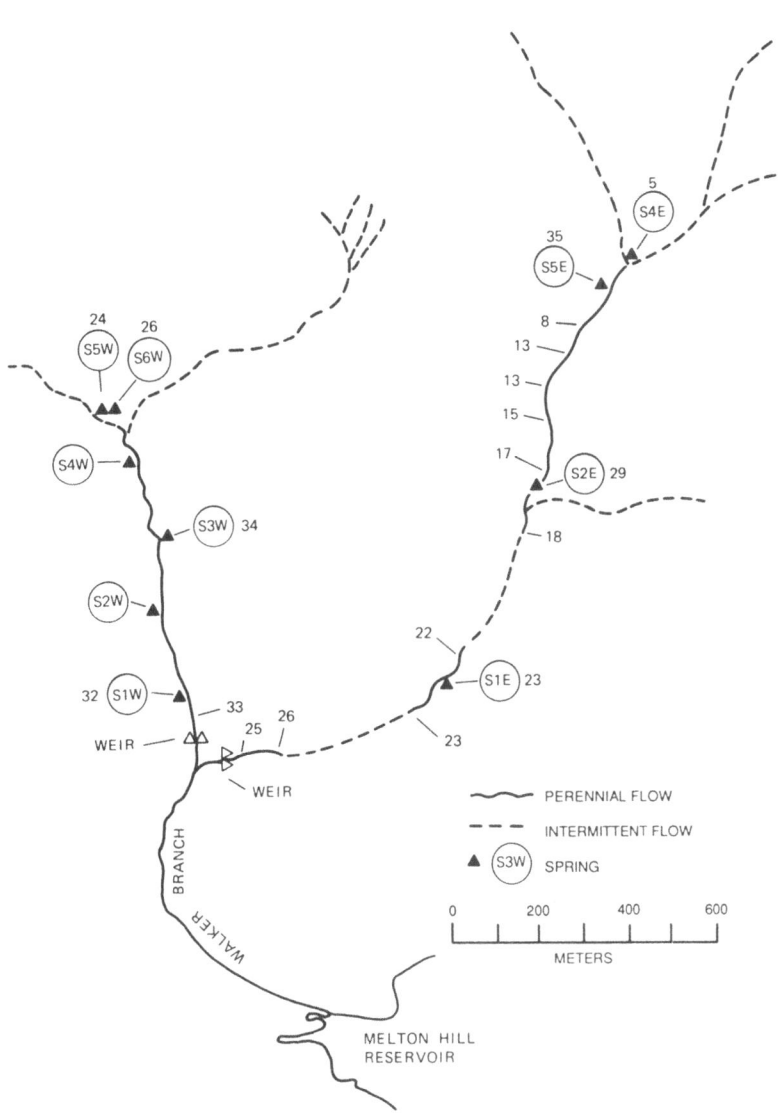

Figure 8.2. Spatial variation in the calcium concentration (in mg/L) in spring water and stream water at different locations in the East Fork and West Fork of Walker Branch (data from Curlin and Nelson 1968).

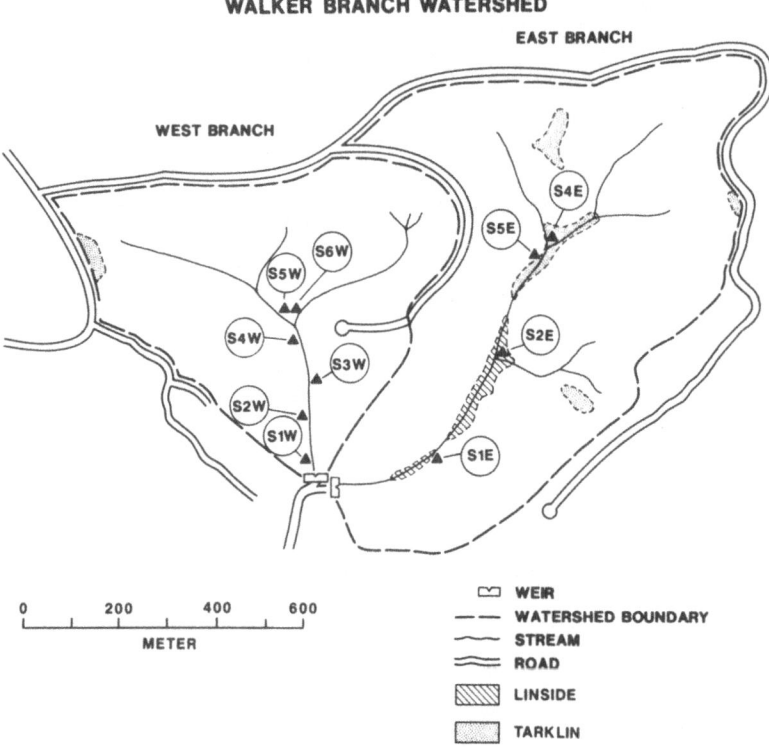

ORNL-DWG 86-1513

Figure 8.3. Distribution of Tarklin and Linside soils in relation to springs feeding the East Fork and West Fork of Walker Branch. The distribution of these two soil types is restricted almost exclusively to the stream margins on the East Fork. Source: Adapted from J.W. Curlin and D.J. Nelson. 1968. Walker Branch Watershed Project: Objectives, Facilities, and Ecological Characteristics. ORNL/TM-2271. Oak Ridge National Laboratory, Oak Ridge, Tennessee.

Fork, dissolved manganese can exhibit sharp spatial (longitudinal) gradients. Elevated manganese concentrations occur immediately below large organic debris dams and at points of reemergence of the stream below dry reaches. The stream bed substrate in these areas is typically characterized by a black coating of hydrous oxides of manganese, which precipitates primarily in response to reaeration of stream water. As shown by Cerling and Turner (1982) for nearby White Oak Creek, the formation and destruction of these coatings can be very rapid (weeks to months) and appears to be regulated by gradients in stream water oxidation-reduction potential. The occurrence of elevated manganese concentrations in zones below organic debris dams and the lack of elevated manganese levels in spring flow suggest that the proximate source of manganese is

ORNL-DWG 86-9532

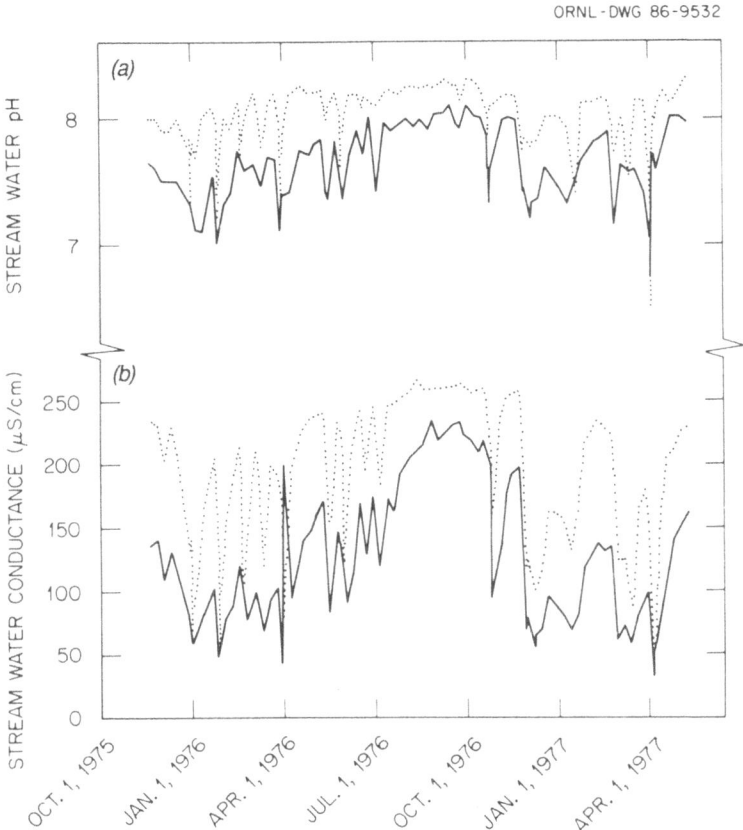

Figure 8.4. (a) Stream water pH and (b) conductivity (in μs/cm at 25°C) in the East Fork and West Fork of Walker Branch. The solid line is the East Fork; the dotted line is the West Fork. Lowest conductivity and pH occur during high discharges, the decrease in the former probably due to dilution, and the decrease in the latter probably due to both dilution of alkalinity and titration of alkalinity by strong mineral acidity (H_2SO_4).

decomposing organic matter and manganese precipitated on stream bed deposits, not bedrock or soils.

In summary, the stream water in the West Fork appears to emanate predominantly from deep groundwater sources that are in contact with the underlying dolomite. In contrast, a greater proportion of the stream water in the East Fork emanates as shallow groundwater from a perched water table caused by a fragipan in the Tarklin and Linside soils. Because this fragipan effectively reduces or prevents contact between the water moving through the soil and the underlying dolomite bedrock before it reaches the stream channel, the spring water emanating from these perched water tables is more dilute than the deeper groundwater in contact with the dolomite. Thus, weathering of dolomite is relatively less important in controlling the chemistry of stream water in the East Fork than in the West Fork of Walker Branch. This is reflected in the lower conductivity, lower concentrations of most base cations (e.g., Ca, Mg, K), and lower pH in stream water in the East Fork than in the West Fork.

8.3.3 Episodic Patterns in Stream Water Chemistry

Temporal variations in solute concentrations in streamflow occur on both short (episodic) and long (seasonal, annual) time scales. The processes controlling solute concentrations may be chemically and physically based, such as kinetic limitations on mineral dissolution, or they may be biologically based, such as nutrient uptake and retention or release by vegetation in the watershed. Changes in solute concentrations during storms fall into one of three classes: (1) dilution effect, where the solute concentration decreases with increasing discharge; (2) concentration effect, where the solute concentration increases with increasing discharge; and (3) no effect, where there is little or no change in concentration with changes in discharge. For Walker Branch, calcium and magnesium exhibit a dilution response (e.g., Fig. 8.5); total nitrogen, sulfate, and soluble reactive phosphorus show a concentration response (e.g., Figs. 8.6–8.8); and sodium concentrations show little or no response to changes in flow rate during storms (Fig. 8.9). The response of potassium varies, depending on the time of year. For example, a concentration response occurs during the late fall-winter period, but there appears to be little or no response to changes in discharge during the remainder of the year (Fig. 8.10).

For the trace metals examined to date, bulk concentrations (dissolved plus particulate) in Walker Branch stream water exhibit a concentration response during storms, with a hysteresis effect where the concentration at a given discharge is greater during the ascending limb of the hydrograph than during the descending limb (Figs. 8.11, 8.12). Trace metals in solution, however, exhibit either a dilution (Fig. 8.11) or a concentration (Fig. 8.12) response, depending on the element.

Calcium and Magnesium

Calcium and magnesium concentrations in streamflow appear to be controlled predominantly by the extent and duration of contact of the stream water with the dolomite bedrock underlying Walker Branch. During storms when flow is increasing, base flow, which has a long contact time with the bedrock, constitutes a smaller proportion of the streamflow and is diluted by water entering the stream channel from other flow paths. The extent to which calcium and magnesium concentrations decline during storms thus depends on the proportion of base flow in total streamflow.

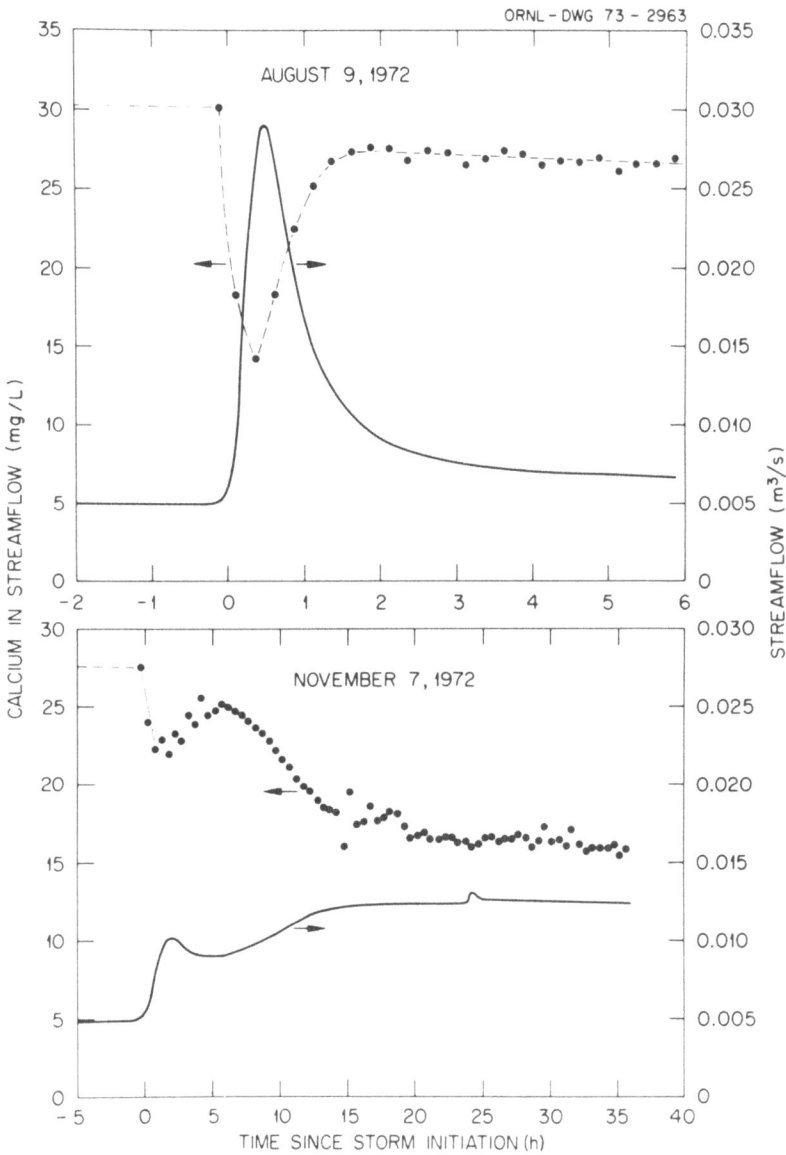

Figure 8.5. Relationship between calcium concentration in stream water and discharge rate during and after a storm on the West Fork of Walker Branch Watershed. Source: G.S. Henderson, A. Hunley, and W. Selvidge. 1977. Nutrient discharge from Walker Branch Watershed. pp. 307–320. IN D.L. Correll (ed.), Watershed Research in Eastern North America: A Workshop to Compare Results. Chesapeake Bay Center for Environmental Studies, Smithsonian Institution, Edgewater, Maryland.

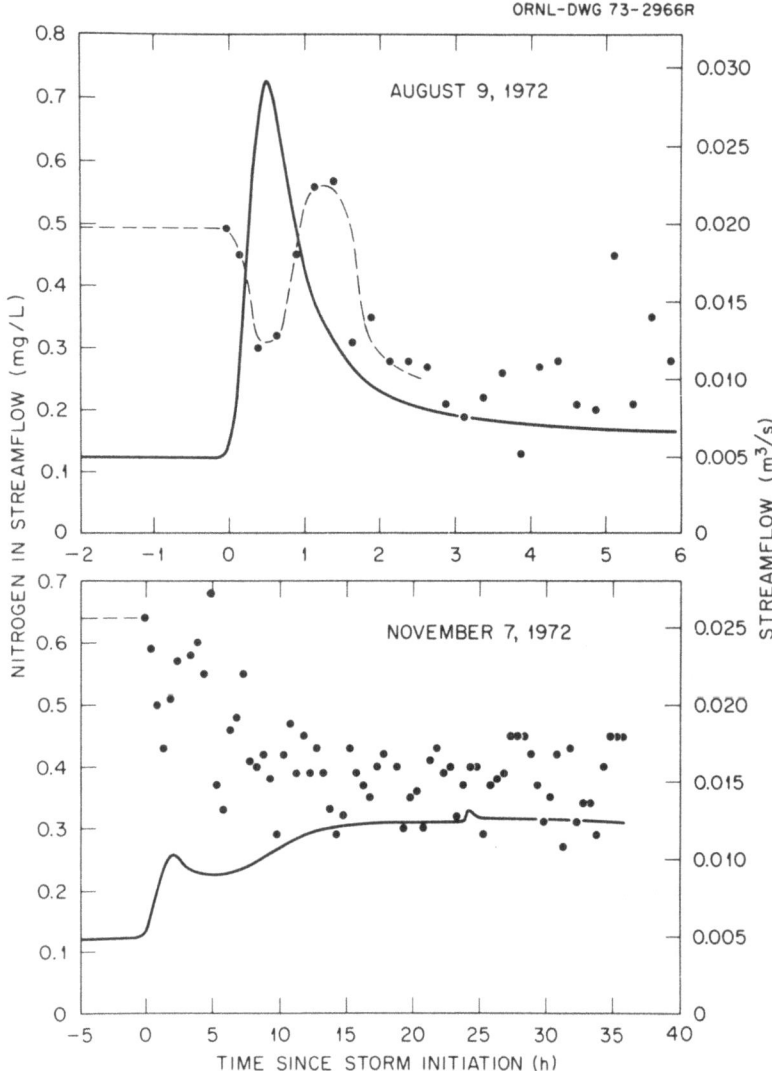

Figure 8.6. Relationship between total nitrogen (dissolved organic nitrogen plus particulate organic nitrogen plus ammonium-nitrogen) and discharge rate during and after two storms on the West Fork of Walker Branch Watershed. Source: G.S. Henderson and W.F. Harris. 1975. An ecosystem approach to characterization of the nitrogen cycle in a deciduous forest watershed. pp. 179–193. IN B. Bernier and C.H. Winget (eds.), Forest Soils and Forest Land Management. Les Presses de L'Université Laval, Quebec, Canada.

ORNL-DWG 76-18645

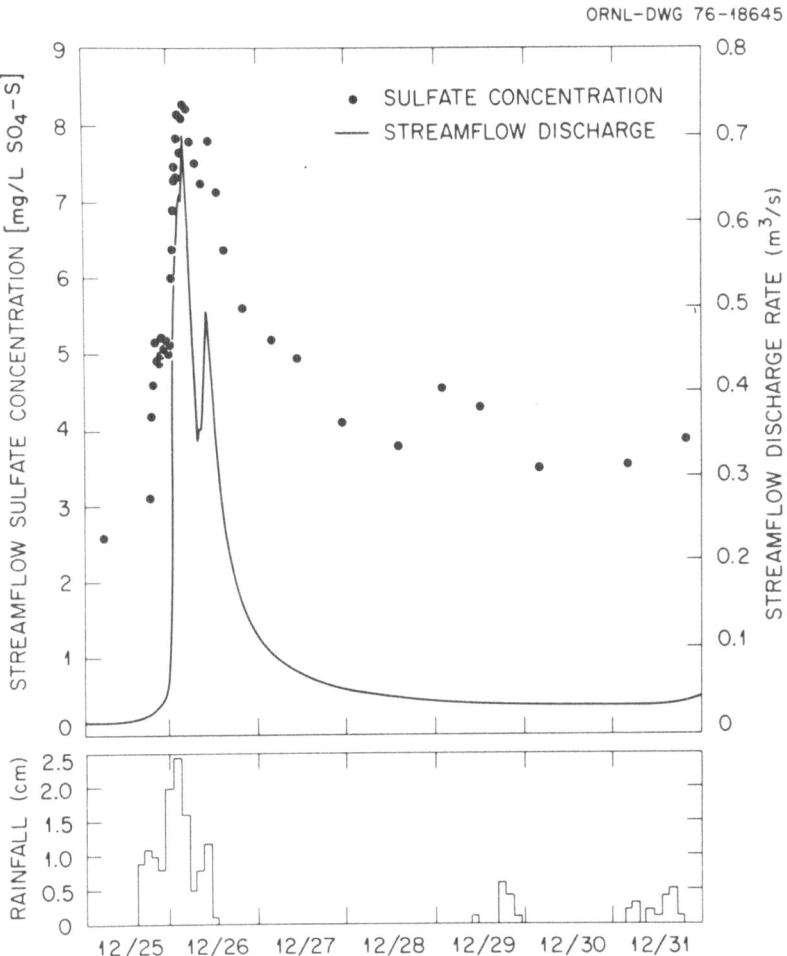

Figure 8.7. Relationship between sulfate concentration in stream water, discharge rate, and precipitation during and after a storm in December on the West Fork of Walker Branch Watershed. Source: D.W. Johnson and G.S. Henderson. 1979. Sulfate adsorption and sulfur fractions in a highly weathered soil under a mixed deciduous forest. Soil Sci. 128:34–40.

The dilution effect of increasing flows on calcium and magnesium concentrations appears to hold during all seasons, although during winter storms which produce exceptionally high flows, calcium concentration reaches a minimum and does not decrease further even though streamflow continues to increase (Henderson etal. 1977). This minimum concentration is 10 to 20% of the normal base flow concentration and is similar to values in soil water at a depth of 75 cm in the soil profile, suggesting that exchange and weathering processes in the soil are more important than bedrock

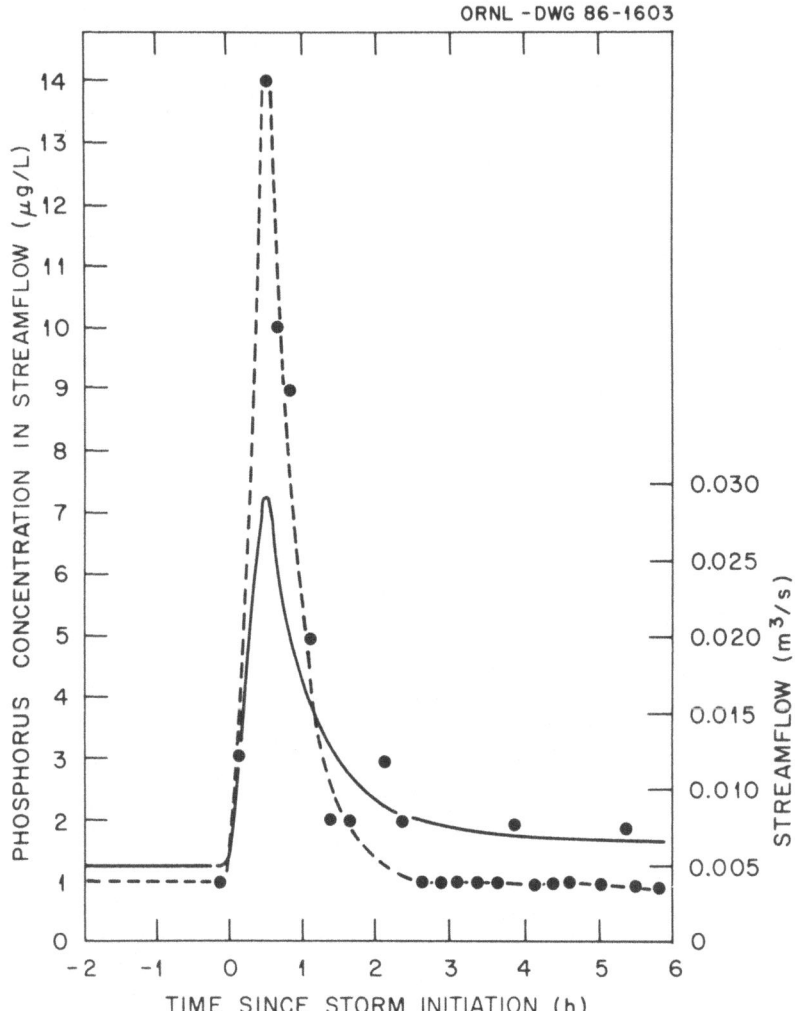

Figure 8.8. Relationship between the concentration of soluble reactive phosphorus (SRP) and discharge rate in the West Fork of Walker Branch during and after a storm in August.

weathering in controlling the concentration of the divalent base cations in stream water during these relatively short periods of high flow.

Sulfate and Total Nitrogen

Sulfate and total nitrogen (ammonium-nitrogen plus dissolved organic nitrogen plus particulate organic nitrogen) exhibit increases in concentration with increasing discharge. These increases are attributed to expansion of

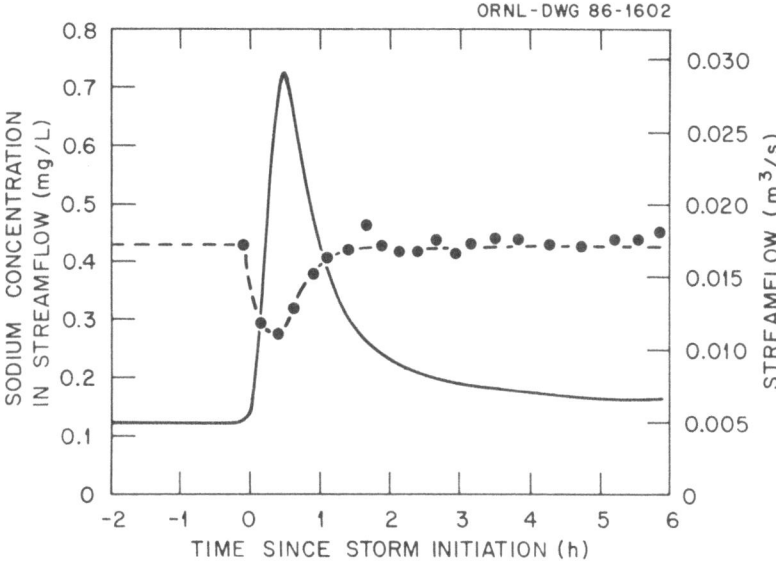

Figure 8.9. Relationship between sodium concentration and discharge rate in the West Fork of Walker Branch during and after a storm in August.

the stream into hydrologic source areas and transport of materials from intermittent drainages. For total nitrogen, the most commonly observed pattern is an initial decrease in concentration followed by an increase and then another decrease to concentrations below those prior to storm initiation (e.g., Fig. 8.6). The initial decrease in concentrations is attributed to dilution from direct channel interception by throughfall that has a low concentration of total nitrogen, with a subsequent increase in total nitrogen due to increases in organic nitrogen. The causes of the increase are probably channel expansion and the resulting leaching of soluble organic nitrogen and the suspension and entrainment of particulate organic materials (e.g., leaf litter) along the stream margin and in intermittent channels. Once the supply of materials subject to leaching and entrainment in these source areas is depleted, the concentrations of total nitrogen decline unless there is further channel expansion into new source areas.

The importance of the concentration effect on total nitrogen transport is that a significant fraction of the seasonal and annual nitrogen transport from Walker Branch occurs in relatively few hydrologic events. Henderson etal. (1977), for example, estimated that 80 to 90% of the annual particulate nitrogen loss from Walker Branch occurs during a 5- to 10-h period during the three or four largest storms of the year. Estimating annual, seasonal, or episodic nitrogen loss from the watershed and nitrogen transfer to downstream systems thus depends on sampling during these relatively infrequent hydrologic events.

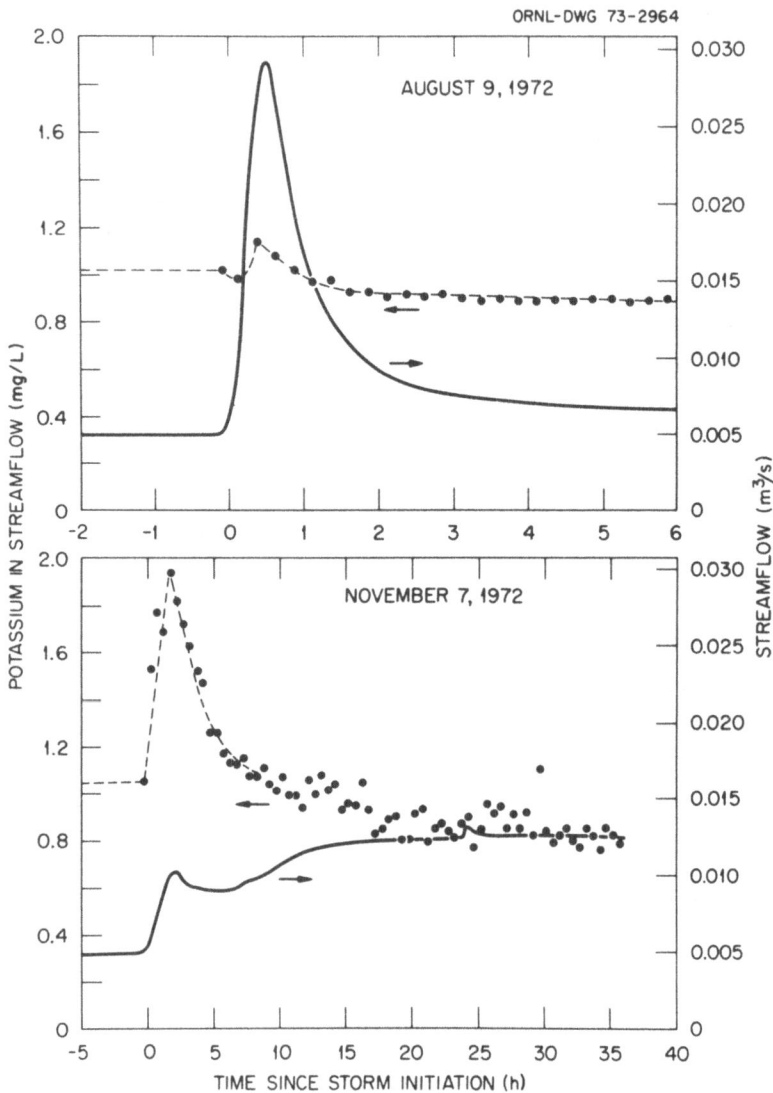

Figure 8.10. Relationship between potassium concentration and discharge rate in the West Fork of Walker Branch during and after two storms. Source: G.S. Henderson, A. Hunley, and W. Selvidge. 1977. Nutrient discharge from Walker Branch Watershed. pp. 307–320. IN D.L. Correll (ed.), Watershed Research in Eastern North America: A Workshop to Compare Results. Chesapeake Bay Center for Environmental Studies, Smithsonian Institution, Edgewater, Maryland.

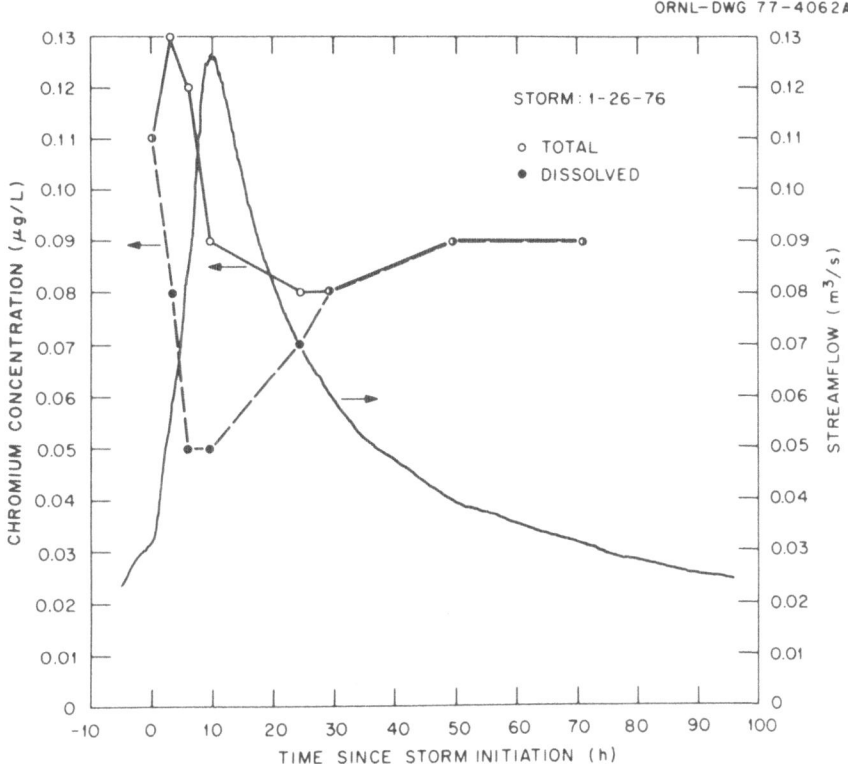

Figure 8.11. Relation between total and dissolved chromium concentration and discharge rate in the West Fork of Walker Branch during a storm in January. Source: R.R. Turner, S.E. Lindberg, and K. Talbot. 1977. Dynamics of trace element export from a deciduous watershed, Walker Branch, Tennessee. p. 661–679. IN D.L. Correll (ed.), Watershed Research in Eastern North America: A Workshop to Compare Results. Chesapeake Bay Center for Environmental Studies, Smithsonian Institution, Edgewater, Maryland.

Sulfate concentrations in streamflow also increase with increasing discharge. Unlike total nitrogen, however, which is diluted when stormflow first increases, sulfate increases as stormflow begins, and this increase continues until peak flow is reached (Fig. 8.7). Thereafter, sulfate concentration decreases as stormflow recedes. The plot of sulfate concentration against flow rate for the entire storm shows that during the first 5 to 6 h, sulfate concentrations increase rapidly relative to the increase in flow rate (Fig. 8.13). From that time until peak flow, sulfate concentration continues to increase, but at a rate less than the initial rate and less than the rate of flow increase. During the descending part of the storm hydrograph, sulfate concentrations decline at a slower rate than streamflow.

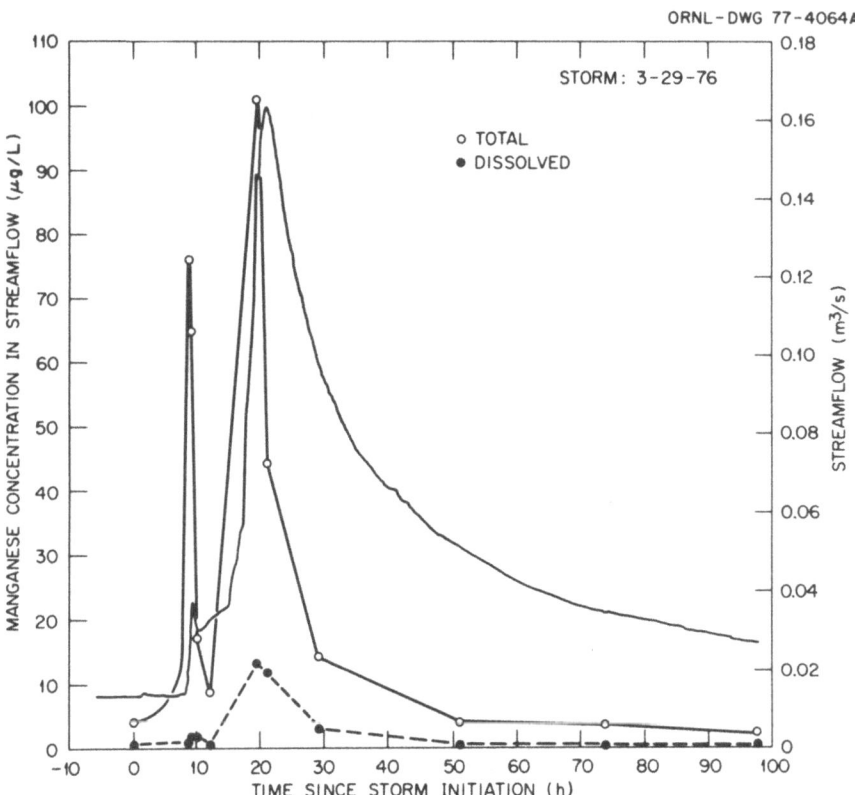

Figure 8.12. Relationship between total and dissolved manganese concentration and discharge rate in the West Fork of Walker Branch Watershed during a storm in March. Source: R.R. Turner, S.E. Lindberg, and K. Talbot. 1977. Dynamics of trace element export from a deciduous watershed, Walker Branch, Tennessee. p. 661–679. IN D.L. Correll (ed.), Watershed Research in Eastern North America: A Workshop to Compare Results. Chesapeake Bay Center for Environmental Studies, Smithsonian Institution, Edgewater, Maryland.

Thus, for a given flow rate, the sulfate concentration is greater when flow is receding than when it is increasing.

The sources of the sulfate that cause concentrations in streamflow to increase during storms include incoming precipitation and throughfall, previously deposited sulfate in soil, and sulfate from mineralized organic sulfur. The initial increase in sulfate concentration in streamflow is most likely due to direct interception by the stream channel of sulfate in rainfall and throughfall. Sulfate concentrations in composite throughfall samples from Walker Branch range from 5 to 10 mg/L, levels that exceed the observed sulfate concentrations in streamflow during base flow. Peak concentrations of sulfate in stream water during long winter storms, however,

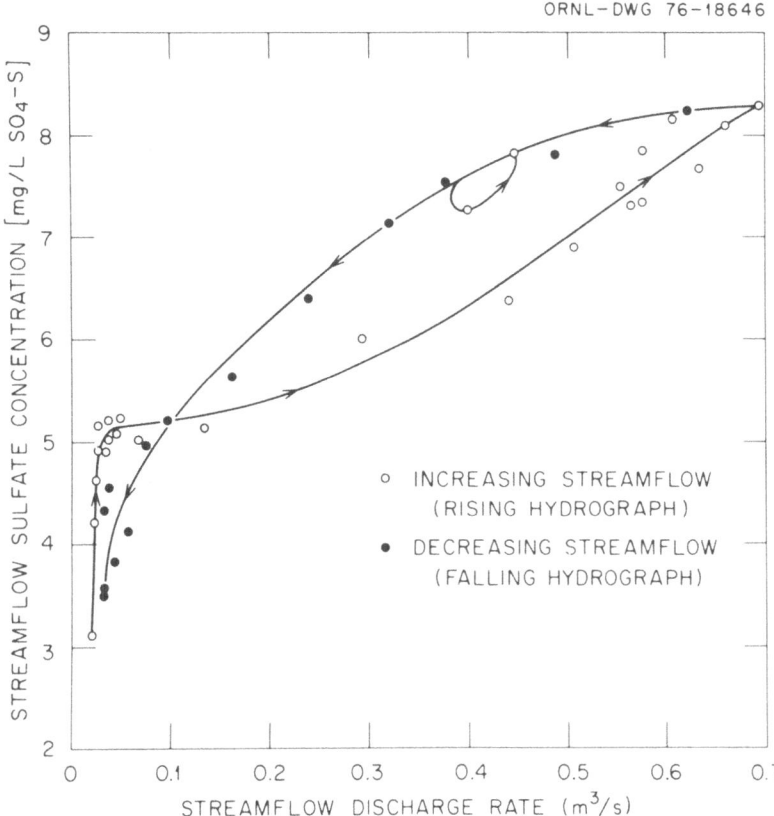

ORNL-DWG 76-18646

Figure 8.13. Relationship between sulfate concentration and discharge rate in the West Fork of Walker Branch during and after a December rainstorm. Data plotted are the same as those in Fig. 8.7, showing the hysteresis effect of flow rate on sulfate concentration. Open circles are for samples collected during increasing flow; solid circles are for samples collected during receding flow.

cannot be accounted for entirely by channel interception, since sulfate discharges during these storms exceed inputs from direct channel interception by two- to threefold. The most likely source of this sulfate in streamflow is water-soluble sulfate transported from the upper soil horizons as they become saturated during storms (Johnson and Henderson 1979; Johnson and Henderson Ch. 7, this volume). Johnson et al. (1982) estimated that 16 to 27% of the annual sulfate export from Walker Branch is due to transport of water-soluble sulfate from shallow subsurface soils, whereas only 7 to 8% is due to direct channel interception of sulfate. The episodic export of sulfate from these two flow paths combined, however, is estimated to account for 23 to 37% of the annual sulfate yield from Walker Branch, demonstrating the importance of stormflow to sulfate export from this, and perhaps other, watersheds for which a concentration response would not be expected, based on the measured sulfate adsorption capacity and hydraulic conductivity of the soils.

Potassium

In contrast to the concentration or dilution response of some solutes, potassium remains relatively constant during storms in late winter, summer, and spring. During fall and early winter, however, potassium concentrations increase markedly during the initial period of a storm and then rapidly decline to the prestorm base flow level (Fig. 8.10). During the fall-winter period, leaching of potassium from senescent leaves and litter in intermittent channels and along the unwetted stream margin appears to be responsible for the increased potassium concentrations in streamflow during the early part of storms. The pool of mobile potassium leached from leaves during storms appears to be limited, however, since the concentration in streamflow declines rapidly even though flow rate continues to increase.

Sodium

Sodium exhibits a slight dilution response (Fig. 8.9) or no response to increases in discharge. This reflects the low abundance of sodium in bedrock, soils, and vegetation on the watershed; the relatively high mobility within the watershed of sodium deposited in precipitation; and most likely the effect of evapotranspiration in concentrating sodium in soil and groundwater, which may be subsequently diluted during storms. The extent of dilution most likely depends on the fraction of stream discharge emanating from quick flow (e.g., rill flow in intermittent channels) and channel interception of throughfall. Sodium outputs from Walker Branch in streamflow closely approximate inputs from atmospheric deposition, with bedrock weathering accounting for only 5% of the annual sodium loss (Henderson et al. 1977). Sodium levels in streamflow thus appear to be controlled primarily by evaporative processes within the watershed, which concentrate sodium in soil solutions and groundwater that was previously deposited in wetfall.

Phosphorus

Soluble reactive phosphorus exhibits the same concentration pattern as sulfate during storms, increasing in concentrations during the ascending limb of the hydrograph and decreasing as flow rate declines (Fig. 8.8). This concentration response is probably a result of both leaching and wash-off of soluble phosphorus associated with soil and litter in the upper horizons and channel interception of throughfall containing an elevated concentration of phosphorus compared with that in stream water (Segars et al. 1985). Water-extractable phosphorus in the Tarklin and Linside soils on Walker Branch is highest in the A horizon and decreases in the B horizon (Johnson et al. 1981). Leaching or exchange of this water-soluble phosphorus as water moves laterally through the upper soil horizons during storms could thus account for the observed concentration response of phosphorus in stream water. In addition, leaching of phosphorus from litter in intermittent channels and along the unwetted stream channel would also contribute phosphorus to the stream during storms. Soluble reactive phosphorus (SRP) is leached from dried leaves, and fragmentation of the leaves can increase the amount leached out as much as threefold (Cowen and Lee 1973). Because the method for measuring SRP is not specific for inorganic orthophosphate, it is unclear what percentage of the increase in phosphorus concentration during storms is due to increased inputs of soluble organophosphorus that becomes hydroloyzed during the analysis and what percentage is due to increased inputs of inorganic orthophosphate.

Trace Metals

The pattern in trace metal concentrations during storms is illustrated by data for chromium and manganese, two elements that exhibit contrasting responses. The bulk concentration (dissolved plus particulate) of both trace metals increases sharply as discharge increases, with the peak concentration often preceding the maximum discharge rate during storms (Figs. 8.11, 8.12). This early peak in the bulk concentration of trace metals is most likely due to direct channel interception of precipitation and throughfall. The early peak in discharge and the ascending limb of the main storm hydrograph are typically characterized by high concentrations of suspended matter.

The occurrence of the highest particulate (bulk minus dissolved) metal concentration coincident with these high suspended matter concentrations is consistent with significant contributions of metals from suspended particles in stream water. Concentrations of Cd, Cu, Zn, Pb, Mn, Fe, and Cr associated with suspended matter in stream water from Walker Branch are significantly correlated with the organic matter content of suspended matter (Turner et al. 1977), suggesting that a sizable fraction of the suspended load of these trace metals is associated with particulate organic carbon.

In contrast to the increase in particulate concentration during the early part of storms, dissolved chromium exhibits a dilution effect (Fig. 8.11), whereas dissolved manganese exhibits a concentration response during storms (Fig. 8.12). The higher concentration of dissolved Mn and other trace metals, including Cu, Zn, and Fe, during early peaks in stormflow and during the rising limb of storm hydrographs appears to be related to early wash-off and leaching of soluble metals from watershed surfaces (e.g., canopy, streamside litter) (Turner et al. 1977). That the dissolved organic carbon (DOC) concentration in Walker Branch stream water also exhibits a concentration response during storms (Comiskey 1978) and that Mn, Zn, Cu, and Fe concentrations in stream water are positively correlated with DOC concentration (Turner et al. 1977) suggests that at least a fraction of the dissolved load of these metals is bound to soluble organic matter.

The importance of storms to the export of trace metals in dissolved and particulate phases from Walker Branch is illustrated in Table 8.3. The export of water, suspended matter, sulfate, and total and dissolved Cd, Pb, Zn, and Mn in streamflow over a 21-month period is compared with that exported during a single 2-d storm (with a recurrence interval of 2–3 years) within the same 21-month period. The storm accounted for 12% of the total water export and 64% of the total suspended matter export during the period. For all the trace metals except zinc, the fraction of total trace metal export during the storm was greater than that for water, a result of the fact that most of the trace metals in water are not in solution

Table 8.3. Export in streamflow of water, sulfate, suspended solids, and dissolved and total Cd, Pb, Zn, and Mn from Walker Branch Watershed during a 21-month period compared with the export during a large 2-d storm[a] that occurred during the same period

	21-month[b] export	2-d storm export	Percent export by storm
	——— 10^6 m^3 ———		
Water	1.098	0.128	12
	——— kg ———		
Sulfate	5.8 x 103	1.1 x 103	19
Suspended matter	36.9 x 103	23.8 x 103	64
	——— g ———		
Total cadmium	28	5.1	18
Dissolved cadmium	21	1.1	5.3
Total lead	1,050	740	66
Dissolved lead	204	25	12
Total manganese	100,000	72,200	72
Dissolved manganese	5,910	3,100	37
Total zinc	2,020	233	12
Dissolved zinc	550	49	9.1

[a]April 4–5, 1977; recurrence interval = 2–3 years; 18% of water yield for calendar year 1977.
[b]September 1975 to May 1977.

but rather are associated with suspended particles (see Table 8.2). Although zinc is also associated predominantly with particles in Walker Branch stream water (see Table 8.2) and the storm accounted for a greater fraction of the suspended solid export during the 21-month period than that for water, the fraction of total zinc export during this one storm was the same as that for water (Table 8.3). This difference between the export of total zinc and total export of the other three trace metals is apparently a result of significant dilution of dissolved zinc during storms. Dissolved cadmium also exhibits a dilution response, whereas the concentration of dissolved lead remains constant during storms. For dissolved trace metals that exhibit a concentration response (e.g., manganese) (see Fig. 8.12), the fraction of total export that occurs during storms is, as expected, greater than that for water. These results demonstrate the importance of storms in exporting materials from Walker Branch and emphasize the need to sample storms with relatively short recurrence intervals (i.e., <1 year to 2–3 years) in order to obtain accurate chemical and hydrologic losses in streamflow.

8.3.4 Seasonal Patterns in Stream Water Chemistry

The concentration of several solutes in stream water also exhibits distinct seasonal patterns, reflecting the effects of geochemical, hydrologic, and biological processes in the watershed. Elements exhibiting distinct seasonal patterns in concentration include calcium, magnesium, nitrate-nitrogen, and soluble reactive phosphorus; those exhibiting no seasonal pattern include potassium and chloride.

Calcium and Magnesium

Calcium and magnesium concentrations reach a maximum in the summer, when streamflow is lowest, and a minimum in winter and spring, when streamflow is highest (Fig. 8.14). This inverse relationship between the concentrations of calcium and magnesium and streamflow is probably due to a dilution-concentration mechanism related to the duration and extent to which the water is in contact with soil colloids and the bedrock (i.e., dolomite), which supply cations to drainage water through weathering and exchange reactions. In addition, the ability of groundwater to weather dolomite may vary seasonally because of differences in the concentration of weak acids (carbonic, humic) resulting from seasonal differences in biological activity in the litter and organic soil horizons.

Some springs on the watershed exhibit a seasonal pattern in calcium and magnesium concentrations similar to that found in the streams, and others exhibit no such pattern (Fig. 8.15). Springs exhibiting the seasonal pattern are probably fed by a conduit-flow aquifer in which water flows through relatively large solution openings in the bedrock, resulting in short residence times in the groundwater and minimal contact with the dolomite

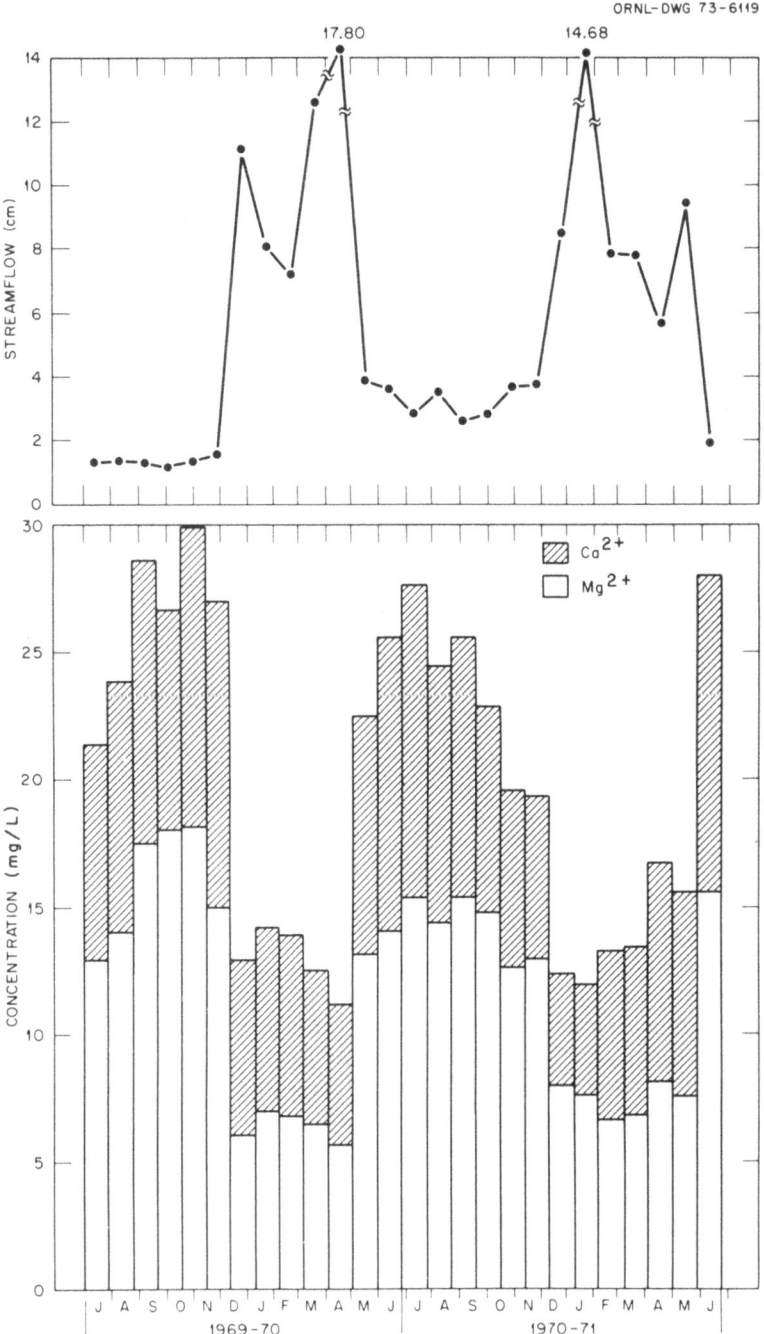

bedrock. The temperature of the spring on the East Fork also follows the expected seasonal pattern for a conduit aquifer, with the maximum temperature occurring in the summer and the minimum in the winter. In contrast, the major spring on the West Fork (S3W, Fig. 8.2) shows no seasonal pattern in calcium and magnesium concentrations (as indicated by the calcium plus magnesium hardness, Fig. 8.15) or in temperature. The lack of a seasonal pattern in this spring suggests that it is fed by a diffuse-flow aquifer in which the contact time of water with the dolomite is longer, resulting in a more constant hardness and temperature.

Seasonal differences in the capacity of drainage water to weather dolomite due to differences in carbonic acid concentrations are well documented (Langmuir 1971; Shuster and White 1971). However, quantitative data are lacking on the relative importance of the capacity of drainage water to weather base cations (as reflected by its pH) vs. its opportunity to weather (as reflected by the contact time of drainage water with weatherable and exchangeable pools of base cations) in causing seasonal variations. Shuster and White (1972) reported distinct seasonal patterns in the CO_2 partial pressure in water from calcite and dolomite springs in Pennsylvania, with the maximum occurring in summer and the minimum in winter. The higher levels of CO_2 in spring water in summer were presumably due to greater metabolic release of CO_2 in the soil from the decomposition of litter and soil organic matter.

Organic acids in the springs feeding Walker Branch have not been measured. Concentrations of dissolved organic carbon in the spring on the West Fork (S3W, Fig. 8.2), which exhibits no seasonal pattern in divalent base cations, are low (<1 mg/L DOC) and also exhibit no seasonal pattern (Comiskey 1978). This suggests that organic acids are not an important factor in bedrock weathering of base cations in this watershed. However, the concentrations of DOC in springs that exhibit seasonal patterns in base cation concentrations have not been determined.

Evidence of seasonal differences in the concentration of carbonic acid in groundwater, which can influence weathering rates, is inconclusive. Calculated CO_2 concentrations (from alkalinity and pH measurements) in the water from the major spring on the West Fork (S3W) and on the East Fork (S1E) exhibit no seasonal pattern. Because spring water can be supersaturated with CO_2, the lack of a seasonal pattern may be due to degasing of some of the excess CO_2 from the spring water samples before

◁————————————————————————————————————

Figure 8.14. Flow-weighted mean concentrations of calcium and magnesium and discharge over 4-week intervals in Walker Branch over 2 water years. Concentrations are volume-weighted means for the East Fork and West Fork combined, and discharge is for the East Fork and West Fork combined. Source: J.W. Elwood and G.S. Henderson. 1975. Hydrologic and chemical budgets at Oak Ridge, Tennessee. pp. 31–51. IN A.D. Hasler (ed.), Coupling of Land and Water Systems. Ecological Studies 10. Springer-Verlag, New York.

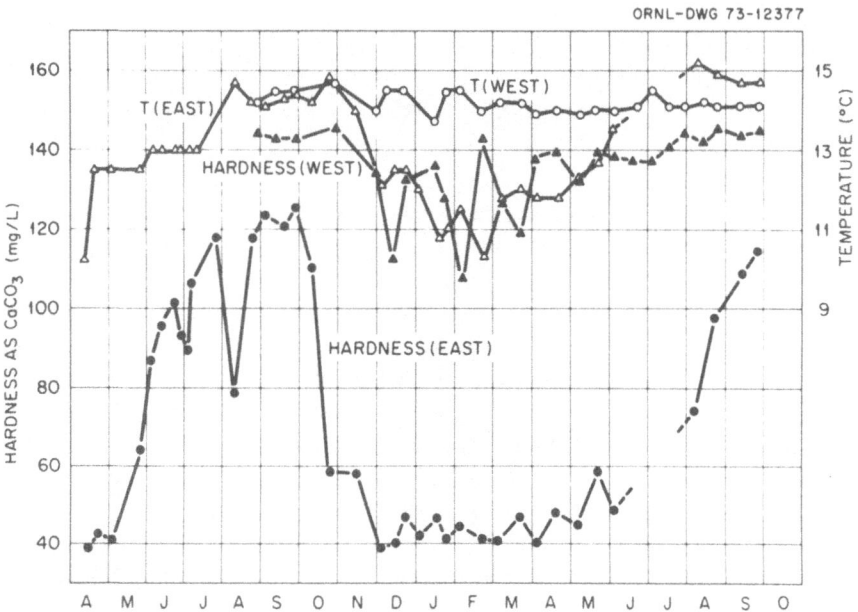

ORNL-DWG 73-12377

Figure 8.15. Seasonal variation in the calcium plus magnesium hardness (expressed as CaCO$_3$) and temperature in two springs on Walker Branch Watershed. The spring on the East Fork is S1E, and that on the West Fork is S3W (see Fig. 8.2).

measurement of pH. Direct measurements of pCO$_2$ and DOC in springs with seasonal patterns in base cation concentrations will be needed to determine the relative importance of weak inorganic and organic acids and hydrologic contact time in causing the seasonal pattern in the concentration of divalent base cations in the springs and streams draining Walker Branch Watershed.

Nitrate-Nitrogen and Phosphorus

The seasonal pattern of nitrate concentration in stream water is opposite to that for Ca^{2+} and Mg^{2+}, with the minimum occurring in summer and fall and the maximum in winter and early spring (Fig. 8.16). This seasonal pattern reflects the seasonal variation in both the uptake of nitrogen by vegetation and microorganisms associated with litter on the watershed and the increase in the transport of remineralized nitrogen through the soil to the stream channels because of less evapotranspiration during the late fall–winter period.

Similar seasonal patterns in nitrate concentrations in streams draining forested watersheds have been reported (e.g., Likens et al. 1970; Vitousek and Reiners 1975; Leonard et al. 1979). Vitousek and Reiners (1975) hy-

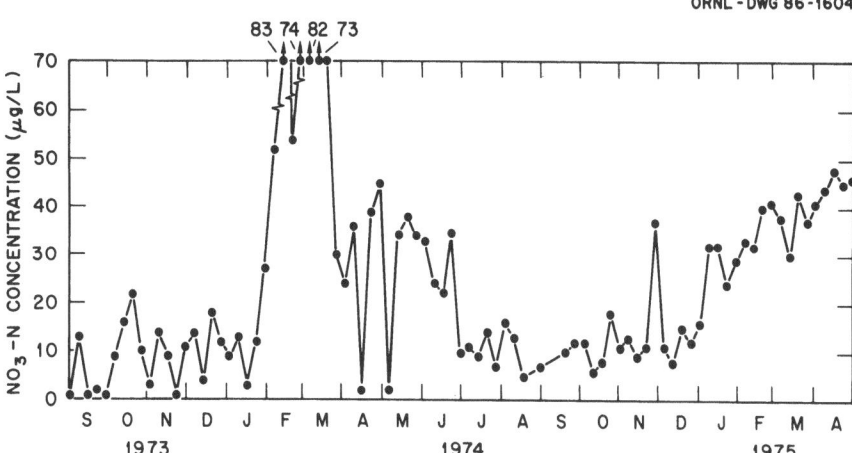

Figure 8.16. Seasonal variation in the concentration of nitrate-nitrogen in the West Fork of Walker Branch over ~2 water years.

pothesized that there would be pronounced seasonal cycles in the concentration of elements that are retained in watersheds primarily through uptake by organisms. They suggested that in an environment milder than that in New England, where soils are frozen during winter, the seasonal pattern in nitrate levels in drainage waters might be less marked or even absent because of continued root uptake of nitrate. For Walker Branch Watershed, the increase in nitrate concentration in stream water occurs in December-January (Fig. 8.14), later than in watersheds in New England (a seasonal pattern consistent with root uptake being the major mechanism of nitrate retention in these watersheds), with root uptake declining sooner in NewEngland than in the southeastern United States.

The seasonal pattern in SRP concentration varies spatially in the West Fork of Walker Branch, depending on the distance from major springs. Immediately downstream of S3W (see Fig. 8.1), for example, there is no distinct seasonal pattern in SRP levels, whereas 100 m farther downstream of this spring, SRP concentrations are highest during the summer and lowest during the fall-winter period (Fig. 8.17). The lack of a distinct seasonal pattern immediately below the spring appears to be a result of input of spring water containing a relatively high, constant concentration of SRP. Since SRP levels in soil water in the B horizon of Walker Branch Watershed are lower than in spring water (Segars et al. 1985), most of the SRP in spring water appears to result from weathering of phosphorus-bearing minerals (e.g., apatite) in the dolomite. Dillon and Kirchner (1975) found that the phosphorus loads in streams were significantly greater in sedimentary than in igneous watersheds of plutonic origin, whereas Omernik (1977) found no effect of geology on phosphorus loads in streams.

ORNL-DWG 86-9534

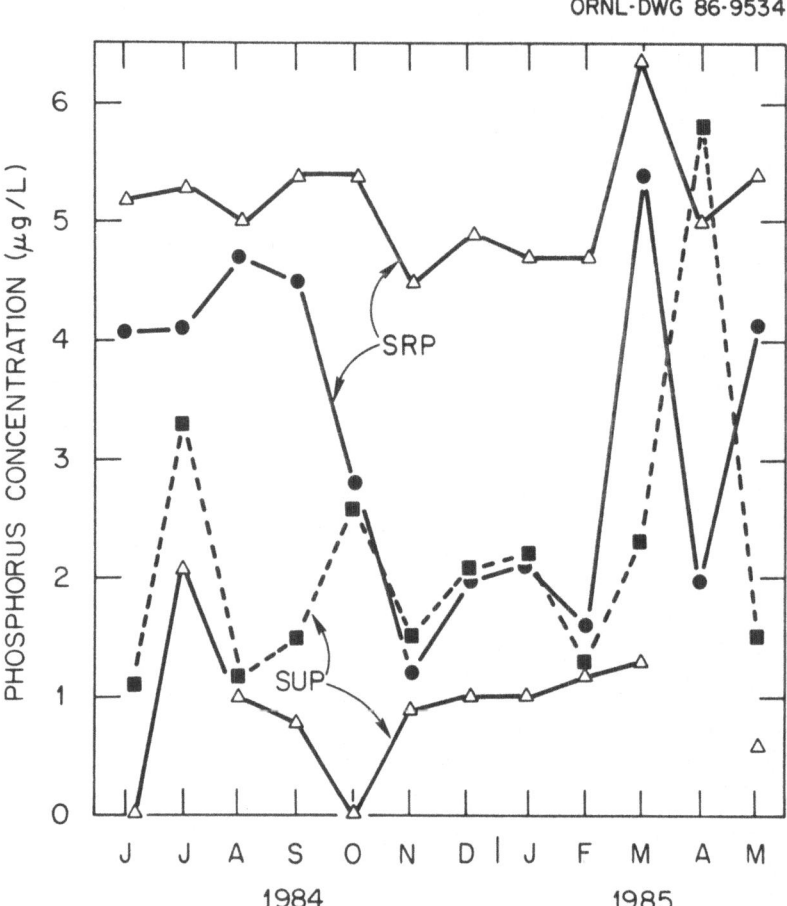

Figure 8.17. Seasonal pattern in the concentration of soluble reactive phosphorus (SRP) and soluble unreactive phosphorus (SUP) in spring water (triangles) and stream water (circles) in the West Fork of Walker Branch. Spring water samples were collected ~120 m downstream of spring S3W (data from Segars et al. 1985).

Unfortunately, comparative data on SRP concentrations in groundwater, soil water, and stream water from forested watersheds with contrasting geology are not available.

The occurrence of a seasonal pattern in SRP farther downstream of the spring on the West Fork is attributed to both a dilution-concentration effect associated with seasonal variation in the flow path of water and variation in the biological uptake of SRP within the stream channel. The dilution effect for SRP appears to be similar to that for calcium and magnesium, where base flow derived from groundwater that is richer in some solutes (calcium, magnesium, SRP) is diluted by interflow and surface flow, which have shorter residence times in the catchment and hence a shorter period in which solution reactions can take place.

Seasonal variation in the net uptake of SRP, or net conversion of SRP to soluble unreactive phosphorus (SUP) and/or particulate phosphorus, by organisms growing on substrata on the stream bottom would account for some of the observed seasonal variation in SRP levels in stream water at the downstream site. The uptake of orthophosphate (which is an unknown fraction of the SRP in Walker Branch) is maximum in the autumn, when the standing stock of detritus, consisting predominantly of leaves from riparian vegetation, is at its annual maximum owing to leaf fall (Mulholland et al. 1985). This input of detritus provides substrata for colonization by bacteria and algae, which remove orthophosphate from stream water. As these leaves decompose and/or are transported out of the stream during storms, the uptake of orthophosphorus declines. Despite having the highest mass-specific uptake rate of orthophosphorus from stream water, the aufwuchs community (algae and bacteria growing on rocks) in Walker Branch generally accounts for <10% of the total uptake due to the low standing stock of these organisms (Newbold et al. 1983a; Mulholland et al. 1985). Edwards (1973), however, reported that SRP and silicon levels in several rivers in England were lowest during the period of diatom blooms, indicating that algae can also contribute to the seasonal variation in SRP levels in stream water. The relative importance of dilution vs. biological uptake in explaining the seasonal variation in SRP concentrations in the West Fork of Walker Branch awaits further research.

Despite the potential contribution of bedrock weathering to phosphorus fluxes from Walker Branch Watershed, data on soluble phosphorus fluxes in throughfall, precipitation (wetfall), and streamflow show a net annual retention of phosphorus by this watershed of ~500 g/ha (Henderson et al. 1977). Because of the importance of SUP and particulate phosphorus to phosphorus fluxes in streams (e.g., Johnson et al. 1976; Omernik 1977; Hobbie and Likens 1973; Rigler 1979; Segars et al. 1985), the phosphorus loss is probably underestimated from measurement of SRP alone. In addition, the measured fluxes of phosphorus from Walker Branch do not include those during storms, when much of the annual flux of particulate phosphorus, and perhaps the SUP, occurs. Thus, the phosphorus fluxes from Walker Branch are most likely greater than the SRP fluxes alone; however, it seems unlikely that the losses of total phosphorus would exceed the inputs of total phosphorus, given the retention of soluble phosphorus by litter and soils in this watershed (Segars et al. 1985).

The tight conservation of phosphorus in Walker Branch is similar to that observed for northern hardwood forest watersheds at Hubbard Brook (Wood etal. 1984). Based on changes in the concentration of both SRP and SUP (assumed to be organic phosphorus) along a flow path of water through Walker Branch, the primary sites of phosphorus uptake and retention in this catchment appear to be the litter and mineral soil in the terrestrial environment and sediments in the aquatic environment (Fig. 8.18). The decrease in SRP and SUP between throughfall and the B horizon occurs primarily in the litter (Segars et al. 1985), suggesting that sorption of phosphorus by microorganisms associated with the litter is the primary cause of this decline. The low concentration of phosphorus in soil solutions in the mineral soil (B horizon) in Walker Branch is presumably a result of both efficient uptake and cycling of phosphorus by vegetation and microorganisms associated with soil organic matter and sorption of phosphorus by sesquioxides of iron and aluminum associated with soil particles.

The decrease in concentration of SRP between spring water and stream water reflects uptake of phosphorus in the stream channel. Laboratory measurements of phosphorus uptake, using fine-grained sediments and rocks from Walker Branch and $^{32}PO_4$, have shown that biotic sorption

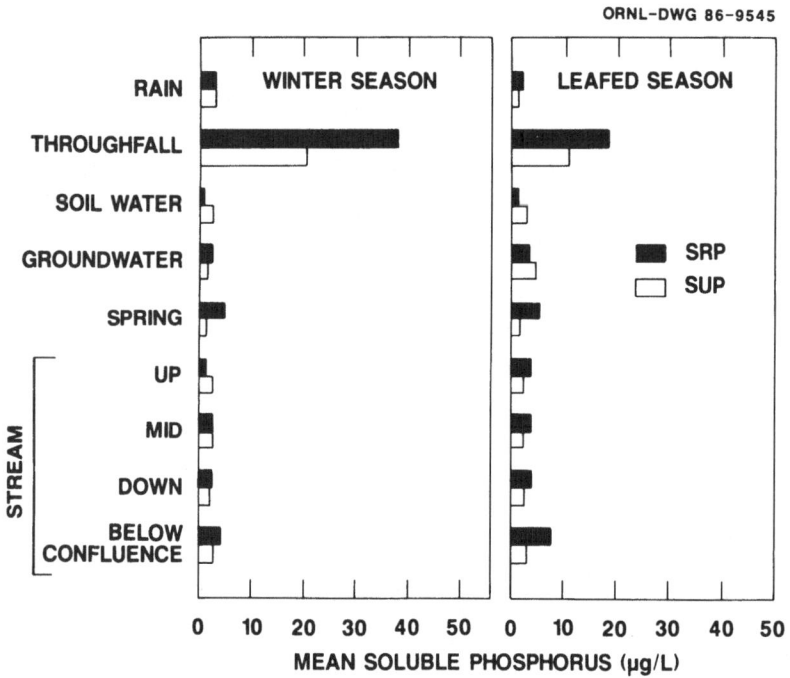

Figure 8.18. Annual arithmetic average concentration of soluble reactive phosphorus (SRP) and soluble unreactive phosphorus (SUP) at different locations along a hydrologic flow path from precipitation to soil water, groundwater, and stream water in the West Fork of Walker Branch Watershed. Spring water samples are from S3W (see Fig. 8.2). Source: Adapted from J.E. Segars, J.W. Elwood, and R.A. Minear. 1985. Chemical Characterization of Soluble Phosphorus Forms Along a Hydrologic Flowpath of a Forested Stream Ecosystem. ORNL/TM-9737. Oak Ridge National Laboratory, Oak Ridge, Tennessee.

accounts for most of the uptake of inorganic phosphorus from stream water (Elwood et al. 1981b; Mulholland et al. 1984). The importance of biotic uptake in regulating SRP levels in Walker Branch contrasts with that in streams at Hubbard Brook, where abiotic uptake appears to be the dominant process regulating SRP levels in stream water. Meyer (1979) concluded that microbial uptake played a relatively minor role in phosphorus sorption by inorganic sediments in Bear Brook, a stream draining one of the undisturbed watersheds at Hubbard Brook watershed in New Hampshire.

While the West Fork of Walker Branch appears to be in steady state with respect to the concentration of total soluble phosphorus (TSP), there is evidence that some of the SRP entering the stream in groundwater is converted to SUP in the stream channel. During the spring-summer period, for example, the SUP concentration increases from upstream to downstream, suggesting that SRP is being converted to organic phosphorus. In addition, laboratory experiments at Oak Ridge National Laboratory (ORNL) in which carrier-free $^{32}PO_4$ was added to batch reactors containing stream water and sediments from Walker Branch have verified that soluble orthophosphorus is converted to soluble organic phosphorus (SOP) in the stream channel. Identification of SOP is based on separation of the labeled inorganic and organic phosphorus by gel permeation chromatography. Rigler (1979) postulated that soluble phosphorus in groundwater accumulated on the bed of a Dartmoor stream during base flow periods but was resuspended and transported as particulate phosphorus during storms. Mass balances of total and soluble phosphorus fractions over reaches of Walker Branch have not been measured; thus, the importance of phosphorus retention and conversion in the stream channel to the measured fluxes of SRP from this watershed are unresolved. The cycling and transformations of phosphorus in the stream channel are discussed in more detail later in this chapter.

Organic Carbon Dynamics

Results of studies on low-order (first- to fourth-order) streams draining forested watersheds have shown that these systems are heterotrophic. Further, after reviewing the literature on stream productivity, Hynes (1963) concluded that "at least a very large part of the productivity of all running water is based on photosynthesis which takes place elsewhere." Research was conducted to examine the organic carbon dynamics in the West Fork of Walker Branch, with emphasis on (1) inputs of DOC and particulate organic carbon (POC) from the terrestrial to the aquatic system, (2) in situ primary productivity, (3) the hydrologic transport of DOC and POC, and (4) in-stream processing of particulate organic carbon by heterotrophic organisms.

Organic Carbon Inputs

Inputs of coarse particulate organic matter (>1 mm) to the West Fork of Walker Branch occur primarily as direct leaf fall and blow-in to the stream channel (Comiskey 1978). As expected, leaf fall inputs are highly seasonal (Fig. 8.19), with maximum inputs generally occurring in October and November. Annual total leaf fall for the total drainage approaches 400 g (dry weight)/m^2 (Comiskey et al. 1977), with direct leaf fall input of 345 g (dry weight)/m^2 to the reach of the West Fork that flows perennially.

Blow-in of detritus from the stream banks is heavily influenced by the slope and aspect of the surrounding terrain (Comiskey et al. 1977). The steepest southwest-facing slopes generally have the highest inputs, with peak inputs for leaves and twigs occurring during the winter-early spring period when winds are the strongest (Fig. 8.20). Total yearly input of organic detritus from blow-in to the stream channel is estimated to be 100 g/m^2, most of which (82%) is leaves. Of the total annual input of allochthonous leaf detritus to the West Fork of Walker Branch of 472 g/m^2, 21% is from blow-in and 79% is from direct leaf fall (Comiskey et al. 1977).

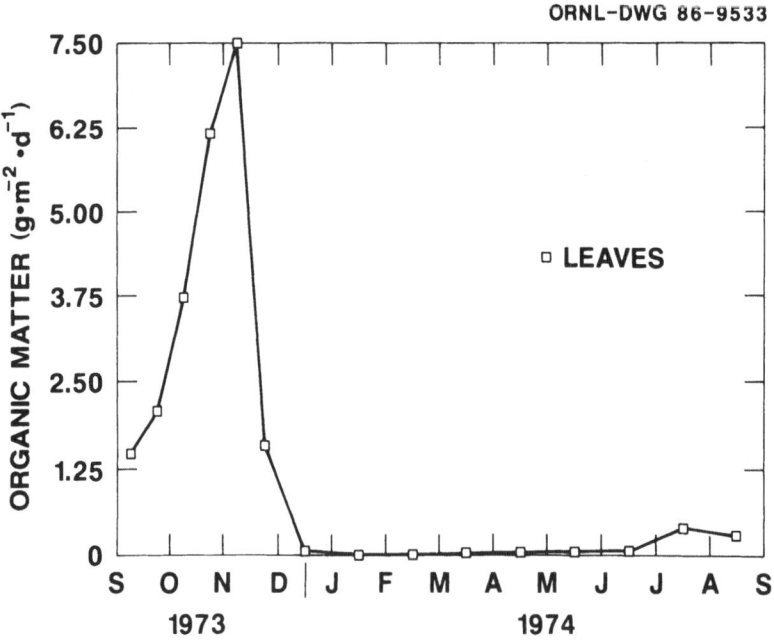

Figure 8.19. Mean daily input rates of leaf litter via litterfall to the West Fork of Walker Branch for the 1973–1974 water year. Source: C.E. Comiskey. 1978. Aspects of the organic carbon cycle on Walker Branch Watershed: A study of land/water interaction. Ph.D. Thesis, University of Tennessee, Knoxville.

ORNL-DWG 86-9629

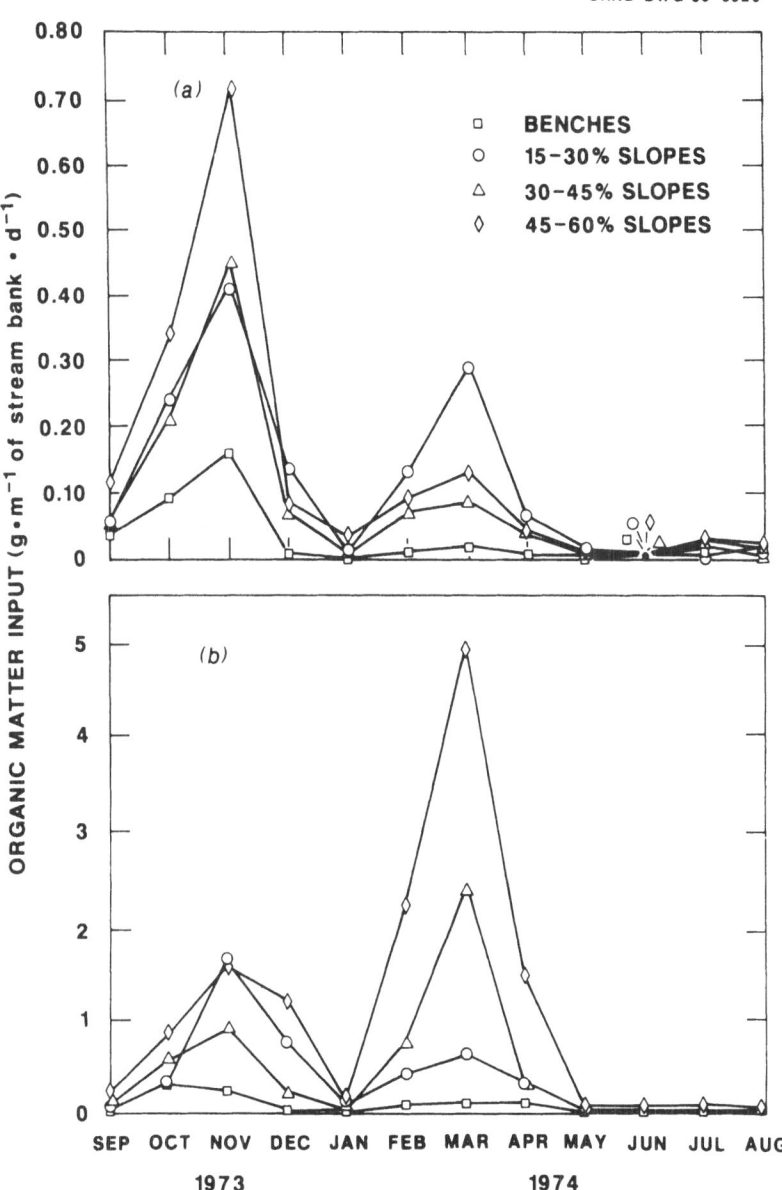

Figure 8.20. Seasonal variation in the blow-in of leaf detritus from (a) northeast-facing and (b) southwest-facing slopes to the West Fork of Walker Branch. Source: C.E. Comiskey. 1978. Aspects of the organic carbon cycle on Walker Branch Watershed: A study of land/water interaction. Ph.D. Thesis, University of Tennessee, Knoxville.

While direct leaf fall input to the stream channel on a unit area basis is relatively constant over the entire drainage area of Walker Branch Watershed, the blow-in varies, depending on the steepness and aspect of the adjoining slopes and the width of the stream channel. As a consequence, the loading rate of allochthonous detritus per unit area of stream channel is not uniform. In addition, blow-in of detritus to the stream channel continues after leaf fall has ended (Fig. 8.20). Blow-in thus represents a potentially important source of carbon and physical substrate for stream organisms, particularly following the winter-spring period, when scouring floods often remove much of the allochthonous detritus in the stream channel that entered as leaf fall during the autumn.

The standing stock of coarse (>1 mm) particulate organic matter (CPOM) in the stream channel exhibits significant seasonal and habitat variation, reflecting the influence of allochthonous inputs, hydrologic transport, and in-stream processing. The standing stock generally reaches a maximum in the autumn after peak leaf fall and a minimum in winter or summer, depending on whether there are scouring floods (which generally occur during winter and spring) that transport detritus from the continuously wetted portion of the stream channel (Figs. 8.21, 8.22).

This seasonal pattern in the standing stock of CPOM is influenced in part by benthic macroinvertebrates, which use the detritus as a food source, converting some of the CPOM to fine particulate organic matter (FPOM) and some to living biomass. The macroinvertebrate community in Walker Branch is dominated by shredders and scrapers. Measurements

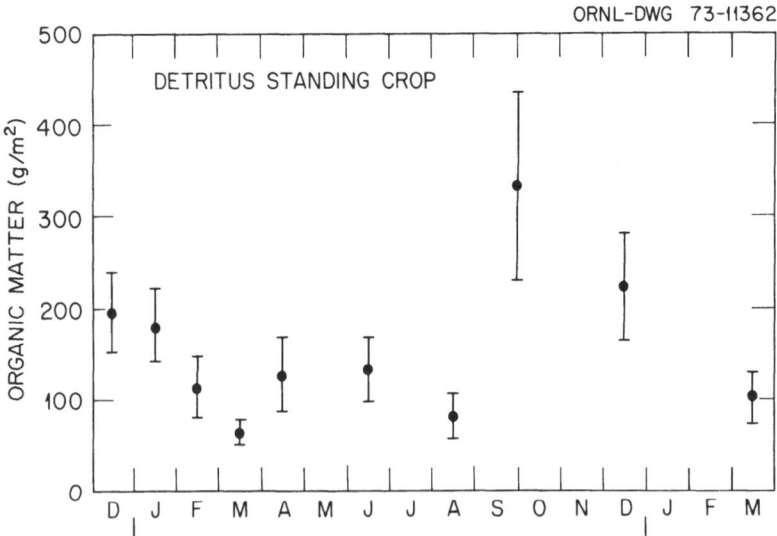

Figure 8.21. Seasonal variation over 2 years in the standing stock of coarse particulate organic matter in the West Fork of Walker Branch.

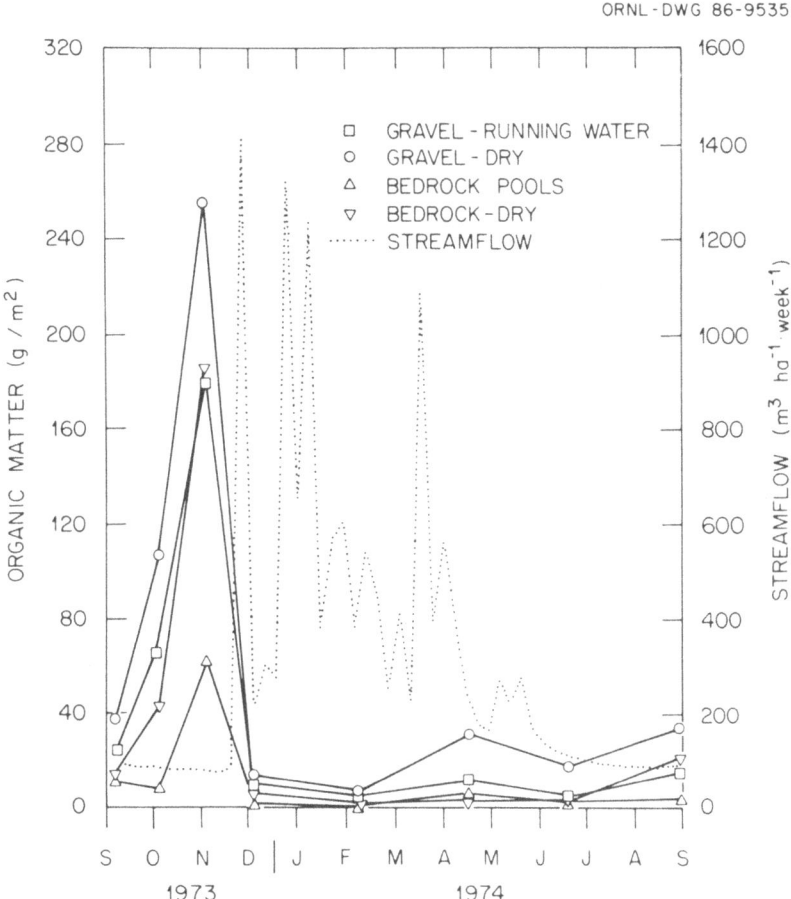

Figure 8.22. Seasonal variation in the standing stock of coarse particulate organic matter (CPOM) [in grams (ash-free dry mass) per square meter] in relation to stream discharge in the West Fork of Walker Branch. This figure illustrates the effect of large storms in reducing the standing stock of CPOM within the perennially wetted stream channel. Source: C.E. Comiskey. 1978. Aspects of the organic carbon cycle on Walker Branch Watershed: A study of land/water interaction. Ph.D. Thesis, University of Tennessee, Knoxville.

of the decomposition of red oak leaves in Walker Branch showed that microbial respiration accounted for only 33% of the weight loss in winter (Elwood et al. 1981a); identical measurements for leaves from less recalcitrant species (e.g., tulip poplar, dogwood) that enter Walker Branch have not been made. The remaining fraction of the weight loss (69%) was due to shredding by consumers (shredders and scrapers) and to physical fragmentation by the current, indicating that for species that decompose relatively slowly, fragmentation by biological and physical processes is more important than microbial oxidation to the loss of CPOM mass.

Because the available supply of leaf detritus in streams is dependent on seasonally pulsed inputs and on storms that export it from the stream channel, the life history patterns of macroinvertebrates that are dependent on allochthonous detritus are likely to be related to the standing stock of their available food supply. Figure 8.23 shows that the diversity of the macroinvertebrate community in Walker Branch exhibits a seasonal pattern similar to that of the standing stock of CPOM. Diversity decreases in the spring as a result of the emergence of several species of aquatic insects, including several species of shredders. Diversity remains low throughout the summer, a period of low standing stocks of detritus and hence low supplies of available food for obligate shredders. In the autumn, after leaf fall begins, the diversity of macroinvertebrates increases, corresponding to the increases in benthic organic matter. The similarity between the seasonal patterns in the standing stock of CPOM and in macroinvertebrate diversity does not prove that a causal relationship exists between these two variables. However, it is reasonable to expect that the life history patterns of aquatic insects dependent onallochthonous sources of CPOM would evolve so that hatching of detritus-feeding larvae would be somewhat synchronous with the seasonally pulsed inputs of detritus,

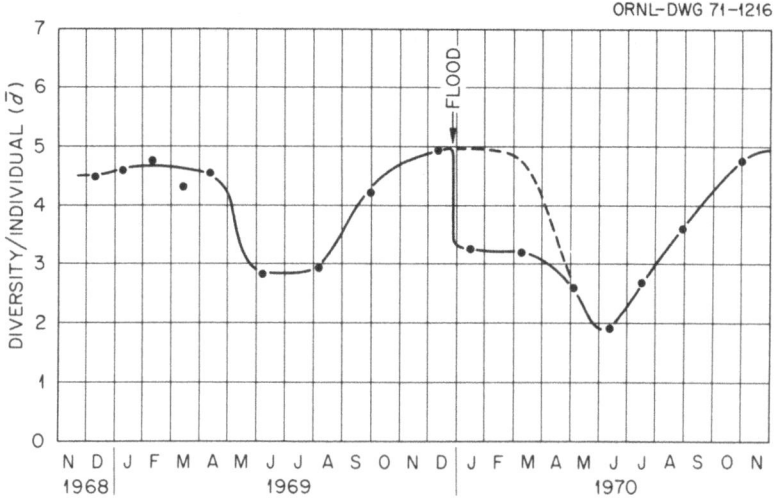

Figure 8.23. Seasonal variation in the diversity per individual ($\bar{d}$) of benthic macroinvertebrates in the West Fork of Walker Branch Watershed. Diversity per individual was computed using the index derived from information theory (Shannon and Weaver 1963). Values plotted are asymptotic diversity at each time, calculated according to Wilhm (1970). The dashed line is the hypothesized pattern without a flood, based on the pattern observed in the preceding year.

and emergence would occur before an extended period of minimum food supplies when the energy requirements of the larvae are at their annual maximum owing to the higher stream water temperatures during summer.

Peltoperla maria, a leaf-shredding stonefly, has a life cycle of ~2 years, with the eggs hatching in December and January and emergence occurring in April and May of the following year (Elwood and Cushman 1975). This species has an extended egg diapause such that the larvae avoid one potentially unfavorable period when food supplies are at a minimum and temperatures are maximum. Hynes (1970) suggested that the egg stage of many aquatic insects may allow the species to avoid unfavorable periods. Thus, for *Peltoperla maria*, each cohort endures only one food-poor period, rather than two, in its 2-year life cycle. Life history patterns of predatory species in Walker Branch indicate that higher trophic levels also have evolved to avoid periods of food scarcity (Cushman et al. 1977).

Dissolved organic carbon (DOC) is another source of organic matter for stream organisms in Walker Branch. The concentration DOC in the West Fork is generally <1 mg/L, except during storms, when the concentration increases with increasing discharge. The increase in DOC concentration during the rising limb of storm hydrographs (Fig. 8.24) is presumably due to direct channel interception by throughfall with a high concentration of DOC relative to stream water, and to leaching of DOC from litter as the stream expands into hydrologic source areas during

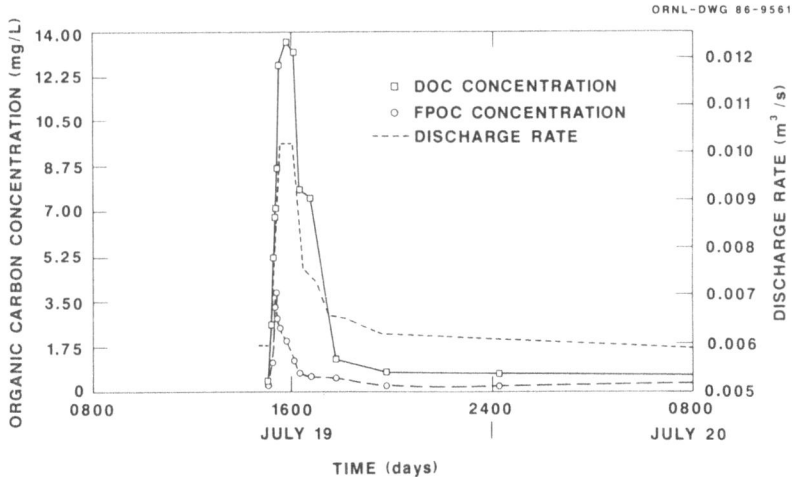

Figure 8.24. Concentration of dissolved organic carbon (DOC) and fine particulate organic carbon (FPOC) in the West Fork of Walker Branch vs. discharge rate during and after a storm on July 19–20, 1974. Source: C.E. Comiskey. 1978. Aspects of the organic carbon cycle on Walker Branch Watershed: A study of land/water interaction. Ph.D. Thesis, University of Tennessee, Knoxville.

storms. This DOC concentration effect is significant because of the importance of storms to DOC loading of the stream and of downstream systems. In addition, storms are important in the export from the watershed of nutrients (e.g., phosphorus and nitrogen) associated with soluble organic carbon. Comiskey (1978) estimated that a single storm that accounted for ~9% of the annual discharge from the West Fork of Walker Branch contributed ~31% of the annual output of DOC. This same storm accounted for 58% of the annual output of fine particulate organic carbon (FPOC) from the West Fork. While the weighted mean concentration of DOC in streamflow is approximately one-half of that for FPOC, DOC accounted for only 32.5% of the combined annual output of organic carbon in the two fractions, reflecting the increased importance of FPOC export during storms relative to base flow conditions.

Base flow concentrations of DOC in the West Fork are low, ranging from 0.1 to 0.4 mg/L during most of the year, with no apparent seasonal pattern (Comiskey 1978). Concentrations of DOC in stream water are comparable to those in spring water during base flow, suggesting either that DOC levels in the stream simply reflect groundwater inputs or that the stream is in steady state with respect to DOC (with inputs by exudation and leaching of DOC within the stream channel and groundwater inputs of DOC equal to uptake, adsorption, and oxidation of DOC). Only during the autumn, after leaf fall, are base flow levels of DOC >0.5 mg/L (Comiskey 1978). Since DOC concentrations in spring water during this period ranged from 0.1 to 0.3 mg/L, the autumn increase in DOC in stream water appears to be a result of in-stream processes probably associated with leaching of soluble organic matter from leaf detritus recently deposited in the stream channel. Thomas (1970) reported that freshly fallen leaves from maple and tulip poplar lost ~25% of their dry mass because of leaching within 24 h after entering Walker Branch.

Another source of organic matter in Walker Branch is autotrophic production by algae associated with solid substrates (rocks, gravel, leaf detritus) on the stream bottom. Indirect estimates of primary production in the West Fork of Walker Branch, using an isotope dilution method to measure biomass turnover rates, indicate that primary production rates are in the range of 16 to 22 mg (ash-free dry mass)$\cdot$m$^{-2}\cdot$d^{-1} (Elwood and Nelson 1972). Grazing rates on periphyton were comparable to the estimated production rates, suggesting that primary production rates in Walker Branch during certain periods may be controlled by grazers, which maintain a low biomass of algae on the stream bottom through their feeding activity. Attempts to eliminate grazers in order to examine the response of periphyton have so far proved unsuccessful. Laboratory stream experiments, using water with nutrient concentrations comparable to those in Walker Branch, indicate that the snails in Walker Branch can regulate primary production rates by regulating algal biomass on the stream bottom (Mulholland et al. 1983).

8.4 Nutrient Limitation and Phosphorus Spiraling

Unlike lentic systems where biomass can increase to the point of exhaustion of the available nutrient supply, streams receive a continuous supply of nutrients from their surrounding watershed. The concentration of many nutrients in water entering streams, however, is low owing to the conservative retention of nutrients by biotic and geochemical processes in the terrestrial system. The question arises whether the available supply of nutrients in streams is so low as to exert a primary limitation on autotrophic (i.e., algal) and heterotrophic (bacteria, fungi) biomass and productivity through a concentration limitation on uptake kinetics. To examine whether phosphorus was limiting algal biomass and leaf decomposition in Walker Branch, two reaches of the stream were continuously enriched for 95 d with orthophosphorus at two levels (60 and 450 μg P/L) over the upstream control level of 4 μg P/L (Elwood et al. 1981a). In the following year, this enrichment experiment was repeated, except that the stream was enriched with one level of ammonium-nitrogen (100 μg/L compared with 20 μg/L in the upstream control reach) in order to test for nitrogen limitation during the winter (Newbold et al. 1983b). Results of these enrichment experiments showed evidence of phosphorus limitation but no evidence of nitrogen limitation during the winter. The decomposition rate of red oak leaves (as reflected by the loss of mass), for example, was 24% greater in the phosphorus-enriched sections than in the control reach, whereas there was no significant difference in decomposition rates between the ammonium-enriched reach and the control. The nitrogen content of leaves in the phosphorus-enriched reaches was 60% greater than that of leaves in the control section, indicating that, at least during the winter, nitrogen immobilization by microbes associated with the leaves in Walker Branch is phosphorus-limited.

The respiration rate of microbes associated with the leaves was significantly higher than that for the control only at the high level of phosphorus enrichment, and microbial respiration accounted for only 10 and 34% of the increased mass loss at the low and high levels of enrichment, respectively. There was no significant difference in the respiration rate of microbes on leaves between the ammonium-enriched reach and the control reach, however (Newbold et al. 1983b). Results for the phosphorus enrichment suggest that the added phosphorus resulted in increased mechanical breakdown of the leaves through faster microbial conditioning, increases in macroinvertebrate feeding (shredding, scraping), or both. Densities of *Goniobasis clavaeformis*, a facultative grazing-shredding snail, were significantly higher in the phosphorus-enriched sections than in the control reach, supporting the hypothesis that feeding rates on leaf detritus were greater in the phosphorus-enriched reaches than in the control reach because of the greater number of consumers and the higher quality of the detritus as a food source (i.e., lower carbon:nitrogen ratio). In the nitrogen-enrichment experiment, the nitrogen concentrations of the leaves were not significantly different between the treatment and the control, indicating that microbes were unable to utilize the increased concentration of ammonium-nitrogen presumably because of phosphorus limitation.

Results for algal biomass, based on chlorophyll standing stock on glass slides, showed that the phosphorus-enriched reaches had a faster rate of increase in algal biomass initially, but after 4 weeks, there was no significant difference in the rate of biomass accumulation between the enriched and the control reaches. Algal biomass showed no significant response to the ammonium enrichment. Because of differences in the density of grazing invertebrates between the phosphorus-enriched and control reaches, the lack of a significant enrichment effect on algal biomass after 4 weeks may have been due to greater grazing rates in the phosphorus-enriched reaches resulting from significantly higher densities of snails. Previous work on grazing rates in Walker Branch demonstrated that snails (which dominate both the density and biomass of benthic macroinvertebrates in Walker Branch) can effectively control the primary productivity of benthic algae in the streams by controlling algal biomass through grazing (Elwood and Nelson 1972).

The finding of phosphorus limitation for leaf decomposition and primary production in Walker Branch suggested that phosphorus cycling has an important role in these streams and indicated a need for a better understanding of the controlling mechanisms, pathways, and rates. For streams where transport is an integral feature of the ecosystem, it is useful to address the question of nutrient cycling with an approach that combines the transport and cycling processes while recognizing that the two processes occur simultaneously in streams. The term "spiraling" has been used to describe the combined processes of nutrient cycling and downstream transport in a stream ecosystem. An index of spiraling, referred to as "spiraling length," was developed to quantitatively describe the cycling of nutrients in stream systems (Newbold et al. 1981; Elwood et al. 1983). Spiraling length is defined as the expected downstream distance traveled by a nutrient atom as it completes a cycle. A cycle, illustrated schematically in Fig. 8.25, consists of the passage of a nutrient atom from a dissolved inorganic form in the water, through various components of the ecosystem, and back to the water. The number of times a nutrient is expected to cycle in a reach of stream of length X is thus the ratio of the reach length to the spiraling length.

To analyze the spiraling of phosphorus in Walker Branch, $^{32}PO_4$ was released to a reach of the West Fork over 30-min periods. The uptake of ^{32}P from water was measured during the release, and the concentration of ^{32}P in coarse (>1 mm) particulate organic matter (CPOM), fine (<1 mm) particulate organic matter (FPOM), aufwuchs, grazers, shredders, collectors, net-spinning filter feeders, and predators was measured over

ORNL-DWG 80-11216 ESD

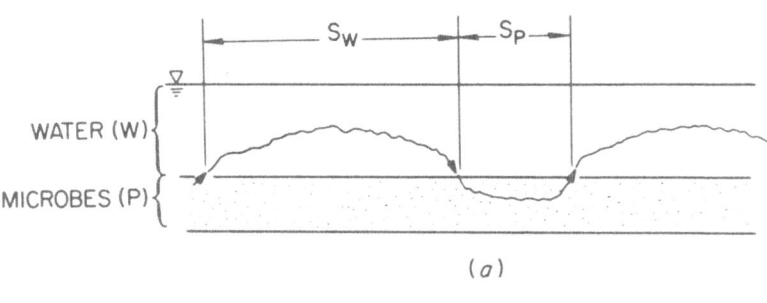

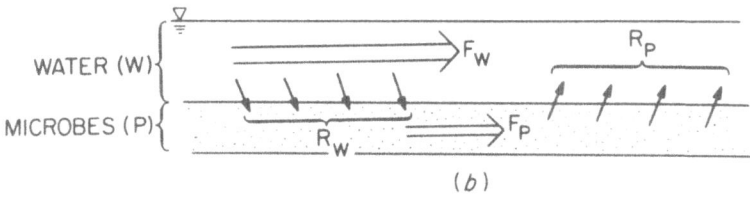

Figure 8.25. Conceptual diagram of (a) spiraling pathways in which nutrient exchanges are assumed to involve only the uptake and release of nutrients by microorganisms growing on substrata on the stream bottom and (b) the fluxes used to estimate the spiraling length of a nutrient from water: S_w is the uptake length of the nutrient from water, S_p is the turnover length of the nutrient associated with microbes, F_w is the downstream flux of the available nutrient supply in water, F_p is the downstream flux of nutrient associated with microbes, R_w is the uptake flux of the nutrient from water by microbes on the stream bottom, and R_p is the release flux of the nutrient from microbes back to water.

time for several weeks following the release. The tracer data were then modeled for a system of coupled partial equations and ordinary differential equations to obtain parameter estimates for phosphorus exchange among ecosystem compartments and for downstream transport.

In July, phosphorus was found to move downstream at an average velocity of 10.4 m/d, cycling once every 18.4 d. The spiraling length was thus 190 m (the product of the cycling time and the downstream transport velocity of phosphorus). While CPOM accounted for 60% of the uptake of orthophosphorus from water, the aufwuchs community was the metabolically most active component, as reflected by the highest concentration of ^{32}P following the release. However, the aufwuchs accounted for only 5% of the phosphorus uptake from water because of its low standing stock. FPOM accounted for the remaining 35% of the $^{32}PO_4$ uptake from stream water.

Of the phosphorus uptake by particulates on the stream bottom, 2.8% was transferred to consumers, and the remainder was recycled directly back to the water without being ingested by consumers. About 30% of the uptake by consumers was transferred to predators. The spiraling length for phosphorus was partitioned into (1) an uptake length of 165 m, defined as the expected travel distance of soluble inorganic phosphorus in stream water, (2) a particulate turnover length of 25 m, defined as the expected travel distance of exchangeable phosphorus associated with the transport of FPOM and CPOM, and (3) a consumer turnover length of 0.05 m, defined as the expected travel distance of phosphorus associated with drifting animals. In July, FPOM accounted for most (99%) of the particulate turnover length. The low consumer turnover length reflects the low uptake of phosphorus from particulates by consumers and the low drift velocity of invertebrates in Walker Branch.

Measures of phosphorus spiraling at other times of the year showed that the uptake length was shortest in November (22 m) and longest in August (97 m) (Mulholland et al. 1985). The standing stock of CPOM, which exhibits large seasonal variation due to seasonally pulsed inputs and storm transport, appeared to be the major determinant of the uptake length of orthophosphorus from stream water (Fig. 8.26). At all times of the year, the spiraling length of phosphorus in Walker Branch is dominated by the uptake length, with the turnover length ranging from 1 to 3 m (Mulholland et al. 1985). This reflects the low export of exchangeable phosphorus in the particulate phase and the rapid recycling of phosphorus back to water.

Considering that microbial decomposition of leaf detritus is phosphorus-limited during at least part of the year (Elwood et al. 1981a), it is surprising that the uptake length of phosphorus is the major component of the spiraling length at all times of the year. Newbold et al. (1982) suggested that in streams that were strongly limited by phosphorus, the turnover length would dominate the spiraling length because of the nutrient demand by phosphorus-limited bacteria and algae. These results for Walker Branch suggest that decomposition rates in this stream are not strongly limited by phosphorus and that detrital organic matter influences phosphorus spiraling more than phosphorous spiraling influences the microbial decomposition of organic matter. The extent to which nutrient uptake by organisms growing on substrates on a stream bottom can control the uptake length is, however, physically constrained to some degree by the extent of contact of soluble nutrients in the overlying water with the stream bottom. Thus, increasing the standing stock of CPOM may increase the surface area in contact with the soluble nutrient supply and thereby reduce the uptake length, whereas reducing the soluble concentration of a nutrient that is already in limiting supply may not increase the uptake length because of this physical constraint on the rate of nutrient removal from the overlying water being transported downstream.

ORNL-DWG 83-16217

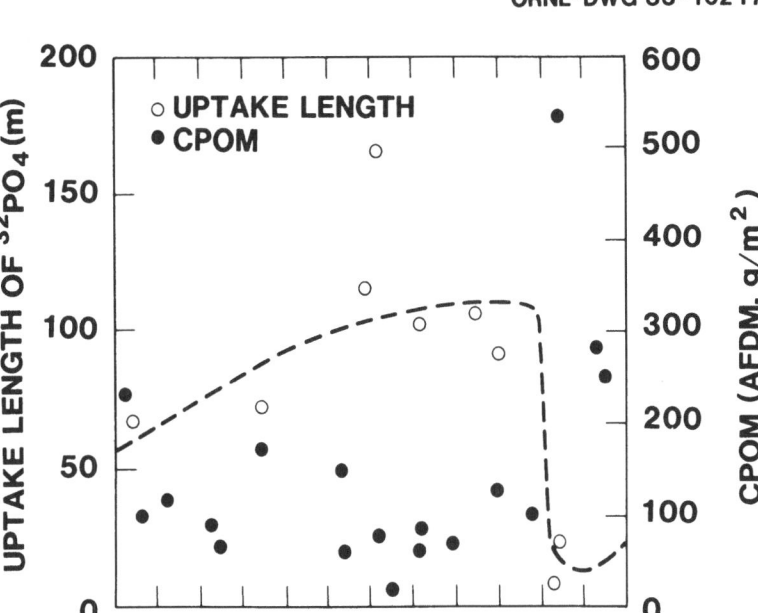

Figure 8.26. Seasonal variation in the uptake length of phosphorus and in the standing stock of coarse particulate organic matter (CPOM) [in grams of ash-free dry mass (AFDM) per square meter]. Source: P.J. Mulholland, J.D. Newbold, J.W. Elwood, L.A. Ferren, and J.R. Webster. 1985. Phosphorus spiraling in a woodland stream: Seasonal variations. Ecology 66:1012–1023.

8.5 Summary

The streams draining Walker Branch Watershed are viewed both as conduits that export nutrients and other products of weathering, erosion, and mineralization and as subsystems of the watershed with internal structure and function that influence the timing, form, and magnitude of material transport in stream water. Calcium and magnesium are the dominant cations, and bicarbonate is the dominant anion in streamflow, reflecting the influence of weathering of the dolomite bedrock underlying Walker Branch Watershed. The spatial and temporal (seasonal, episodic) variations in the concentration of these ions are attributed to differences in the proportion of streamflow emanating from groundwater that has been in contact with the dolomite.

Soluble calcium and magnesium concentrations exhibit a dilution response to increasing flow rate; sulfate, SRP, and total nitrogen (ammonium-nitrogen plus organic nitrogen) exhibit a concentration response to changes in flow rate during storms, whereas sodium shows little or no response. Potassium concentration increases with flow rate during winter but remains relatively constant during the remainder of the year, reflecting the influence of biotic processes on potassium retention and transport in Walker Branch. Seasonal variations in nitrate and SRP concentrations appear to be due, respectively, to variation in biological uptake by terrestrial vegetation and to biotic uptake by microorganisms associated with stream sediments.

Concentrations of trace metals (Cd, Cu, Cr, Pb, Zn, Mn, and Fe) in stream water are very low, consistent with the relatively undisturbed nature of Walker Branch Watershed and the low solubility of solid phases of these metals at alkaline pH's typical of the stream water. The sources of temporal variation in the dissolved metal concentrations are not completely known, but stream discharge and related hydrologic variables appear to be important factors. Dissolved manganese and zinc generally exhibit a concentration response to increasing flow rate, whereas chromium and lead are diluted with increasing flow rate. Sharp increases in the concentrations of dissolved Mn, Zn, Cu, and Fe during early increases in stormflow rate appear to result from the flushing of metals from watershed surfaces (forest canopy, litter). Because concentrations of these same dissolved metals are positively correlated with dissolved organic carbon, it appears that a significant fraction of the load of these dissolved metals in stream water is associated with soluble organic matter.

The concentrations of metal associated with the suspended matter (seston) show a significant positive correlation with the percent of organic matter in the seston, suggesting that a sizable fraction of the suspended trace metal load in streamflow is associated with particulate organic matter. Hydrous iron–manganese oxide coatings on stream sediments also appear to be important carriers of some trace metals.

Particulate organic matter enters the streams from the surrounding watershed via direct leaf fall and blow-in from the stream banks, and soluble organic matter enters the streams via groundwater and soil water. Inputs of leaf detritus are seasonal, with 79% occurring as direct leaf fall, primarily during the autumn, and 21% as blow-in. Loss of detritus mass in the stream is influenced by microbial activity, macroinvertebrate feeding, physical fragmentation, and leaching. Microbial respiration was shown to account for only 33% of the mass loss during winter, indicating that invertebrate consumption and physical fragmentation are collectively more important than microbial oxidation.

The diversity of species of benthic macroinvertebrates exhibits a seasonal pattern similar to that of the standing stock of CPOM in the stream bottom. This suggests that the life history patterns of aquatic insects in woodland streams such as Walker Branch have evolved such that the feeding stages (larvae) of species dependent on allochthonous detritus are not present in the stream during periods of minimum food abundance.

Life history patterns of some predatory species also reflect this seasonal pattern, suggesting that the hatching and emergence of higher trophic levels have evolved to avoid periods of low food supply.

Nutrient enrichment experiments with orthophosphate and ammonium-nitrogen demonstrated that the decomposition rate of detritus in Walker Branch during winter is phosphorus-limited but not nitrogen-limited, a result consistent with the high nitrogen:phosphorus ratio in the stream water. The nitrogen content of detritus in phosphorus-enriched sections was greater than in the unenriched control section, indicating that nitrogen immobilization by microbes associated with detritus in Walker Branch is phosphorus-limited.

Algal biomass (based on chlorophyll standing stock) showed a positive response to the phosphorus enrichment initially, but no difference between the enriched reaches and the upstream control was evident after 4 weeks. However, the higher density of grazers in the enriched reaches may have obscured the differences in algal biomass between the enriched and control reaches resulting from phosphorus enrichment.

Studies of the spatially dependent cycling (referred to as spiraling) of phosphorus in Walker Branch showed a seasonal pattern in the spiraling length, or cycling distance, defined as the distance required for a nutrient atom to pass through the cycle once. The minimum spiraling length (23 m) occurs in the autumn, and the maximum spiraling length (99–190 m) occurs in the summer. The major determinant of the spiraling length of phosphorus appears to be the standing stock of CPOM. At all times of the year, the expected transport distance of soluble orthophosphate in water (defined as the uptake length of a nutrient) accounts for most of the spiraling length of phosphorus, whereas the turnover length of phosphorus is low, ranging from 1 to 3 m. This pattern is consistent with that predicted for streams that are moderately phosphorus-limited, suggesting that the productivity of algae and bacteria in Walker Branch is never strongly limited by the supply of available phosphorus.

References

Cerling, T.E., and R.R. Turner. 1982. Formation of freshwater Fe-Mn coatings on gravel and the behavior of ^{60}Co, ^{90}Sr, and ^{137}Cs in a small watershed. Geochim. Cosmochim. Acta 46:1333–1344.

Comiskey, C.E. 1978. Aspects of the organic carbon cycle on Walker Branch Watershed: A study of land/water interaction. Ph.D. Thesis, University of Tennessee, Knoxville.

Comiskey, C.E., G.S. Henderson, R.H. Gardner, and F.W. Woods. 1977. Patterns of organic matter transport on Walker Branch Watershed. pp. 439–467. IN D.L. Correll (ed.), Watershed Research in Eastern North America: A Workshop to Compare Results. Chesapeake Bay Center for Environmental Studies, Smithsonian Institution, Edgewater, Maryland.

Cowen, W.F., and G.F. Lee. 1973. Leaves as source of phosphorus. Environ. Sci. Technol. 7:853–854.

Curlin, J.W., and D.J. Nelson. 1968. Walker Branch Watershed Project: Objectives, Facilities, and Ecological Characteristics. ORNL/TM-2271. Oak Ridge National Laboratory, Oak Ridge, Tennessee.

Cushman, R.M., J.W. Elwood, and S.G. Hildebrand. 1977. Life history and production dynamics of *Alloperla mediana* and *Diplectrona modesta* in Walker Branch, Tennessee. Am. Midl. Nat. 98:354–364.

Dillon, P.J., and W.B. Kirchner. 1975. The effects of geology and land use on the export of phosphorus from watersheds. Water Res. 9:135–148.

Edwards, A.M.C. 1973. The variation of dissolved constituents with discharge in some Norfolk rivers. J. Hydrol. 18:219–242.

Elwood, J.W., and R.M. Cushman. 1975. The life history and ecology of *Peltoperla maria* (Plecoptera: Peloperlidae) in a small spring-fed stream. Verh. Int. Verein. Limnol. 19:3050–3056.

Elwood, J.W., and G.S. Henderson. 1975. Hydrologic and chemical budgets at Oak Ridge, Tennessee. pp. 31–51. IN A.D. Hasler (ed.), Coupling of Land and Water Systems. Ecological Studies 10. Springer-Verlag, New York.

Elwood, J.W., and D.J. Nelson. 1972. Periphyton production and grazing rates in a stream measured with a ^{32}P material balance method. Oikos 23:295–303.

Elwood, J.W., J.D. Newbold, A.F. Trimble, and R.W. Sark. 1981a. The limiting role of phosphorus in a woodland stream ecosystem: Effects of P enrichment on leaf decomposition and primary producers. Ecology 62:146–158.

Elwood, J.W., J.D. Newbold, R.V. O'Neill, R.W. Stark, and P.T. Singley. 1981b. The role of microbes associated with organic and inorganic substrates in phosphorus spiralling in a woodland stream. Verh. Int. Verein. Limnol. 21:850–856.

Elwood, J.W., J.D. Newbold, R.V. O'Neill, and W. Van Winkle. 1983. Resource spiralling: An operational paradigm for analyzing lotic ecosystems. pp. 3–27. IN T.D. Fontaine III and S.M. Bartell (eds.), Dynamics of Lotic Ecosystems. Ann Arbor Science, Ann Arbor, Michigan.

Henderson, G.S., and W.F. Harris. 1975. An ecosystem approach to characterization of the nitrogen cycle in a deciduous forest watershed. pp. 179–193. IN B. Bernier and C.H. Winget (eds.), Forest Soils and Forest Land Management. Les Presses de l'Université Laval, Quebec, Canada.

Henderson, G.S., A. Hunley, and W. Selvidge. 1977. Nutrient discharge from Walker Branch Watershed. pp. 307–320. IN D.L. Correll (ed.), Watershed Research in Eastern North America: A Workshop to Compare Results. Chesapeake Bay Center for Environmental Studies, Smithsonian Institution, Edgewater, Maryland.

Hobbie, J.E., and G.E. Likens. 1973. Output of phosphorus, dissolved organic carbon, and fine particulate carbon from Hubbard Brook watersheds. Limnol. Oceanogr. 18:734–742.

Hynes, H.B.N. 1963. Imported organic matter and secondary production in streams. Proc. Int. Congr. Zool. 16:324–329.

Hynes, H.B.N. 1970. The Ecology of Running Waters. University of Toronto Press, Toronto, Ontario, Canada.

Johnson, A.H., Jr., R. Bouldin, E.A. Goyette, and A.M. Hedges. 1976. Phosphorus loss by stream transport from a rural watershed: Quantities, processes, and sources. J. Environ. Qual. 5:148–157.

Johnson, D.W., and G.S. Henderson. 1979. Sulfate adsorption and sulfur fractions in a highly weathered soil under a mixed deciduous forest. Soil Sci. 128:34–40.

Johnson, D.W., D.W. Cole, F.W. Horng, H. Van Miegroet, and D.E. Todd. 1981. Chemical Characteristics of Two Forested Ultisols and Two Forested Inceptisols Relevant to Anion Production and Mobility. ORNL/TM-7646. Oak Ridge National Laboratory, Oak Ridge, Tennessee.

Johnson, D.W., G.S. Henderson, D.D. Huff, S.E. Lindberg, D.D. Richter, D.S. Shriner, D.E. Todd, and J. Turner. 1982. Cycling of organic and inorganic sulfur in a chestnut oak forest. Oecologia 54:141–148.

Johnson, D.W., D.D. Richter, H. Van Miegroet, D.W. Cole, and J.M. Kelly. 1986. Sulfur cycling in five forest ecosystems. Water Air Soil Pollut. 30:965–979.

Langmuir, D. 1971. The geochemistry of some carbonate ground water in central Pennsylvania. Geochim. Cosmochim. Acta 35:1023–1045.

Leonard, R.L., L.A. Kaplan, J.F. Elder, R.N. Coats, and C.R. Goldman. 1979. Nutrient transport in surface runoff from a subalpine watershed, Lake Tahoe Basin, California. Ecol. Monogr. 49:281–310.

Likens, G.E., F.H. Bormann, N.M. Johnson, D.W. Fisher, and R.S. Pierce. 1970. Effects of forest cutting and herbicide treatment on nutrient budgets in the Hubbard Brook watershed ecosystem. Ecol. Monogr. 40:23–47.

McSwain, M.R., and R.J. Watrous. 1974. Improved methylthymol blue procedure for automated sulfate determinations. Anal. Chem. 46:1329–1331.

Messer, J.J., C.W. Ariss, J.R. Baker, S.K. Drouse, K.N. Eshleman, P.R. Kaufmann, R.R. Linthurst, J.M. Omernik, W.S. Overton, M.J. Sale, R.D. Schonbrod, S.M. Stambaugh, and J.R. Tuschall, Jr. 1987. National Surface Water Survey: National Stream Survey, Phrase I—Pilot Survey. EPA/600/4-87-026. U.S. Environmental Protection Agency, Environmental Research Laboratory, Corvallis, Oregon.

Meyer, J.L. 1979. The role of sediments and bryophytes in phosphorus dynamics in a headwater stream ecosystem. Limnol. Oceanogr. 24:365–375.

Mulholland, P.J., J.D. Newbold, J.W. Elwood, and C.L. Hom. 1983. The effect of grazing intensity on phosphorus spiralling in autotrophic streams. Oecologia 58:358–366.

Mulholland, P.J., J.W. Elwood, J.D. Newbold, J.R. Webster, L.A. Ferren, and R.E. Perkins. 1984. Phosphorus uptake by decomposing leaf detritus: Effect of microbial biomass and activity. Verh. Int. Verein. Limnol. 22:1899–1905.

Mulholland, P.J., J.D. Newbold, J.W. Elwood, L.A. Ferren, and J.R. Webster. 1985. Phosphorus spiralling in a woodland stream: Seasonal variations. Ecology 66:1012–1023.

Murphy, J., and J.P. Riley. 1962. A modified single solution method for the determination of phosphate in natural waters. Anal. Chim. Acta 27:31–36.

Menzel, D.W., and N. Corwin. 1965. The measurement of total phosphorus in sea water based on the liberation of organically bound fractions by persulfate oxidation. Limnol. Oceanogr. 10:280–282.

Newbold, J.D., J.W. Elwood, R.V. O'Neill, and W. Van Winkle. 1981. Measuring nutrient spiralling in streams. Can. J. Fish. Aquat. Sci. 38:860–863.

Newbold, J.D., R.V. O'Neill, J.W. Elwood, and W. Van Winkle. 1982. Nutrient spiralling in streams: Implications for nutrient limitation and invertebrate activity. Am. Nat. 120:628–652.

Newbold, J.D., J.W. Elwood, R.V. O'Neill, and A.L. Sheldon. 1983a. Phosphorus dynamics in a woodland stream: A study of nutrient spiralling. Ecology 64:1249–1265.

Newbold, J.D., J.W. Elwood, M.S. Schulze, R.W. Stark, and J.C. Barmeier. 1983b. Continuous ammonium enrichment of a woodland stream: Uptake kinetics, leaf decomposition, and nitrification. Freshwater Biol. 13:193–204.

Omernik, J.M. 1977. Non-Point Source-Stream Nutrient Level Relationships: A Nationwide Survey. EPA-600/3-77-105. U.S. Environmental Protection Agency, Environmental Research Laboratory, Corvallis, Oregon.

Peters, L.N., D.F. Grigal, J.W. Curlin, and W.J. Selvidge. 1970. Walker Branch Watershed Project: Chemical, Physical, and Morphological Properties of the Soils of Walker Branch Watershed. ORNL/TM-2968. Oak Ridge National Laboratory, Oak Ridge, Tennessee.

Rigler, F.H. 1979. The export of phosphorus from Dartmoor catchments: A model to explain variations of phosphorus concentrations in streamwater. J. Mar. Biol. Assoc. U.K. 59:659–687.

Segars, J.E., J.W. Elwood, and R.A. Minear. 1985. Chemical Characterization of Soluble Phosphorus Forms Along a Hydrologic Flowpath of a Forested Stream Ecosystem. ORNL/TM-9737. Oak Ridge National Laboratory, Oak Ridge, Tennessee.

Shannon, C.E., and W. Weaver. 1963. The Mathematical Theory of Communication. University of Illinois Press, Urbana.

Shuster, E.T., and W.B. White. 1971. Seasonal fluctuations in the chemistry of limestone springs: A possible means for characterizing carbonate aquifers. J. Hydrol. 14:93–128.

Shuster, E.T., and W.B. White. 1972. Source areas and climatic effects in carbonate groundwaters determined by saturation indices and carbon dioxide pressures. Water Resour. Res. 8:1067–1073.

Stainton, M.P. 1980. Errors in molybdenum blue methods for determining orthophosphate in freshwater. Can. J. Fish. Aquat. Sci. 37:472–478.

Tarapchak, S.J. 1983. Soluble reactive phosphorus measurements in lake water: Evidence for molybdate-enhanced hydrolysis. J. Environ. Qual. 12:105–108.

Technicon Industrial Systems. 1971. Technicon Autoanalyzer Methodology, Methods 108-71W, 100-70W, 93-70W. Technicon Corporation, Tarrytown, New York.

Thomas, W. A. 1970. Weight and calcium losses from decomposing tree leaves on land and in water. J. Appl. Ecol. 7:237–241.

Turner, R.R., S.E. Lindberg, and K. Talbot. 1977. Dynamics of trace element export from a deciduous watershed, Walker Branch, Tennessee. p. 661–679. IN D.L. Correll (ed.), Watershed Research in Eastern North America: A Workshop to Compare Results. Chesapeake Bay Center for Environmental Studies, Smithsonian Institution, Edgewater, Maryland.

Vitousek, R.M., and W.A. Reiners. 1975. Ecosystem succession and nutrient retention: A hypothesis. BioScience 25:376–381.

Wilhm, J.L. 1970. Effects of sample size on Shannon's formula. Southwest. Nat. 14:441–445.

Wood, T., F.H. Bormann, and G.K. Voigt. 1984. Phosphorus cycling in a northern hardwood forest: Biological and chemical control. Science 223:391–393.

Chapter 9
Modeling Chemical Transport, Uptake, and Effects in the Soil-Plant-Litter System

R.J. Luxmoore

9.1 Introduction

The processes of chemical transport, uptake, and effects in forested land-scapes interact with the water and carbon (growth, decomposition) dynamics of the soil-plant-litter-atmosphere system. The development of a unified mechanistic approach to modeling the relationships of water, carbon, and chemicals (nutrients, pollutants) in forests was undertaken as a means of evaluating chronic pollutant impacts on terrestrial environments and the associated streams. Low loading rates of gaseous pollutants, trace metals, and acidifying chemicals (SO_4^{2-}, NO_3^-) could have unexpected and subtle effects in an ecosystem over the long term that may not be experimentally apparent in the near term. Simulation modeling offers a reasoned approach to the evaluation of mechanisms of system response to impacts from diverse sources. The broad scope of the research on Walker Branch Watershed has provided a strong incentive for the development of mechanistic simulation models that can be used for synthesis of diverse sources of data and as a means of evaluating the significance of the physical, chemical, and biological processes operating in forested landscapes.

Historically, the unified modeling approach was initiated and a first-version model was developed during the 1972–1976 period in a project directed toward quantifying trace metal transport in watersheds including Walker Branch (Baes et al. 1976). A hydrologic transport code (see Ch. 5) formed the bookkeeping framework for coupling codes for growth and chemical transport in vegetated landscapes. An underlying concept in the approach was to represent flow processes with concentration-dependent gradient equations in which empirical pathway characteristics (e.g., con-

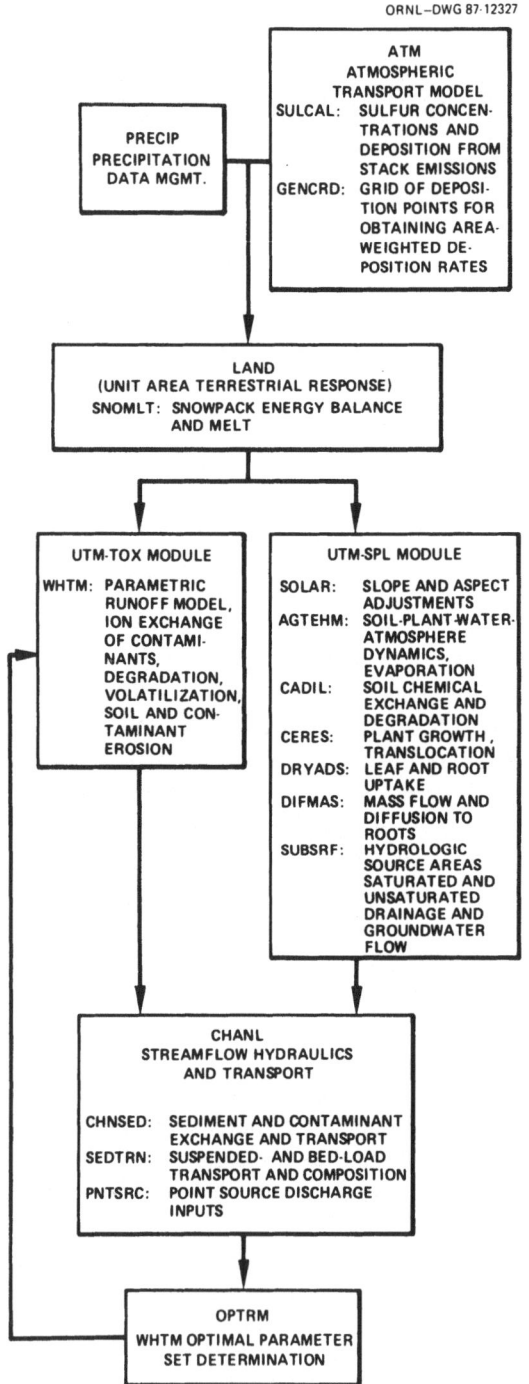

Figure 9.1. The submodels that link to form the two versions of the Unified Transport Model (UTM-TOX, UTM-SPL).

ductivities) could be readily changed to represent alternative landscape units of interest. The hydrologic model operated on time steps of 15 min for precipitation periods and 1h for other periods. The coupled models for carbon and chemical processes also operated with these time steps. All of the meteorological data (e.g., irradiance, temperature) necessary for simulation of coupled processes was conveniently available from the hydrologic code.

9.2 Landscape Context

The unified approach to transport processes includes the coupling of atmospheric, terrestrial, and aquatic processes, and these mechanisms are represented by subprograms of the Unified Transport Model (UTM) (Baes et al. 1976). The PRECIP module reads rainfall and meteorological data and provides the appropriate values at each time period of the simulation to the other modules. The Atmospheric Transport Model (ATM) simulates the movement of chemicals from a point (smoke stack), line (road, railroad), or area (landfill, pond), using a Gaussian plume model, and estimates deposition on a selected area (watershed). Chemical transport in the terrestrial system can be represented with either a parametric approach (UTM-TOX) or the coupled-processes routines (soil-plant-litter) of UTM-SPL. The CHANL module represents streamflow transport processes (Fig. 9.1). The surface and subsurface flows from each land segment are assigned to a particular stream section called a reach, and the cumulative flows from all reaches are used to generate the stream hydrograph (Fig. 9.2).

In the next section, the coupled-processes models for transport processes in the soil-plant-litter system are described.

9.3 Modeling the Soil-Plant-Litter System

The physical, chemical, and physiological mechanisms of water, carbon, and chemical dynamics in the soil-plant-litter system of forested landscapes form a complex network of relationships that are well suited to investigation by simulation methods. Sufficient knowledge is available to mathematically characterize the mechanisms involved (Fig. 9.3).

Five FORTRAN codes have been developed, and they are linked within the overall framework of the UTM. The general attributes of these five codes (Table 9.1) include gradient-dependent flow equations; empirical relationships for soil, plant, and litter characteristics; and simplified relationships for complex processes (e.g., macropore flow, plant chemical demand). The underlying philosophy of this modeling development has been to use a consistent level of detail. In some instances, a more complex

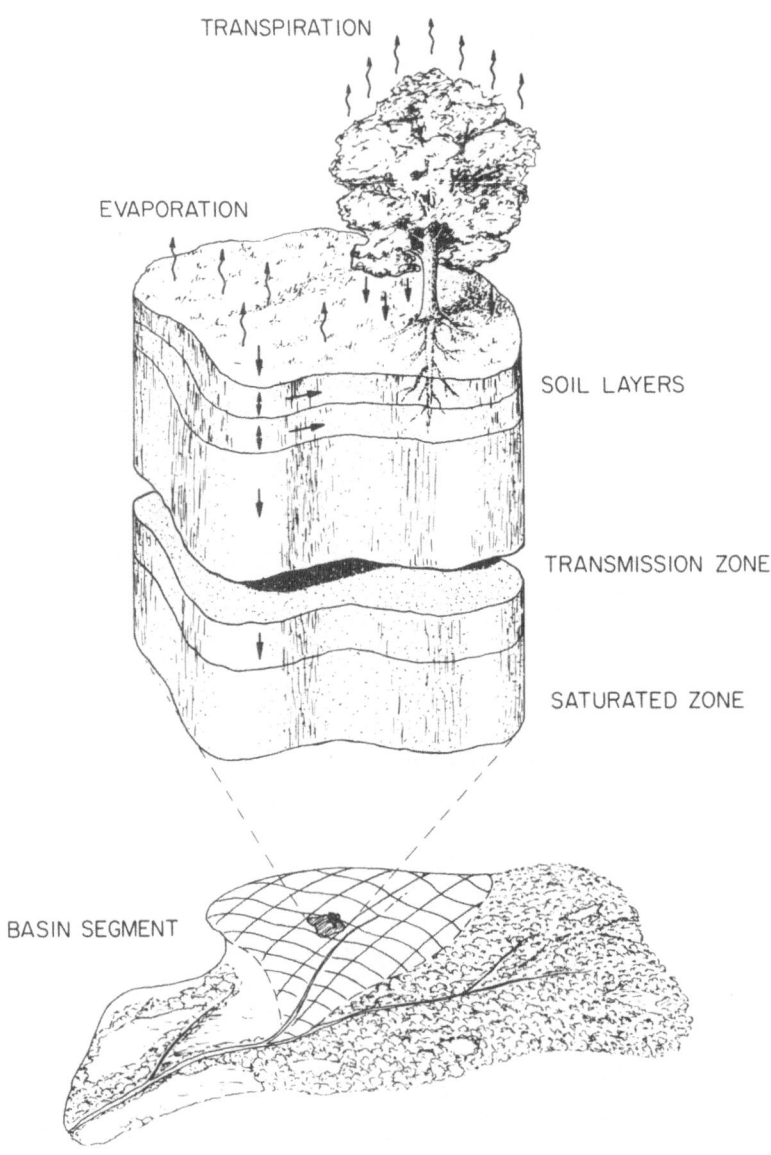

ORNL-DWG 74 - 4167

TRANSPIRATION

EVAPORATION

SOIL LAYERS

TRANSMISSION ZONE

SATURATED ZONE

BASIN SEGMENT

Figure 9.2. Land segments in a landscape context.

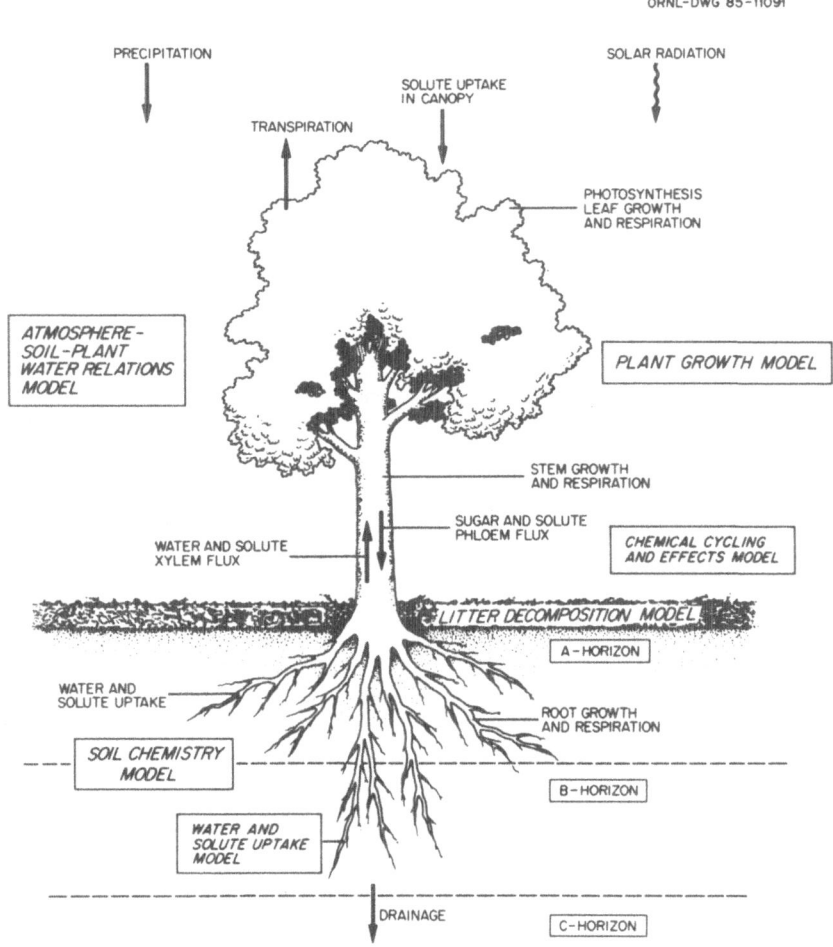

ORNL-DWG 85-11091

Figure 9.3. Component processes in the soil-plant-litter system. Source: R.J Luxmoore, C.L. Begovich, and K.R. Dixon. 1978. Modeling solute uptake and incorporation into vegetation and litter. Ecol. Modell. 5:137–171.

treatment of a process has been avoided. For example, the CO_2 gradient equation used for net photosynthesis is a more simplified approach than is available in many photosynthesis models; nevertheless, the selected algorithm is suited to the context of the UTM-SPL.

9.3.1 Soil-Plant-Water System

An introduction to the water flow model was given in Ch. 5, along with results from model applications to a yellow-poplar stand in a valley and an oak stand on a ridgetop. One aspect of soil water flow not presented

Table 9.1. Some attributes of coupled models describing carbon, water, and chemical dynamics in the soil-plant-litter system

	Component model				
	Soil-plant-atmosphere water flow	Soil chemistry	Vegetation and litter carbon dynamics	Diffusion and mass flow of solutes	Solute dynamics in vegetation and litter
Name	PROSPER (AGTEHM)	CADIL	CERES	DIFMAS	DRYADS
Time step	15 or 60 min	15 or 60 min	60 min	15 or 60 min	15 or 60 min
Characteristics	Evapotranspiration by combination method; Soil water flow by Darcy flow equation; Macropore flow; Empirical relationship between surface resistance and surface water potential; Empirical data for hydraulic properties of soil	Freundlich adsorption isotherm; First-order degradation equation; Chemical transport in soil macropores; A constant soil nitrogen solution concentration was introduced in one study (Luxmoore et al. 1981)	CO_2 diffusion equation for net photosynthesis; Substrate gradient equation for translocation; Input values for potential growth of plant components; Empirical litter decomposition relationships	Implements model of diffusion and mass flow of solutes to roots by Baldwin et al. (1973)	Solute uptake by roots and leaves; Diffusive gas uptake by leaves; Gradient equation for translocation; Transpiration flux used for xylem transport; Plant demand function determined by potential solute concentration input values
Reference	Huff et al. (1977); Hetrick et al. (1982)	Emerson et al. (1984)	Dixon et al. (1976, 1978)	Luxmoore et al. (1976, 1978)	Luxmoore et al. (1976, 1978)

Source: R.J. Luxmoore, T. Grizzard, and R.H. Strand. 1981. Nutrient translocation in the outer canopy and understory of an eastern decidous forest. For. Sci. 27:505–518.

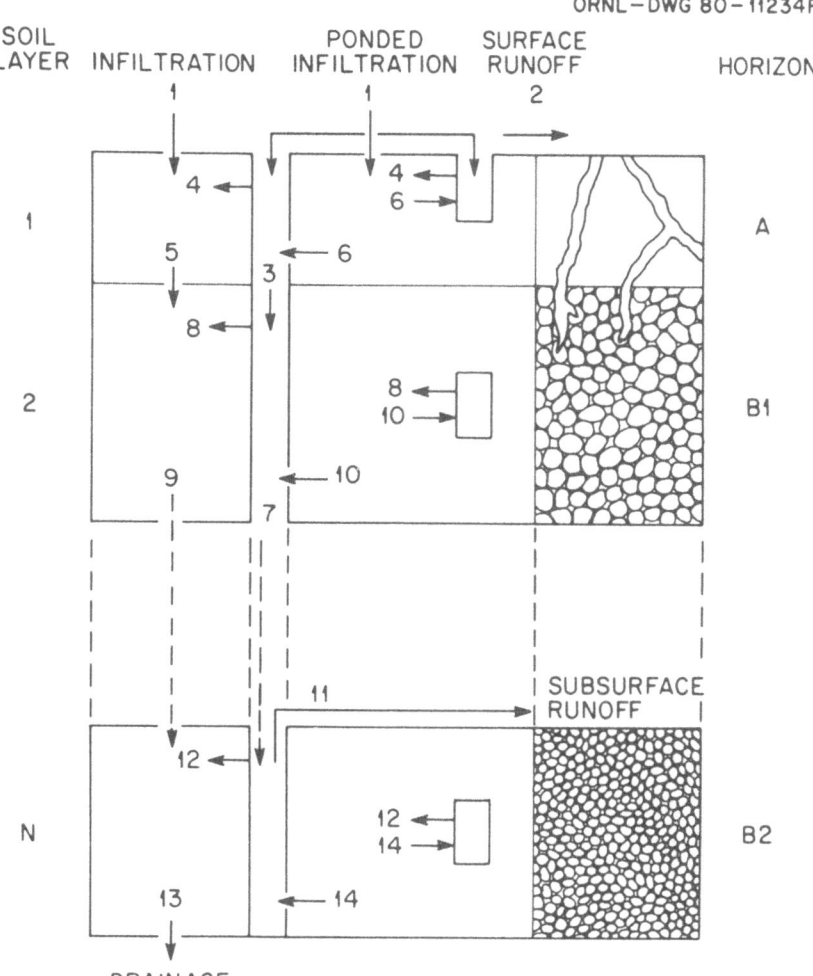

Figure 9.4. Water flow relationships in a soil with macropores in a mesopore matrix.

earlier is the hydrologic modeling of macropore effects. Cracks between soil aggregates, root and worm channels, and animal burrows can facilitate deep percolation as well as rapid subsurface flow in a lateral direction in sloping terrain (Beven and Germann 1982). A simplified concept of macropore effects on subsurface flow (Fig. 9.4) has been incorporated into the soil water subroutines. The basis for these calculations, described by Hetrick et al. (1982), are given below:

1. Flow rate *within* macropores is rapid and not controlled by geometrical factors. At each time step, macropore flow between soil layers is controlled by the lesser of two values: (1) the macropore water content or (2) the capacity of the lower layer to accept additional macropore water.
2. Flow from macropores to the smaller mesopores is controlled by the hydraulic properties of the mesopore matrix, using cylindrical and linear flow equations for cylinders and cracks, respectively.
3. Flow from mesopores to macropores results from the excess water entering mesopores above the saturation value.

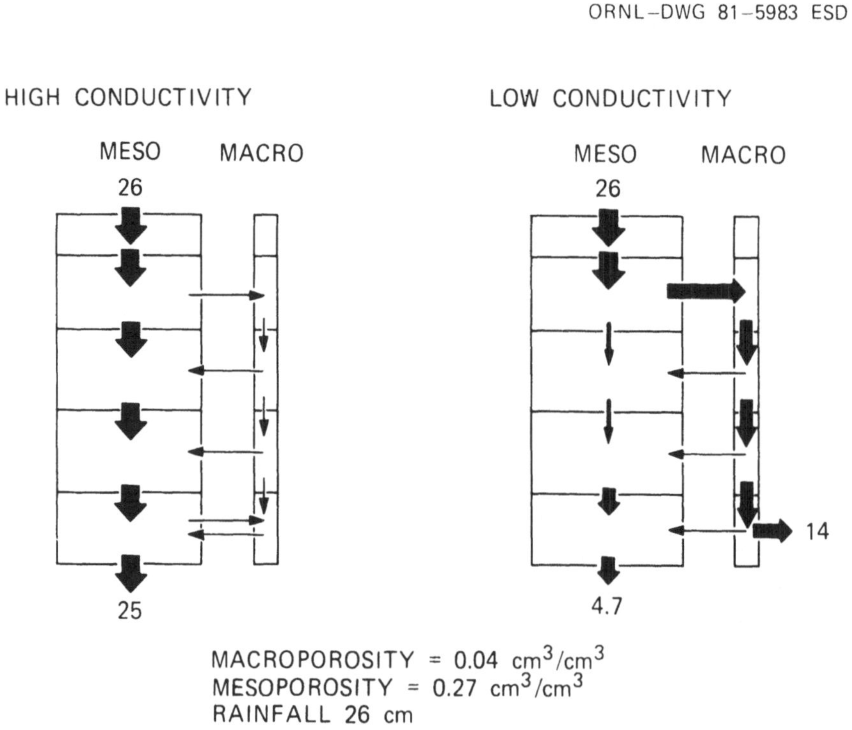
ORNL–DWG 81–5983 ESD
HIGH CONDUCTIVITY
LOW CONDUCTIVITY
MESO
MACRO
26
25
MESO
MACRO
26
14
4.7
MACROPOROSITY = 0.04 cm3/cm3
MESOPOROSITY = 0.27 cm3/cm3
RAINFALL 26 cm

Figure 9.5. Macropore water flow in soils of high vs low hydraulic conductivity (*K*) in the soil mesopore matrix. Source: After L. Fong and H.R. Appelbaum. 1980. Macropore-Mesopore Model of Water Flow Through Aggregated Porous Media. ORNL/MIT-312. Oak Ridge National Laboratory, Oak Ridge, Tennessee.

4. Excess water in the continuous macropore system becomes subsurface runoff and enters the stream channel.
5. Drainage from the mesopores enters the deep soil transmission zones and eventually percolates into groundwater.

Macropore effects on soil water flow are more pronounced in areas with high rainfall and in soils with a low hydraulic conductivity of its matrix materials. An illustration of this, in Fig. 9.5, shows much greater macropore flow in a soil profile with lower subsoil hydraulic conductivity. The macropore transport mechanism of subsurface flow is the most reasonable current explanation for the rapid transport of chemicals (e.g., sulfate; Fig. 7.16c of Ch. 7) through Walker Branch Watershed during storm events.

9.3.2 Soil Chemistry

The initial soil chemistry model (SCEHM) developed by Begovich and Jackson (1975) described the solubility and adsorption of trace metals in soil and the movement of these chemicals between soil layers. The code represented chemical adsorption onto soil as a linear function of soil solution concentration (K_d) and was suitable for low-chemical-concentration conditions. It has since been expanded in a new version called CADIL (Emerson et al. 1984). In CADIL, adsorption is represented by a Freundlich isotherm, which can be applied to all ranges of organic and inorganic chemical loading to soil.

Chemical degradation and soil temperature algorithms are included in CADIL. A concentration-dependent (first-order) degradation function with a temperature-dependent rate coefficient is used to represent a range of nutrient or pollutant transformation reactions in soil.

The CADIL code also describes chemical transport between layers of the soil matrix or through the matrix bypass mechanism provided by macropores. An example simulation (Fig. 9.6) shows that as soil chemical adsorption decreases with an increase in the power term of the Freundlich adsorption equation, there is an increase in chemical transport in both the vertical and lateral drainage (Emerson et al. 1984). In the case shown, chemical transport in the vertical drainage water is ~7 times greater than that in the lateral flow.

9.3.3 Vegetation and Litter Carbon Dynamics

The plant compartments of the code CERES (Dixon et al. 1978), which describes the dynamics of carbon in vegetation and litter, include structural carbon (storage) and mobile carbon (substrate) (Fig. 9.7). Photosynthesis is described by a CO_2 gradient-resistance equation (Table 9.2) in which the internal CO_2 concentration increases as the leaf substrate (sucrose)

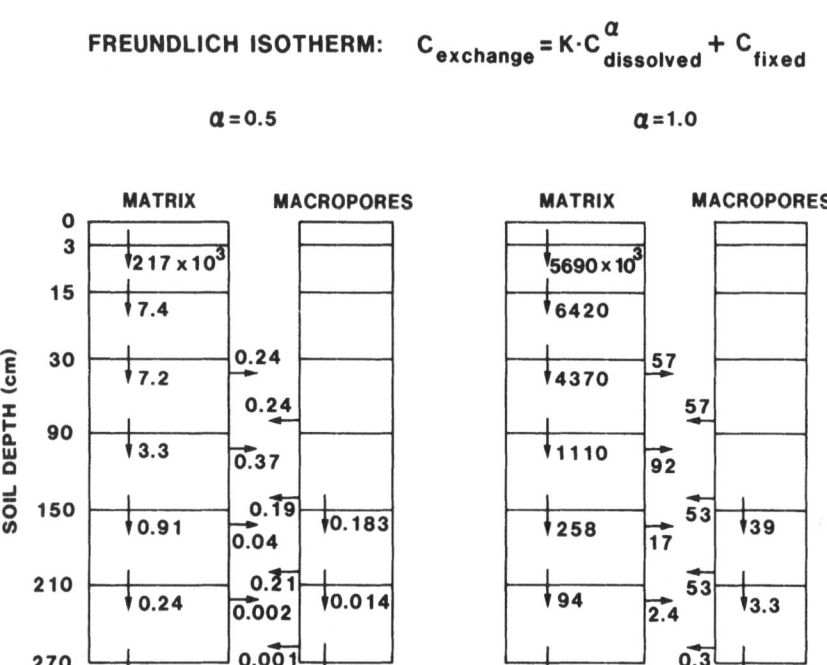

Figure 9.6. Chemical fluxes through soil matrix and macropores for two values of the Freundlich isotherm power term, α. K is an empirical constant; C represents concentration. Source: Adapted from C.J. Emerson, B. Thomas, Jr., and R.J. Luxmoore. 1984. CADIL: Model Documentation for Chemical Adsorption and Degradation in Land. ORNL/TM-8922. Oak Ridge National Laboratory, Oak Ridge, Tennessee.

level increases. Formation of storage tissue (growth) is dependent on plant water potential and is much more sensitive to water stress than is stomatal resistance. Sucrose translocation is represented by the difference in sucrose concentration between tissues (e.g., leaf to stem) divided by the phloem resistance (input parameter). Tissue respiration is a function of tissue mass and temperature.

Photosynthesis, leaf sugar, and growth vary with leaf water potential during several diurnal cycles (Fig. 9.8). All the simulation results from each of the component models of the UTM-SPL are given on a per-unit-soil-area basis, usually square meters, unless otherwise stated. The predicted annual growth of leaf area and stem heartwood (tree ring) for a deciduous forest (Fig. 9.9) shows a seasonal cycle in the application of Dixon et al. (1978).

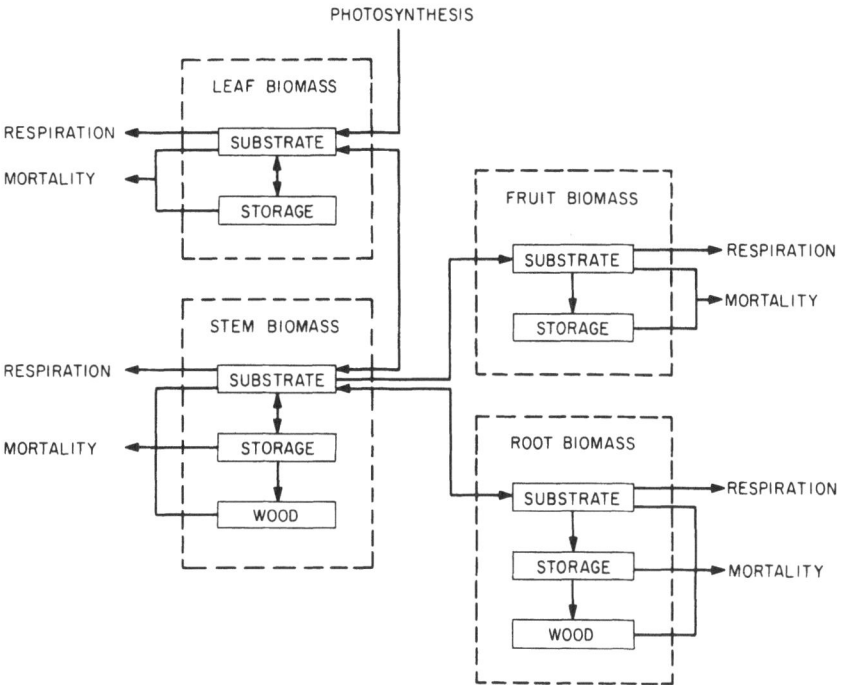

ORNL-DWG 75-15811R

Figure 9.7. Leaf, stem, root, and fruit components and processes represented in CERES. Source: K.R. Dixon, R.J. Luxmoore, and C.L. Begovich. 1978. CERES—a model of forest stand biomass dynamics for predicting trace contaminant, nutrient, and water effects. I. Model description. Ecol. Modell. 5:17–38.

Table 9.2. Characteristics of carbon uptake, translocation, and utilization used in CERES[a,b]

Photosynthesis (Penman and Schofield 1951)

$$\text{Photosynthate production} = A\,\frac{(CO_2 \text{ in air } - CO_2 \text{ in chloroplast})}{r_a^* + r_s^* + r_i}$$

Allocation (Thornley 1972)

$$\text{Sugar transport} = \frac{(\text{sugar in leaf } - \text{ sugar in stem or root})}{\text{pathway resistance}}$$

Utilization
 Sugar utilization by respiration + growth:
 Respiration = f(temperature,* amount of tissue)
 Growth = f(temperature,* water potential,*
  sugar supply, phenology)

---

[a]Values marked with an asterisk (*) are from the soil-plant-atmosphere water flow model.
[b]A = conversion factor; r_a = boundary-layer resistance; r_s = surface resistance; r_i = internal resistance.

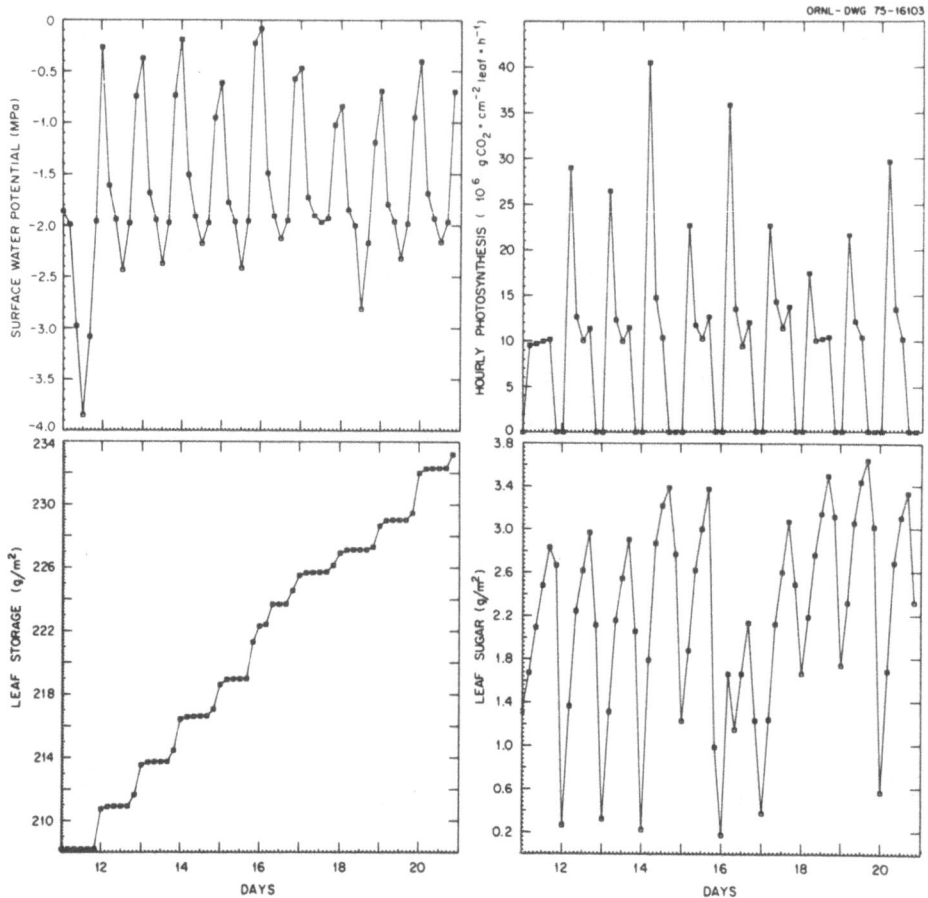

Figure 9.8. Hourly patterns of surface water potential, photosynthesis, leaf sugar, and leaf storage from the CERES-PROSPER models, showing reduced leaf growth (storage) during periods of low water potential.

Litter decomposition is represented by temperature-dependent rate coefficients for each component of the litter [leaf, stem, fruit (e.g., acorns), and root].

9.3.4 Solute Uptake by Roots

A model that describes solute uptake by roots was developed by Baldwin et al. (1973) and has been incorporated into the UTM-SPL in the code called DIFMAS (Luxmoore et al. 1978). Baldwin et al. (1973) recommended their model for inclusion in whole plant-soil models. DIFMAS uses the transpiration rate from PROSPER for calculating the mass flow component of solute movement from bulk soil solution to root cylinders

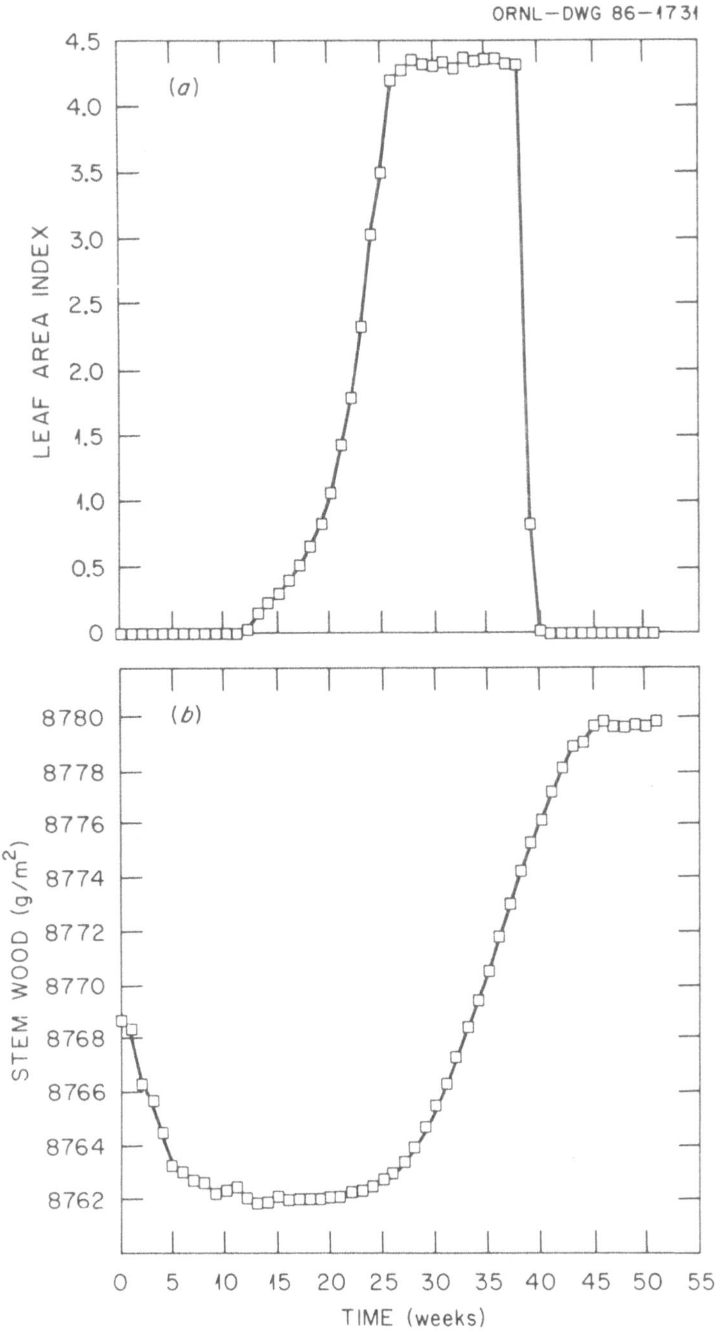

Figure 9.9. Simulated leaf area index and stem heartwood weight for a deciduous forest: (a) leaf area index (cm^2 leaf/cm^2 land) during an annual cycle; (b) stem heartwood (g/m^2 land) during an annual cycle. Source: K.R. Dixon, R.J. Luxmoore, and C.L. Begovich. 1978. CERES—a model of forest stand biomass dynamics for predicting trace contaminant, nutrient, and water effects. II. Model application. Ecol. Modell. 5:93–114.

in an algorithm combining mass flow and diffusion processes. During periods without transpiration (e.g., rainfall, night), solute uptake is calculated using a diffusion gradient equation. The solute concentration in the root is determined by the DRYADS code. A simulation of lead uptake by roots in a deciduous forest (Luxmoore et al. 1978) shows a strong diurnal dependence on transpiration (Fig. 9.10).

9.3.5 Chemical Dynamics in Vegetation and Litter

The compartments for chemical incorporation in the DRYADS model (Fig. 9.11) correspond to the compartments for carbon incorporation in CERES.

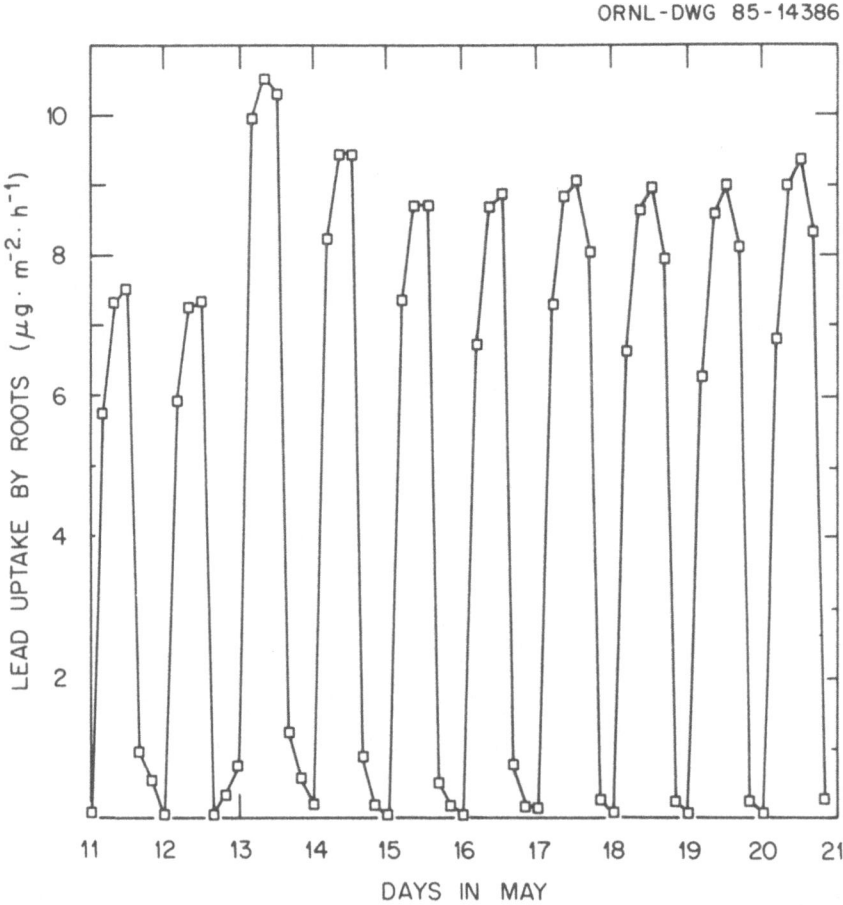

Figure 9.10. Lead uptake by roots simulated for a deciduous forest. Source: R.J. Luxmoore, C.L. Begovich, and K.R. Dixon. 1978. Modelling solute uptake and incorporation into vegetation and litter. Ecol. Modell. 5:137–171.

ORNL−DWG 76−3672R

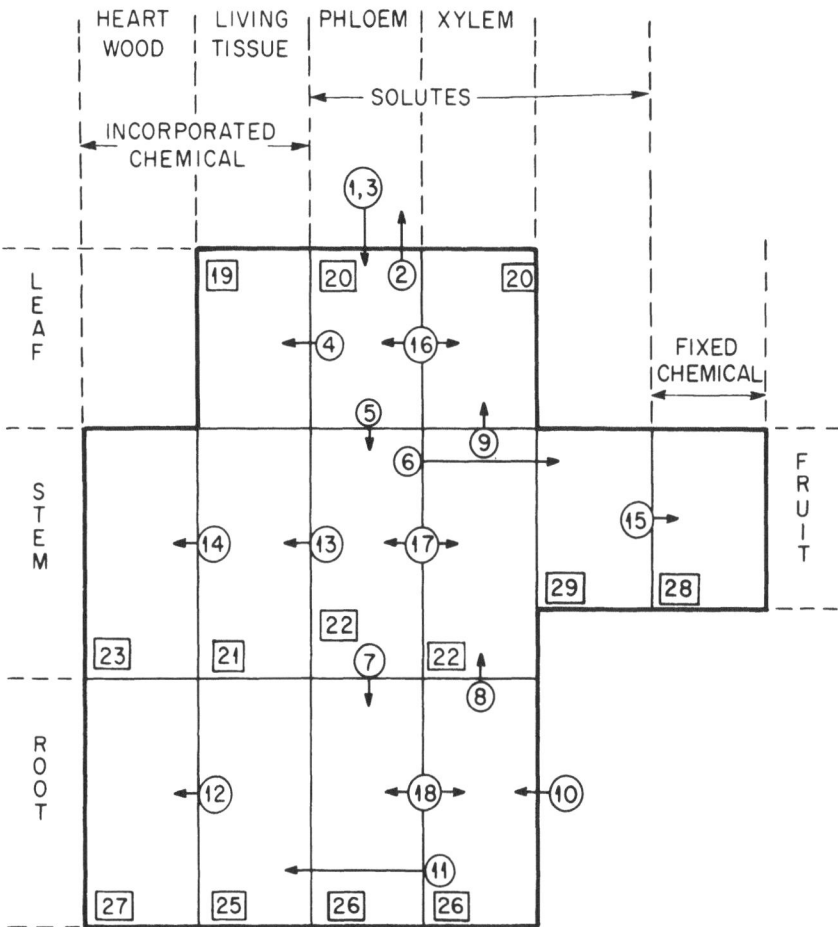

THE NUMBERS INDICATE THE SEQUENCE OF CALCULATIONS IN DRYADS

PROCESSES
LEAF SOLUTE UPTAKE OR LEACHING 1–2
LEAF GASEOUS UPTAKE 3
PHLOEM TRANSLOCATION 5–7
XYLEM TRANSPORT 8–9
ROOT SOLUTE UPTAKE 10
CHEMICAL INCORPORATION
    IN LIVING TISSUE 4, 11, 13, 15
    IN HEARTWOOD 12, 14
SOLUTE EQUILIBRATION 16–18
MORTALITY LOSSES 19–29

Figure 9.11. Vegetation compartments and sequence of chemical transport calculations used in DRYADS. Source: R.J. Luxmoore, C.L. Begovich, and K.R. Dixon. 1978. Modelling solute uptake and incorpation into vegetation and litter. Ecol. Modell. 5:137–171.

The movement of chemicals between compartments depends on linkages with the water and carbon models. The DRYADS model is based on the Munch hypothesis of plant transport processes. Munch (1930) introduced a simple theory which describes the nature of material movement in plants in terms of a protoplasmic continuum (symplasm) and a nonprotoplasmic continuum (apoplasm). Symplasm consists of cellular cytoplasm interconnected by plasmodesmata and includes phloem tissues. Movement of solutes in apoplasm occurs in the water flow within the xylem from roots to leaves and within cell walls. Cellular metabolites and photosynthates move in phloem (Nelson 1963), and this is the main pathway of material movement from leaves to roots represented in DRYADS. This simple view of material movement in plants forms one of the two main concepts of the DRYADS model.

The second concept is that of plant solute demand. Plants require many minerals for normal development and growth. These requirements are met by foliar absorption or uptake of solutes from the soil. Plants show preferential uptake of particular elements and can exclude others at their membranes. In the DRYADS model, vegetation is hypothesized to absorb solutes up to a predetermined maximum concentration in each of its leaf, stem, fruit, and root tissues. Changes in the amount of plant tissue caused by growth and senescence cause temporal changes in the chemical content. The deficit between the maximum concentration and the current tissue concentration forms the basis of the demand function for solutes by the vegetation.

Leaves and roots are the major interfaces of material exchange between vegetation and the environment. Foliar uptake of gaseous materials by diffusion through stomata has been shown to occur for sulfur dioxide, oxides of nitrogen, and ozone (Hill 1971). Movement of solutes from wet and dry deposition through the leaf surface can also be an important pathway. Radioactive isotopes of lead (^{210}Pb) and cadmium (^{109}Cd) applied to lettuce and radish leaves have been shown by Hemphill and Rule (1975) to be absorbed into the leaves and translocated. It is generally thought that heavy metals that enter plants become chelated and move in the xylem and phloem tissue (Lagerwerff 1972). Beauford et al. (1975) have established that radioactive zinc (^{65}Zn) taken up from nutrient solution by pea (*Pisum sativum*) and broad bean (*Vicia faba*) can be released to the atmosphere. They also obtained a similar result for Pb, Cu, Hg, and Mn. Leaching of solutes out of leaves during rainfall has also been shown to occur for many elements (Tukey and Morgan 1962). Equations describing the bidirectional flux of solutes through the leaf surface and the uptake of gases by diffusion have been incorporated into the leaf physiology algorithms of the DRYADS model (Fig. 9.11).

Transfer of solutes from leaves to stems and from stems to roots is determined by phloem resistances and solute gradients. The flux of solutes from the roots to stems and leaves is a function of the transpiration flux, the water content of roots and stems, and the solute content in roots and stems. One of the major consequences of water movement in plants is solute redistribution (Crafts and Crisp 1971).

The soluble (mobile) forms of chemicals in leaves, stems, roots, and fruits are fixed to immobile forms in association with the formation of insoluble carbon (growth) calculated in CERES (Dixon et al. 1976). The levels of immobile chemical plus mobile solute in the tissues are compared with input values to determine the deficiency, sufficiency, and toxicity effects of the element on tissue growth. Indexes of these effects are used in the CERES growth model.

The "absorbing power" of vegetation for chemicals is a function of the permeability of its roots to solutes and its capacity for solutes. It has been demonstrated that the uptake of solutes requires metabolic energy, and various carrier mechanisms that influence solute uptake across cellular membranes have been proposed (Gauch 1972). The details of these processes are not included in DIFMAS or DRYADS, and the assumption is made that solute uptake is limited neither by metabolic energy nor by carrier sites. The uptake of solutes by roots is determined by both the vegetation's demand for solutes (absorbing power) and the supply and transport of solutes in the soil to the root surface.

The chemicals associated with plant tissue mortality are added to the chemical pool in the litter (Fig. 9.12). The mineralization rate of the chemicals in litter is proportional to the litter decomposition rate obtained from CERES. The amount of a chemical in the litter is used to determine deficiency, sufficiency, and toxicity feedback effects on decomposition rates in the forest phytomass model.

The DRYADS model of solute movement in vegetation and litter and the DIFMAS model of solute movement and uptake by roots are coupled to three other models: a forest phytomass model (CERES), the soil-plant-atmosphere water flow model (PROSPER), and the soil chemistry model (CADIL). The coupling between models (Fig. 9.13) shows that every model has information transfer with at least two other models, and these transfers take place on either an hourly time step or every 15 min during rainfall. Hourly values of stomatal resistance and plant water potential from PROSPER are used in CERES to determine photosynthesis and growth, respectively. Leaf growth and root growth, in turn, influence transpiration and thus soil water flow. During rainfall, infiltration and the movement of water between soil layers (calculated in PROSPER) are used in the soil chemistry model (CADIL) to calculate chemical fluxes. Chemical concentration and root water uptake information are used in DIFMAS to calculate chemical uptake into root by diffusion and mass flow. Chemicals within the plant are moved up in the transpiration stream and down in the phloem pathway. These models are executed as subroutines in the TEHM hydrology model (Huff et al. 1977; Hetrick et al. 1982).

Ramp functions (Fig. 9.14) are used to determine the effects of pollutants

ORNL–DWG 76–3671R

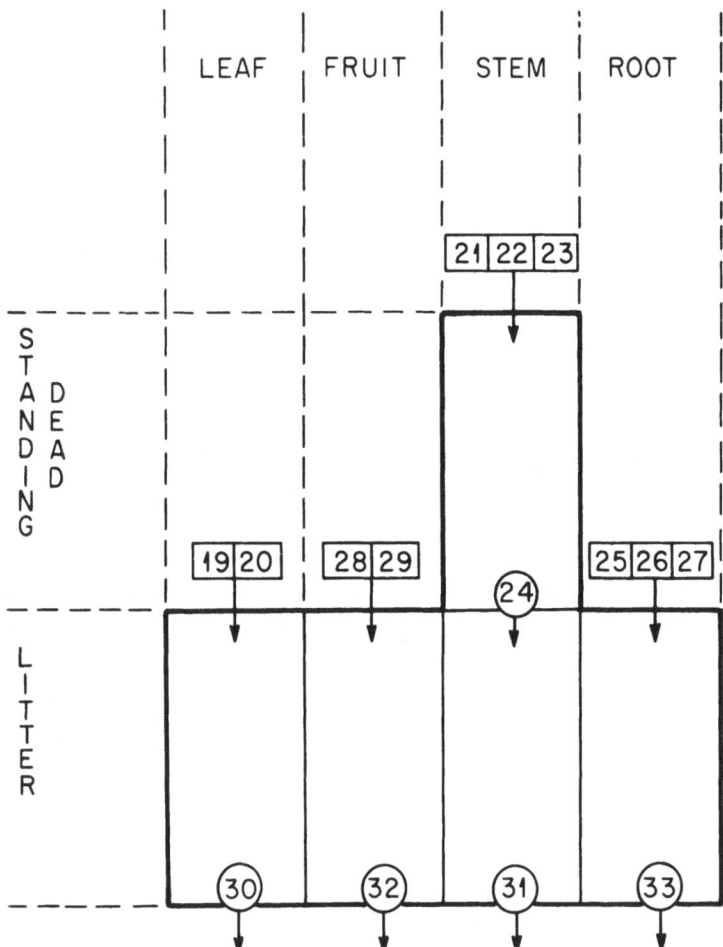

THE NUMBERS INDICATE THE SEQUENCE OF CALCULATIONS IN DRYADS

PROCESSES
MORTALITY INPUTS 19–29
WIND FALL 24
MINERALIZATION 30–33

Figure 9.12. Litter compartments and sequence of chemical transport calculations used in DRYADS. This sequence follows the vegetation calculations (Fig. 9.11). Source: R.J. Luxmoore, C.L. Begovich, and K.R. Dixon. 1978. Modelling solute uptake and incorporation into vegetation and litter. Ecol. Modell. 5:137–171.

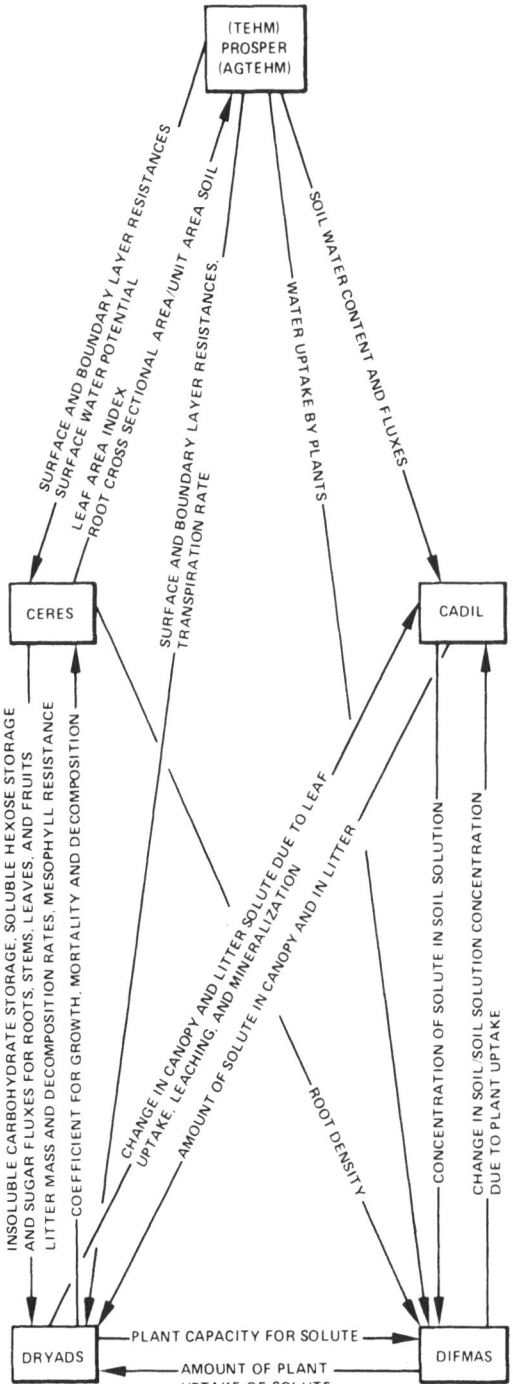

Figure 9.13. Coupling of five process models that describe hourly carbon, water, and solute dynamics of the soil-plant-litter system. Source: Adapted from R.J. Luxmoore, C.L. Begovich, and K.R. Dixon. 1978. Modelling solute uptake and incorporation into vegetation and litter. Ecol. Modell. 5:137–171.

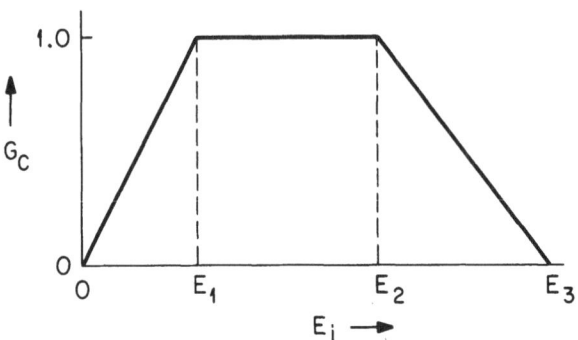

ORNL-DWG 76-3275R

Figure 9.14. The relationship between the growth coefficient (G_c) and the amount of chemical in tissue (E_i) that is used to represent deficiency $(E_i < E_1)$, sufficiency $(E_1 < E_i < E_2)$, and toxicity $(E_i > E_2)$ effects. Source: R.J. Luxmoore, C.L. Begovich, and K.R. Dixon. 1978. Modelling solute uptake and incorporation into vegetation and litter. Ecol. Modell. 5:137–171.

on the growth and decomposition of leaf, stem, root, and fruit components. Separate ramp functions for either growth effects (Fig. 9.14) or control of decomposition in the litter (same form as for growth effects) represent ranges of chemical deficiency, sufficiency, and toxicity as the chemical concentration increases. Empirical data are needed to define these ranges in any model application. Hypotheses concerning the beneficial effects (e.g., fertilizer) as well as the toxic effects of pollutants can thus be examined. The product of the growth coefficient and tissue growth rate (from CERES) provides a growth rate modified by pollutant effects.

A 6-year simulation of the deposition, transport, and uptake of several heavy metals in an oak-hickory forest showed that the lead accumulation was greatest in the litter (Luxmoore et al. 1978). Lead uptake by roots increased through the 6-year period, whereas uptake by leaves remained constant for the repetitive annual deposition of 25 g Pb/m^2 land area. Because of the buildup of lead in the plant tissues, the mortality of plant parts returned increasingly greater amounts of dry matter and lead to the litter. Consequently, the dry weight of the litter increased through the 6-year period by 949 g/m^2. This compares reasonably well with field studies conducted in the oak-hickory forest. A difference of 1130 g/m^2 between the litter mass at a control site and that at a site exposed to equivalent heavy metal deposition over a 6-year period was reported by Watson et al. (1976). The results of the simulation provided an alternative explanation for the experimental inference of reduced rates of litter decomposition at elevated levels of heavy metal accumulation (Jackson and Watson 1977) by showing that increased litter mass could be obtained through increased mortality of plant parts.

9.4 Model Applications

9.4.1 Simulation of Nitrogen Dynamics

Simulation of nitrogen dynamics was used to evaluate the relative importance of soil nitrogen and the internal plant storage pool in supplying the nitrogen requirements for spring foliar growth in an oak forest. The set of five models describing the hourly carbon, water, and solute dynamics in a deciduous forest (Table 9.1) was implemented for the soil (Peters et al. 1970), vegetation, and weather conditions (1969 meteorological data) on Walker Branch Watershed.

The complexity of soil nitrogen dynamics was simplified by assuming that the concentration of nitrogen in soil solution was maintained at a constant value. The annual nitrogen uptake by vegetation as a function of the concentration of nitrogen in soil solution (Fig. 9.15) showed a large response for the two cases representing a mechanism of either diffusion (root conductivity of 8×10^{-6} cm/s) or mass flow (root conductivity of 2×10^{-5} cm/s) across root membranes (Luxmoore et al. 1981). Chemical budget estimates of the annual nitrogen uptake by roots range from 95 kg

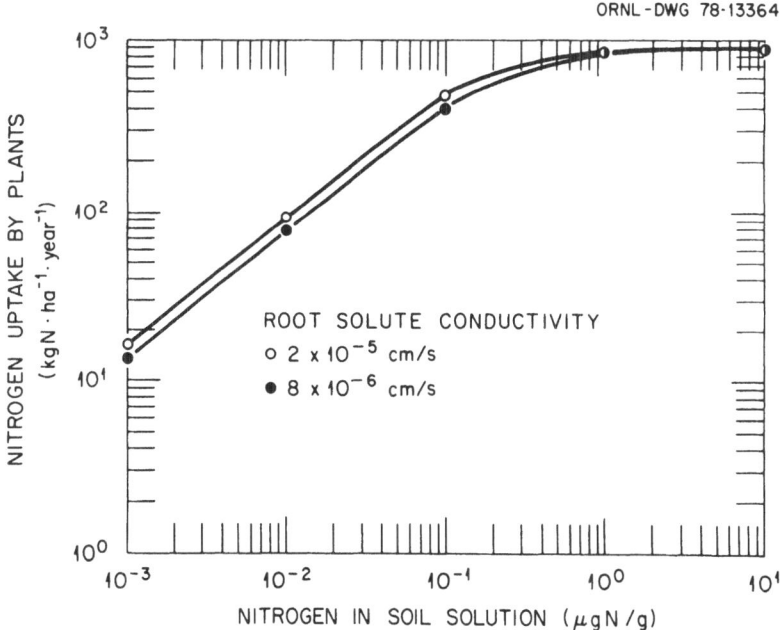

Figure 9.15. Simulated annual nitrogen uptake by plants for two values of root conductivity over a range of nitrogen concentrations in soil solution. Source: R.J. Luxmoore, T. Grizzard, and R.H. Strand. 1981. Nutrient translocation in the outer canopy and understory of an eastern deciduous forest. For. Sci. 27:505–518.

N·ha^{-1}·year^{-1} at Coweeta, North Carolina (Mitchell et al. 1975) to 124 kg N·ha^{-1}·year^{-1} at Walker Branch (Henderson and Harris 1975). The latter estimate was based on yellow-poplar (*Liriodendron tulipifera* L.) at a fertile site and is likely to be an upper estimate. It is shown that an annual uptake of 100 kg N·ha^{-1}·year^{-1} can be obtained with a soil solution concentration near 0.01 μgN/mL. Data from soil solution sampling within the root zone have shown very low nitrogen concentrations throughout the growing season, and a value of 0.01 μgN/mL is representative for Walker Branch (Johnson et al. 1985).

The monthly pattern of nitrogen dynamics was examined for the case of 100 kg N·ha^{-1}·year^{-1} uptake to evaluate the soil as a source of nitrogen during the spring flush period. According to the simulation, nitrogen uptake by roots was initiated in March and reached a maximum of 25 kg N·ha^{-1}·month^{-1} during June (Fig. 9.16); the uptake rate then declined through the summer to a negligible rate after leaf fall in November. Cumulative nitrogen uptake by roots in the March-May period amounted to 18.6kgN/ha (Fig.9.16), whereas the cumulative nitrogen translocation to leaves for the overstory deciduous vegetation by the end of May was 7.74 mg N/leaf (Luxmoore et al. 1981). Leaf population was estimated, from foliar and leaf fall data, to be 10^7/ha during the spring period, giving a total nitrogen translocation to leaves of 77.4 kg N/ha. Because the leaf population cannot be expected to be constant throughout this development period, the estimated translocation can be viewed as an upper limit value. Simulated soil nitrogen uptake can account for only 24% (18.6/77.4) of the observed foliar nitrogen influx up to the end of May. This comparison leads to the inference that internal nitrogen storage reserves were the major source of foliar nitrogen and that soil nitrogen uptake provides a replenishment through the 8-month growing season.

Nitrogen uptake from soil solution, even at low concentration, was greater during the day (91%) than at night (9%), according to the simulation (Luxmoore et al. 1981). This was due in part to the fact that the transpiration stream that moves nitrogen from the roots to the trunk and leaves maintained a favorable nitrogen gradient for continuing uptake by the roots during the day. The mass flow transport of nitrogen to the root from the bulk soil solution does not account for the increased uptake during the day. This is understandable, since the product of the quantity of water transpired during the growing season (58.0 cm) and the nitrogen concentration in soil solution (0.01 mg N/kg) amounted to 0.06 kg N/ha. This is a trivial part of annual uptake. Essentially all of the daytime nutrient supply could be accounted for by diffusive transport to the root surfaces. The model of nitrate-nitrogen uptake by plant roots reported by Phillips et al. (1976) suggests that the relative importance of diffusion over mass flow is large when the transpiration rate is small and the absorbing power of the roots is high. In the DRYADS model, the absorbing power of the roots reduces with nutrient accumulation within the roots: thus, at night when transpiration rate is small, the absorbing power becomes small as nutrients

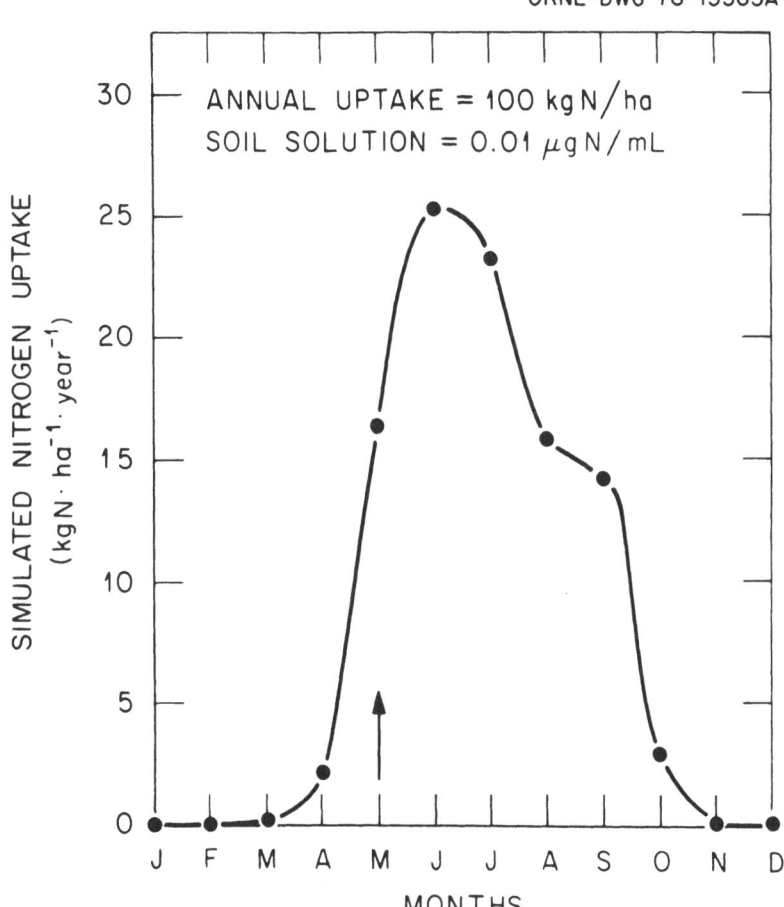

ORNL-DWG 78-13365A

Figure 9.16. Monthly nitrogen uptake during a year in which the concentration of nitrogen in soil solution was fixed at 0.01 μgN/ml. Source: R.J. Luxmoore, T. Grizzard, and R.H. Strand. 1981. Nutrient translocation in the outer canopy and understory of an eastern deciduous forest. For. Sci. 27:505–518.

accumulate, and the diffusive uptake becomes smaller than would be predicted by the model of Phillips et al. (1976).

9.4.2 Air Pollutant Uptake

The uptake of air pollutants by vegetation may occur directly through leaves (gaseous and particulate) or indirectly through roots after the pollutants have been incorporated into the soil. Gaseous uptake is represented by a diffusion equation (same form as the photosynthesis equation). Thus,

$$U_g - \frac{(g_e - g_i)g_d}{r_a + r_s + r_m} \tag{9.1}$$

where g_e = the atmospheric pollutant concentration (ml/ml), g_i = the internal pollutant concentration (ml/ml), g_d = gas density (mg/ml), r_a = boundary layer diffusion resistance (s/cm), r_s = stomatal resistance (s/cm), r_m = mesophyll resistance (s/cm), U_g = uptake ($\mu g \cdot cm^{-2} leaf \cdot s^{-1}$). The value of g_i is made to vary between zero and g_e, depending on the level of pollutant in leaf storage (E_i), as follows:

$$g_i = g_e \left(1 - \frac{E_m - E_i}{E_m}\right) \qquad (9.2)$$

E_m is the maximum allowable level of pollutant in leaf storage, an input parameter. Operationally, this is the pollutant level at which the leaf tissue ceases growth.

Sulfur dioxide uptake by an oak-hickory forest in the vicinity of a lead mine and smelter complex in southeastern Missouri was simulated, and the results illustrate the behavior of the model. The cumulative sulfur

ORNL-DWG 80-11126

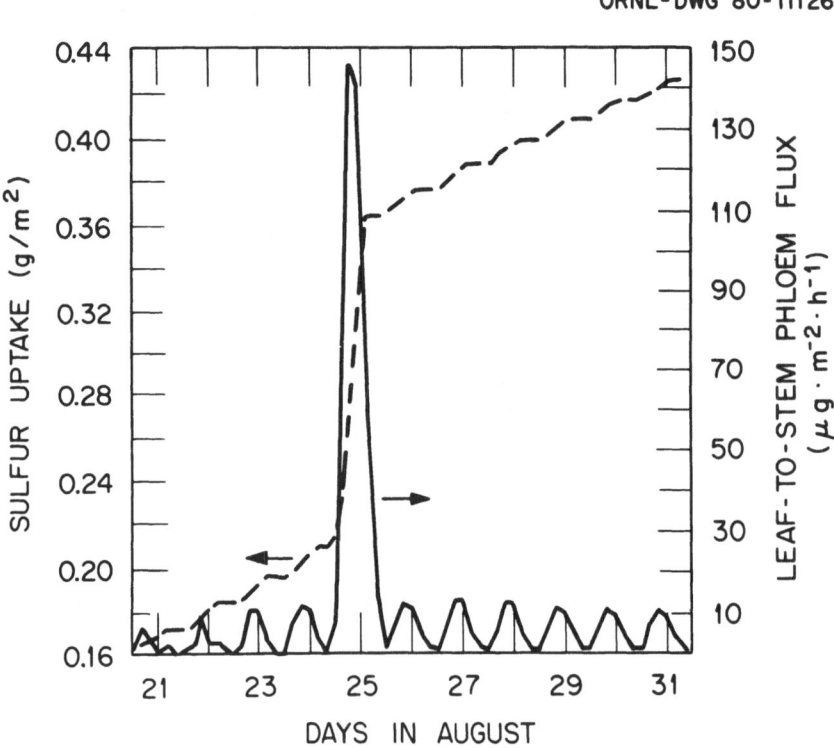

Figure 9.17. Simulated cumulative sulfur uptake by vegetation and rate of leaf-to-stem translocation for an 11-d period. Source: R.J. Luxmoore. 1980. Modeling pollutant uptake and effects on the soil-plant-litter system. pp. 174–180. IN P.R. Miller (ed.), Effects of Air Pollutants on Mediterranean and Temperate Forest Ecosystems. Report PSW-43. Pacific Southwest Forest and Range Experiment Station, Berkeley, California.

levels in leaves (Fig. 9.17) show a rapid increase on August 25, a day when the atmospheric SO_2 level increased 10-fold above ambient. The translocation of sulfur from leaf to stem (Fig. 9.17) clearly shows a diurnal pattern and elevated rates on August 25. Some of the sulfur that was transported to the roots leaked into the transpiration stream and returned from the roots to the stem, albeit in trace amounts. The phloem and xylem transport pathways allow considerable mobility of solutes between plant tissues, according to the simulation. The cumulative sulfur levels in the leaf, stem, and roots show that most of the sulfur remained in the leaves.

9.4.3 Chemical Uptake from Particle Deposition

The uptake of pollutants from particles deposited on leaves (U_i) is represented by a gradient equation, using empirical input values for the conductivity (k_l) and thickness (W) of the cuticle. Thus,

$$U_i = k_l \frac{(S_e - S_i)}{W} \tag{9.3}$$

where S_e is the quantity of pollutant on all leaf surfaces above a unit land area (g/m² land), and S_i is the quantity of pollutant within the foliage (g/m² land).

The amount of dissolved pollutant on leaf surfaces is calculated as the lesser of either (1) the product of the pollutant's solubility and the water volume on the surface of the leaves (interception) or (2) the amount of pollutant on the surface of the leaves. The soluble pollutant within the leaves (S_i) is assumed to be uniformly distributed and has one of two fates: it may be transported to other plant parts, or it may be incorporated into the leaves in an immobile form. The cuticular uptake process is reversible in the model. Thus, during rainfall, the pollutant is washed off, and when S_e becomes less than S_i, the pollutant is leached out of the leaves.

Sensitivity analysis of the leaf cuticle conductivity (Fig. 9.18) shows that greater conductivity is associated with greater uptake of chemical (zinc in the example) by the leaves and a slightly reduced uptake of zinc from the soil solution (Begovich and Luxmoore 1979). The latter, more subtle effect is induced by the higher zinc level in the plant at higher leaf cuticular conductivity, which feeds back a reduced chemical demand in the root uptake algorithm. Cuticular conductivity and the equivalent property at the root-soil interface (root conductivity, k_r) were shown to be sensitive parameters in the model, and yet these are perhaps the least well characterized experimentally. Results from an analysis of the sensitivity of lead uptake to root conductivity (Table 9.3) show large increases in lead uptake by roots and in lead concentration in tree tissues with an increase in k_r from 10^{-10} cm/s to 10^{-6} cm/s. The simulations also show that pollutants accumulate preferentially in the leaf and root, the sites of pollutant entry. A modification subsequently added to the model allows

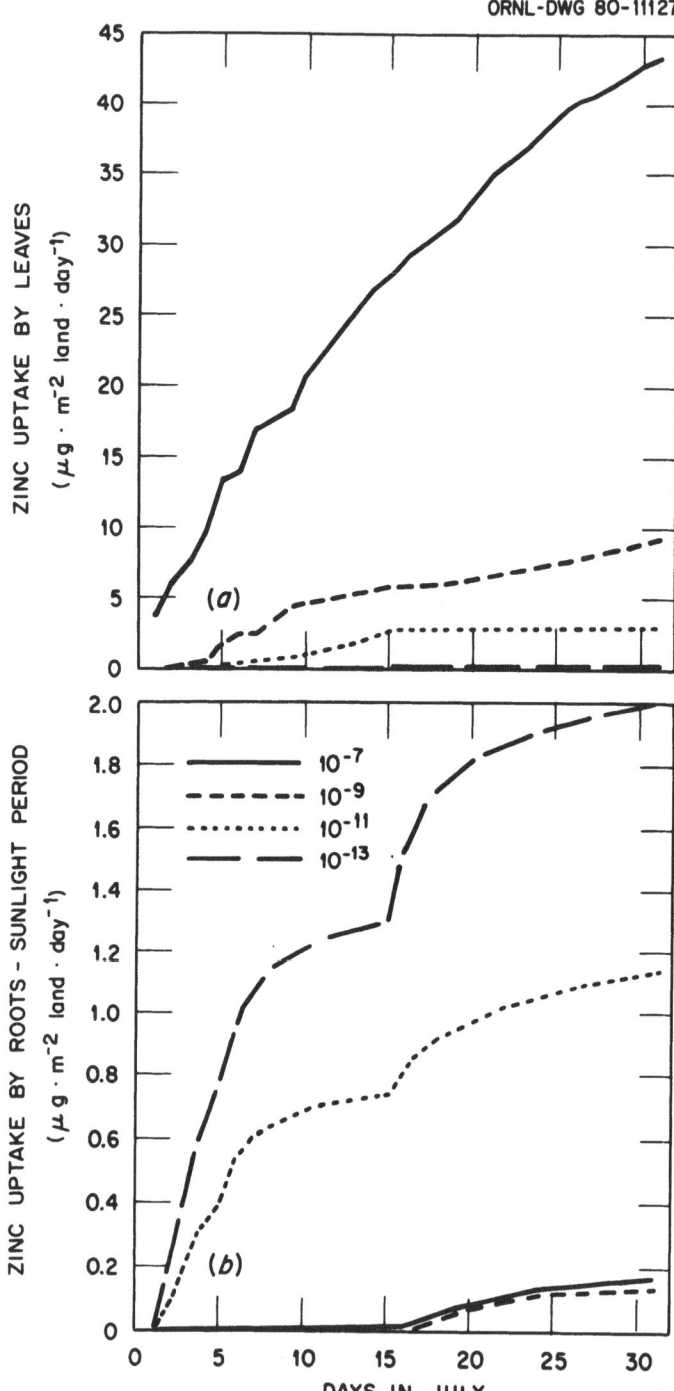

Figure 9.18. Influence of leaf cuticle conductivity on zinc uptake by (a) leaves and (b) roots. Source: R.J. Luxmoore. 1980. Modeling pollutant uptake and effects on the soil-plant-litter system. pp. 174–180. IN P.R. Miller (ed.), Effects of Air Pollutants on Mediterranean and Temperate Forest Ecosystems. Report PSW-43. Pacific Southwest Forest and Range Experiment Station, Berkeley, California.

Table 9.3. Sensitivity of annual lead uptake by roots and concentration of lead in tissue (prior to leaf fall) in an oak forest to a change in the root conductivity parameter (k_r).

k_r (cm/s)	Lead uptake by roots ($\mu g \cdot cm^{-2}$ land$\cdot$year^{-2})	September tissue concentration ($\mu g/g$)						
			Stem		Root		
		Leaf	Sapwood	Heartwood	Sapwood	Heartwood	Fruit
10^{-4}	156.1	309	854.0	3.30	967	10.2	264.0
10^{-5}	143.3	258	774.0	2.80	961	9.9	191.0
10^{-6}	128.0	242	683.0	2.32	993	11.7	251.0
10^{-8}	20.7	113	5.06	0.01	432	3.1	1.2
10^{-10}	0.52	112	0.48	0.003	11	0.1	0.1

Source: R.J. Luxmoore. 1980. Modeling pollutant uptake and effects on the soil-plant-litter system. pp. 174–180. IN P.R. Miller (ed.), Effects of Air Pollutants on Mediterranean and Temperate Forest Ecosystems. Report PSW-43. Pacific Southwest Forest and Range Experiment Station, Berkeley, California.

Table 9.4. Simulated lead uptake by roots and leaves (mg Pb•m^{-2} land•month^{-1}) in an oak forest in the vicinity of a lead mine-smelter complex

Month	Roots Day	Roots Night	Leaves
January	0	0	0
February	0	0	0
March	57	4	250
April	213	13	410
May	243	28	890
June	196	45	1190
July	286	41	1130
August	285	37	890
September	315	40	420
October	226	38	20
November	0	0	0
December	0	0	0
Total	1821	246	5200

Source: R.J. Luxmoore. 1980. Modeling pollutant uptake and effects on the soil-plant-litter system. pp. 174–180. IN P.R. Miller (ed.), Effects of Air Pollutants on Mediterranean and Temperate Forest Ecosystems. Report PSW-43. Pacific Southwest Forest and Range Experiment Station, Berkeley, California.

chelation of chemical pollutants within the plant (Luxmoore and Begovich 1979), which has the effect of increasing the mobility of pollutants within the plant. Thus, the site of pollutant entry may not be the site of accumulation (Jackson et al. 1978).

The monthly pattern of lead uptake by roots and foliage simulated for an oak forest near a lead mine-smelter complex during the first year of operation shows that uptake corresponds with the growing season (Table 9.4). The major proportion (88%) of lead uptake by the roots occurred during the day chiefly because of two complementary transport processes: (1) the mass flow of pollutant to the roots and (2) the mass flow of pollutant from the roots to the shoots. The latter was the controlling process in the simulations. Overall, lead uptake by leaves was more than double that simulated for roots in the first year of smelter operation.

9.5 Pollutants and the Diurnal Cycle

The modeling of water, carbon, and chemicals as coupled components in soil-plant-litter systems has stimulated the development of a conceptual framework for the diurnal cycle in plants (Fig. 9.19) that can be used to derive hypotheses of pollutant effects on whole plants (Luxmoore 1980). Plants alternate between two *relative* states during each diurnal cycle: (1)

ORNL-DWG 76-3670R

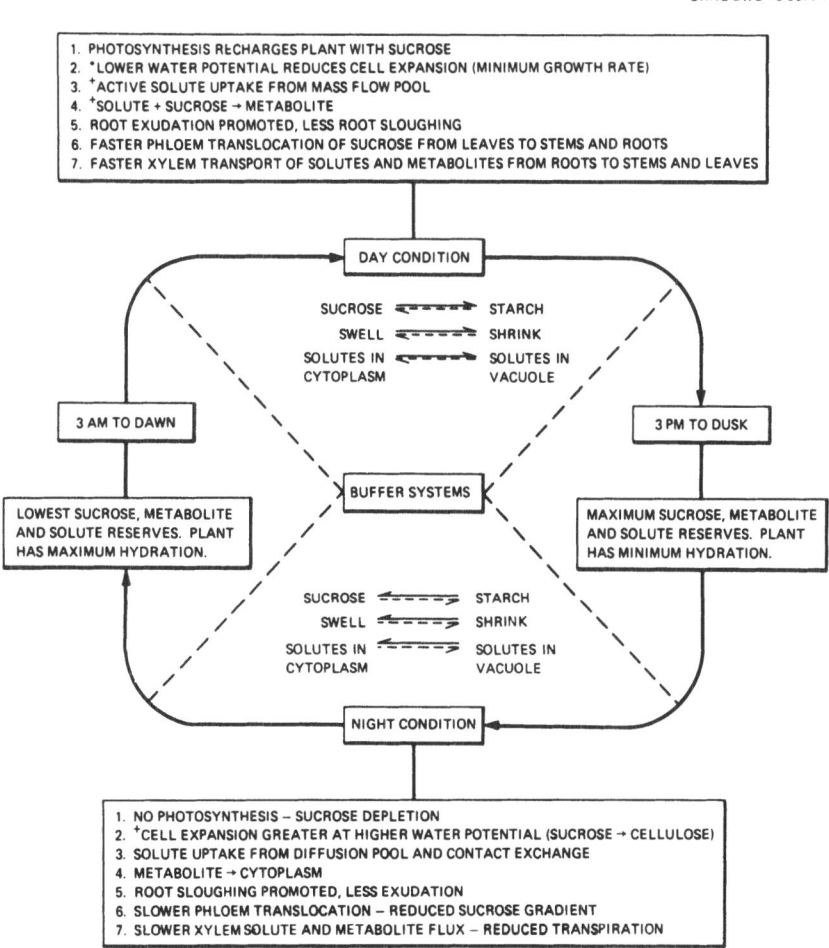

Figure 9.19. Diurnal pattern of carbon, water, and solute dynamics, showing *relative tendencies* and *relative states* in vegetation.

lowest sucrose, metabolite, and solute reserves and maximum hydration at dawn; and (2) highest sucrose, metabolite, and solute reserves and minimum hydration at dusk. These states are relative for any given day. Photosynthesis recharges the plant with sucrose and increases starch storage (or equivalent) during the day. At the same time, the plant is absorbing nutrients and undergoing dehydration. The loss of water can reduce the rate of cell expansion during the day. Thus, there is an imbalance between carbon gain and utilization, which is buffered by changes in internal storage. The higher internal carbon content in leaves during the afternoon may reduce the significance of pollutant impacts on them during that part of the day. At that time photosynthesis may be already slowed by the accumulation of photosynthetic products; alternatively, detoxification mechanisms, using readily available carbon metabolites and/or energy, may more easily cope with pollutant exposure during the afternoon than during early morning, when internal sucrose content is lower.

The diurnal pattern of carbon dynamics (Fig. 9.19) suggests that the exudation of carbon compounds by the roots and the transport of carbon to mycorrhizae and root nodules could be facilitated during the day. Disruption of these processes by air pollutants may have detrimental effects on nutrient uptake. Pollutants that cause reduced photosynthesis and/or greater respiration in forest ecosystems may decrease the carbon supply to the mycorrhizae, potentially decreasing the supply of nutrients and water to the tree. Thus, phytotoxic air pollutants may cause forest ecosystems to become less efficient in nutrient retention. A similar response was proposed by O'Neill et al. (1977) in ecosystems with trace metal pollution. Conversely, beneficial air pollutants may increase nutrient retention in forest ecosystems. Elevated levels of atmospheric CO_2 have been shown to be an example of the latter (Luxmoore et al. 1986).

9.6 Walker Branch Watershed Perspective

Data from research at Walker Branch Watershed have been used extensively in the development and testing of the hydrologic algorithms in the UTM-SPL (see Ch. 5). Some testing of the forest growth and nitrogen cycling models has been conducted (Luxmoore et al. 1981); however, much remains to be attempted. The quantitative linkages between the carbon dynamics (Ch. 6), deposition in the forest canopy (Ch. 4), and nutrient cycling (Ch. 7) can be evaluated with UTM-SPL. The synthesis of these diverse sources of data is an appropriate and feasible challenge for computer simulation, given the consistent focus of many experimental studies conducted over a relatively long period of time at Walker Branch Watershed.

Nutrient cycling processes, subsurface transport mechanisms, and forest biomass productivity can all be linked with the mechanistic modeling of the UTM-SPL. Ongoing research at Walker Branch Watershed and the testing and development of the component models of UTM-SPL will remain a robust combination for quantifying the interacting physical, chemical, and biological processes of forested landscapes. These studies will also continue to be a basis for addressing issues of environmental concern in other forested areas.

There is a large range in the transport rates of energy-related contaminants in a forested landscape from those that tend to accumulate (e.g., heavy metals) to those that can be readily leached (e.g., acidifying ions such as SO_4^{2-}). The effect of these contaminants and their possible interactions are of national and international concern (McLaughlin 1985). The decline in forest vigor in parts of Europe and in the Appalachian Mountains in the United States is hypothesized to be caused by multiple sources of stress, including natural weather variability as well as chemical pollutants. Modeling is a suitable tool for evaluating the significance of various mechanisms of forest response to multiple stresses. The development and testing of mechanistic models, such as the UTM-SPL, through Walker Branch Watershed experiments provide a basis from which to address questions and concerns that go beyond the specific bounds of the Walker Branch Watershed.

9.7 Summary

The coupled system of models having hourly resolution of carbon, water, and solute dynamics in terrestrial ecosystems is a "mathematical research facility" of ordered knowledge combined with selected hypotheses. The development, testing, and application of the UTM-SPL in combination with results from Walker Branch Watershed, the "experimental research facility," is an evolving and essentially never-ending process that leads to new knowledge and new hypotheses. The combination of modeling and experimental research provides a robust approach to quantifying complex phenomena and evaluating multiple hypotheses of cause and effect.

References

Baes, C.F., C.L. Begovich, W.M. Culkowski, K.R. Dixon, D.E. Fields, J.T. Holdeman, D.D. Huff, D.R. Jackson, N.M. Larson, R.J. Luxmoore, J.K. Munro, M.R. Patterson, R.J. Raridon, M. Reeves, O.C. Stein, J.L. Stolzy, and T.C. Tucker. 1976. The Unified Transport Model. pp. 13–62. IN R.I. Van Hook and W.D. Shults (eds.), Ecology and Analysis of Trace Contaminants Progress Report, October 1974–December 1975. ORNL/NSF/EATC-22, Oak Ridge National Laboratory, Oak Ridge, Tennessee.

Baldwin, J.P., P.B. Nye, and P.B. Tinker. 1973. Uptake of solutes by multiple root systems from soil. III. A model for calculating the solute uptake by a randomly dispersed root system developing in a finite volume of soil. Plant Soil 38:621–635.

Beauford, U., J. Barber, and A.R. Barringer. 1975. Heavy metal release from plants into the atmosphere. Nature 256:35–36.

Begovich, C.L., and D.R. Jackson. 1975. Documentation and application of SCEHM. A model for soil chemical exchange of heavy metals. ORNL/NSF/ EATC-16. Oak Ridge National Laboratory, Oak Ridge, Tennessee.

Begovich, C.L., and R.J. Luxmoore. 1979. Some Sensitivity Studies of Chemical Transport Simulated in Models of the Soil-Plant-Litter System. ORNL/TM-6791. Oak Ridge National Laboratory, Oak Ridge, Tennessee.

Beven, K., and P. Germann. 1982. Macropores and water flow in soils. Water Resour. Res. 18:1311–1325.

Crafts, A.S., and C.E. Crisp. 1971. Phloem Transport in Plants. W.H. Freeman and Co., San Francisco.

Dixon, K.R., R.J. Luxmoore, and C.L. Begovich. 1976. CERES—A Model of Forest Stand Biomass Dynamics for Predicting Trace Contaminant, Nutrient, and Water Effects. ORNL/NSF/EATC-25. Oak Ridge National Laboratory, Oak Ridge, Tennessee.

Dixon, K.R., R.J. Luxmoore, and C.L. Begovich. 1978. CERES—A model of forest stand biomass dynamics for predicting trace contaminant, nutrient, and water effects. I. Model description, II. Model application. Ecol. Modell. 5:17–38, 93–114.

Emerson, C.J., B. Thomas, Jr., and R.J. Luxmoore. 1984. CADIL: Model Documentation for Chemical Adsorption and Degradation in Land. ORNL/TM-8922. Oak Ridge National Laboratory, Oak Ridge, Tennessee.

Fong, L., and H.R. Appelbaum. 1980. Macropore-Mesopore Model of Water Flow Through Aggregated Porous Media. ORNL/MIT-312. Oak Ridge National Laboratory, Oak Ridge, Tennessee.

Gauch, H.G. 1972. Inorganic Plant Nutrition. Dowden, Hutchinson and Ross, Inc., Stroudsburg, Pennsylvania.

Hemphill, D.D., and J. Rule. 1975. Foliar uptake and translocation of ^{210}Pb and [109]Cd. Paper presented at International Conference on Heavy Metals in the Environment, Toronto, Canada, October 27–31, 1975 (pp. C239–C240 in program abstracts).

Henderson, G.S., and W.F. Harris. 1975. An ecosystem approach to characterization of the nitrogen cycle in a deciduous forest watershed. pp. 179–193. IN B. Bernier and C.H. Winget (eds.), Forest Soils and Forest Land Management. International Scholarly Books Service, Portland, Oregon.

Hetrick, D.M., J.T. Holdeman, and R.J. Luxmoore. 1982. AGTEHM: Documentation of Modifications to the Terrestrial Ecosystem Hydrology Model (TEHM) for Agricultural Applications. ORNL/TM-7856. Oak Ridge National Laboratory, Oak Ridge, Tennessee.

Hill, A.C. 1971. Vegetation: A sink for atmospheric pollutants. J. Air Pollut. Control Assoc. 21:341–346.

Huff, D.D., R.J. Luxmoore, J.B. Mankin, and C.L. Begovich. 1977. TEHM: A Terrestrial Ecosystem Hydrology Model. ORNL/NSF/EATC-27. Oak Ridge National Laboratory, Oak Ridge, Tennessee.

Jackson, D.R., and A.P. Watson. 1977. Description of nutrient pools and transport of heavy metals in a forested watershed near a lead smelter. J. Environ. Qual. 6:331–338.

Jackson, D.R., W.J. Selvidge, and B.S. Ausmus. 1978. Behavior of heavy metals in forest microcosms. I. Transport and distribution among components. Water Air Soil Pollut. 10:3–11.

Johnson, D.W., D.D. Richter, G.M. Lovett, and S.E. Lindberg. 1985. The effects of atmospheric deposition on potassium, calcium, and magnesium cycling in two deciduous forests. Can. J. For. Res. 15:773–782.

Lagerwerff, J.V. 1972. Lead, mercury, and cadmium as environmental contaminants. pp. 593–636. IN J.J. Mortvedt, P.M. Giordano, and W.L. Lindsay (eds.), Micronutrients in Agriculture. Soil Science Society of America, Inc., Madison, Wisconsin.

Luxmoore, R.J. 1980. Modeling pollutant uptake and effects on the soil-plant-litter system. pp. 174–180. IN P.R. Miller (ed.), Effects of Air Pollutants on Mediterranean and Temperate Forest Ecosystems. Report PSW-43. Pacific Southwest Forest and Range Experiment Station, Berkeley, California.

Luxmoore, R.J., and C.L. Begovich. 1979. Simulated heavy metal fluxes in tree microcosms and a deciduous forest. Int. Soc. Ecol. Modell. J. 1:48–60.

Luxmoore, R.J., C.L. Begovich, and K.R. Dixon. 1976. DRYADS and DIFMAS: FORTRAN Models for Investigating Solute Uptake and Incorporation into Vegetation and Litter. ORNL/NSF/EATC-26. Oak Ridge National Laboratory, Oak Ridge, Tennessee.

Luxmoore, R.J., C.L. Begovich, and K.R. Dixon. 1978. Modelling solute uptake and incorporation into vegetation and litter. Ecol. Modell. 5:137–171.

Luxmoore, R.J., T. Grizzard, and R.H. Strand. 1981. Nutrient translocation in the outer canopy and understory of an eastern deciduous forest. For. Sci. 27:505–518.

Luxmoore, R.J., E.G. O'Neill, J.M. Ells, and H.H. Rogers. 1986. Nutrient-uptake and growth responses of Virginia pine to elevated atmospheric CO_2. J. Environ. Qual. 15:244–251.

McLaughlin, S.B. 1985. Effects of air pollution on forests: A critical review. J. Air Pollut. Control Assoc. 35:512–534.

Mitchell, J.E., J.B. Waide, and R.L. Todd. 1975. A preliminary compartment model of the nitrogen cycle in a deciduous forest ecosystem. pp. 41–57. IN F.G. Howell, J.B. Gentry, and M.H. Smith (eds.), Mineral Cycling in Southeastern Ecosystems. CONF-740513. National Technical Information Service, Springfield, Virginia.

Munch, E. 1930. Die Stoffbewegungen in der Pflanze. Fisher, Jena, East Germany.

Nelson, C.D. 1963. Effect of climate on the distribution and translocation of assimilates. pp. 149–174. IN L. T. Evans (ed.), Environmental Control of Plant Growth. Academic Press, New York.

O'Neill, R.V., B.S. Ausmus, D.R. Jackson, R.I. Van Hook, P. Van Voris, C. Washburne, and A.P. Watson. 1977. Monitoring terrestrial ecosystems by analysis of nutrient export. Water Air Soil Pollut. 8:271–277.

Penman, H.L., and R.K. Schofield. 1951. Some physical aspects of assimilation and transpiration. pp. 115–129. IN Carbon Dioxide Fixation and Photosynthesis. Academic Press, New York.

Peters, L.N., D.F. Grigal, J.W. Curlin, and W.J. Selvidge. 1970. Walker Branch Watershed Project: Chemical, Physical and Morphological Properties of the Soils of Walker Branch Watershed. ORNL/TM-2968. Oak Ridge National Laboratory, Oak Ridge, Tennessee.

Phillips, R.E., T. NaNagara, R.E. Zartman, and J.E. Leggett. 1976. Diffusion and mass flow of nitrate-nitrogen to plant roots. Agron. J. 68:63–66.

Thornley, J.H.M. 1972. A model to describe the partitioning of photosynthate during vegetative plant growth. Ann. Bot. 33:419–430.

Tukey, H.B., and J.V. Morgan. 1962. The occurrence of leaching from aboveground plant parts and the nature of the material leached. 16th Int. Hortic. Congr. Rep., (Brussels, Belgium) 4:153–160.

Watson, A.P., R.I. Van Hook, D.R. Jackson, and D.E. Reichle. 1976. Impact of a Lead Mining-Smelting Complex on the Forest Floor Litter Arthropod Fauna in the New Lead Belt Region of Southeast Missouri. ORNL/NSF/EATC-30. Oak Ridge National Laboratory, Oak Ridge, Tennessee.

Chapter 10
Implications of Walker Branch Watershed Research

R.I. Van Hook

The original objectives of the Walker Branch Watershed Project, as defined in 1967, were to provide base-line values for unpolluted natural waters, to contribute to our knowledge of cycling and loss of chemical elements in natural ecosystems, and to enable construction of models for predicting the effects of man's activities on the landscape. These objectives were established prior to identification of specific environmental issues such as acidic precipitation, increasing levels of atmospheric carbon dioxide, and hydrologic transport of both organic and inorganic trace contaminants through forested landscapes. The watershed concept encompasses the coupling of both water and chemical budgets as they are influenced by the biological, chemical, and physical systems operating in the landscape. Integrating plot-level studies of the specific processes that govern transport and distribution of materials with watershed-level mass balances of water and chemicals permits evaluation of the effects of future perturbations on the landscape.

As indicated in earlier chapters, there are only a limited number of long-term watershed-level studies. The studies on Hubbard Brook Watershed in New Hampshire, the Coweeta watersheds in North Carolina, and the H.J. Andrews Forest in Oregon are the major contributors, along with those on Walker Branch Watershed, to our understanding of forest processes at landscape levels. These studies have addressed, in varying degrees, hydrology, meteorology, productivity, nutrient cycling, and land-water interactions while at the same time evaluating the effects of perturbations such as harvesting, fertilization, air pollution stress, and fire. Comparison of the results from this core group of watershed studies with numerous stand-level studies in the United States and abroad, along with the synthesis of the data by means of simulation models, has led to a quantified understanding of forest system behavior. This has permitted comparisons of ecological processes across biomes through participation in the International Biological Program in the early 1970s and exchange and integration of data with Man and the Biosphere Biosphere Reserves, the National Science Foundation's Long-Term Ecological Research Sites, and the Department of Energy's National Environmental Research Parks. These links are critical to developing a nationwide assembly of data and models to serve as base lines for measuring changes resulting from future perturbations. Additionally, Walker Branch Watershed research, along with other studies, will be integral in the development of programs on global habitability and geosphere-biosphere interactions.

Significant accomplishments of the Walker Branch Watershed Project over its 18-year history include scientific advancements in (1) atmospheric deposition to forest canopies, (2) radiation transfer in the canopy, (3) the role of macropores in hydrology, (4) biogeochemical cycles and biomass dynamics of forest ecosystems, (5) factors regulating nutrient and trace element concentrations in stream water, (6) nutrient limitation and cycling (spiraling) in streams, and (7) terrestrial-aquatic interactions in the transport and cycling of nutrients and organic carbon. These areas of research, which have been discussed in detail in earlier chapters, have permitted significant extrapolation of Walker Branch Watershed research results to other watershed sites as well as to larger regional landscapes.

Tests of contemporary phytoactinometric theory in the Walker Branch forest have shown that extant models, developed largely for agronomic crops, are deficient in terms of deciduous forest canopies. Because of foliage clumping in the deciduous Walker Branch forest, the traditional approaches fail to simulate the penetration and scattering of beam radiation in the canopy sufficiently accurately for many applications. A canopy radiation transfer theory based on the negative binomial probability distribution function has been developed that permits more accurate simulation of the partitioning of beam radiation in structurally heterogeneous forest canopies. A canopy-level stomatal resistance model has also been developed that allows accounting for physiological control of canopy stomatal resistance to be evaluated on the basis of distributions of photosynthetically active irradiances and of leaf areas, temperatures, and water potentials within the canopy. This stomatal resistance model forms a submodel of a "big leaf"–type model of dry deposition to vegetation canopies (Hicks et al. 1985), which is used to assess regional dry deposition of acidic pollutants as part of the National Acid Precipitation Assessment Program.

Atmospheric deposition research has established the importance of deposition to element cycles in forests and the significant role played by dry deposition of both trace and major constituents. Both of these findings were used in developing the National Plan for Acid Deposition in the early 1980s, and more recently in developing the long-term multisite Forest Effects Study of the Electric Power Research Institute. As a result of our earlier work, new efforts were initiated here and elsewhere to refine sampling methods for dry deposition of large particles, including inert surface, foliar extraction, and rain and throughfall chemistry approaches. In addition, using results from Walker Branch Watershed research, we have developed formal and informal joint efforts with researchers in Sweden, the United Kingdom, and the Federal Republic of Germany in studies of the role of the atmosphere in forest diebacks, and with scientists in the United States in studies of deposition methodology and deposition of trace organics. The Metropolitan Area Power Plant Pollution Study site at Walker Branch Watershed continues to provide wetfall-only data on rainfall events to this network for analysis of factors influencing rain chemistry in the Southeast, and the National Atmospheric Deposition Program site was recently designated as a National Trends Network site for continued comparison of spatial and temporal trends in rain chemistry.

Throughout the history of the Walker Branch Watershed Project, model development, testing, and implementation have been the central theme. The Atmosphere-Soil-Plant Water Flow Model (PROSPER) initiated in 1974 was incorporated into the Terrestrial Ecosystem Hydrologic Model (TEHM) in 1977, and this paved the way for mechanistic hydrologic analysis and provided the framework for coupling forest growth and chemical transport models. These models, collectively called the Unified Transport Model (UTM), have played key roles in Walker Branch research and in the application of the research results to other ecosystems. The applications of Walker Branch results to research projects and environmental concerns on the Oak Ridge Reservation have included the areas of resource utilization, forest management, water management, and contaminant transport.

In the area of resource utilization and forest management, the evaluation of the effects of total removal of aboveground forest biomass (clear-cutting) compared with the effects of conventional forest management practices was significantly aided by direct transfer of data on soils, vegetation, and hydrology from Walker Branch Watershed to an adjacent watershed. This significantly reduced the costs associated with sample collection and analysis, since critical ecosystem components requiring measurement had already been identified. The knowledge developed in Walker Branch research also played a major role in initial site selection for the experiment.

The results from Walker Branch research have been used in several studies associated with groundwater management on the Oak Ridge Reservation. The watershed approach to understanding landscapes has been utilized in evaluating the White Oak Creek drainage basin for sources of contaminants transported to the Clinch River. Basic geological and hydrological data on Knox dolomite derived from Walker Branch Watershed research have been used in site characterization of potential shallow land burial sites on the Oak Ridge Reservation. The TEHM has been utilized for the estimation of water budgets at several burial grounds at Oak Ridge National Laboratory; the specific concern here was to minimize the groundwater levels by either increasing evapotranspiration through the use of vegetative cover or redirecting the groundwater away from the burial grounds.

Studies of nutrient spiraling in Walker Branch were among the first to document the fact that stream ecosystems can be nutrient-limited. In addition, the studies showed that benthic communities associated with organic and inorganic substrata control the downstream flux of the limiting nutrient and that seasonal variation in the spiraling length of phosphorus in the stream is primarily due to variation in the standing stock of organic matter derived from riparian vegetation. While most watershed studies have treated streams as conduits that transport materials (water nutrients) exported from the terrestrial portion of the watershed, the studies on Walker Branch show that in-stream processes can modify both the form and flux of nutrients in streamflow. The results from Walker Branch research have demonstrated that to distinguish between the influences of terrestrial and aquatic processes in regulating both nutrient concentrations in streamflow and areal fluxes of nutrients and organic carbon from the watershed, and to correctly interpret temporal variation in element concentrations and fluxes, it is necessary to include studies of in-stream processes as an integral subsystem of a watershed.

Following the development and testing of the TEHM with data from Walker Branch Watershed research, a series of simulation studies was undertaken for two grassland watersheds in Chickasha, Oklahoma. These applications demonstrated the influence of spatial variability in the hydraulic properties of soil on hydrologic processes and showed that average soil properties could not represent the range of streamflow or evapotranspiration responses of the watersheds.

In a more extensive application of Walker Branch Watershed research results to an off-site situation, Walker Branch scientists were requested to assist the Environmental Protection Agency, National Science Foundation, and National Park Service in evaluating the impacts of a lead mine-mill-smelter operation on an oak-pine forest in southeastern Missouri. The viburnum trend, or "new lead belt," in the Clark National Forest near Salem, Missouri, became the site of lead mining, milling, and smelting operations in 1968. Walker Branch investigators joined scientists from the University of Missouri–Rolla in identifying and instrumenting Crooked Creek watershed, which was approximately the same size as Walker Branch Watershed and contained similar vegetation. Experience and data developed on the Walker Branch Watershed Project facilitated the selection of key components on Crooked Creek to be monitored for the impacts of Cd, Pb, Zn, and Cu on ecosystem processes, significantly reducing the manpower and analytical costs of the research. Soils, litter, vegetation, stream water, and stream sediments were involved, as were deposition measurements. The UTM was applied to Crooked Creek watershed, and budgets of water and the four trace contaminants, including off-site transport, were calculated for this system. Guidelines that were established from these results and from measurements of SO_x and particulates assisted the AMAX Mining Corporation in controlling both point source and fugitive emissions from their operations and in evaluating various procedures for meeting the water quality standards of the National Pollutant Discharge Elimination System.

The results of Walker Branch Watershed studies have also been used in developing regional-scale assessments of the effects of atmospheric pollution on forested landscapes. In 1985 the National Acid Precipitation Assessment Program assessed soil sensitivity to acidic precipitation in the eastern United States through utilization of specific information on anion mobility and cation leaching from Walker Branch watershed coupled with regional soils data and chemical composition data from other eastern stand and watershed studies. Research on Walker Branch and elsewhere has led to the hypothesis that sulfate adsorption is a major factor controlling soil sensitivity to acid deposition. We have found general relationships between soil classification and sulfate adsorption, based on laboratory tests, allowing regional mapping of soil sensitivity to leaching by atmospheric sulfate inputs.

In addition to developing maps of soil sensitivity from Walker Branch information, we are using the 18-year histories of soil water and vegetation dynamics to verify our stand succession model (FORET), which will be used in developing an assessment of the future effects of acidic precipitation on forest systems in the eastern United States. The availability of these data sets, together with data sets on atmospheric deposition and streamflow, is critical to the interpretation of long-term trends in these ecosystems. Long-term data sets, combined with verified models of material transport in forested systems, represent a rational mechanism for predicting the future effects of man's activities on the environment. Continuance of these types of studies, including those on the Walker Branch, Hubbard Brook, Coweeta, and H.J. Andrews sites, along with the newly established Long-Term Ecological Research Sites, Biosphere Reserves, and National Environmental Research Parks, will ensure the development of data bases against which future changes may be assessed.

Future directions on the Walker Branch Watershed include (1) continued long-term monitoring of atmospheric input, soil solution, and stream outputs, along with soil and vegetation inventories; (2) increased emphasis on decomposition processes in belowground and stream systems; (3) continued emphasis on the development of a thorough understanding of biogeochemical cycles in this system; and (4) regionalization of watershed results, including the areas of soils, hydrology, and vegetation, utilizing modeling and remote sensing as major tools. The objectives of the project will continue to focus on providing base-line data, contributing to our knowledge of element cycling in natural systems, and refining models for simulating material transport and effects in forested landscapes. Throughout these efforts, the development of a basic understanding of the biological and physicochemical processes governing the behavior of watersheds will be emphasized.

References

Hicks, B.B., D.D. Baldocchi, R.P. Hosker, Jr., B.A. Hutchison, D.R. Matt, R.T. McMillen, and L.C. Satterfield. 1985. On the Use of Monitored Air Concentration to Infer Dry Deposition. NOAA Technical Memorandum ERL ARL-141. National Oceanic and Atmospheric Administration, Air Resources Laboratory, Silver Spring, Maryland.

Index